Inositol and Phosphoinositides

Metabolism and Regulation

Experimental Biology and Medicine

INOSITOL
AND
PHOSPHOINOSITIDES

Metabolism and Regulation

Edited by

John E. Bleasdale, Joseph Eichberg, and George Hauser

Humana Press · Clifton, New Jersey

Library of Congress Cataloging in Publication Data
Main Entry under title:

Inositol and phosphoinositides.

    "The Chilton Conference on Inositol and
Phosphoinositides, held on January 9–11, 1984 at
Southwestern Medical School, University of Texas Health
Science Center, Dallas, Texas, was the third in a
series of conferences on cyclitols and
phosphoinositides"—Pref.
    Includes index.
    1. Inositol—Metabolism—Congresses.
2. Phosphoinositides—Metabolism—Congresses.
I. Bleasdale, John E.   II. Eichberg, Joseph, 1935–
III. Hauser, George.   IV. Chilton Conference on
Inositol and Phosphoinositides (1984 : Southwestern
Medical School) [DNLM: 1. Inositol—metabolism—
congresses. QU 93 I58 1984]
QP772.I5I56   1985   599'.019'24   85-131
ISBN 0-89603-074-1

© 1985 The Humana Press Inc.
Crescent Manor
PO Box 2148
Clifton, NJ 07015

Printed in the United States of America

PREFACE

The Chilton Conference on Inositol and Phosphoinositides, held on January 9–11, 1984 at Southwestern Medical School, University of Texas Health Science Center, Dallas, Texas, was the third in a series of conferences on cyclitols and phosphoinositides. The first took place in 1968 in New York [*Ann. New York Acad. Sci.* (1969), *165,* 508–819] and the second was held in 1977 in East Lansing, Michigan [*Cyclitols and Phosphoinositides,* Wells, W. W. and Eisenberg, F., eds., (1978) Academic press, New York, pp. 1–607.]

In the interim since the previous conference, not only has the pace of research in the field accelerated markedly, but the physiological importance of phosphoinositide metabolism has become apparent to an increasing number of investigators from diverse fields in the life sciences. Thus it seemed to us timely for both recent and established workers in this area, as well as others whose interests impinged on it, to meet in order to disseminate new information, to review, and perhaps arrive at, a consensus of our current understanding of the role of inositol and phosphoinositides, and to establish new directions for research for the next few years.

The expansion of the field since the last meeting made it mandatory to restrict the scope of the topics covered at the conference, primarily to aspects dealing with mammalian systems. We sincerely regretted the exclusion of recent research on cyclitols and phosphoinositides in microbes and plants and hope that these areas will be included in future conferences.

The attendance and active participation of nearly 200 scientists, including representation from ten countries outside of the United States, provided compelling testimony to the current interest in the field. Leading researchers reported on advances in our knowledge of *myo*-inositol biosynthesis, the factors and processes that affect *myo*-inositol homeostasis, phospho-inositide enzymology, receptor-mediated alterations in phos-phoinositide metabolism, and the effects these changes evoke in Ca^{2+} mobilization and protein phosphorylation, arachidonic acid release, and abnormalities associated with diabetes in the biochemistry of inositol and phosphoinositides. The most clearcut accomplishment to emerge from the meeting was the general agreement that two products of stimulated phosphoinositide hydrolysis, diacylglycerol and *myo*-inositol 1,4,5-trisphosphate, are serious candidates for second messenger molecules intimately involved in the regulation of a wide variety of cellular phenomena. The conference was divided into six sessions and the manuscripts in this volume are arranged in the order in which they were presented. In addition, the abstracts of poster presentations are also included.

The success of the conference was due to the efforts of many people to whom thanks are due. We acknowledge the help of a number of individuals at the University of Texas Health Science Center at Dallas, most particularly the enthusiastic support provided by Dr. Jack Johnston of the Department of Biochemistry. We are especially grateful to The Chilton Foundation and its officers Mr. Andy Bell and Mr. Sam Winstead for providing the financial support that made the conference possible. The support of Ms. Evelyn Whitman of Southwestern Medical Foundation is also gratefully acknowledged. We thank American Cyanamid Company, Burroughs-Wellcome Company, Pfizer, Inc., Sandoz, Inc., and The Upjohn Company for additional financial support. Last, we would like to thank Mr. Thomas Lanigan of Humana Press for his interest in publishing the proceedings and Ms. Dolly Tutton for her devoted service in retyping all manuscripts into camera-ready form and for her invaluable assistance with other editorial tasks.

The editors of the volume arising from the previous conference stated in their Preface that "perhaps by the time of the next conference the exact role played by the inositol phospholipids in the physiology of the cell membrane will be revealed." Although this goal is still some distance off, progress has been such that the outlines of that role are emerging, and the excitement now being generated over the importance of these compounds assures that intense study of the functions of inositol and phosphoinositides will continue well into the future.

We feel confident that the onrush of new findings will result in the convening of a fourth conference much sooner than the seven years between the previous two conferences on this subject.

John E. Bleasdale
Joseph Eichberg
George Hauser

CONTENTS

vii

LIST OF CONTRIBUTORS

Abdel-Latif, A. A. • *Department of Cell and Molecular Biology, Medical College of Georgia, Augusta, Georgia*

Agranoff, B. W. • *Neuroscience Laboratory, Mental Health Research Institute,063University of Michigan, Ann Arbor, Michigan*

Akhtar, R. A. • *Department of Cell and Molecular Biology, Medical College of Georgia, Augusta, Georgia*

Allan, D. • *Department of Experimental Pathology, School of Medicine, University College, London, UK*

Alving, C. R. • *Department of Membrane Biochemistry, Walter Reed Army Institute of Research, Washington, D.C.*

Anceschi, M. • *Departments of Biochemistry, Obstetrics-Gynecology, Cecil H. and Ida Green Center for Reproductive Biology Sciences, University of Texas Health Science Center, Dallas, Texas*

Anderson, R. E. • *Cullen Eye Institute and Program in Neuroscience, Baylor College of Medicine, Houston, Texas*

Aub, D. L. • *Department of Pharmacology, Medical College of Virginia-VCU, Richmond, Virginia*

Ban, C. • *Departments of Biochemistry, Obstetrics-Gynecology, Cecil H. and Ida Green Center for Reproductive Biology Sciences, University of Texas Health Science Center, Dallas, Texas*

Bazan, N. G. • *LSU Eye Center, Louisiana State University Medical Center, New Orleans, Louisiana*

Bell, M. E. • *Department of Biochemical and Biophysical Sciences, University of Houston, Houston, Texas*

Benton, H. • *Cancer Research Campaign, Cell Proliferation Unit, Royal Postgraduate Medical School, London, UK*

Berridge, M. J. • *AFRC Unit of Insect Neurophysiology and Pharmacology, Department of Zoology, University of Cambridge, Cambridge, UK*

Berry, G. • *Division of Metabolism, Children's Hospital, Philadelphia, Pennsylvania*

Berti-Mattera, L. • *Department of Biochemical and Biophysical Sciences, University of Houston, Houston, Texas*

Blackmore, P. F. • *Howard Hughes Medical Institute, Vanderbilt University School of Medicine, Nashville, Tennessee*

Bleasdale, J. E. • *Department of Biochemistry, University of Texas Health Science Center, Dallas, Texas*

Bone, E. A. • *Department of Biochemistry, University of Birmingham, Birmingham, UK*

Broekman, M. J. • *Divisions of Hematology/Oncology, Department of Medicine, N.Y. Veterans Administration Medical Center and Cornell University Medical College, New York, New York*

Burgess, G. M. • *Department of Pharmacology, Medical College of Virginia-VCU, Richmond, Virginia*

Charest, R. • *Howard Hughes Medical Institute, Vanderbilt University School of Medicine, Nashville, Tennessee*

Cockcroft, S. • *Department of Experimental Pathology, School of Medicine, University College, London, UK*

Daniel, L. W. • *Department of Biochemistry, Bowman Gray School of Medicine of Wake Forest University, Winston-Salem, North Carolina*

Dawson, R. M. C. • *Department of Biochemistry, AFRC Institute of Animal Physiology, Babraham, Cambridge, UK*

DeGraan, P.N.E. • *Division of Molecular Neurobiology, Rudolf Magnus Institute of Pharmacology and Institute of Molecular Biology, State University of Utrecht, Utrecht, The Netherlands*

Downes, C. P. • *ICI Pharmaceuticals Division, Bioscience Department II, Alderley Park, Macclesfield, Cheshire, UK*

Eichberg, J. • *Department of Biochemical and Biophysical Sciences, University of Houston, Houston, Texas*

Eisenberg, F. Jr. • *National Institute of Arthritis, Diabetes, Digestive and Kidney Diseases, Bethesda, Maryland*

Exton, J. H. • *Howard Hughes Medical Institute, Vanderbilt University School of Medicine, Nashville, Tennessee*

Fain, J. N. · Section of Biochemistry, Brown University, Providence, Rhode Island

Farese, R. V. · Department of Medicine, Veterans Administration Hospital, University of South Florida College of Medicine, Tampa, Florida

Farrell, L. E. · Department of Biochemistry, Michigan State University, East Lansing, Michigan

Filburn, C. R. · Laboratory of Molecular Aging, National Institute on Aging, Geronotology Research Center, Baltimore City Hospital, Baltimore, Maryland

Foudin, L. · Sinclair Comparative Medicine Research Farm and Biochemistry Department, University of Missouri, Columbia, Missouri

Fretten, P. · Department of Biochemistry, University of Birmingham, Birmingham, UK

Gillon, K. R. · Department of Biochemistry, Medical School, Queen's Medical Centre, Nottingham, UK

Girard, A. · Department of Pharmacology, Hôpital Necker, Paris, France

Gispen, W. H. · Division of Molecular Neurobiology, Rudolf Magnus Institute of Pharmacology and Institute of Molecular Biology, State University of Utrecht, Utrecht, The Netherlands

Giusto, N. M. · Instituto de Investigaciones Bioquimicas, Universidad Nacional del Sur-Consejo Nacional de Investigaciones Cientificas y Technicas, Bahia Blanca, Argentina

Godfrey, P. P. · Department of Pharmacology, Medical College of Virginia-VCU, Richmond, Virginia

Greene, D. A. · Diabetes Research Laboratories, Department of Medicine, School of Medicine, University of Pittsburgh, Pittsburgh, Pennsylvania

Hanley, M. R. · Department of Biochemistry, Imperial College of Science and Technology, London, UK

Hauser, G. · Ralph Lowell Laboratories, McLean Hospital, Harvard Medical School, Belmont, Massachusetts

Hawthorne, J. N · Department of Biochemistry, Medical School, Queen's Medical Centre, Nottingham, UK

Hokin-Neaverson, M. • *Departments of Physiological Chemistry and Psychiatry and the Wisconsin Psychiatric Institute, University of Wisconsin, Madison, Wisconsin*

Hollyfield, J. G. • *Cullen Eye Institute and Program in Neuroscience, Baylor College of Medicine, Houston, Texas*

Holub, B. J. • *Department of Nutrition, University of Guelph, Guelph, Ontario, Canada*

Honchar, M. P. • *Department of Psychiatry, Washington University School of Medicine, St. Louis, Missouri*

Horwitz, J. • *University of Illinois College of Medicine, Chicago, Illinois*

Huang, S. F-L. • *Sinclair Comparative Medicine Research Farm and Biochemistry Department, University of Missouri, Columbia, Missouri*

Ilincheta de Boschero, M. G. • *Instituto de Investigaciones Bioquimicas, Universidad Nacional del Sur-Consejo Nacional de Investigaciones Cientificas y Tecnicas, Bahia Blanca, Argentina*

Irvine, R. F. • *Department of Biochemistry, AFRC Institute of Animal Physiology, Babraham, Cambridge, UK*

Janowsky, A. • *National Institute of General Medical Sciences, Bethesda, Maryland*

Jett, M. • *Department of Membrane Biochemistry, Walter Reed Army Institute of Research, Washington, DC*

Johnston, J. M. • *Departments of Biochemistry, Obstetrics-Gynecology, Cecil H. and Ida Green Center for Reproductive Biology Sciences, University of Texas Health Science Center, Dallas, Texas*

Joseph, S. K. • *Department of Biochemistry and Biophysics, University of Pennsylvania School of Medicine, Philadelphia, Pennsylvania*

Kaibuchi, K. • *Department of Biochemistry, Kobe University School of Medicine, Kobe and Department of Cell Biology, National Institute for Basic Biology, Okazaki, Japan*

Katakami, Y. • *Department of Biochemistry, Kobe University School of Medicine, Kobe and Department of Cell Biology, National Institute for Basic Biology, Okazaki, Japan*

Kikkawa, U. • *Department of Biochemistry, Kobe University School of Medicine, Kobe and Department of Cell Biology, National Institute for Basic Biology, Okazaki, Japan*

Kirk, C. J. • *Department of Biochemistry, University of Birmingham, Birmingham, UK*

Koutouzov, S. • *Department of Pharmacology, Hôpital Necker, Paris, France*

Labarca, R. • *Clinical Neuroscience Branch, National Institute of Mental Health, Bethesda, Maryland*

Lander, D. J. • *Department of Biochemistry, AFRC Institute of Animal Physiology, Babraham, Cambridge, UK*

Lapetina, Eduardo G. • *The Wellcome Research Laboratories, Research Triangle Park, North Carolina*

Laposata, M. • *Division of Hematology-Oncology, Departments of Medicine and Biological Chemistry, Washington University School of Medicine, St. Louis, Missouri*

Lattimer, S. A. • *Diabetes Research Laboratories, Department of Medicine, School of Medicine, University of Pittsburgh, Pittsburgh, Pennsylvania*

Letcher, A. J. • *Department of Biochemistry, AFRC Institute of Animal Physiology, Babraham, Cambridge, UK*

Lightman, S. L. • *Medical Unit, Charing Cross and Westminister Medical School Hospital, London, UK*

Lin, S.-H. • *Section of Biochemistry, Brown University, Providence, Rhode Island*

Litosch, I. • *Section of Biochemistry, Brown University, Providence, Rhode Island*

Mack, S. E. • *Department of Biochemistry, Dalhousie University, Halifax, Nova Scotia, Canada*

Maeda, T. • *National Institute of Arthritis, Diabetes, Digestive and Kidney Diseases, Bethesda, Maryland*

Majerus, P. W. • *Division of Hematology-Oncology, Departments of Medicine and Biological Chemistry, Washington University School of Medicine, St. Louis, Missouri*

Marche, P. • *Department of Pharmacology, Hôpital Necker, Paris, France*

Michell, R. H. • *Department of Biochemistry, University of Birmingham, Birmingham, UK*

Millar, F. A. • *Department of Biochemistry, Medical School, Queen's Medical Centre, Nottingham, UK*

Munsell, L. Y. · *Department of Psychiatry, Washington University School of Medicine, St. Louis, Missouri*

Neufeld, E. J. · *Division of Hematology-Oncology, Departments of Medicine and Biological Chemistry, Washington University School of Medicine, St. Louis, Missouri*

Nishizuka, Y. · *Department of Biochemistry, Kobe University School of Medicine, Kobe and Department of Cell Biology, National Institute for Basic Biology, Okazaki, Japan*

Oestreicher, A. B. · *Division of Molecular Neurobiology, Rudolf Magnus Institute of Pharmacology and Institute of Molecular Biology, State University of Utrecht, Utrecht, The Netherlands*

Palmer, F. B. St. C. · *Department of Biochemistry, Dalhousie University, Halifax, Nova Scotia, Canada*

Palmer, S. · *Department of Biochemistry, University of Birmingham, Birmingham, UK*

Parries, G. S. · *Departments of Physiological Chemistry and Psychiatry and the Wisconsin Psychiatric Institute, University of Wisconsin, Madison, Wisconsin*

Paul, S. M. · *Clinical Neuroscience Branch, National Institute of Mental Health, Bethesda, Maryland*

Perlman, R. L. · *University of Illinois College of Medicine, Chicago, Illinois*

Peterson, R. G. · *Department of Anatomy, Indiana University School of Medicine, Indianapolis, Indiana*

Prpic, V. · *Howard Hughes Medical Institute, Vanderbilt University School of Medicine, Nashville, Tennessee*

Putney, J. W., Jr. · *Department of Pharmacology, Medical College of Virginia-VCU, Richmond, Virginia*

Rhodes, D. · *Howard Hughes Medical Institute, Vanderbilt University School of Medicine, Nashville, Tennessee*

Rittenhouse, S. E. · *Brigham and Women's Hospital, Harvard Medical School, Boston, Massachusetts*

Roccamo de Fernandez, A. M. · *Instituto de Investigaciones Bioquimicas, Universidad Nacional del Sur-Consejo, Nacional de Investigaciones Cientificas y Technicas, Bahia Blanca, Argentina*

Rubin, R. P. • *Department of Pharmacology, Medical College of Virginia, Richmond, Virginia*

Sawamura, M. • *Department of Biochemistry, Kobe University School of Medicine, Kobe and Department of Cell Biology, National Institute for Basic Biology, Okazaki, Japan*

Segal, S. • *Division of Metabolism, Children's Hospital, Philadelphia, Pennsylvania*

Sherman, W. R. • *Department of Psychiatry, Washington University School of Medicine, St. Louis, Missouri*

Seyfred, M. A. • *Department of Biochemistry, Michigan State University, East Lansing, Michigan*

Smith, C. D. • *Department of Biochemistry, Michigan State University, East Lansing, Michigan*

Smith., E. M. • *Department of Biochemistry, Medical School, Queen's Medical Centre, Nottingham, UK*

Smith, J. P. • *Department of Cell and Molecular Biology, Medical College of Georgia, Augusta, Georgia*

Sun, G. Y. • *Sinclair Comparative Medicine Research Farm and Biochemistry Department, University of Missouri, Columbia, Missouri*

Takai, Y. • *Department of Biochemistry, Kobe University School of Medicine, Kobe and Department of Cell Biology, National Institute for Basic Biology, Okazaki, Japan*

Tang, W. • *Sinclair Comparative Medicine Research Farm and Biochemistry Department, University of Missouri, Columbia, Missouri*

Todd, K. • *Medical Unit, Charing Cross and Westminister Medical School, London, UK*

Thomas, A. P. • *Department of Biochemistry and Biophysics, University of Pennsylvania, School of Medicine, Philadelphia, Pennsylvania*

Tsymbalov, S. • *University of Illinois College of Medicine, Chicago, Illinois*

Uchida, T. • *Laboratory of Molecular Aging, National Institute on Aging, Gerontology Research Center, Baltimore City Hospital, Baltimore, Maryland*

Van Dongen, C. J. • *Division of Molecular Neurobiology, Rudolf Magnus Institute of Pharmacology and Institute of*

Molecular Biology, State University of Utrecht, Utrecht, The Netherlands

Van Rooijen, L. A. A. · *Neuroscience Laboratory, Mental Health Research Institute, University of Michigan, Ann Arbor, Michigan*

Vu, N-D. · *Laboratory of Molecular and Developmental Biology, National Eye Institute, Bethesda, Maryland*

Wallace, M. · *Section of Biochemistry, Brown University, Providence, Rhode Island*

Wells, W. W. · *Department of Biochemistry, Michigan State University, East Lansing, Michigan*

Williamson, J. R. · *Department of Biochemistry and Biophysics, University of Pennsylvania, School of Medicine, Philadelphia, Pennsylvania*

Yandrasitz, J. R. · *Division of Metabolism, Children's Hospital, Philadelphia, Pennsylvania*

Zelenka, P. S. · *Laboratory of Molecular and Developmental Biology, National Eye Institute, Bethesda, Maryland*

Zwiers, H. · *Division of Molecular Neurobiology, Rudolf Magnus Institute of Pharmacology and Institute of Molecular Biology, State University of Utrecht, Utrecht, The Netherlands*

COMMENT ON ABBREVIATIONS

The Chilton Conference on Inositol and Phospho-inositides held in Dallas, Texas, January 9-11, 1984, brought together almost 200 investigators who are actively pursuing research in this area. While some of the participants used the IUPAC-IUB recommended abbreviations, most did not. The lack of acceptance relates primarily to the fact that the recommended abbreviations are much too long to be adapted to spoken language. The several alternative systems which have arisen are quite different. For example, for PtdIns(4,5)P_2 (phosphatidylinositol 4,5-bisphosphate), one finds PIP_2, $PhIP_2$, PIPP, and TPI. In an effort to standardize both our written and our spoken terminology, we agreed upon a single new trivial system. This volume then uses only two terminologies: the IUPAC-IUB recommended abbreviations, and the new, more pronounceable ones outlined here. While one common abbreviation system would be even better than two, perhaps we can take heart in Ralph Waldo Emerson's expression--"foolish consistency is the hobgoblin of little minds".

B.W. Agranoff
F. Eisenberg, Jr.
G. Hauser
J.N. Hawthorne
R.H. Michell

Footnotes to table:

a   IUPAC-IUB Commission on Biochemical Nomenclature, Hoppe-Seyler's Z. Physiol. Chem. 358:599-616, 1977; Proc. Natl. Acad. Sci. USA 74:2222-2230, 1977; Biochem. J. 171:1-19, 1978. The reader should note, however, that there are inaccuracies in designations and structures in the foregoing. See Agranoff, B.W., Trends in Biochem. Sci. 3:N283-285, 1978.

b   By inference from a.

c   The positions of phosphorus substituents in this and the following degradation products, if not specified, are assumed to be those of the parent lipid. Thus, IP_3 is D-myo-inositol 1,4,5-trisphosphate, $GPIP_2$ is sn-glycero-3-phospho-D1-myo-inositol 4,5-bisphosphate, etc.

Name	IUPAC-IUB Abbreviation[a]	"Chilton Convention"
Phosphatidylinositol	PtdIns	PI
Phosphatidylinositol 4-phosphate	PtdIns4P	PIP
Phosphatidylinositol 4,5-bisphosphate	PtdIns(4,5)P_2	PIP$_2$
Phosphatidate	PtdOH[b]	PA
Diacylglycerol	Ac$_2$Gro	DG
Cytidine diphospho-diacylglycerol	PtdCMP[b]	CDP-DG (or CMP-PA)
Phosphatidylserine	PtdSer	PS
Phosphatidylcholine	PtdCho	PC
Phosphatidylethanolamine	PtdEtn	PE
Phosphatidylglycerol	PtdGro	PG
Glycerophospho-inositol[c]	Gro$P$Ins[b]	GPI
Glycerophospho-inositol phosphate	Gro$P$Ins$P$[b]	GPIP
Glycerophospho-inositol bisphosphate	Gro$P$InsP_2[b]	GPIP$_2$
Inositol 1-phosphate	Ins$P$[b]	IP
Inositol 1,4-bisphosphate	InsP_2[b]	IP$_2$
Inositol 1,4,5-tris-phosphate	InsP_3[b]	IP$_3$

PART I

MYO-INOSITOL METABOLISM AND HOMEOSTASIS

THE MECHANISM OF ENZYMATIC ISOMERIZATION

OF GLUCOSE 6-PHOSPHATE TO

L-*MYO*-INOSITOL 1-PHOSPHATE

F. Eisenberg, Jr. and T. Maeda

National Institutes of Health, NIADDK
Building 10, Room 9B07, Bethesda, Maryland 20205

SUMMARY

By the addition of NaB^3H_4 to a system of homogeneous L-*myo*-inositol 1-phosphate synthase, glucose 6-phosphate, and NAD^+, inosose-2 1-phosphate was trapped as a mixture of tritiated *myo*-inositol and *scyllo*-inositol. Based on the concentration of the cyclitols and the specific radioactivity of H_1 of glucitol produced by concomitant reduction of glucose 6-phosphate we estimate that 2 moles of inosose-2 1-phosphate are bound per mole of enzyme. In controls without enzyme or without substrate the cyclitols were unlabeled, showing that incorporation of tritium was related to the actively synthesizing system. Iditol and glucitol H_5, epimeric alditols representing 5-ketoglucose 6-phosphate, the other postulated intermediate, were not significantly labeled. We conclude that 5-ketoglucose 6-phosphate has no finite existence, its cyclization to inosose-2 1-phosphate occurring as quickly as its formation from glucose 6-phosphate. Since inosose-2 1-phosphate is sufficiently long-lived to be trapped by borotritide reduction, its reduction by putative NADH in the enzymatic reaction must be the rate limiting step, and thus the site of hydrogen isotope effects observed previously.

INTRODUCTION

In a search for intermediates in the reaction catalyzed by partially purified L-*myo*-inositol 1-phosphate synthase from rat testis, Chen and Eisenberg (1975) identified inosose-2 1-phosphate bound to both active and boiled enzyme. The greater amount found associated with active enzyme was suggestive that inosose-2 1-phosphate is an intermediate in the catalyzed reaction (Fig. 1). With the purification of the enzyme to homogeneity by Maeda and Eisenberg (1980) the same intermediate has now been identified and its formation unequivocally correlated with the enzymatic reaction. The calculated steady state concentration of the intermediate suggests that 2 moles per mole of enzyme are bound during the reaction. If 2 available sites per molecule of enzyme are assumed, then the other postulated intermediate, 5-keto-glucose 6-phosphate, should not be demonstrable. Accordingly we have found no evidence for that intermediate and conclude that 5-ketoglucose 6-phosphate is too short-lived to be detectable. Experiments leading to these and other conclusions are described in this paper.

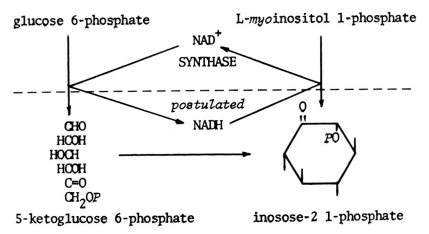

Figure 1. The postulated mechanism of isomerization of glucose 6-phosphate to L-<u>myo</u>-inositol 1-phosphate by L-<u>myo</u>-inositol 1-phosphate synthase. Putative enzyme-bound intermediates are shown below the broken line.

MATERIALS AND METHODS

Enzyme Preparation

L-*myo*-Inositol 1-phosphate synthase (EC 5.5.1.4) was purified from 250 g of rat testis by the procedure described by Maeda and Eisenberg (1980). A solution of 1 ml of homogeneous enzyme containing 3.2 mg of protein (800 mU, M_r 210,000) was prepared.

Incubation and Borotritide Reduction

One ml of solution containing NAD^+ (1 mM), glucose 6-phosphate (5 mM), NH_4Cl (2 mM), 1.6 mg (0.00762 μmol) of enzyme, and Tris-HCl (50 mM), pH 7.4 was incubated at 37°C. A control without enzyme and a second control without substrate were incubated at the same time. After 2 min the solutions were chilled quickly and to each was added 1 ml of the following mixture: 0.7 mg, 250 mCi NaB^3H_4 (Amersham); 3 mg $NaBH_4$, and 60 mg Na_2CO_3 in 3 ml H_2O. After 15 min, excess borotritide was destroyed with glacial acetic acid. The two incubations that contained enzyme were deproteinized with warm 70% ethanol, followed by centrifugation. All solutions were then assayed for inositol 1-phosphate (Barnett *et al.*, 1970); only the complete system contained the product, (0.462 μmol), neither control contained detectable inositol 1-phosphate.

Each solution was passed through a small cation exchange column (Fisher, Rexyn 101H), the eluates adjusted to pH 3.5 with $NaHCO_3$, evaporated to dryness, and then evaporated several times with MeOH to remove boric acid. The pH of each solution was raised to 8.0 and the solutions were then incubated with bacterial alkaline phosphatase (Worthington) and $MgCl_2$ (2 mM) to convert phosphates into free alcohols. One mg each of D-glucitol, L-iditol, *myo*-inositol, and *scyllo*-inositol were added to the three solutions. They were then de-ionized with Amberlite MB3 and evaporated to 200 μl for paper chromatographic separation.

Chromatographic Separation of Carrier Alcohols

The solutions were loaded on sheets of Whatman No. 1 paper and subjected to descending chromatography for 13 h in 80% acetone/H_2O. Cyclitols were found at 15 cm from the origin and alditols at 30 cm by silver nitrate staining (Trevelyan et al., 1950) of an edge-strip of the chromatogram. The bands were cut out and compounds were recovered from the paper by descending elution with water. The mixed alditol eluate was evaporated to dryness in preparation for acetylation; the mixed cyclitol eluate was concentrated and loaded on Whatman No. 1 previously washed with 80% phenol/H_2O and subjected to descending chromatography for 96 h in the same solvent. myo-Inositol and scyllo-inositol ($R_{myo\text{-}inositol} = 0.675$) bands were eluted with water and 10 mg of respective carrier were added. The solutions were evaporated to dryness and further dried over P_2O_5 in preparation for silylation.

Purification of Cyclitols as Trimethylsilyl Ethers

The cyclitols were silylated, the products sublimed, and hydrolyzed to free crystalline compounds as described earlier (Chen and Eisenberg, 1975). The specific radioactivity became constant after two repetitions of this sequence. Radioactivity of each product (as the hexakistrimethylsilyl ether) was assayed by liquid scintillation spectrometry in Aquasol.

Separation and Purification of Alditols as Hexaacetates

The dried alditol mixtures were further diluted with 100 mg each of D-sorbitol (Fisher) and L-iditol [prepared by borohydride reduction of L-sorbose (Fisher) at 0°C then separated from glucitol as described below, and deacetylated with NaOMe]. The mixtures were then acetylated with acetic anhydride in pyridine and the acetates precipitated with the addition of ice. Iditol hexaacetate was separated from glucitol hexaacetate by crystallization from methanol at room temperature. It was recrystallized with unlabeled glucitol hexaacetate to eliminate contaminating glucitol radioactivity and then several times more until it was homogeneous by gas-liquid chromatography (ESSG, Applied Science Labs.). Radio-activity of the recrystallized material was assayed in Aquasol. Glucitol hexaacetate, contaminated with iditol hexaacetate, was deacetylated with NaOMe in MeOH. After neutralization with acetic acid, 10 g of sorbitol was added and the mixture dried *in vacuo* in preparation for separation of H_1 and H_5.

Separation of H_1 and H_5 of Glucitol

A procedure for the isolation of H_5 was sought when it was found that iditol, representing putative 5-keto-glucose 6-phosphate, was not labeled by borotritide, raising the possibility that the reduction of the intermediate was exclusively directed toward glucitol formation. The amount of radiolabel in H_5 of glucitol thus became of vital interest. For this determination, glucitol was converted first into 1,3,2,4-diethylideneglucitol (Hockett and Schaefer, 1947) followed by catalytic oxidation to 3,5,4,6-diethylidene-L-gulonic acid (D'Addieco, 1958) and hydrolysis to L-gulonolactone (D'Addieco, 1958). The lactone was degraded by HIO_4 to formaldehyde (H_1 of glucitol), formic acid (H_{2-4}), and glyoxylic acid (H_5). After destruction of HIO_3 and excess HIO_4, the solution was lyophilized to remove volatile formaldehyde and formic acid. Non-volatile glyoxylic acid was crystallized as the semicarbazone and radioactivity in its sodium salt was assayed in Aquasol. For recovery of the compound from Aquasol, formic acid was added causing immediate precipitation of crystalline glyoxylic acid semicarbazone which was then assayed again for radioactivity as its sodium salt. In this way, glyoxylic acid could be purified to constant specific radioactivity with little loss of material. Details of this degradation procedure, including a test with $[1-{}^3H]-$ and $[5-{}^3H]$glucitol will be published elsewhere.

RESULTS AND DISCUSSION

Table I shows the radiolabeling found in the alditols expected from the reduction of 5-ketoglucose 6-phosphate and in the cyclitols from inosose-2 1-phosphate (Fig. 1). The observed values were normalized to constant carrier dilution by multiplying by the dilution factor (F); both observed and normalized (a) values are shown in the

7

complete system but only normalized values are given for controls without enzyme (b) and without substrate (c).

As expected, the bulk of the radioactivity was found in H_1 of glucitol, derived from the reduction of unused substrate (a, 4.54 µmol; b, 5.00 µmol). $H_1(c)$ was a measure of contamination of carrier glucitol by a radioactive impurity. The same amount of radiolabel appeared in $H_5(a)$ and was unrelated to the enzymatic reaction since there was as much radiolabel in $H_5(b)$. Because $H_1(c)$ was relatively small there was no need to examine $H_5(c)$. We concluded that the radiolabel in glucitol $H_5(a)$ was spurious and therefore not the result of reduction of 5-ketoglucose 6-phosphate. The same conclusion applies to iditol; the radiolabel present was contamination from glucitol H_1 and likewise unrelated to the enzymatic reaction. We were thus unable to detect 5-ketoglucose 6-phosphate and conclude that it was cyclized to inosose-2 1-phosphate as rapidly as it was formed from glucose 6-phosphate.

As seen also in the earlier experiment (Chen and Eisenberg, 1975) both cyclitols were radiolabeled in the complete system (a); the unequal radiolabeling is the result of stereospecific reduction in favor of *myo*-inositol. That this radiolabeling is unequivocally related to the enzymatic reaction is shown in the absence of radioactivity in the controls (b and c). Control (b) eliminates the possibility that radiolabeling of the cyclitols is the result of contamination by glucitol H_1. Control (c) eliminates the reduction of free inosose derived from the pathway described by Sherman *et al.* (1968) as the source of labeled cyclitols. A direct test for this enzyme system as a contaminant in purified synthase proved negative (Maeda and Eisenberg, 1980). We conclude that inosose-2 1-phosphate is an intermediate in the synthase reaction.

Besides showing that the cyclitols are radiolabeled in association with the enzymatic reaction, we can correlate the extent of radiolabeling with the amount of intermediate bound to the enzyme. For this purpose it is essential first to ascertain accurately the amount of tritium available for reduction. Ordinarily this is done by adding a reducible compound to borotritide, recovering the reduced product, and measuring its specific radioactivity. In the

TABLE I

Radioactivity of Carriers Isolated after Borotritide Reduction

Inositol 1-Phosphate Synthase Intermediates

Carrier	Dilution*	F**	Observed Complete	Specific Activity cpm/mmol carrier Normalized (Obs. x F)		
				Complete (a)	-Enzyme (b)	-Substrate (c)
glucitol $\diagdown$ H$_1$ $\diagup$ H$_5$	$1{:}7.2 \times 10^6$	1	275,000	275,000	350,000	400
			600	600	550	--
iditol	$1{:}7.2 \times 10^4$	1/100	2,160	22	18	36
scyllo-inositol	$1{:}7.3 \times 10^3$	1/1014	396,000	391***	4	1
myo-inositol	$1{:}7.3 \times 10^3$	1/1014	505,000	498***	4	0

* equal to ratio of concentrations of added carrier to enzyme-bound intermediate, assuming 1 mol intermediate bound per mol enzyme. ** equal to ratio of dilutions with respect to glucitol. *** sum, 889 cpm/mmol.

9

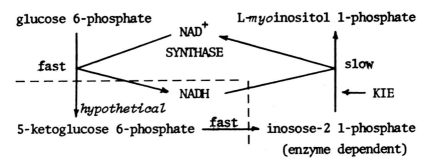

Figure 2. Diagrammatic summary of conclusions. Hypothetical and post-ulated intermediates and inferred fast reactions are shown within the right angle. KIE, kinetic isotope effect.

present experiment H_1 of substrate glucose 6-phosphate (4.54 μmol) met this requirement; the specific radio-activity of glucitol H_1 (a, 275,000 cpm/mmol) is readily measurable as diethylideneglucitol or diethylidenegulonic acid, both crystalline derivatives synthesized from glucitol prior to its degradation as gulonolactone (see Separation of H_1 and H_5 of Glucitol). Assuming 1 mol of either intermediate per mol of enzyme (0.00762 μmol), the specific radioactivity of that intermediate will then be 0.00762 x 275,000/4.54 = 462 cpm/mmol. Since this value is about one half the total observed cyclitol specific radioactivity (889 cpm/mmol, Table 1), we conclude that there are two binding sites per mol of enzyme, both occupied by inosose-2 1-phosphate. This argument is consistent with our earlier finding of two like subunits of the three that comprise the synthase molecule (Maeda and Eisenberg, 1980). Because both sites are presumed to be occupied by inosose-2 1-phosphate, the chance of demon-strating 5-ketoglucose 6-phosphate is reduced to nil, also consistent with our inability to detect that intermediate.

Previously Eisenberg and Bolden (1968), Loewus (1977) and Eisenberg (1978), have shown kinetic hydrogen isotope effects pointing to the involvement of H_5

of glucose 6-phosphate in the synthase reaction. Our demonstration that inosose-2 1-phosphate is sufficiently long-lived to be trapped, while 5-ketoglucose 6-phosphate is not, suggests that the reduction of inosose-2 1-phosphate to inositol 1-phosphate by putative NADH is the rate-limiting step. The isotope effect must then be localized to the transfer of H to the product rather than from the substrate. Fig. 2 summarizes these conclusions.

ACKNOWLEDGEMENTS

The authors acknowledge the help of Mr. Charles Paule in the development of the degradation procedure for glucitol.

REFERENCES

Barnett, J.E.G., Brice, R.E. and Corina, D.L. (1970) *Biochem. J.* *119*, 183-186.

Chen, C.H.-J. and Eisenberg, F. Jr. (1975) *J. Biol. Chem.* *250*, 2963-2967.

D'Addieco, A.A. (1958) *U.S. Patent* 2,847,421; (1959) *Chem. Abstr.* *53*, 3084.

Eisenberg, F. Jr. (1978) in: "*Cyclitols and Phosphoinositides*" (W.W. Wells and F. Eisenberg, Jr., eds.) pp. 269-278, Academic Press, New York.

Eisenberg, F. Jr. and Bolden, A.H. (1968) *Fed. Proc.* *27*, 595.

Hockett, R.C. and Schaefer, F.C. (1947) *J. Am. Chem. Soc.* *69*, 849-851.

Loewus, M.W. (1977) *J. Biol. Chem.* *252*, 7221-7223.

Maeda, T. and Eisenberg, F. Jr. (1980) *J. Biol. Chem.* *255*, 8458-8464.

Sherman, W.R., Stewart, M.A., Kurien, M.M. and Goodwin, S.L. (1968) *Biochim. Biophys. Acta* *158*, 197-205.

Trevelyan, W.E., Procter, D.P. and Harrison, J.E. (1950) *Nature* *166*, 444-445.

REGULATION OF THE LIPID COMPOSITION

OF LUNG SURFACTANT

John E. Bleasdale

Department of Biochemistry, University of Texas
Health Science Center, Dallas, Texas 75235

SUMMARY

During lung development, the phosphatidylinositol
(PI)-rich surfactant that is produced by type II
pneumonocytes of immature lungs is replaced by the
phosphatidylglycerol (PG)-rich surfactant that is
characteristic of mature lungs. Synthesis of
phosphatidylcholine (PC) (the most abundant lipid of
surfactant) increases before the change in the relative
amounts of PG and PI and was found to be accompanied
by an increase in the intracellular amount of CMP.
CMP-dependent synthesis of PG in lung was demonstrated
and its characteristics were consistent with the proposition that CMP promotes the reverse reaction catalyzed
by PI synthase (EC 2.7.8.11) thus decreasing net synthesis of PI and increasing the availability of CDP-
diacylglycerol for PG synthesis. When the amount of CMP
in isolated type II pneumonocytes was increased, directly
or indirectly, synthesis of PG was increased. Synthesis
of PG by these cells was also influenced by extracellular
myo-inositol at concentrations similar to those found in
fetal serum. Availability of *myo*-inositol to lungs, from
either synthesis *in situ* or uptake from the blood, was
found to decline in late gestation. This developmental
decline in *myo*-inositol availability is delayed in pregnancies that are complicated by diabetes mellitus and this
may account for the delayed appearance of PG-rich fetal
lung surfactant during such pregnancies.

INTRODUCTION

In several species, the surfactant that is produced by the type II pneumonocytes of mature lungs contains PC and PG as the two most abundant glycerophospholipids. The surfactant produced by immature lungs, however, contains PI in place of PG. Late in human fetal lung development, PI-rich surfactant is replaced by surfactant rich in PG (Hallman et al., 1977). Although the function of PG in surfactant is controversial (Hallman et al., 1977; Beppu et al., 1983), infants who are born before the appearance of PG-rich surfactant are apparently at increased risk of succumbing to respiratory distress syndrome (Hallman et al., 1977). Furthermore, the appearance of PG-rich surfactant is delayed in the fetuses of women with the less severe forms of diabetes mellitus (Kulovich and Gluck, 1979). This developmental change in surfactant composition, in addition to being clinically important, provides a model for investigating the regulation of the synthesis of PI and PG from their common precursor, CDP-diacylglycerol.

The increase in PG synthesis at the expense of PI synthesis in type II pneumonocytes does not appear to be due to either an induction of the synthesis of CDP-diacylglycerol:glycerol 3-P phosphatidyltransferase (PGP-synthase, EC 2.7.8.5) or a repression of the synthesis of CDP-diacylglycerol:inositol 3-phosphatidyltransferase (PI synthase, EC 2.7.8.11) (reviewed by Possmayer, 1984). The latter enzyme, however, was found to catalyze a reversible reaction in rabbit lung (Bleasdale et al., 1979) as it does in other tissues (Petzold and Agranoff, 1965; Hokin-Neaverson et al., 1977). It was proposed that in vivo the reaction catalyzed by PI synthase is near equilibrium and consequently the net synthesis of PI is influenced by changes in the intracellular concentrations of reactants, viz., CDP-diacylglycerol, myo-inositol, PI, cytidine monophosphate (CMP) (Bleasdale et al, 1979). The evidence presented below supports the proposition that in late gestation, changes in the concentration of CMP and myo-inositol within the type II pneumonocyte favor PG synthesis at the expense of PI synthesis.

REGULATORY FUNCTION OF CMP

When the production of surfactant by fetal type II pneumonocytes is initiated, the synthesis of PC increases and the limited amount of CDP-diacylglycerol that is available is used primarily to synthesize PI. This results in the formation of a surfactant that is characteristic of immature lungs. Later in lung development, because of greatly increased PC synthesis, there is an increase in the intracellular concentration of CMP (a co-product in the reaction catalyzed by cholinephosphotransferase). CMP promotes the reverse reaction catalyzed by PI synthase so that net synthesis of PI diminishes and more CDP-diacylglycerol (synthesized from PA) is now available for PG synthesis. The evidence of such a regulatory function of CMP is as follows: (1) the amount of CMP in lung tissue increases at the time in rabbit lung development when the synthesis of PC for surfactant increases, (2) rabbit lung microsomes catalyze CMP-dependent synthesis of PG, (3) stimulation of PC synthesis by isolated type II pneumonocytes results in an increase in intracellular amounts of CMP and a selective increase in the synthesis of PG, (4) addition of CMP to permeabilized type II pneumonocytes stimulates selectively the incorporation of [^{14}C]glycerol 3-P into PG.

Changes in the amounts of various cytidine nucleotides in rabbit lung during development are summarized in Figure 1 (Quirk et al., 1980). Early in gestation, the most abundant cytidine nucleotide in fetal lung tissue was CTP. Later in gestation, amounts of CTP fell and the amount of CDP-choline increased. On approximately the 26th day of gestation, the synthesis of PC for surfactant was stimulated and this was accompanied by a decrease in the amount of CDP-choline and a doubling of the amount of CMP in lung tissue. Amounts of other nucleoside monophosphates did not increase during this period and an increase in the amount of CMP in liver was not observed (Quirk et al., 1980). These data are indicative that between the 23rd and the 26th day of gestation it is likely that PC synthesis is restricted by the limited availability of diacylglycerol and not CDP-choline. Furthermore, stimulation of PC synthesis is accompanied by an increase in the intracellular concentration of CMP.

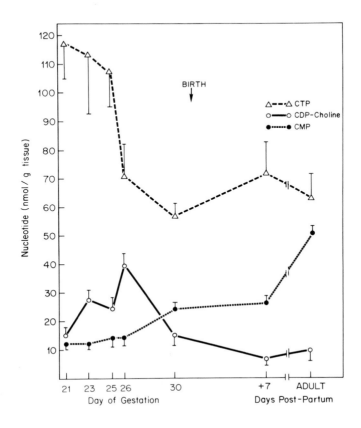

<u>Figure 1</u>. Amounts of CTP, CDP-choline and CMP in rabbit lung tissue during development. Lung tissue was removed rapidly from rabbits at various stages of development and was freeze-clamped in liquid N_2. Nucleotides were extracted from tissue samples, separated by HPLC and then quantitated as described elsewhere (Quirk <u>et</u> <u>al</u>., 1980). Data are mean values ± S.E. derived from 5-12 animals.

The effect of an increase in the intracellular concentration of CMP on the synthesis of PG was investigated. When rabbit lung microsomes were incubated in the presence of [^{14}C]glycerol 3-P, the amount of endogenous CDP-diacylglycerol was too small to support appreciable incorporation of ^{14}C into PG. This incorporation could be increased greatly by the addition of

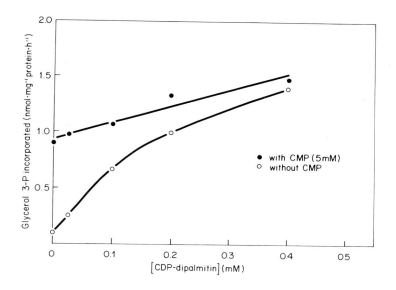

Figure 2. CDP-diacylglycerol-dependent and CMP-dependent incorporation of [^{14}C]glycerol 3-P into phosphatidylglycerol by rabbit lung microsomes. Rabbit lung microsomes (300 µg protein) were suspended in 0.2 ml of Tris-maleate buffer (50 mM, pH 7.4) that contained [^{14}C]glycerol 3-P (0.5 mM, 5 Ci/mol), 2-mercaptoethanol (5 mM), $MnCl_2$ (0.5 mM) and AMP (5 mM). CDP-dipalmitoylglycerol was added at various concentrations and CMP, when added was at a concentration of 5 mM. Reaction mixtures were incubated at 37°C for 30 min. Reactions were terminated and radiolabeled products were quantitated as described by Bleasdale and Johnston (1982). Phosphatidylglycerol accounted for more than 95% of all [^{14}C]glycerol 3-P incorporated. CMP-dependent incorporation of [^{14}C]glycerol 3-P was half-maximal at a CMP concentration of 0.2 mM.

either CDP-diacylglycerol or CMP (Fig. 2). The characteristics of CMP-dependent synthesis of PG described previously (Bleasdale and Johnston, 1982) and summarized below are consistent with the proposed involvement of CMP in the regulation of the lipid composition of surfactant. (1) CMP-dependent incorporation of glycerol 3-P into PGP represents net synthesis and not glycerol 3-P exchange, (2) CMP-dependent and CDP-diacylglycerol-dependent synthesis of PG are not additive, (3) the

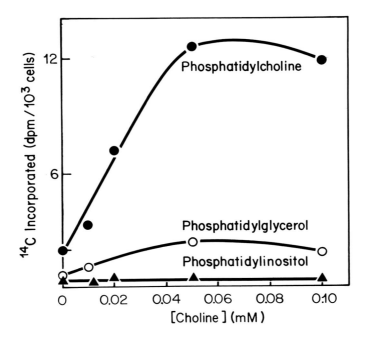

Figure 3. Effect of choline on [^{14}C]glycerol incorporation into glycerophospholipids by choline-depleted type II pneumonocytes. Type II pneumonocytes were depleted of choline (Anceschi et al., 1984) and then incubated for 12 h at 37°C in Dulbecco's modified Eagle's medium that contained [^{14}C]glycerol (0.07 mM, 30 Ci/mol) and choline at various concentrations. Cells were then harvested and the incorporation of ^{14}C into glycerophospholipids was measured (Bleasdale et al., 1983).

concentration of CMP required to support PG synthesis is similar to that found in type II pneumonocytes, (4) inhibitors of PI synthase (that do not affect PGP synthase) inhibit CMP-dependent synthesis of PG, (5) under *in vitro* conditions where there is no synthesis of CDP-diacylglycerol from PA, CMP-dependent synthesis of PG requires microsomal PI. Depletion of microsomal PI using PI-specific phospholipase C resulted in a loss of CMP-dependent synthesis of PG (Bleasdale and Johnston, 1982).

Regulation of the Lipid Composition of Lung Surfactant

Additional evidence of the proposed involvement of CMP came from experiments in which the synthesis of PC by isolated type II pneumonocytes was altered experimentally. Type II pneumonocytes were isolated from adult rat lungs and the incorporation of [^{14}C]glycerol into glycerolipids was measured. Incorporation of [^{14}C]glycerol into PC was relatively insensitive to choline in the extracellular medium, presumably because intracellular amounts of choline were large. It was found, however, that type II pneumonocytes could be depleted of choline by incubation in a medium that lacked choline but contained choline oxidase and catalase (Anceschi *et al.*, 1984). Intracellular choline was reduced from 250 ± 75 to 84 ± 20 pmol/10^6 cells (mean values ± S.D., 7 experiments). PC synthesis by choline-depleted type II pneumonocytes was extremely sensitive to choline in the extracellular medium (Fig. 3). Incorporation of [^{14}C]glycerol into PC was stimulated approximately 7-fold as the extracellular concentration of choline was increased. Incorporation into other glycerophospholipids was largely unaffected except for a 3-fold increase in incorporation into PG. The extracellular concentration of choline that supported maximal incorporation into PG was that which supported maximal PC synthesis. Associated with the choline-induced increase in synthesis of PC and PG was an increase in the amount of CMP in type II pneumonocytes (from 48 ± 9 to 76 ± 16 pmol CMP/10^6 cells, mean values ± S.D., 4 determinations).

The direct effect of CMP on PG synthesis by type II pneumonocytes also has been investigated. Type II pneumonocytes were permeabilized by incubation in calcium-free medium as described by Streb and Schulz (1983). Permeabilized cells were incubated in a potassium-rich buffer (Streb and Schulz, 1983) that was supplemented with ATP (5 mM), [^{14}C]glycerol 3-P (0.1 mM) and CMP at various concentrations. After 30 min, lipids were extracted from the cells and the incorporation of ^{14}C into various lipids was measured (Fig. 4). CMP increased the incorporation of [^{14}C]glycerol 3-P into PG but not into other glycerophospholipids. This effect of CMP was not observed with nonpermeabilized cells and was not reproduced in permeabilized cells if CMP was replaced by cytosine-β-D-arabinofuranoside 5'-monophosphate. Raetz *et al.* (1977) found that cytosine-β-D-

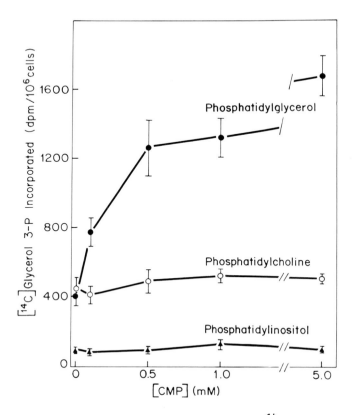

Figure 4. Effect of CMP on the incorporation of [^{14}C]glycerol 3-P into glycerophospholipids by permeabilized type II pneumonocytes. Type II pneumonocytes (approx. 2×10^6 cells/ml) that were permeabilized by exposure to calcium-free medium were incubated in a potassium-rich buffer (Streb and Schulz, 1983) that was supplemented with ATP (5 mM), [^{14}C]glycerol 3-P (0.1 mM) and CMP at various concentrations. After 30 min at 37°C, lipids were extracted from the cells and incorporation of ^{14}C into various glycerophospholipids was measured (Bleasdale et al., 1983). Data are mean values ± S.E. derived from 4 experiments.

arabinofuranoside 5'-diphosphate-1,2-dipalmitin supported the synthesis of PI but not PG by subcellular fractions of rat liver. As in the case of choline-stimulated PG synthesis, CMP increased the incorporation of [^{14}C]glycerol 3-P into PG without decreasing the incorporation into PI. Therefore, the total utilization (and presumably synthesis) of CDP-diacylglycerol was apparently increased by CMP. The effect of CMP on the synthesis of PA or CDP-diacylglycerol by type II pneumonocytes is unknown. Possmayer (1974), however, reported that synthesis of PA from glycerol 3-P by rat brain microsomes was greatest when the cytidylate energy charge was low.

These data support the hypothesis that an increase in the intracellular concentration of CMP that accompanies stimulated synthesis of PC for surfactant, influences the reaction catalyzed by PI synthase such that net synthesis of PI is decreased and PG synthesis is increased.

EFFECT OF *MYO*-INOSITOL ON SURFACTANT COMPOSITION

The relative rates of utilization of CDP-diacylglycerol for the synthesis of PG and PI are also influenced by the availability of *myo*-inositol. A decreasing availability of *myo*-inositol during development would favor the synthesis of PG-rich surfactant. For this reason, fetal homeostasis of *myo*-inositol in the rabbit, rat, and human was investigated. Fetal rabbit lungs expressed activity of L-*myo*-inositol 1-P synthase, the enzyme that catalyzes the putative rate-limiting reaction in *myo*-inositol synthesis. The specific activity of L-*myo*-inositol 1-P synthase in fetal rabbit lung on day 28 of gestation was greater than that in adult lung [81.0 ± 9.0 (7) and 23.2 ± 1.0 (3) nmol x h^{-1} x g^{-1} tissue respectively, mean values ± S.E. for the number of determinations in parentheses] (Bleasdale *et al.*, 1982).

Fetal rabbit lungs, in addition to having the potential for synthesis of *myo*-inositol, were found to take up *myo*-inositol from the extracellular medium. The uptake of *myo*-inositol by slices of fetal rabbit lung was found to be specific, energy-dependent and required Na^{+} ions (Bleasdale *et al.*, 1982). *Myo*-inositol uptake was saturable and half-maximal uptake occurred at a concentration of *myo*-inositol of approximately 0.1 mM. A similar affinity for uptake was found for slices of adult rabbit lung tissue. The significance of this latter finding is that the concentration of *myo*-inositol in fetal rabbit serum is initially high enough to support maximal uptake of *myo*-inositol by the lungs but then, in late gestation,

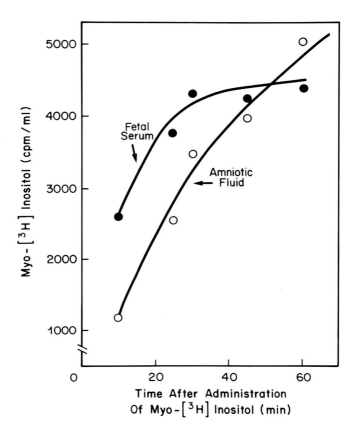

Figure 5. Placental transfer of myo-[³H]inositol to the fetal rabbit. Pregnant New Zealand white rabbits were anesthetized on the 28th day of gestation and myo-[³H]inositol (40 µCi, 12.5 Ci/mmol) in 2 ml of sorbitol solution (311 mM) was administered via the marginal vein of the ear. After various periods of time, fetuses were removed individually and ³H in fetal serum and amniotic fluid was quantitated. myo-Inositol in serum and amniotic fluid was separated by paper chromatography as described (Bleasdale et al., 1982) and was found to account for more than 95% of total ³H in the samples.

falls toward adult levels that are insufficient to support maximal uptake. A similar developmental decline in the concentration of *myo*-inositol in the serum of other species has been described (Campling and Nixon, 1954; Burton and Wells, 1974). Some of the *myo*-inositol in fetal rabbit serum may originate from the maternal circulation by transplacental transfer since it was found that when *myo*-[^{3}H]inositol was administered intravenously to pregnant rabbits there was a time-dependent appearance of *myo*-[^{3}H]inositol in the serum and amniotic fluid of their fetuses (Fig. 5). When carrier-free *myo*-[^{3}H]inositol (1 nmol) was administered intravenously to pregnant rabbits the half-time for disappearance of ^{3}H from the serum was 22.8 min and this was increased to 55.8 min if an excess of nonradiolabeled *myo*-inositol (0.6 mmol) was injected simultaneously. When the larger dose of *myo*-inositol was administered to pregnant rabbits, the absolute amount of *myo*-inositol in fetal serum after 60 min was 2.5-times greater than with the lower dose.

Although transplacental transfer may account for some of the *myo*-inositol in fetal rabbit serum, it is also likely that much of the *myo*-inositol is synthesized in the placenta where a high specific activity of L-*myo*-inositol 1-P synthase was measured (527 ± 64 nmol x h^{-1} x g^{-1} tissue, mean ± S.E., 6 determinations) (Bleasdale *et al.*, 1982). It was concluded that fetal rabbit lungs obtain *myo*-inositol by synthesis *in situ* and by uptake from the blood. The availability of *myo*-inositol from both sources, however, apparently declines in late gestation. Evidence for a declining availability of *myo*-inositol to the fetal lungs of the rat (Quirk *et al.*, 1984) and the human (Quirk and Bleasdale, 1983) also was obtained.

The influence of extracellular *myo*-inositol on the relative rates of synthesis of PG and PI by type II pneumonocytes was examined (Bleasdale *et al.*, 1983). Cells were incubated for 24 h in medium that contained [^{14}C]-glycerol (0.1 mM) and *myo*-inositol at various concentrations. Incorporation of ^{14}C into various lipids was measured. As the concentration of *myo*-inositol in the medium was increased, the rate of *myo*-inositol uptake increased and there was a decreased incorporation of ^{14}C into PG relative to that into PI (Fig. 6). The concentration at which *myo*-inositol had the most pronounced effect on the relative rates of synthesis of PG and PI by type II pneumonocytes were those found in fetal serum in late gestation. During development, therefore, a declining availability of *myo*-inositol to the fetal lungs favors PG synthesis at the expense of PI synthesis.

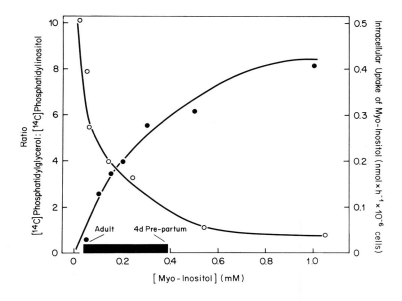

<u>Figure 6</u>. Dependence of <u>myo</u>-inositol uptake and the synthesis of phos-phatidylinositol and phosphatidylglycerol by type II pneumonocytes upon extracellular <u>myo</u>-inositol. Type II pneumonocytes from adult rats were incubated in Dulbecco's modified Eagle's medium that contained <u>myo</u>-inositol at various concentrations. Either [^{14}C]glycerol (0.1 mM, 10 Ci/mol) or <u>myo</u>-[^{3}H]inositol was added to the cell suspensions. Incorp-oration of [^{14}C]glycerol into glycerolipids after 24 h at 37°C and the uptake of <u>myo</u>-[^{3}H]inositol after 20 min at 37°C were measured as described (Bleasdale <u>et al</u>., 1983). The data are expressed as the ratio of ^{14}C incorporated into PG to that incorporated into PI (O) and as nmol of <u>myo</u>-inositol taken up per h by 10^{6} cells (O). A half-maximal rate of uptake of <u>myo</u>-inositol occurred at a <u>myo</u>-inositol concentration of 0.3 mM. Superimposed on the abscissa is the range of concentrations of <u>myo</u>-inositol that were measured in rat serum at various stages of development between the 18th day of gestation (4 days before birth) and adulthood (Bleasdale <u>et al</u>., 1982).

We (Bleasdale *et al.*, 1982) and Hallman and Gluck (1982) proposed that the imbalance of *myo*-inositol homeo-stasis that is known to occur in the patient with diabetes mellitus persists in the pregnant diabetic woman and this affects adversely *myo*-inositol homeostasis in her fetus.

Consequently, the normal developmental decline in *myo*-inositol availability to the fetal lungs is retarded and this accounts for the observed delay in the appearance of PG-rich surfactant in the lungs of such fetuses. This possibility has been tested using experimental animals. Female rabbits were rendered hyperglycemic with alloxan and 5 days later were mated. On the 29th day of gestation, pregnant rabbits were sacrificed. The concentrations of glucose in the serum of alloxan-treated rabbits and their fetuses were approximately twice those in saline-treated control animals (Table I). The concentration of *myo*-inositol in the serum of adult rabbits was only marginally increased by alloxan treatment but was greatly increased in the serum and amniotic fluid of their fetuses (Table I). Slices of lung tissue from adult rabbits and their fetuses were prepared and then incubated in medium that contained either [^{3}H]choline or [^{14}C]glycerol. Incorporation of [^{3}H]choline into PC, lyso-PC and sphingomyelin per mg of tissue was greater in fetal lung than in adult lung, but in neither case was there an effect of alloxan treatment. Likewise, the incorporation of [^{14}C]glycerol into lipids was greater in fetal lung than in adult lung and incorporation into most lipids was unaffected by alloxan. One exception, however, was the incorporation of [^{14}C]glycerol into PG by fetal lung which was decreased as a result of alloxan treatment (Fig. 7). When the specific activities of PGP synthase and PI synthase in the lungs of fetuses of alloxan-treated rabbits were assayed under optimal *in vitro* conditions, no effect of alloxan treatment was observed (data not shown). Therefore, the likely explanation for the decreased incorporation of [^{14}C]glycerol into PG by lungs of fetuses of alloxan-treated rabbits is the hyperinositolemia of such fetuses.

The possibility that hyperinositolemia itself can alter the lipid composition of lung surfactant was tested in experiments in which *myo*-inositol was administered intraperitoneally to pregnant rats that were given access to drinking water that contained *myo*-inositol. On day 21 of gestation, after four days of treatment, the rats were sacrificed and surfactant was lavaged from their lungs. In the surfactant of rats that were treated with *myo*-inositol, PG was almost entirely replaced by PI. The concentration of *myo*-inositol in the serum of treated rats

TABLE I

Serum Concentrations of Glucose and *Myo*-Inositol
in Normal and Alloxan-Treated Pregnant Rabbits

	Saline-Pretreated	Alloxan-Pretreated
[Glucose] (mg/dl) on day 29 of gestation:		
Adult Serum	103 ± 10	203 ± 46
Fetal Serum	66 ± 5	138 ± 24
[*Myo*-Inositol] (μM) on day 29 of gestation:		
Adult Serum	30 ± 4	42 ± 7
Fetal Serum	142 ± 14	263 ± 37
Amniotic Fluid	38 ± 10	70 ± 13

New Zealand white rabbits were treated with alloxan (50 mg/kg) intravenously and 5 days later were mated. Pregnant animals were killed on the 29th day of gestation and the concentrations of glucose and myo-inositol in the serum of the pregnant animals and their fetuses and in amniotic fluid were determined as described (Bleasdale et al., 1982). Data are mean values ± S.E. derived from 6 alloxan-pretreated pregnant animals and 5 saline-pretreated pregnant animals. All changes due to alloxan pretreatment, except the change in the concentration of myo-inositol in adult rabbit serum, were statistically significant (p<0.05, student t test).

was increased greatly (from 0.06 to 0.67 mM). The concentration of *myo*-inositol in serum of fetuses of these rats also was abnormally high (0.52 mM compared to 0.23 mM in saline-treated animals). Furthermore, the normal developmental decline in the concentration of *myo*-inositol in fetal serum was altered (Quirk and Bleasdale, 1984). Treatment of pregnant rats with *myo*-inositol significantly decreased the molar ratio of PG to PI in the maternal lung tissue, maternal lung lavage and fetal lung tissue on day 21 of gestation (Fig. 8). Thus, an alteration in fetal homeostasis of *myo*-inositol was accompanied by a failure to synthesize PG-rich surfactant.

Regulation of the Lipid Composition of Lung Surfactant

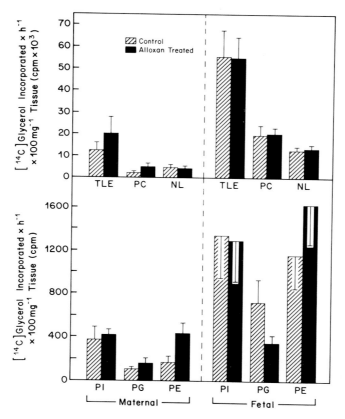

Figure 7. Incorporation of [14C]glycerol into lipids by slices of lung tissue from fetuses and from adult pregnant rabbits. Either NaCl solution (0.154 M) or alloxan (50 mg/kg) was administered intravenously to adult female New Zealand white rabbits. Five days later, the rabbits were mated and on the 29th day of gestation they were killed. Slices (0.3 mm thick) of fetal and maternal lung tissue were prepared using a McIlwain tissue chopper and were incubated in Krebs-Henseleit medium that contained [14C]glycerol (0.1 mM, 30 Ci/mol) for 1 h at 37°C. Slices were then collected on glass-fiber filters, washed twice with ice-cold NaCl solution (0.154 M), blotted dry and weighed. Lipids were extracted from slices and 14C incorporated into various lipids was measured as described (Bleasdale et al., 1983). TLE, total lipid extract; NL, neutral lipids. Alloxan pre-treatment caused statistically significant (p<0.05 student t test) changes in the incorporation of [14C]glycerol into PG in fetal lung slices and into PE in maternal lung slices.

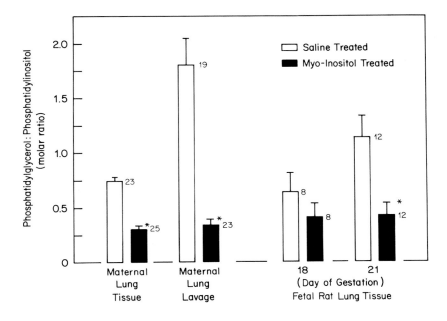

Figure 8. Effect of <u>myo</u>-inositol administration to pregnant rats on the molar ratio of phosphatidylglycerol to phosphatidylinositol in lung tissue and surfactant. Pregnant rats were treated with <u>myo</u>-inositol [180 mg intraperitoneally twice daily plus access to drinking water that contained 0.39 M <u>myo</u>-inositol] beginning on the 14th or 17th day of gestation. Control animals received NaCl solution (0.154 M) intraperitoneally and were given access to tap-water for drinking. After 4 days of treatment, rats were killed and the molar ratios of PG to PI in maternal lung tissue, maternal lung lavage (day 21 of gestation) and fetal lung tissue (day 18 and day 21 of gestation) were measured using methods described elsewhere (Bleasdale <u>et</u> <u>al</u>., 1983). Data are mean values ± S.E. derived from the number of determinations shown. Statistically significant differences from controls ($p < 0.05$, student t test) are indicated with asterisks.

CONCLUSION

The data discussed above are consistent with the hypothesis that CMP and *myo*-inositol are involved in the regulation of the PG-content of surfactant. Late in gestation when the synthesis of PC for surfactant increases there is a concomitant increase in the intracellular amount of CMP. During this time, the availability of *myo*-inositol to the fetal lungs is decreasing. CMP in increasing amounts and *myo*-inositol in decreasing amounts reduce the net synthesis of PI and favor increased util-ization of CDP-diacylglycerol for PG synthesis. This mechanism provides a biochemical explanation for the regulation of the PI and PG content of surfactant and for the integration of the synthesis of these lipids with that of PC. In addition, abnormal homeostasis of *myo*-inositol in the fetus of the pregnant woman with diabetes mellitus can account for the delayed appearance of fetal surfactant that is rich in PG.

ACKNOWLEDGEMENTS

These investigations were supported by USPH grant HD14373. The technical assistance of Ms. N. Tyler, Mr. J. Head and Mr. G. Rader and the editorial assistance of Mrs. D. Tutton is gratefully acknowledged.

REFERENCES

Anceschi, M.M., Di Renzo, G.C., Venincasa, M.D. and Bleasdale, J.E. (1984) *Biochem. J. 224*, 253-262.

Beppu, O.S., Clements, J.A. and Goerke, J. (1983) *J. Appl. Physiol. 55*, 496-502.

Bleasdale, J.E. and Johnston, J.M. (1982) *Biochim. Biophys. Acta 710*, 377-390.

Bleasdale, J.E., Maberry, M.C. and Quirk, J.G. (1982) *Biochem. J. 206*, 43-52.

Bleasdale, J.E., Tyler, N.E., Busch, F.N. and Quirk, J.G. (1983) *Biochem. J. 212*, 811-818.

Bleasdale, J.E., Wallis, P., MacDonald, P.C. and Johnston, J.M. (1979) *Biochim. Biophys. Acta 575*, 135-147.

Burton, L.E. and Wells, W.W. (1974) *Develop. Biol. 37*, 35-42.

Campling, J.D. and Nixon, D.A. (1954) *J. Physiol. (London) 126*, 71-80.

Hallman, M., Feldman B.H., Kirkpatrick, E. and Gluck, L. (1977) *Pediatr. Res. 11*, 714-720.

Hallman, M., Wermer, D., Epstein, B.L. and Gluck, L. (1982) *Am. J. Obstet. Gynecol. 142*, 877-882.

Hokin-Neaverson, M.R., Sadeghian, K., Harris, D.W. and Merrin, J.S. (1977) *Biochem. Biophys. Res. Commun.* *78*, 364-371.
Kulovich, M.V. and Gluck, L. (1979) *Am. J. Obstet. Gynecol.* *135*, 64-70.
Petzold, G.L. and Agranoff, B.W. (1965) *Fed. Proc.* *24*, 426.
Possmayer, F. (1974) *Biochem. Biophys. Res. Commun.* *61*, 1415-1426.
Possmayer, F. (1984) in: *"Pulmonary Surfactant"* (B. Robertson, L.M.G. van Golde and J.J. Batenburg, eds.) pp. 295-355, Elsevier Biomedical Press, Amsterdam.
Quirk, J.G., Baumgarten, B. and Bleasdale, J.E. (1984) *J. Perinat. Med.* (in press).
Quirk, J.G. and Bleasdale, J.E. (1983) *Obstet. Gynecol.* *62*, 41-44.
Quirk, J.G., Bleasdale, J.E., MacDonald, P.C. and Johnston, J.M. (1980) *Biochem. Biophys. Res. Commun.* *95*, 985-992.
Raetz, C.R.H., Chu, M.Y., Srivastava, S.P. and Turcotte, J.G. (1977) *Science* *196*, 303-305.
Streb, H. and Schulz, I. (1983) *Am. J. Physiol.* *245*, G347-G357.

NUTRITIONAL REGULATION OF THE

COMPOSITION, METABOLISM, AND

FUNCTION OF CELLULAR PHOSPHATIDYLINOSITOL

Bruce J. Holub

Department of Nutrition, University of Guelph,
Guelph, Ontario, Canada N1G 2W1

SUMMARY

Molecular species analyses have revealed that in most mammalian cells the 1-stearoyl 2-arachidonoyl species of phosphatidylinositol predominates greatly over other species and this may be significant in the functioning of this phospholipid. Tracer studies conducted *in vivo* and enzyme studies have suggested the importance of deacylation-reacylation reactions for the formation of the 1-stearoyl 2-arachidonoyl species and the enrichment of phosphatidylinositol with arachidonic acid via acyl-CoA:acyl-*sn*-glycero-(3)phosphoinositol acyltransferase activity. Our laboratory has also been interested in the influence of nutritional status on the physiological levels of *myo*-inositol and phosphatidylinositol in health and disease and on the fatty acid composition of phosphatidylinositol since such dietary-induced alterations could modify its cellular functioning. Inositol-deficient diets have been shown to produce a reduction in phosphatidylinositol levels and a marked accumulation of hepatic triglyceride in several animal species. The consumption of certain types of dietary fatty acids (*e.g.*, fish oils containing eicosapentaenoic acid) produces a lowering of the arachidonoyl phosphatidylinositol in human platelets and provides for the introduction of novel molecular species (eicosapentaenoyl phosphatidylinositol) which may be related to the accompanying reduced platelet aggregation.

Dietary cholesterol supplementation was found to cause an elevation in the concentration of total and arachidonoyl phosphatidylinositol in the platelets of gerbils. Human subjects with hypercholesterolemia also have elevated levels of phosphatidylinositol and cholesterol (expressed as $\mu g/10^8$ platelets) which may be associated with the increased platelet aggregability. Patients with chronic renal failure exhibit a dramatic hyperinositolemia without a marked alteration in the circulating levels of lipoprotein-bound phosphatidylinositol although its fatty acid composition is abnormal. These results indicate that the concentration of cellular phosphatidylinositol and the relative amounts of the various molecular species, including the arachidonoyl species, are subject to significant nutritional modification which may offer clinical applications in the control of phosphatidylinositol-mediated physiological responses.

INTRODUCTION

Considerable interest has arisen recently in the composition, metabolism, and function of phosphatidylinositol (PI) at the level of individual molecular species of PI in various animal tissues and cells. Attention to this area was stimulated by observations that different molecular species exhibit a metabolic heterogeneity which may relate to their cellular functioning. Our laboratory has studied the composition of individual molecular species of PI in various mammalian tissues and cells and has attempted to elucidate and characterize the metabolic pathways and enzyme-catalyzed reactions leading to their formation. Various cellular functions and responses appear to be mediated by the enhanced turnover of arachidonoyl and other species of membrane PI including the generation of 1,2-diacylglycerol, phosphatidic acid, PI 4-phosphate (PIP), PI 4,5-bisphosphate (PIP$_2$), and free arachidonic acid for prostaglandin synthesis. Therefore, we have also investigated the influence of nutrition (e.g., dietary myo-inositol levels, type of dietary fat, dietary cholesterol levels) on the amount and fatty acyl compositions of PI, with some emphasis on the complement of arachidonoyl species, in an attempt to provide opportunities for non-pharmacological control of PI-mediated cell responses. In addition, assessment of the levels of serum myo-inositol (inositol,) the amounts of lipoprotein-PI and cell membrane-associated PI and their component molecular species in certain human disease states (*e.g.*, chronic renal failure, hypercholesterolemia) have revealed biochemical abnormalities which may possibly be related to certain aspects of the abnormal cellular functions and responses which exist in many of these patients.

MOLECULAR SPECIES OF PHOSPHATIDYLINOSITOL

It is of considerable interest that stearic and arachidonic acids represent the major fatty acids in phosphatidylinositol isolated from almost all mammalian tissues and that they are nearly exclusively located in the *sn*-1- and *sn*-2 positions, respectively, of the *sn*-glycero(3)phosphoinositol backbone (Holub, 1982). Molecular species analyses using gas-liquid chromatography after argentation thin-layer chromatography of intact PI or its 1,2-diacylglycerol acetate derivatives, in combination with specific enzyme hydrolyses (phospholipases C and A_2 and pancreatic lipase), have revealed that PI from sources such as rat liver (Holub and Kuksis, 1971) and human platelets (Mahadevappa and Holub, 1982) is much more enriched in tetraenoic species than are phosphatidylcholine, phosphatidylethanolamine, or phosphatidylserine isolated from the same sources. As seen in Fig. 1, which depicts the molecular species composition of human platelet PI, the 1-stearoyl 2-arachidonoyl species greatly predominated in this phospholipid (71 mol% of total) followed by the 1-stearoyl 2-oleoyl species (12%). This predominance of the 1-stearoyl 2-arachidonoyl species is characteristic of PI isolated from rat liver (77 mol%) as well as PI, PIP and PIP_2 isolated from bovine brain (Holub and Kuksis, 1971; Holub *et al.*, 1970) although exceptions do exist in mammalian sources (*e.g.*, lamb liver where the monoenes predominate, Luthra and Sheltawy, 1972). It is of interest that, in all of the above analyses, the preferential pairing of arachidonate in the 2-position with stearate (over palmitate) in the 1-position has been found to be much more restrictive than when oleate resides in the 2-position of PI. It is of interest that the existence of supraenoic molecular species (more unsaturated than hexaenes) have been reported in bovine retina PI where they represent 12% of the molecular species in rod outer segment PI (Aveldano and Bazan, 1983).

33

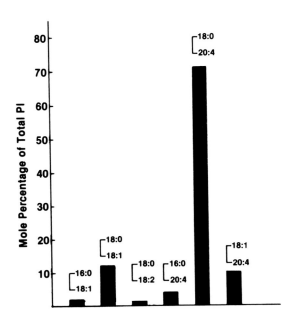

<u>Figure 1</u>. The molecular species composition of phosphatidylinositol from the human platelet. The upper and lower fatty acids in each pairing represent those in the <u>sn</u>-1 and <u>sn</u>-2-positions, respectively. Molecular species contributing less than 1 mol% to the total PI have been omitted from the figure.

Comparative studies on platelet PI isolated from different mammalian sources have revealed significant differences; the arachidonate-containing species account for approximately 76, 71, 70, 56, 52, and 52% of the total PI in the case of human, pig, rat, gerbil, seal, and horse, respectively; much more dramatic differences were found also across animal species in the percentages of oleate and linoleate in PI (Holub *et al.*, unpublished observations). These results suggest that, before extrapolating to the human platelet, caution needs to be exerted in evaluating the contribution of arachidonoyl PI to phospholipid turnover and arachidonic acid release when the data are derived from experiments with non-human platelets. It remains to be established to what extent the cellular functioning of PI may be attributed to its domination by 1-stearoyl 2-arachidonoyl species.

METABOLIC ORIGIN OF ARACHIDONOYL PHOSPHATIDYLINOSITOL

It has been of continued interest for us during the past several years to evaluate if the *de novo* pathway alone, as operative *in vivo*, could account for the preponderance of tetraenoic molecular species of PI in rat liver PI (Holub, 1978, 1982). One approach was to study the distribution of [^{3}H]inositol among the various classes of PI soon after the intraperitoneal injection of [^{3}H]inositol. Work from our laboratory had indicated that the [^{3}H]inositol which enters rat liver PI under physiological conditions does so mainly by the *de novo* pathway and not by the exchange reaction. The specific radioactivities of the total and individual species of PI were well below their eventual maxima at early times following injection such that the distribution of radioactivity likely reflects the species of PI formed by CDP-diacylglycerol:inositol phosphatidyltransferase (PI synthase) under physiological conditions. Using this approach, we observed that the monoenoic plus dienoic species represented 37% and the tetraenoic species accounted for 41% of the PI formed *in vivo* via the PI synthase. In contrast, 83% of the total PI in liver is represented by tetraenoic (1-saturated 2-arachidonoyl) species on a molar basis.

The potential importance of deacylation-reacylation reactions for the eventual enrichment of PI in arachidonate in the 2-position was suggested by the observed transfer of [^{3}H]inositol (and radiolabeled orthophosphate and glycerol in other experiments) from monoenoic plus dienoic to tetraenoic molecular species over extended time periods following injection of these radiolabeled precursors for *de novo* synthesis. These combined results suggested, therefore, that the reaction of free inositol with endogenous arachidonoyl CDP-diacylglycerol catalyzed by PI synthase can account for only approximately one-half of the natural abundance of tetraenoic PI whereas the other half originates by a retailoring cycle involving lyso-PI intermediates as depicted in Fig. 2.

Evidence for such a biochemical sequence has been forthcoming from *in vitro* studies using liver preparations which have demonstrated the predominant formation of tetraenoic species (74% of total) of PI from radiolabeled

35

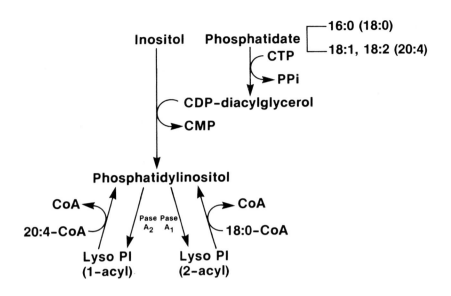

<u>Figure 2</u>. Schematic representation of the interconversion of different molecular species of phosphatidylinositol by deacylation-reacylation reactions and the formation of the 1-stearoyl 2-arachidonoyl species. Abbreviations: lyso PI, lysophosphatidylinositol; Pase, phospholipase.

1-acyl-*sn*-glycero(3)phosphoinositol (1-acyl-GPI) in the absence of added acyl-CoA (incubations containing ATP, CoA, and MgCl$_2$) as well as the existence of microsomal acyl-CoA:1-acyl-GPI and acyl-CoA:2-acyl-GPI acyltransferases with affinities for arachidonoyl-CoA and stearoyl-CoA, respectively, as substrates (Holub, 1976; Holub and Piekarski, 1979). Thus, the acylation reactions given in Fig. 2 could provide for the formation of a 1-stearoyl 2-arachidonoyl species of PI which cannot be readily accounted for by the *de novo* pathway alone. The acyl-CoA:1-acyl-GPI acyltransferase in rat brain and human platelet microsomes has also been found to exhibit a selectivity towards arachidonoyl-CoA (Baker and Thompson, 1973; Kameyama *et al.*, 1983). The existence of phospholipases A$_1$ and A$_2$ which can deacylate PI in mammalian tissues and thereby generate the intermediary (1-acyl) and (2-acyl)lysoPI depicted in Fig. 2 have been reported (Gray and Strickland, 1982; Hirasawa *et al.*,

1981). In accordance with evidence that the *de novo* pathway alone cannot totally account for the enrichment of liver PI with 1-stearoyl 2-arachidonoyl species, CDP-di-acylglycerol is derived from phosphatidate which is rich in palmitate in the 1-position and oleate and linoleate in the 2-position but limited with respect to stearate and arachidonate in the 1- and 2-positions, respectively (Possmayer *et al.*, 1969) as represented in Fig. 2. It has been reported that rat liver CDP-diacylglycerol has a stearate and arachidonate content which is intermediary between that of phosphatidate and CDP-diacylglycerol (Thompson and MacDonald, 1977).

NUTRITIONAL REGULATION OF CELLULAR LEVELS OF PHOSPHATIDYLINOSITOL AND CONSTITUENT MOLECULAR SPECIES

Effect of Dietary Inositol

Despite the capacity of certain animal tissues (*e.g.*, testis, mammary gland, liver, kidney) to provide for the endogenous biosynthesis of inositol, it has been well documented that the level of dietary inositol can have a significant influence on the level of free inositol in the circulation as well as the tissue level of its lipid-bound form, PI. The feeding of an inositol-deficient diet to the neonatal and developing rat was found to result in lower free inositol levels in plasma and all tissues studied (testis, liver, lung, heart, lens, kidney, and small intestine) than those found in the inositol-supplemented controls with the exception of the cerebrum and cerebellum (Burton *et al.*, 1976). Interestingly, only the liver exhibited lower levels of lipid-bound inositol. Recently, Chu and Geyer (1983) have reported dramatically lower plasma inositol levels (by 85%) and more moderately lower levels of inositol in the intestine, liver, kidney, and pancreas, but not in the brain, of inositol-deficient female gerbils as compared to supplemented controls; lipid-bound inositol levels were lower only in the intestine and liver. Normal rats consuming a standard diet containing 1% inositol have been found by Greene and Lattimer (1983) to have plasma and sciatic nerve concentrations of free inositol which were greater by 560% and 30%, respectively, than those of rats ingesting inositol at a dietary level of only 0.01%. A significant elevation of the mean concentration of inositol in serum (by 80%) has been observed in diabetic patients given two 500 mg supplements of inositol daily (Salway *et al.*, 1978); this supplementary amount (1 g) is similar to the estimated daily intake of humans (Clements and Reynertson, 1977).

It has been recognized for several years that dietary inositol can function as a lipotropic factor. Depending upon the overall composition of the experimental diet employed and other factors, it has been possible to produce a significant accumulation of hepatic lipid in the rat (Andersen and Holub, 1980), gerbil (Hoover *et al.*, 1978), and other animals with the omission of supplementary inositol despite the presence of dietary choline, another lipotrope, at levels which are considered to be nutritionally adequate. The rainbow trout (*Salmo gairdneri*) is a particularly sensitive animal with regard to the development of hepatic lipid abnormalities in association with a depletion in membrane PI when deprived of dietary inositol (Holub *et al.*, 1982). In a collaborative study with Professor B. Woodward, we fed to rainbow trout semi-purified diets containing choline with or without a supplement of inositol (500 mg/kg diet). Despite the absence of any difference in liver weights between the two groups after eight weeks, lipid analyses revealed that inositol-deprived fish had total triglyceride and cholesterol levels (in mg/g liver) which were greater by 488% and 56%, respectively, than those for control animals.

In striking contrast to the neutral lipid responses, total phospholipid, phosphatidylcholine, and phosphatidylethanolamine concentrations were moderately lower (by 24–30%) in deficient fish whereas the hepatic level of PI was 68% lower in the inositol-deficient group. When the data is calculated on a fat-free liver weight basis, PI is the only phospholipid type to exhibit significantly lower concentrations in the inositol-deprived group. A depletion in membrane PI may impede lipoprotein formation and triglyceride transport from liver in inositol-deficient animals as observed in gerbils (Hoover *et al.*, 1978) and rats (Burton and Wells, 1979). Alternatively, Hayashi *et al.* (1978) have attributed the triglyceride accumulation to a stimulated lipolysis in adipose tissue that is mediated by an excitation of the sympathetic nerve terminals causing

38

an enhanced mobilization of fatty acids to liver. As a third potential mechanism, Beach and Flick (1982) have suggested that the liver lipodystrophy occurring during inositol-deficiency may arise from elevated levels of lipogenic enzymes (acetyl-CoA carboxylase and fatty acid synthetase). The observation that the percentage of oleic acid in intestinal PI is increased in inositol deficiency, in gerbils consuming diets containing coconut oil, has led to the suggestion that an altered complement of PI molecular species may possibly be related to the resulting intestinal lipodystrophy (Woods and Hegsted, 1979; Chu and Hegsted, 1980).

Effect of Dietary Fat Type

Manipulation of the fatty acid profile of the dietary fat has provided an approach whereby the molecular species composition of cellular PI, including the arachidonoyl PI, can be significantly altered without an accompanying change in total PI concentrations. In addition, it may provide for the introduction of essentially novel molecular species into naturally-occurring PI. Our interest in this area of research has focused upon the effect of dietary fish oils containing eicosapentaenoic acid (20:5ω3) and other fatty acids on the level and fatty acid composition of PI in human platelets in relation to platelet reactivity (Ahmed and Holub, 1983).

This work was stimulated by knowledge that dietary fish oils containing eicosapentaenoic acid diminish platelet aggregation and prolong bleeding times in human subjects, that PI may be closely involved in platelet aggregation, and that previous research had not studied the effect of fish oil supplementation on the fatty acid composition of human platelet PI separated from other phospholipids. In our experimental design, bleeding times, platelet aggregability (thrombin-induced), and biochemical analyses were conducted on three different occasions from each of seven volunteers – on day 0, after receiving a cod-liver supplement for 14 days, and after having terminated supplementation for a period of 14 days (day 28). The PI levels ($\mu g/10^9$ platelets) and the percentage contribution of PI to the total phospholipid were not significantly altered by the dietary manipulation. By day 14, however, the eicosapentaenoic acid level in

39

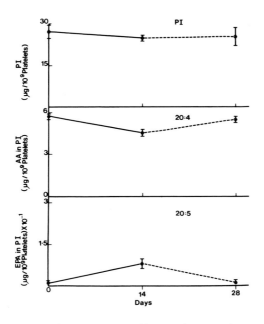

Figure 3. Effect of cod-liver oil supplementation on phosphatidyl-inositol in human platelets. Blood samples were taken before receiving the supplement (day 0), after receiving the supplement for 14 days, and 14 days after terminating the supplement (day 28). Values are mean values ± S.E. (n = 7). Abbreviations: AA, arachidonic acid (20:4); EPA, eicosapentaenoic acid (20:5).

platelet PI rose markedly over the very minimal basal (day 0) values, whereas a moderate concomitant drop in arachidonoyl PI was observed (Fig. 3).

The subjects also exhibited decreased platelet aggregability and increased bleeding times at day 14 relative to basal values. Interestingly, the absolute level of eicosapentaenoic acid accumulation in PI was less than that in the other diacyl phospholipids. These alterations, as well as other minor changes in the fatty patterns of PI, approached normal values by day 28 (following term-ination of supplementation). It is possible that the observed moderate diet-induced perturbations in the fatty acid composition of PI may be related to the altered platelet aggregability and platelet-vessel wall interactions.

It is of interest to note that, although the level of accumulation of eicosapentaenoic acid in platelet PI is moderate with dietary supplementation, our laboratory has found (Ahmed *et al.*, 1983) exogenous [^{14}C]eicosapentaenoic acid to be readily incorporated into PI in isolated human platelets and lost from this phospholipid upon stimulation with an agonist such as thrombin.

Effect of Dietary Cholesterol

In view of recognized physico-chemical interactions between cholesterol and hydrophobic regions of phospholipids in cellular membranes, we have investigated the influence of dietary cholesterol levels on the level of PI and possible compensations in its fatty acid composition which may accompany the induced-changes in the cholesterol content of such membranes. For such studies, we have chosen the Mongolian gerbil as our animal model for studies of the relationship between dietary cholesterol and platelet PI because plasma cholesterol levels can be significantly altered in this animal even when it consumes cholesterol at a level of only 0.1% by weight in the diet (comparable to North American human intakes). Compositional analyses have revealed substantially higher levels (up to 64%) of PI ($\mu g/10^8$ platelets) when gerbils were fed semi-purified diets containing 1% cholesterol relative to unsupplemented controls (Agwu and Holub, 1983) so as to produce a five-fold elevation in plasma cholesterol levels. The PI response was accompanied by a 47% elevation in platelet cholesterol and increases in the other individual and total phospholipids ranging from 25-56%. In addition, the percentage contribution of arachidonate to the total fatty acids in platelet PI was significantly higher in the supplemented group such that the absolute concentration ($\mu g/10^8$ platelets) of the arachidonoyl PI was elevated by 80%. At much lower levels of dietary cholesterol (0.1%), which produced only a doubling of plasma cholesterol levels, the cholesterol content, percentage of arachidonate in PI, and the arachidonoyl PI-content were higher by 19, 9, and 43%, respectively, than levels found in platelets isolated from the unsupplemented control animals. These results suggest that the dietary cholesterol-induced alterations in the level and fatty acyl composition of platelet PI may be intimately associated with the platelet hyperaggregability found in response to hypercholesterolemia.

41

ABNORMALITIES IN INOSITOL METABOLISM AND PHOSPHATIDYLINOSITOL COMPOSITION IN HUMAN DISEASE STATES

There has been considerable movement in interest during the past few years towards understanding the etiology of certain disease states in relation to possible inositol- and PI-related disorders and in applying our basic information on inositol and PI metabolism and their regulation to clinical situations. Clinical interest in our laboratory in abnormalities of inositol and PI metabolism accompanying human disease have been partly directed towards individuals with chronic renal failure undergoing regular maintenance hemodialysis (Chapkin *et al.*, 1983; Holub, 1984). Measurement of free serum inositol levels in the patient group and in matched controls was conducted by temperature-programmed gas-liquid chromatography using trimethylsilyl derivatives of inositol and methyl-α-D-mannopyranoside as the internal standard. The patient group exhibited greatly elevated levels (pre-dialysis) of circulating inositol (240 ± 34 μM, mean $\pm$ S.E., n = 10) which were greater by 630% than those for the matched controls (33 ± 1 μM, n = 9).

These findings are supportive of the work by Clements and Diethelm (1979) who emphasized the importance of the kidney as a major regulator of circulating inositol concentrations in man based on its participation, along with other tissues, in inositol synthesis and as the key organ responsible for the oxidation and disposal of this cyclitol. Interestingly, quantitation of free inositol in the early effluent perfusate appears to be a reliable method for evaluating the viability of preserved kidneys and predicting graft functioning after transplantation (Kuzuhara *et al.*, 1982). The hyperinositolemia of patients with advanced renal disease likely arises from a decreased catabolism of inositol, a decreased glomerular filtration rate, and a disturbed tubular reabsorption (Pitkanen, 1976). The potential toxicity of such greatly elevated serum inositol levels is under further investigation. In this regard, Liveson and colleagues (1977) have observed the development of ultrastructural abnormalities following the exposure of cultures of dorsal root ganglion cells to *myo*-inositol. Neurotoxicity was

42

revealed at concentrations as low as 109 µM which is considerably below the inositol levels (240 µM) found in the uremic patients which we have studied.

Work from our laboratory has revealed that, despite their dramatic hyperinositolemia and abnormal lipoprotein profiles [particularly notable are the depressed high-density lipoprotein (HDL) levels], the patients with renal failure exhibited only a modest elevation in PI in total serum lipoprotein (Chapkin *et al.*, 1983) which was of borderline significance and no difference in HDL-PI concentrations (mg/100 ml). However, the patient group had higher percentages of linoleate and lower arachidonate/linoleate ratios in the PI. In view of the role of free inositol as a precursor for the biosynthesis of PI and likely for the function of PI in cellular responses such as platelet aggregation and neurophysiological functioning, the question arises as to whether or not an abnormal metabolism of these compounds might contribute to the hemorrhagic diathesis (including platelet dysfunction) and polyneuropathy which many of the renal patients experience. It is possible that the hyperinositolemia may result in abnormal turnover rates of cellular PI without necessarily causing a significant change in the absolute levels of membrane PI. It is of interest to determine if a diminution in the level of circulating inositol through dietary restriction might improve the overall clinical status of subjects with chronic renal failure.

In contrast to the patients with renal failure, evidence for abnormalities in inositol metabolism in patients with diabetes has led to the suggestion that hyperglycemia in the untreated diabetic may impair inositol transport thereby resulting in an intracellular deficiency which may contribute to the pathogenesis of diabetic neuropathy (Clements and Reynertson, 1977). Dietary inositol supplementation, which elevates plasma inositol levels and improves the sub-normal nerve levels found in experimental diabetes, has proved beneficial in ameliorating the defect in motor nerve conduction velocities in the acute streptozotocin-diabetic rat (Greene *et al.*, 1982). In addition to elevating circulating inositol concentrations, oral doses of inositol increased the amplitude of the evoked action potentials of the median, sural, and popliteal nerves in diabetic patients (Salway *et al.*, 1978). Thus, sub-normal and excessive levels of tissue inositol may bear relationship to the clinical peripheral polyneuropathy found as a complication of diabetes and uremia, respectively.

Since the platelets from hypercholesterolemic human subjects exhibit a heightened sensitivity to pro-aggregating agents which may contribute to their increased risk for arterial thrombosis, our laboratory has studied the levels and fatty acid compositions of PI in such platelets relative to values obtained on normocholesterolemic control subjects. This research was conducted in collaboration with Dr. M.A. Mishkell of the McMaster Lipid Clinic. As found also in the dietary cholesterol-induced hypercholesterolemic gerbil (see earlier section and also Agwu and Holub, 1983), platelets isolated from the patient group (showing serum cholesterol levels twice those for matched control subjects) had significantly higher concentrations ($\mu g/10^8$ platelets) of both cholesterol and phospholipids than controls. Shastri *et al.* (1980) have reported that only the PI (percentage of total phospholipid) was significantly higher in the platelets from type II hypercholesterolemic patients relative to normal subjects. As observed also in the gerbil, we have found that the percentage elevation of platelet PI levels in the hypercholesterolemic human platelet over controls was greater than that for any other single or total phospholipid. This contributed to the considerably higher concentration of arachidonoyl PI ($\mu g/10^8$ platelets) in the hypercholesterolemic platelet since the percentage of total fatty acids in PI represented by arachidonate was similar in the two groups.

It is possible that the higher quantity of arachidonoyl PI per platelet may permit a greater yield of arachidonoyl derivatives (1,2-diacylglycerol and phosphatidic acid), and unesterified arachidonic acid, upon stimulation as well as influencing membrane fluidity and its responsiveness to platelet agonists. In this regard, Stuart *et al.* (1980) have suggested that an enhanced release of arachidonic acid from phospholipid and its increased conversion to thromboxane A_2 may contribute to the increased aggregation of cholesterol-enriched platelets. It remains to be determined whether a restoration of platelet PI levels to normal values by cholesterol-lowering dietary modification will be directly associated with a decreased aggregability of platelets from the hypercholesterolemic patient.

ACKNOWLEDGEMENTS

The author's research described herein was supported by grants from the Medical Research Council of Canada, Ontario Heart Foundation, Natural Sciences and Engineering Research Council, and the Human Nutrition Research Council of Ontario. The assistance of Mrs. Frances Graziotto and Miss Alice Wilson in the preparation of this manuscript is gratefully acknowledged.

REFERENCES

Agwu, D.E. and Holub, B.J. (1983) *Fed. Proc.* *42*, 1186.

Ahmed, A.A. and Holub, B.J. (1983) *Can. Fed. Biol. Soc.* *26*, 41.

Ahmed, A.A., Mahadevappa, V.G. and Holub, B.J. (1983) *Nutr. Res.* *3*, 673-680.

Andersen, D.A. and Holub, B.J. (1980) *J. Nutr.* *110*, 488-495.

Aveldano, M.I. and Bazan, N.G. (1983) *J. Lipid Res.* *24*, 620-627.

Baker, R.R. and Thompson, W. (1973) *J. Biol. Chem.* *248*, 7060-7065.

Beach, D.C. and Flick, P.K. (1982) *Biochim. Biophys. Acta* *711*, 452-459.

Burton, L.E. and Wells, W.W. (1979) *J. Nutr.* *109*, 1483-1491.

Burton, L.E., Ray, R.E., Bradford, J.R., Orr, J.P., Nickerson, J.A. and Wells, W.W. (1976) *J. Nutr.* *106*, 1610-1616.

Chapkin, R.S., Haberstroh, B., Liu, T. and Holub, B.J. (1983) *J. Lab. Clin. Med.* *101*, 726-735.

Chu, S.W. and Geyer, R.P. (1983) *J. Nutr.* *113*, 293-303.

Chu, S.W. and Hegsted, D.M. (1980) *J. Nutr.* *110*, 1217-1233.

Clements, R.S. Jr. and Diethelm, A.G. (1979) *J. Lab. Clin. Med.* *93*, 210-219.

Clements, R.S. Jr. and Reynertson, R. (1977) *Diabetes* *26*, 215-221.

DETECTION OF RECEPTOR-LINKED

PHOSPHOINOSITIDE METABOLISM IN BRAIN OF

LITHIUM-TREATED RATS

William R. Sherman, Michael P. Honchar and
Ling Y. Munsell

Department of Psychiatry, Washington University
School of Medicine, 4940 Audubon Ave.,
St. Louis, MO   63110

SUMMARY

The metabolism of phosphoinositides that results from receptor-coupled processes in the CNS can be readily detected in rats treated with low doses of lithium chloride.   Brain levels of *myo*-inositol 1-phosphate (IP) rise in response to agonists of this class of receptor and decrease on treatment with antagonists.   For example, peripheral administration of the direct-acting muscarinic cholinergic agonist pilocarpine causes a rapid (<30 min) increase in the IP levels of cerebral cortex.   Subsequent administration of the muscarinic antagonist atropine reverses the effect.   The anticholinesterase physostigmine produces a similar IP elevation.   Both of these agents produce seizures when given to lithium-treated rats. Because diazepam reduces the increase in IP while arresting seizures and because peripherally-administered kainic acid causes both seizures and an elevation in IP level that is partially reversed by atropine, we believe that the seizure activity itself contributes to the increase in IP observed.   Nicotine, when administered subcutaneously to the lithium-treated rat, suppresses IP levels, while the two nicotinic cholinergic ganglionic blockers mecamylamine and pempidine give rise to atropine-reversible increases in IP that are comparable to the increases obtained with muscarinic agonists but without producing seizures.   The effect of the nicotinic blockers is ascribed to disinhibition of Renshaw cell-like circuits.

INTRODUCTION

In 1981 we showed that the lithium-induced increase in cerebral cortex levels of *myo*-inositol 1-P (IP) was composed principally of the D- rather than the L-enantiomer of IP (Sherman *et al.*, 1981). This distinction had to be made since D-IP is a product of phospho-inositide catabolism while the L- form is produced during the synthesis of *myo*-inositol in brain and other tissues (Eisenberg, 1967; Leavitt and Sherman, 1982). The lithium effect on cortical IP levels was originally dis-covered by Allison *et al.* in 1976, being an outgrowth of his original finding (Allison and Stewart, 1971), that lithium caused a reversible depletion of up to 30% of *myo*-inositol levels in rat cerebral cortex. During this period, Naccarato *et al.* (1974) showed that 250 mM Li$^+$ completely inhibited *myo*-inositol 1-phosphatase. We subsequently found that lithium was a potent uncom-petitive inhibitor of that enzyme, with a K$_i$ near 0.8 mM, which is at the low end of the plasma lithium concen-tration sought in the treatment of mania (Hallcher and Sherman, 1980). In 1982 Berridge *et al.* showed that lithium could be a powerful tool in the study of recep-tor-dependent phosphoinositide metabolism *in vitro*, as revealed by a large increase in the tissue level of IP in the presence of lithium. At that time we reported that rats treated with modest doses of lithium, followed by the muscarinic agonist pilocarpine or by the anticholinesterase physostigmine, underwent large increases in cortical IP levels and simultaneously suffered limbic seizures that were accompanied by brain damage (Honchar *et al.*, 1983; Olney *et al.*, 1983). The further study of the effects of those agonists, and of other agents, on phosphoinositide metabolism is the subject of this report.

METHODS

Male Sprague-Dawley rats bred at Washington Uni-versity and weighing 250-300 g were used in all experi-ments. Drugs were obtained from Sigma Chemical Co., St. Louis, MO, except pempidine (1,2,2,6,6-pentamethyl-piperidine) which was purchased from Kodak Laboratory and Specialty Chemicals, Rochester, NY. Solutions of the following drugs were prepared at the concentrations shown in distilled water except as noted: LiCl (0.75 M); pilocarpine (10 mg/ml); physostigmine (0.2 mg/ml); atropine (75 mg/ml); nicotine (1.8 mg/ml in saline); mecamylamine (10 mg/ml); pempidine (14.8 mg/ml, dissolved in dilute HCl); kainic acid (4.74 mg/ml, disolved in dilute NaOH); metrazole (35 mg/ml); diazepam (5 mg/ml). All amine weights are as the free base. Experiments were terminated by decapitating the rats and immediately freezing the heads in liquid nitrogen. The methods of dissection at -17°, lyophilization, weighing and direct tissue derivatization, followed by gas chromatography, are described in Allison *et al.* (1976) and in Leavitt and Sherman (1982). Gas chromatography was carried out on 6 ft × 1/4 in glass columns packed with 3% OV-17 on Chromosorb W-HP prepared by the Ohio Valley Specialty Chemical, Inc., Marietta, OH. The packing was tested by us for its ability to chromatograph inositol phosphates as described in Leavitt and Sherman (1982). The Supelco packing described in that reference was subsequently *not* found to be satisfactory for inositol phosphates (it is not warranted by Supelco to be useful for that purpose). In some recent experiments we have used *myo*-inositol 2-phosphate (M2P) as an internal standard and as a guide to column quality. A fresh, satisfactory, packing should give the same chromatographic peak (area/mole) for each of these substances. An unsatisfactory packing will give a high IP/M2P ratio. Flame photometric detection was used for IP measurements, and a flame ionization detector for *myo*-inositol.

myo-Inositol 1-P was either prepared enzymatically as described by Burton and Wells (1974) or purchased from Elastin Products, St. Louis, MO. *myo*-Inositol 2-P is available from Sigma.

Lithium levels were measured by atomic absorption measurement of supernatants of homogenates or after acid digestion.

RESULTS

Lithium-Treated Rats Undergo A Rapid and Large Increase in Cortical *myo*-Inositol 1-P upon Systemic Administration of Pilocarpine and Physostigmine

Pilocarpine. When rats are administered LiCl alone (3 meq/kg s.c.), cerebral cortex levels of IP reach a maximum of 1.25 ± 0.10 mmol/kg of dry tissue at 18 h and fall to 0.75 ± 0.14 mmol/kg by 24 h (Fig. 1). If the muscarinic cholinergic agonist pilocarpine (30 mg/kg) is then administered subcutaneously, and cortex examined in 30 min, the level of IP is found to have risen to 7.51 ± 0.93 mmol/kg where it remains, insignificantly changed, for an additional 2.5 h. Five h after the injection of pilocarpine the levels of IP again approach those obtained with lithium alone. Pilocarpine, by itself, produces only a small increase in cortical IP concentration, from a control value of 0.26 ± 0.11 mmol/kg to 0.53 ± 0.06 mmol/kg after 1 h in the presence of the drug. *myo*-Inositol levels in the animals treated with 3 meq/kg of LiCl are not significantly depressed at 18 or 24 h following the lithium injection, but fall precipitously after pilocarpine is administered to the lithium-treated rats. The amount of *myo*-inositol withheld in the form of IP is about 7 mmol/kg (dry wt); thus there is close stoichiometry between the decrease in *myo*-inositol and the increase in IP. Pilocarpine does not affect Li uptake by cortex when administered at levels of 3 and 5 meq LiCl/kg (data not shown).

Physostigmine. Physostigmine (0.4 mg/kg) has an effect similar to that of pilocarpine, causing an increase in the levels of IP with a simultaneous decrease in *myo*-inositol (Table I). The degree of the change is a function of the cortical levels of lithium, levels which are not affected by the presence of the anticholinesterase. Physostigmine, administered to lithium-treated rats in this way, also brings about seizures and consequent brain damage.

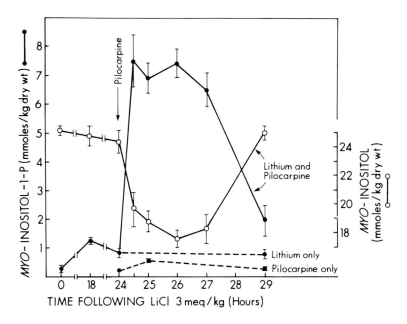

Figure 1. When pilocarpine (30 mg/kg) is administered subcutaneously to rats that have received LiCl (3 meq/kg, s.c.) 24 h earlier, myo-inositol-1-P levels in cerebral cortex rapidly increase (closed circles) and myo-inositol levels (open circles) fall (note the ordinate scale for myo-inositol is compressed by a factor of two relative to the IP scale). The course of IP levels with the same dose of lithium alone and with pilocarpine alone are also shown. Rats thus treated undergo limbic seizures and suffer severe brain damage (Honchar et al., 1983).

Atropine Reverses the Increase in IP Brought About by Pilocarpine in Lithium-Treated Rats

One hour after 150 mg/kg of atropine is given subcutaneously to rats that have previously received LiCl (3 meq/kg) and pilocarpine (30 mg/kg), the level of IP falls to 60% of the level that would result with lithium and pilocarpine and, by another hour, to about 45% of that level (Fig. 2). Administration of atropine 1/2 h prior to pilocarpine completely blocks the IP elevation due to the cholinergic agonist (Honchar et al., 1983) (Fig. 2). In the experiment where atropine was administered after the

TABLE I

The Anticholinesterase Physostigmine, Like Pilocarpine, Causes IP Levels to Increase and *myo*-Inositol Levels to Fall in Cerebral Cortex of Lithium-Treated Rats

Treatment	(n)	IP	*myo*-Inositol	Lithium
		(mmol/kg dry wt)		
None	(5)	0.22 ± 0.004	23.14 ± 0.55	--
Li[3]	(5)	0.85 ± 0.15	24.42 ± 0.81	2.55 ± 0.19
Li[3]+P	(5)	3.81 ± 0.48	20.84 ± 0.12	2.53 ± 0.32
None	(4)	0.22 ± 0.02	26.18 ± 0.64	--
Li[5]	(4)	3.26 ± 0.56	21.11 ± 0.81	5.48 ± 0.46
Li[5]+P	(3)	8.07 ± 0.64	19.64 ± 1.10	5.11 ± 0.16

LiCl was administered s.c. in doses of [3] or [5] meq/kg of body wt 24 h before sacrifice. Physostigmine (P, 0.4 mg/kg) was given s.c. 1 h (Li[3]) or 1.5 h (Li[5]) before sacrifice. The myo-inositol levels in the Li-physostigmine treated animals were lower than those in animals treated with lithium alone, in each case (p < 0.005). Mean values ± S.E.

pilocarpine neither seizures nor brain damage was prevented, whereas *prior* administration of atropine prevented both (Olney *et al.*, 1983).

Nicotinic Cholinergic Receptor Activity is Also Found to Affect Phosphoinositide Metabolism

Because of the apparent muscarinic cholinergic aspect of the action of lithium alone, and because the potentiation of the response to muscarinic agonists in the presence of lithium are both so pronounced, the action of nicotinic cholinergic agents was investigated for comparison.

Nicotine, when administered subcutaneously in doses that induce prostration, causes about a 30% reduction in IP levels in cerebral cortex of lithium-treated rats (Table II). The centrally-active ganglionic nicotinic inhibitors mecamylamine and pempidine each have the opposite effect. They cause an elevation of IP levels, the extent

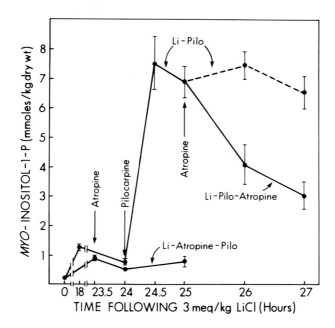

Figure 2. Atropine administered 30 min before pilocarpine completely blocks the stimulation of phosphoinositide metabolism as well as the seizures. When administered 1 h following pilocarpine the level of IP falls at a rate of about 3 mmol/h while seizure activity continues. At the level of lithium in brain at this time (about 0.5 meq/kg dry wt) myo-inositol 1-phosphatase is 43% inhibited.

of which is dependent on the amount of lithium given to the rats and which is reversible by atropine (Table III). Each of the IP increases is accompanied by a significant *myo*-inositol decrease. Thus these nicotinic antagonists have an IP-elevating effect similar to that of muscarinic agonists with the difference that neither seizures nor brain damage accompany the administration of any of the nicotinic drugs to LiCl-treated rats.

TABLE II

Nicotine Reduces the Level of IP in Cerebral Cortex of Lithium-Treated Rats

Treatment[a]	myo-Inositol 1-P (mmol/kg dry wt)
Li[3]	1.90 ± 0.04
Li[3], nicotine [0.15]	1.28 ± 0.20[b]
Li[3], nicotine [1.5]	1.25 ± 0.22[b]
Li[5]	3.20 ± 0.20
Li[5], nicotine [1]	2.14 ± 0.27[c]

[a] LiCl was administered s.c. in doses of [3] or [5] meq/kg 24 h before sacrifice. Nicotine was also administered s.c. at 0.15 and 1.5 mg/kg of body wt, 5 times over a 1 h period, or 1 mg/kg 6 times over 1 h, and the rats sacrificed. In the 3 meq LiCl group 3 rats were used, and in the 5 meq LiCl group 5 rats were used, per experiment. Mean values ± S.E.
[b] p < 0.05 relative to Li[3].
[c] p < 0.02 relative to Li[5].

When mecamylamine and pilocarpine are administered together to rats, in the absence of lithium, a slight IP elevation occurs which is amplified by lithium-pretreatment (Table III). The degree of the increase is, however, not greater than that obtained with pilocarpine alone (Fig. 1). Rats treated with both drugs do, however, undergo seizures and sustain brain damage that is not significantly different from that which occurs on treatment with lithium and pilocarpine.

The Action of Non-Cholinergic Agents and of Seizures on CNS Phosphoinositide Metabolism

Lithium-treated rats receiving diazepam (20 mg/kg, s.c.) prior to pilocarpine do not undergo brain-damaging limbic seizures as do rats unprotected by diazepam (Olney et al., 1983). Furthermore, rats receiving diazepam after pilocarpine have their seizures arrested. In an effort to separate the effects of the direct action of

TABLE III

Mecamylamine and Pempidine Cause an Atropine-Reversible Elevation of IP Levels in Cerebral Cortex of Lithium-Treated Rats

Treatment	myo-Inositol 1-P	myo-Inositol
	(mmol/kg dry wt)	
Expt. 1		
None	0.21 ± 0.01	--
Mecamylamine	0.23 ± 0.10	24.05 ± 0.68
Li[3]	0.73 ± 0.07	
Li[3], mecamylamine	2.37 ± 0.31	20.62 ± 0.44
Li[3], atropine, mecamylamine	0.36 ± 0.02	23.14 ± 0.38
Expt. 2		
Li[5]	3.32 ± 0.55	26.95 ± 1.62
Li[5], mecamylamine	8.14 ± 0.60	22.50 ± 0.87
Expt. 3		
Li[3], pempidine	4.11 ± 0.30	--
Li[3], atropine, pempidine	0.97 ± 0.13	--
Expt. 4		
Li[5], pempidine	6.05 ± 0.49	--
Expt. 5		
Mecamylamine, pilocarpine	0.33 ± 0.04	22.92 ± 0.58
Li[3], mecamylamine, pilocarpine	6.56 ± 0.32	18.80 ± 0.38

Drugs were administered in the order given. LiCl, 3 or 5 meq/kg, Li[3], Li[5], was injected s.c. and rats sacrificed in 24 h; mecamylamine (50 mg/kg) or pempidine (46 mg/kg), was given s.c., 2 h before sacrifice. Atropine (150 mg/kg) was administered 0.5 h before mecamylamine or pempidine. In the pilocarpine (30 mg/kg)-mecamylamine experiments rats were exposed to mecamylamine for 2 h, and to pilocarpine for 1 h. Five rats were used in each experiment, except as noted. myo-Inositol decreases are significant at $p < 0.005$. The 3 meq/kg Li level is a group average of 14 rats from several experiments. This level of Li gives a more variable response than higher doses. Mean values $\pm$ S.E.

pilocarpine on receptor-related phosphoinositide metabolism from the effects of the seizure state itself, we compared the cortical IP levels of lithium-pilocarpine-diazepam-treated rats with those of rats that received the "noncholinergic" seizuregenic agents kainic acid or metrazole. Figure 3 shows that diazepam moderates the increase in the level of IP that occurs in rats whose phosphoinositide metabolism has been stimulated by pilocarpine. Figure 4 shows that administering diazepam to rats before pilocarpine has an even greater effect on IP levels than administration after, although one that is not as profound as that of atropine administered in the same way. Diazepam also suppresses the increase in IP caused by lithium alone (Fig. 4). In another experiment, 24 h after rats received 5 meq/kg LiCl the cortical level of IP was 1.98 ± 0.23 mmol/kg dry wt ($\pm$ SEM), while, 1.5 h following diazepam, 20 mg/kg s.c., the IP level was reduced to 0.78 ± 0.09 mmol/kg ($p < 0.005$). Comparable data with lithium/atropine are 3.26 ± 0.59 mmol/kg for 5 meq/kg lithium alone and 1.43 ± 0.30 for lithium with atropine, 150 mg/kg 2 h before death. The degree of suppression of endogenous cortical phosphoinositide metabolism thus appears to be similar with both drugs.

In an effort to obtain information about the degree to which seizure activity itself contributes to the cortical IP changes, lithium-treated rats were given kainic acid (12.5 mg/kg, s.c.) or metrazole (70 mg/kg, s.c.). Under the conditions used, both drugs brought about seizures which appeared not to be modified by lithium-treatment.

The results (Table IV) show that neither kainate nor metrazole, when given alone, caused changes in IP levels that were statistically significant. However, in the presence of lithium, both drugs caused a readily-detected increase in IP. In the case of kainic acid-lithium, the IP increase was attenuated by atropine (the latter result is significant at the $p < 0.05$ level, $\alpha = 1$, assuming that the IP levels were expected to fall).

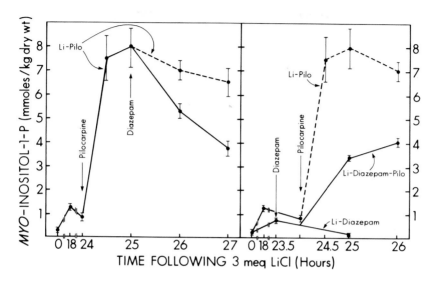

Figure 3 (left). Diazepam depresses cerebral cortex phosphoinositide metabolism that has been stimulated by pilocarpine. Seizures were terminated and brain damage was attenuated. Pilocarpine (30 mg/kg) and diazepam (20 mg/kg) were administered s.c. at the times indicated to rats that had earlier received 3 meq/kg LiCl.

Figure 4 (right). Diazepam also depressed endogenously-generated phosphoinositide metabolism (Li-DIAZEPAM). When pilocarpine was administered 30 min after diazepam the phosphoinositide-stimulating effect of pilocarpine is diminished (Li-DIAZEPAM-PILO). Overt seizures and brain damage were prevented.

DISCUSSION

Figure 1 vividly shows the ability of the lithium-treated rat to reveal muscarinic cholinergic activation of phosphoinositide metabolism *in vivo*. Twenty-four hours after receiving 3 meq/kg of LiCl, the cerebral cortex has a lithium concentration of 2.5 mmol/kg dry wt (Table I), which remains constant for the duration of the experiment (5 h). That level of lithium inhibits *myo*-inositol 1-phosphatase *in vitro* by 43% over the substrate range

59

Sherman et al.

TABLE IV

Kainic Acid has an Atropine-Reversible Effect on IP
Levels in Cerebral Cortex of Lithium-Treated Rat.
Metrazole Also Affects this Metabolism.

Treatment[a]	myo-Inositol 1-P (mmol/kg dry wt)	myo-Inositol
None	0.18 ± 0.02	23.40 ± 0.52
Li	0.59 ± 0.05	23.66 ± 0.49
Kainic acid	0.45 ± 0.23	24.00 ± 0.61
Li, kainic acid	3.99 ± 0.80[b]	18.54 ± 2.89
Li, atropine, kainic acid	1.74 ± 0.64[b]	22.42 ± 0.44
Metrazole	0.16 0.02[c]	24.00 ± 0.51
Li, metrazole	1.05 ± 0.20[c]	23.60 ± 0.43

[a] Li (3 meq/kg) was administered s.c. to rats 24 h before sacrifice,
kainic acid (12.5 mg/kg), 2 h before sacrifice, atropine (150 mg/kg),
2.5 h before sacrifice and metrazole (70 mg/kg), 1 h before sacrifice.
Five rats were used in each experiment.
[b] For these two groups $p < 0.05$, 1-tailed. [c] For these two groups $p <$
0.005, 2-tailed. Mean values ± S.E.

studied, which was 0.1 - 0.5 mM IP (Hallcher and
Sherman, 1980). In the rat treated only with lithium, the
level of IP rises only 4-fold even with this degree of
inhibition. Thus the amount of IP being generated in
cortex under these conditions is hydrolyzed by the
inhibited phosphatase with only a small increase in the
concentration of IP. Without lithium, stimulation by the
muscarinic agonist pilocarpine causes an increase in IP
production which the phosphatase is able to keep up with
by only a doubling of substrate level. That same rate of

Hallcher, L.M. and Sherman, W.R. (1980) *J. Biol. Chem.* *255*, 10896–10901.

Hawthorne, J.N. and Kai, M. (1970) in: *Handbook of Neurochemistry* (A. Lajtha, ed.) Vol. 3, pp. 491–508, Plenum, New York.

Hitzemann, R.J., Natsuki, R. and Loh, H.H. (1978) *Biochem. Pharmacol.* *27*, 2519–2523.

Honchar, M.P. and Sherman, W.R. (1983) *Abstr. Soc. Neurosci.* *9*, 960.

Honchar, M.P., Olney, J.W. and Sherman, W.R. (1983) *Science* *220*, 323–325.

Lapetina, E.G., Brown, W.E. and Michell, R.H. (1976) *J. Neurochem.* *26*, 649–651.

Leavitt, A.L. and Sherman, W.R. (1982) in: *Methods in Enzymology* *89*, (W.A. Wood, ed.) pp. 9–18, Academic Press, New York.

Naccarato, W.F., Ray, R.E. and Wells, W.W. (1974) *Arch. Biochem. Biophys.* *164*, 194–201.

Olney, J.W., Honchar, M.P. and Sherman, W.R. (1983) *Abstr. Soc. Neurosci.* *9*, 401.

Sherman, W.R., Leavitt, A.L., Honchar, M.P., Hallcher, L.M. and Philips, B.E. (1981) *J. Neurochem.* *36*, 1947–1951.

Sherman, W.R., Munsell, L.Y., Gish, B.G. and Honchar, M.P. (1984) *J. Neurochem*, in press.

Simmons, D.A., Winegrad, A.I. and Martin, D.B. (1982) *Science* *217*, 848–851.

Streb, H., Irvine, R.F., Berridge, M.J. and Schulz, I. (1983) *Nature* *306*, 67–69.

PROPRANOLOL-INDUCED MEMBRANE PERTURBATION AND

THE METABOLISM OF PHOSPHOINOSITIDES AND

ARACHIDONOYL DIACYLGLYCEROLS IN THE RETINA

Nicolas G. Bazan, Ana M. Roccamo de Fernandez,
Norma M. Giusto and Monica G. Ilincheta de Boschero

Louisiana State University Medical Center, LSU Eye Center,
136 South Roman St., New Orleans, LA 70116, USA and
Instituto de Investigaciones Bioquimicas, Universidad
Nacional del Sur-Consejo Nacional de Investigaciones
Cientificas y Tecnicas, 8000 Bahia Blanca, Argentina

SUMMARY

The very active metabolic pathway for phosphatidic
acid and phosphoinositides of the retina is affected
greatly by the amphiphilic cation, propranolol. Propranolol, in addition to stimulating the synthesis of
components of this pathway, also stimulates polyphosphoinositide degradation, probably due to the activation
of a phospholipase C. Hence, propranolol stimulates the
phosphoinositide cycle. This effect may be induced by a
membrane perturbation caused by intercalation of the
hydrophobic moiety of the drug in a membrane domain
containing the components of the cycle. Exogenously
added *myo*-inositol stimulates the synthesis of phosphatidylinositol and inhibits the propranolol-induced
degradation of polyphosphoinositides suggesting that
myo-inositol may modulate phosphoinositide metabolism.
The use of propranolol as an experimental probe and the
manipulation of phosphoinositide pathways by *myo*-inositol
will prove useful in the study of the metabolism and
function of phosphoinositides.

INTRODUCTION

The *de novo* biosynthesis of phosphatidic acid and phosphoinositides in retina is highly active when measured *in vivo* or *in vitro* using radiolabeled glycerol (Bazan and Bazan, 1976; Bazan *et al.*, 1976a; Careaga and Bazan, 1981; Aveldano and Bazan, 1983). These high rates of synthesis raise the question of whether the newly synthesized lipids become part of the phosphoinositide cycle or whether they remain as independent pools. When cationic amphiphilic drugs, such as propranolol and phentolamine, were added *in vitro*, *de novo* synthesis via the phosphatidic acid–phosphoinositides pathway in the retina was greatly stimulated (Bazan *et al.*, 1976b, 1977, 1982 a,b; Bazan and Bazan, 1977; Bazan, 1978; Ilincheta de Boschero and Bazan, 1982, 1983; Giusto *et al.*, 1983), as has also been shown in other tissues (Hauser and Eichberg, 1975; Abdel-Latif *et al.*, 1983; Michell *et al.*, 1976). At later incubation times, radiolabeled diacylglycerols were produced in retinas incubated in the presence of radiolabeled glycerol, suggesting that the precursor is incorporated into a phospholipid that is part of the cycle and subsequently released as a diacylglycerol by the action of a phospholipase C (Bazan *et al.*, 1977). Light also has been shown to stimulate glycerol incorporation into phosphatidic acid–phosphatidylinositol (Bazan and Bazan, 1977; Schmidt, 1983a,b), suggesting the physiological relevance of the active *de novo* biosynthesis of these retinal lipids. In other tissues the *de novo* biosynthetic route of phosphatidic acid – phosphoinositides has been linked to hormonal cellular responses (Farese, 1983). In addition, other studies have shown that a classical phosphoinositide cycle also operates in the retina (Schmidt, 1983a,b; Anderson and Hollyfield, 1981) and that rod outer segments of visual cells display an active metabolism of phosphatidylinositol (Anderson *et al.*, 1980) and phosphatidic acid (Bazan *et al.*, 1982a).

The objectives of the work described here were to assess the biosynthesis and cleavage of phosphatidyl-inositol 4-phosphate (PIP) and phosphatidylinositol 4,5-bisphosphate (PIP_2) in the bovine retina incubated *in vitro*. In addition, the metabolism of arachidonoyl-diacylglycerols was studied and correlated with the degradation of polyphosphoinositides. This was done by

a) assessing the content of acyl chains of diacylglycerols; b) conducting pulse-chase experiments with radioactive glycerol or arachidonic acid; c) using the amphiphilic cation, propranolol, as an experimental probe and d) studying the effect of exogenously-added *myo*-inositol.

PROPRANOLOL STIMULATES THE DE NOVO BIOSYNTHESIS OF POLYPHOSPHOINOSITIDES

The effect of *dl*-propranolol on the phosphatidic acid-polyphosphoinositides pathway was explored in experiments using [2-^{3}H]glycerol as a precursor (Tables I and II). Table I confirms that the *de novo* synthesis of phosphatidic acid is enhanced by propranolol, because there is both an increased incorporation of [2-^{3}H]glycerol and an increased pool size of phosphatidic acid. Similar results have been shown with microsomal preparations isolated from intact retinas exposed *in vitro* to propranolol (Giusto *et al.*, 1983). Therefore, the newly synthesized phosphatidic acid is then channeled to CDP-diacylglycerol, since propranolol also stimulates the labeling of this liponucleotide (Table I and Hauser and Eichberg, 1975) and inhibits phosphatidic acid phosphohydrolase (Bazan *et al.*, 1976b; Michell *et al.*, 1976; Eichberg *et al.*, 1979). The addition of 10 mM *myo*-inositol decreased the accumulation of phosphatidic acid induced by propranolol without affecting the incorporation of glycerol. However, *myo*-inositol decreased the labeling of CDP-diacylglycerol both in the presence and absence of propranolol. This was due to the use of *myo*-inositol as substrate for the CDP-diacylglycerol:inositol 3-phosphatidyltransferase that resulted in an enhanced formation of phosphatidylinositol (Table II). Hence, the apparent reduction in the propranolol-induced accumulation of phosphatidic acid by *myo*-inositol (Table II) may be a consequence of the overall stimulation of the pathway by the increased availability of one of the substrates, *myo*-inositol, that enhances the use of the phosphatidate moiety of CDP-diacylglycerol. In fact, *myo*-inositol increased the amount of [^{3}H]glycerol in phosphatidylinositol and potentiated the effect of propranolol.

TABLE I

EFFECT OF PROPRANOLOL ON THE SYNTHESIS
OF PHOSPHATIDIC ACID AND CDP-DIACYLGLYCEROL IN THE RETINA

Myo-inositol (mM)	Phosphatidic Acid		CDP-Diacylglycerol	
	Control	Propranolol (0.5 mM)	Control	Propranolol (0.5 mM)
[³H]Glycerol Incorporation				
0	1.54 ± 0.12	8.03 ± 0.80	0.16 ± 0.01	1.48 ± 0.25
10	1.58 ± 0.17	9.42 ± 1.36	0.10 ± 0.02	1.12 ± 0.02
Content of Phosphatidic Acid				
0	1.20 ± 0.35	3.6 ± 1.45	–	–
10	1.37 ± 0.65	1.51 ± 0.63	–	–

Incorporation is given in pmol glycerol/mg of protein and content is given in nmol phosphatidic acid/mg of protein. Retinas were preincubated for 40 min with the radiolabeled precursor. Lipid extraction, phospholipid separation and quantitation were carried out as described previously (Giusto and Bazan, 1979).

The propranolol effect was also seen in the poly-phosphoinositides where an increased labeling took place as has been reported previously (Aveldano and Bazan, 1983). However, when *myo*-inositol was present, with or without propranolol, polyphosphoinositides exhibited less labeling than the controls. This would seem to indicate that the presence of *myo*-inositol does not increase the synthesis of polyphosphoinositides. However, it was puzzling to observe a decrease in glycerol labeling in these lipids, as if *myo*-inositol was antagonizing the propranolol-stimulated synthesis of polyphosphoinositides.

TABLE II

EFFECT OF PROPRANOLOL AND *MYO*-INOSITOL ON THE METABOLISM OF PHOSPHOINOSITIDES IN THE RETINA

Phosphoinositide		$[2-^3H]$Glycerol Incorporated (nmol/mg protein)	
		minus *myo*-Inositol	(10 mM) *myo*-Inositol
PI	control	18.7 ± 1.6	25.3 ± 1.8[a]
	propranolol	74.3 ± 6.7	139.8 ± 10.7
PIP	control	0.36 ± 0.02	0.24 ± 0.02[b]
	propranolol	4.1 ± 0.5	2.8 ± 0.5
PIP_2	control	0.07 ± 0.02	0.08 ± 0.02[a]
	propranolol	3.4 ± 0.5	2.5 ± 0.5

Experimental conditions were as in Table I. Student t-test for the effect of myo-inositol: [a]$p < 0.005$ and [b]$p < 0.001$.

Several experiments were performed to determine whether an increased formation of polyphosphoinositides induced by propranolol results in their degradation by a phospholipase C. We tested the hypothesis that pro-pranolol was also exerting a stimulatory effect on a polyphosphoinositide-specific phospholipase C. In addition, *myo*-inositol was examined to determine if it was also eliciting another effect by antagonizing the action of propranolol on polyphosphoinositide degradation.

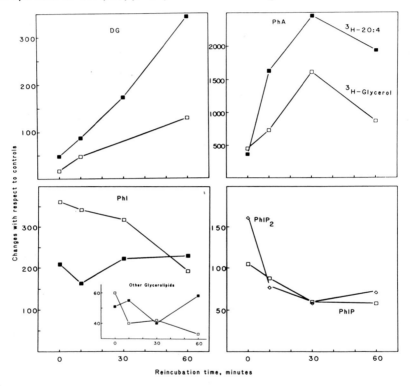

Fig. 1. Effect of dl-propranolol on the metabolism of retinal lipids. Retinas were preincubated for 10 min, exposed to [2-^{3}H]glycerol or [^{3}H]arachidonic acid for 10 min and washed three times in fresh medium. Reincubation was carried out for the given incubation time periods. Propranolol (0.5 mM) was present throughout the incubation and reincubation periods. Open squares represent [^{3}H]arachidonic acid. For PIP$_2$ and PIP only [^{3}H]arachidonic acid data are shown. Values represent changes in incorporation with respect to each control.

The metabolism of diacylglycerols was correlated with the polyphosphoinositides in pulse-chase experiments using [^{3}H]arachidonic acid or [^{3}H]glycerol. Figure 1 shows that in the presence of *dl*-propranolol there was an increased labeling in retinal diacylglycerols with either precursor as a function of reincubation time. At zero time and after 10 min reincubation there was a decreased incorporation of either arachidonic acid or glycerol in the propranolol-treated retinas. This is in agreement with a previous study that used radiolabeled glycerol (Bazan *et al.*, 1977). Phosphatidic acid, phosphatidylinositol and PIP$_2$ displayed an enhanced incorporation of both precursors at zero time due to propranolol. Hence, acidic phospholipids derived from phosphatidic acid through CDP-diacylglycerol showed an enhanced synthesis. In contrast, other glycerolipids showed incorporation with either precursor that was reduced by propranolol to levels below those of the controls. This pattern is in accordance with the redirection of the biosynthetic pathways in retinal glycerolipids that is elicited by cationic amphiphilic drugs, such as propranolol or phentolamine (Bazan *et al.*, 1976b).

PROPRANOLOL STIMULATES THE DEGRADATION OF POLYPHOSPHOINOSITIDES TO ARACHIDONOYL-DIACYLGLYCEROLS

Figure 1 also shows that there is an increased stimulation followed by a decline in radiolabeled arachidonate and glycerol in phosphatidic acid after 30 min reincubation. From 0 to 10 min of reincubation, however, propranolol stimulates the loss of arachidonoyl chains in the polyphosphoinositides. These propranolol-induced changes indicate that the increased turnover of the cycle also involves the polyphosphoinositides with the additional entrance of glycerol through the *de novo* biosynthesis of phosphatidic acid. These changes also indicate that there is a propranolol-stimulated degradation of PIP and PIP$_2$ leading to the accumulation of diacylglycerols.

73

METABOLISM AND FATTY ACID
COMPOSITION OF RETINAL DIACYLGLYCEROLS

The amounts of diacylglycerols in bovine retina increased 34% when the tissue was incubated *in vitro* for 40 min in the presence of 0.5 mM propranolol. This agrees with the previous observation that radiolabeled glycerol increased in retinal diacylglycerols when retinas were incubated for periods in excess of 20 min (Bazan *et al.*, 1977). Figure 2 shows that the content of all the fatty acyl chains in diacylglycerols increased in propranolol-exposed retinas, except the docosahexaenoyl chains. Arachidonoyl chains attained the largest increase (52%) and palmitoyl chains the least (32%). Retinas incubated with 10 mM *myo*-inositol did not differ from controls, although when *myo*-inositol and *dl*-propranolol were added together there was an inhibition of the propranolol-induced production of diacylglycerols.

We have also analyzed the molecular species of diacylglycerols in separate experiments (data not shown) and have found that when propranolol is present the tetraenoic molecular species is the one that exhibits the greatest increase in incorporation of [^{3}H]glycerol. In one set of experiments, retinas were radiolabeled and exposed to propranolol, subcellular fractionation was performed and diacylglycerols were analyzed in microsomal fractions and mitochondrial fractions. These analyses showed no changes in the labeling of the diacylglycerol pool, suggesting that the diacylglycerols may be changing in another fraction, such as the plasma membrane of retinal cells. However, plasma membrane fractions are difficult to prepare and it is not possible at the present time to determine their cellular origin.

In rod outer segments, a membrane fraction that is relatively simple to prepare, we have observed an accumulation of diacylglycerols with a concomitant loss of phosphatidylinositol during incubation (unpublished) and an active metabolism of phosphatidic acid (Bazan *et al.*, 1982a).

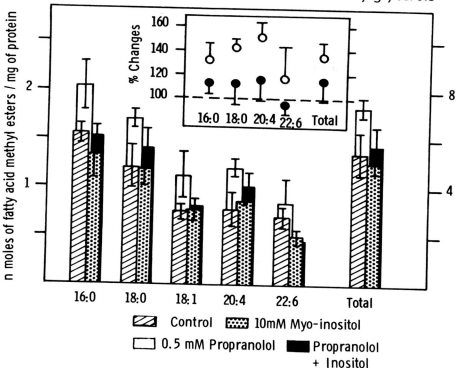

Fig. 2. Effect of dl-propranolol and myo-inositol on the acyl group composition and content of diacylglycerols in the bovine retina. Retinas were preincubated for 10 min and incubated for 40 min. Insert: mean percent change ± S.E. of propranolol-treated/control with respect to (propranolol plus myo-inositol-treated)/(myo-inositol-treated). Student t-test: 16:0 = NS; 18:0 = $p < 0.005$; 20:4 = $p < 0.25$ and 22:6 = NS. Data are mean values ± S.E. of four samples of two retinas each. Diacylglycerols were isolated by thin-layer chromatography and methyl esters were prepared with 14% BF_3-methanol and analyzed by gas-liquid chromatography as described previously (Aveldano de Caldironi and Bazan, 1979).

DISCUSSION

The metabolism of phosphoinositides, particularly PIP and PIP_2, can be studied in the intact bovine retina incubated *in vitro*. The retina has been used in these studies because 1) it is a "nature-made" slice equivalent to a slice of grey matter neural tissue because of its lack of myelin; 2) it is very useful for the study of membrane lipid metabolism (Bazan, 1982, 1983) and 3) there is little data on the molecular basis of information transfer in retinal circuitry which is needed to understand membrane structure and function in retinal degenerative diseases. Despite the lability of polyphosphoinositides to postmortem ischemia (Gonzalez-Sastre *et al.*, 1971), the incubation conditions used with bovine retinas brought to the laboratory a few hours after death have produced measurable amounts of these lipids. Moreover, an increased mass of polyphosphoinositides was observed after *in vitro* incubation (unpublished observations).

Propranolol increases the diacylglycerol content, notably 1-stearoyl-2-arachidonoyl-*sn*-glycerol, in the retina when it is added in concentrations (0.5 mM) far exceeding those that antagonize β-adrenergic receptors. This effect was confirmed by measuring the pool size, fatty acid composition, molecular species and amount of radioactive glycerol or arachidonic acid in diacylglycerols. The time course of the pulse-chase experiments suggests a correlation between the accumulation of diacylglycerols and the degradation of polyphosphoinositides, particularly PIP_2. The propranolol-stimulated accumulation of diacylglycerols also is correlated with very high rates of *de novo* biosynthesis of phosphatidic acid and phosphatidylinositol in the retina.

Propranolol *in vitro* produces a dual effect on the metabolism of diacylglycerols. At early incubation times (5-10 min) there is a marked inhibition of radiolabeled glycerol incorporation into diacylglycerols. Twenty to 40 min after the start of incubation, there is a propranolol-stimulated formation of radiolabeled diacylglycerols. This early inhibitory effect is thought to result from an inhibition of phosphatidate phosphohydrolase and the subsequent redirecting of the *de novo* biosynthetic flow to CDP-diacylglycerols and phosphoinositides (Bazan *et al.*, 1977). The later increase in labeled diacylglycerols is thought to be due to the propranolol-induced activation of a phospholipase C (Bazan *et al.*, 1977; Ilincheta de Boschero and Bazan, 1983) hydrolyzing a phospholipid that already has attained a relatively high specific radioactivity. In addition to phosphatidic acid, phosphoinositides also attained very

Aveldano de Caldironi, M.I. and Bazan, N.G. (1979) *Neurochem. Res. 4,* 213-221.

Bazan, N.G. (1978) in: *"Cyclitols and Phosphoinositides"* (W. Wells and F. Eisenberg, eds.) pp. 563-568, Academic Press, New York.

Bazan, N.G. (1982) *Vis. Res. 22,* 1539-1548.

Bazan, N.G. (1983) *Handbook of Neurochemistry 3,* 17-39.

Bazan, H.E.P. and Bazan, N.G. (1976) *J. Neurochem. 27,* 1051-1057.

Bazan, H.E.P. and Bazan, N.G. (1977) *Adv. Exp. Med. Biol. 83,* 489-495.

Bazan, N.G., Aveldano, M.I., Bazan, H.E.P. and Giusto, N.M. (1976a) in: *"Lipids"* Vol. 1 (R. Paoletti, G. Porcellati and G. Jacini, eds.) pp. 89-97, Raven Press, New York.

Bazan, N.G., Ilincheta de Boschero, M.G., Giusto, N.M. and Bazan, H.E.P. (1976b) *Adv. Exp. Med. Biol. 72,* 139-148.

Bazan, N.G., Ilincheta de Boschero, M.G. and Giusto, N.M. (1977) *Adv. Exp. Med. Biol. 83,* 377-388.

Bazan, H.E.P., Careaga, M.M. and Bazan, N.G. (1981a) *Biochim. Biophys. Acta 666,* 63-71.

Bazan, N.G., Aveldano de Caldironi, M.I. and Rodriguez de Turco, E.B. (1981b) *Progress in Lipid Res. 20,* 523-529.

Bazan, N.G., di Fazio de Escalante, M.S., Careaga, N.M., Bazan, H.E.P. and Giusto, N.M. (1982a) *Biochim. Biophys. Acta 712,* 702-706.

Bazan, N.G., Morelli de Liberti, S.M. and Rodriguez de Turco, E.B. (1982b) *Neurochem. Res. 7,* 839-843.

Bleasdale, J.E., Maberry, M.C., Quirk, J.G. (1982) *Biochem. J. 206,* 43-52.

Brindley, D.N. and Bowley, M. (1975) *Biochem. J. 148,* 461-469.

Careaga, N.M. and Bazan, H.E.P. (1981) *Neurochem. Res. 6,* 1169-1178.

Daum, G., and Kohlwein, S.D., Zinser, E. and Paltauf, F. (1983) *Biochim. Biophys. Acta 753,* 430-438.

Eichberg, J., Gates, J. and Hauser, G. (1979) *Biochim. Biophys. Acta 573,* 90-106.

Esko, J.D. and Raetz, C.R.H. (1980) *J. Biol. Chem. 255,* 4474-4480.

Farese, R.V. (1983) *Metabolism 32,* 628-641.

Freinkel, N., Younsi, C., Dawson, R.M.C. (1975) *Eur. J. Biochem. 59,* 245-252.

Giusto, N.M. and Bazan, N.G. (1979) *Exp. Eye Res. 29,* 155-168.

Giusto, N.M., Ilincheta de Boschero, M.G. and Bazan, N.G. (1983) *J. Neurochem. 40,* 563-568.

Gonzalez-Sastre, F., Eichberg, J. and Hauser, G. (1971) *Biochim. Biophys. Acta 248,* 96-104.

Hallman, M. and Epstein, B.L. (1980) *Biochem. Biophys. Res. Comm. 92,* 1151-1159.

Hauser, G. and Eichberg, J. (1975) *J. Biol. Chem. 250,* 105-112.

Ilincheta de Boschero, M.G. and Bazan, N.G. (1982) *Biochem. Pharmacol. 31,* 1049-1055.

Ilincheta de Boschero, M.G. and Bazan, N.G. (1983) *J. Neurochem. 40,* 260-266.

Michell, R.H., Allan, D., Bowley, M. and Brindley, D.N. (1976) *J. Pharm. Pharmacol. 28,* 331-332.

Pappu, A.S., and Hauser, G. (1981) *Biochem. Pharmacol. 30,* 3243-3246.

Porzig, H. (1975) *J. Physiol. 249,* 27-49.

Quirk, J.G., Jr. and Bleasdale, J.E. (1983) *Obstet. Gynecol. 62,* 41-44.

Rodriguez de Turco, E.B., Morelli de Liberti, S.M., and Bazan, N.G. (1983) *J. Neurochem. 40,* 252-259.

Schmidt, S.Y. (1983a) *J. Biol. Chem. 258,* 6863-6868.

Schmidt, S.Y. (1983b) *J. Neurochem. 40,* 1630-1638.

Sturton, G.R., and Brindley, D.N. (1980) *Biochim. Biophys. Acta 619,* 494-505.

Yamashita S., and Oshima, A. (1980) *Eur. J. Biochem. 104,* 611-616.

NEUROTRANSMITTER RECEPTOR-MEDIATED

MYO-INOSITOL 1-PHOSPHATE ACCUMULATION

IN HIPPOCAMPAL SLICES

Aaron Janowsky*, Rodrigo Labarca and Steven. M. Paul

Sections on Preclinical Studies and Molecular
Pharmacology, Clinical Neuroscience Branch,
National Institute of Mental Health and *National
Institute of General Medical Sciences, Bethesda, MD 20205

SUMMARY

The agonist-dependent accumulation of [^{3}H]inositol
1-phosphate was investigated in rat hippocampal slices
that were preincubated with [^{3}H]inositol. The addition of
lithium to the incubation medium markedly amplified the
effect of all agonists in stimulating [^{3}H]inositol 1-
phosphate accumulation. The response to a number of
neurotransmitter receptor agonists was dependent on the
amount of tissue (mg protein), and the concentrations of
lithium, and [^{3}H]inositol. The accumulation of [^{3}H]-
inositol 1-phosphate in hippocampal slices was stimulated
by carbachol an effect that was potently blocked by
atropine, a muscarinic-cholinergic antagonist. [^{3}H]-
Inositol 1-phosphate also accumulated in response to
norepinephrine and this effect may be mediated by both
α_1 and α_2 adrenoreceptors, since the response was
attenuated by relatively low concentrations of the
selective α_1 and α_2 adrenoreceptor antagonists, phentolamine and yohimbine, respectively. The accumulation of
[^{3}H]inositol 1-phosphate induced by serotonin was blocked
by metergoline, but not by mianserin and therefore,
appeared to be mediated via $5HT_1$ receptors. Intracerebroventricular administration of 6-hydroxydopamine, a
specific neurotoxin of noradrenergic neurons, resulted in
an enhanced response to norepinephrine as well as

carbachol and serotonin. In addition, acute or chronic administration of the tricyclic antidepressant imipramine (10 mg/kg daily) also resulted in an enhanced response to those same neurotransmitter agonists *in vitro*. Our results suggest that the measurement of agonist-induced [^{3}H]inositol 1-phosphate accumulation, in the presence of lithium represents a sensitive method for studying receptor-mediated events in brain slice preparations and should prove useful in studying the neurochemical effects of various psychotropic drugs.

INTRODUCTION

Receptor-mediated phosphatidylinositol turnover has been demonstrated in a variety of tissues including brain (Michell, 1975; Downes, 1982; Janowsky *et al.*, 1984). The agonist-induced response results in an increase in the hydrolysis of phosphatidylinositol catalyzed by phospholipase C. A metabolite in the catabolic pathway to free inositol, inositol 1-phosphate, will accumulate following exposure of tissue, both *in vivo* and *in vitro* to lithium, which inhibits the enzyme *myo*-inositol 1-phosphatase (Allison and Blisner, 1976; Sherman *et al.*, 1981). The accumulated inositol 1-phosphate can be efficiently isolated by Dowex anion exchange chromatography and therefore represents a sensitive measure of receptor-mediated phosphatidylinositol hydrolysis (*cf.* Berridge *et al.*, 1982).

We have used this technique to characterize the effects of various neurotransmitter agonists on phosphatidylinositol hydrolysis in the rat hippocampus, an area of the brain that is well described both neuroanatomically and neurochemically. Our results suggest that the measurement of agonist-induced inositol 1-phosphate accumulation in the presence of lithium provides a sensitive method for studying receptor-mediated events in the CNS.

METHODS

Male Sprague-Dawley rats (125 - 200 g) were killed by decapitation and their brains quickly removed and dissected on ice. Hippocampi were dissected according to Glowinsky and Iversen (1966) and chopped into 350 x 350 µm slices on a McIlwain Tissue Chopper. The slices were placed in 50 ml of Krebs Ringer bicarbonate buffer (KRBB) containing 10 mM glucose, and incubated in a gently shaking water-bath (37°C) under a constant stream of O_2/CO_2 (95:5). After a 30 min incubation, the buffer was gently aspirated and 30 µl (approximately 1 mg of protein) of the tissue preparation was aliquoted into Beckman Bio-Vials containing 250 µl of KRBB buffer containing [³H]inositol (0.5 µM to 4 µM). After a 60 min preincubation at 37°C, LiCl (10 mM) and any antagonists were added. Agonists were added after 10 min, and in most experiments the slices were allowed to incubate for an additional 60 min.

To stop the reaction and isolate the [³H]inositol 1-phosphate, slices were rinsed four times with 3 ml of buffer and 0.94 ml chloroform/methanol (1:2) was added according to the method of Berridge et al. (1982). The tubes were vortexed, and after 10 min, 0.31 ml of chloroform and 0.31 ml of H_2O were added. The tubes were centrifuged for 10 min at approximately 200 x g, and 1 ml of the upper aqueous phase was removed and placed over a BioRad mini-column containing 1 cc of Dowex 1-X8 (BioRad, Richmond, CA) in the formate form. The columns were rinsed 4 times with 5 ml of H_2O containing 250 mM myo-inositol to remove any non-metabolized [³H]inositol. [³H]Inositol 1-phosphate was eluted with 0.2 M ammonium formate in 0.1 M formic acid (Berridge et al., 1982). Radioactivity was measured by conventional scintillation spectroscopy. Protein was determined by the method of Lowry et al. (1951).

RESULTS

Since the tissue and/or protein concentration could affect the agonist-induced response as well as the extraction efficiency and subsequent isolation of [³H]inositol 1-phosphate, vials containing 2 mM carbachol were incubated in the presence of increasing amounts of tissue. Figure 1 indicates that at protein concentrations above 1 mg per assay (300 µl) there was no further increase in the maximal response to carbachol. In subsequent experiments, approximately 1 mg of tissue protein was used.

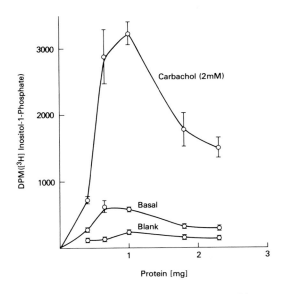

<u>Figure 1.</u> The effect(s) of increasing tissue concentration on carbachol-induced [³H]inositol 1-phosphate accumulation in hippocampal slices. The concentration of carbachol was 2 mM. Each point is the mean value ± S.E. of quadruplicate determinations from a representative experiment repeated three times.

Preliminary experiments, to determine the optimal lithium concentration required for amplification of the agonist-induced response, were carried out using the incubation conditions described above. In the presence of 2 mM carbachol, 10 mM lithium had a maximal effect on [³H]inositol 1-phosphate accumulation, with an EC_{50} value of approximately 1 mM. These results are in agreement with those of Berridge *et al.* (1982), and 10 mM lithium was therefore used in all subsequent experiments. The maximal response to carbachol occurred after a 60 min incubation and the response decreased with further incubation. A 60 min incubation with agonist was therefore used in all subsequent experiments.

Under optimal incubation conditions, carbachol stimulated the accumulation of [³H]inositol 1-phosphate by approximately 800% over the basal (nonstimulated) level. The effect of carbachol was maximal at a concentration of approximately 200 μM, with an EC_{50} value of 50 μM. The

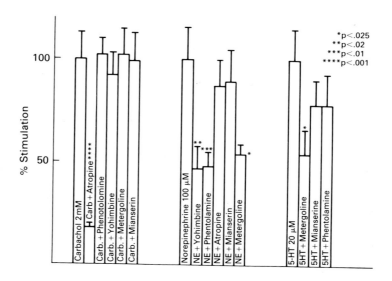

Figure 2. The effect(s) of selective neurotransmitter antagonists on agonist-induced [³H]inositol 1-phosphate accumulation. The concentration of all antagonists was 1 μM. [³H]Inositol 1-phosphate accumulation in the presence of antagonists alone was not significantly different from that observed in the absence of drug. Data are mean values ± S.E. of quadruplicate determinations.

response to norepinephrine appears to plateau at concentrations between 1 and 10 μM and is maximal at 100 μM. The response to serotonin is biphasic since agonist concentrations above 25 μM, results in a decrease in the accumulation of [³H]inositol 1-phosphate.

To determine the subtype of receptor involved in the agonist-induced responses, hippocampal slices were pre-incubated with specific receptor antagonists. Figure 2 shows that the muscarinic-cholinergic antagonist atropine almost completely blocks the [³H]inositol 1-phosphate accumulation elicited by 2 mM carbachol. The response to norepinephrine was partially blocked by both phentolamine (1 μM) and yohimbine (1 μM) suggesting that both α_1 and α_2 adrenoreceptors may play a role in mediating the effect(s) of norepinephrine. The response to serotonin was studied in the presence of the serotonin receptor antagonists metergoline and mianserin. Figure 2 indicates

87

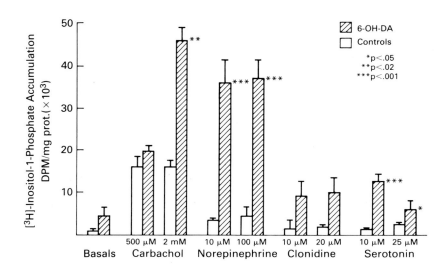

<u>Figure 3</u>. The effect(s) of 6-hydroxydopamine-treatment on agonist-induced accumulation of [³H]inositol 1-phosphate in hippocampal slices. Data are mean values ± S.E. of quadruplicate determinations. See text for details.

that the $5HT_2$ antagonist mianserin (1 μM) had no effect on the serotonin-induced accumulation of [³H]inositol 1-phosphate. Metergoline, however, attenuated the serotonin-induced accumulation, indicating that serotonin exerts its effects through a $5HT_1$ receptor.

In subsequent experiments we examined the effects of selective lesions of the ascending noradrenergic pathway to the hippocampus on the agonist-induced accumulation of [³H]inositol 1-phosphate (Janowsky, A., Labarca, R. and Paul, S.M., in preparation). Hippocampal slices from animals previously treated with intracerebroventricular 6-hydroxydopamine (200 μg in two separate injections over a 7 d period) were incubated in the presence of various agonists. Figure 3 indicates that removal of noradrenergic input to the hippocampus results in an enhanced response to norepinephrine, as well as to carbachol and serotonin. In addition, the response to the $α_2$ agonist clonidine is greatly increased. These results are quite similar to that observed following chronic administration of the tricyclic antidepressant imipramine (10 mg/kg daily for 2 wk) to animals (Labarca, R., Janowsky, A. and Paul, S.M., in preparation) and may indicate an action on post-receptor events in the phosphatidylinositol pathway.

DISCUSSION

The stimulation of [^{3}H]inositol 1-phosphate accumulation in the hippocampus appears to be a receptor-mediated event that is sensitive to a number of neurotransmitter agonists. The bimodal effect of norepinephrine on [^{3}H]inositol 1-phosphate accumulation and the blockade of at least part of the norepinephrine-induced stimulation by yohimbine, indicate that the hippocampus may contain α_2 adrenoreceptors that are coupled to phosphatidylinositol metabolism. However, additional work using more selective α_2 antagonists will be necessary to support this conclusion. The increase in norepinephrine and clonidine stimulated [^{3}H]inositol 1-phosphate accumulation that is observed following lesions with 6-hydroxydopamine implies that the respective receptor-mediated responses are not located presynaptically. In addition, the enhanced response to carbachol may indicate that the noradrenergic innervation to the hippocampus is responsible for tonically inhibiting and modulating neurotransmitter-stimulated, phosphatidylinositol hydrolysis in general. Since there is no change in muscarinic receptor density in the hippocampus (as determined by direct ligand binding) following lesions of the noradrenergic system (Gurwitz, *et al.*, 1980) it is possible that the effects of lesioning, reported here, are due to changes in post-receptor events mediating phosphatidylinositol hydrolysis. Experiments to determine the mechanism(s) responsible for the lesion or drug-induced enhancement of agonist-dependent [^{3}H]inositol 1-phosphate accumulation are in progress.

REFERENCES

Allison, J.H. and Blisner, M.E. (1976) *Biochem. Biophys. Res. Commun. 68*, 1332-1338.
Berridge, M.J., Downes, C.P. and Hanley, M.R. (1982) *Biochem. J. 206*, 587-595.
Downes, C.P. (1982) *Cell Calcium 3*, 413-428.

Janowsky et al.

Glowinski, J. and Iverson, L.L. (1966) *J. Neurochem.* *13*, 655-669.
Gurwitz, D., Kloog, Y., Egozi, Y. and Sokolovsky, M. (1980) *Life Sci.* *26*, 79-84.
Janowsky, A., Labarca, R. and Paul, S.M. (1984) *Life Sci.*, in press.
Lowry, O.H., Rosebrough, N.J., Farr, A.L. and Randall, R.J. (1951) *J. Biol. Chem.* *193*, 265-275.
Michell, R.H. (1975) *Biochim. Biophys. Acta* *415*, 81-147.
Sherman, W.R., Leavitt, A.L., Honchar, M.P., Hallcher, L.M. and Phillips, B.E. (1981) *J. Neurochem.* *36*, 1947-1951.

EARLY EVENTS IN THE SELECTIVE CYTOTOXICITY TO TUMOR CELLS INDUCED BY LIPOSOMES CONTAINING PLANT PHOSPHATIDYLINOSITOL

Marti Jett and Carl R. Alving

Department of Membrane Biochemistry, Walter Reed
Army Institute of Research, Washington, DC 20307-5100

SUMMARY

Recently we reported a cytotoxic effect induced by liposomes comprised of plant phosphatidylinositol (PI) against 10 out of 11 cultured tumor cell lines but not against 4 normal cell lines. The cytotoxicity was not observed with liposomes comprised of other negatively charged phospholipids, including animal PI. Cytotoxicity against tumor cells was most obvious 72 h after incubation of the cells with plant PI liposomes. In this paper, we describe an early biochemical event that occurs prior to the onset of overt cytotoxicity. We have documented a dramatic drop in uptake of inositol that occurs as early as 3 h after incubation of the cells with plant PI liposomes. The decrease in inositol uptake is not seen with normal cells, nor does it occur in tumor cells exposed to liposomes comprised of animal PI. Therefore, we conclude that interference with PI turnover may be an important early event in the development of cytotoxicity against tumor cells.

INTRODUCTION

We previously reported a cytotoxic effect against tumor cells by liposomes comprised of plant phosphatidylinositol and cholesterol in a 2:1 molar ratio (Jett and Alving, 1983). The cytotoxic effect was observed with 10 out of 11 tumor cell lines in tissue culture, but did not occur with 4 normal cell lines. The cytotoxic effect was so pronounced that over a period of a week it caused actual disintegration of the tumor cells. Several other negatively charged phospholipids, including PI of animal origin, were used in the same system and found to be without effect. These data suggested that we were observing a cytotoxic effect of plant phosphatidylinositol that seemed to be unique to tumor cells.

METHODS

Cytotoxic effects of plant PI were measured by 4 methods that yielded identical results. The methods were thymidine uptake, microscopic quantitation, protein analysis, and leucine or glucosamine incorporation. The decrease in thymidine uptake by the tumor cells was most dramatic after 72 h of incubation with the liposomes. In the present study, we investigated the biochemical events that precede the onset of detectable cytotoxicity. One logical system to examine was the intracellular turnover of endogenous PI.

Several cell lines were tested to examine the time course of [^{3}H]inositol uptake by cells in medium that lacked carrier inositol. The total inositol taken up by the different cell lines varied from low uptake by lymphoblastoid cells to approximately 8-fold greater uptake by a neuroblastoma cell line. Steadily increasing inositol uptake was observed over 3 h for all cell lines tested. Pulse-chase experiments showed a typical pattern with approximately one-third of the tritium being released by the cells within 1 h after replacing the medium that contained [^{3}H]inositol with normal culture medium. Pulse-chase experiments in which the pulse was carried out in medium that contained equal amounts of radiolabeled and unlabeled inositol, yielded a similar pattern of uptake parallel to, but with one-half the radioactivity seen when cells were incubated with only radiolabeled inositol. Cellular uptake of [^{3}H]inositol was examined 3, 24 and 48 h after plant PI addition and the uptake pattern was compared with thymidine uptake. For these experiments, cells were cultured in 96-well plates as described in the legend to Figure 1. Liposomes (multilamellar vesicles) were prepared by vortexing dried lipids with a solution of Eagle's minimal essential medium in

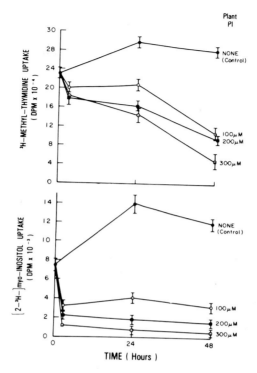

Figure 1. Comparison of thymidine and inositol uptake by SK-N-MC neuro-blastoma cells at various concentrations of plant PI liposomes. (Top) [³H]methyl-thymidine uptake. Cells were plated in 96-well sterile cluster flat bottom plates (100 µl/well) so that they would be 60% confluent by the following day. Eighteen hours later, liposomes comprised of phospholipid:cholesterol in a 2:1 molar ratio (Alving and Swartz, 1984) were added to the cultures (20 µl/well) to give a final concentration of 100, 200, or 300 µM. Three, 24, or 48 h after liposome addition to the cultures, uptake was measured by addition of [³H]methyl-thymidine (1 µCi/120 ng) to each well. The cells were incubated for 3-4 h and harvested by an automatic cell harvesting device (Skatron, Sterling, Va.) using glass-fiber filter mats (Mash). (Bottom) myo-[2-³H]inositol uptake. For inositol uptake, cells were plated and liposomes added as for the measurement of thymidine uptake. At the times of assay, (3, 24, or 48 h) the culture fluid was removed by suction and 100 µl of inositol-free Eagle's minimal essential medium with Earle's balanced salt solution that contained myo-[2-³H]inositol (1 µCi/11 ng/well) was added. Cells were incubated for 2-3 h in this fluid and then were harvested as described above.

93

Earle's balanced salt solution under aseptic conditions (Alving and Swartz, 1984).

RESULTS AND DISCUSSION

Thymidine uptake by plant PI-treated cells (Fig. 1a) deviated slightly from control cells after only 3 h of treatment, but deviation was much more pronounced by 48 h. Dramatic decreases in thymidine uptake occurred even at the lower concentrations of plant PI by 72 h (data not shown). In contrast, inositol uptake (Fig. 1b) was altered maximally by 3 h, with rather slight changes occurring at later time points.

The effect of liposomes comprised of animal or plant PI on inositol uptake was concentration dependent, as shown for a mouse neuroblastoma cell line (N4TG1) and a human melanoma cell line (HT-144) (Fig. 2). Liposomes made with animal PI appeared to stimulate inositol uptake by each of these cell lines. For certain other cell lines, however, animal PI liposomes caused little, if any, change in inositol uptake. In contrast, liposomes comprised of plant PI showed a much different pattern on each of the above cell lines, with inositol uptake being markedly decreased (Fig. 2). Liposomes containing plant phosphatidylcholine did not cause a decrease in inositol uptake (data not shown).

Figure 3 contrasts the effects of plant PI liposomes on inositol uptake by WI-38 normal human embryonic lung fibroblasts and WI-38/SV-40 transformed cells. As shown in Fig. 3a, normal WI-38 lung cells showed little deviation from untreated controls when the cells were treated with plant PI liposomes at 100 μM, 200 μM or 300 μM concentrations of PI. In contrast, the WI-38/SV-40 transformed lung cells responded much differently to plant PI liposomes in their pattern of inositol uptake (Fig. 3b). Approximately 60% suppression of inositol uptake was seen at a PI concentration of 200 μM. Even at a 100 μM concentration of plant PI, obvious strong suppression of inositol uptake was observed.

Our results demonstrate that the cytotoxic effect induced by plant PI against cultured tumor cells is preceded by a very early suppression of [3H]inositol

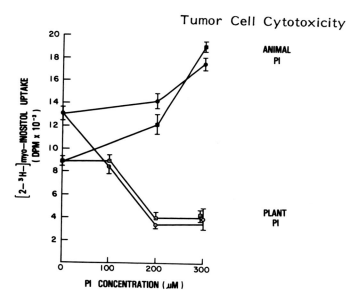

Figure 2. Comparison of the effects of animal and plant PI liposomes on two cell lines. Liposomes containing animal PI (solid symbols), or plant PI (open symbols), were incubated, at the concentrations indicated, with N4TG1 mouse neuroblastoma cells (squares), or HT-144 human melanoma cells (circles). Inositol uptake was measured by the method described in the legend to Fig. 1b.

uptake by the cells. Suppressed uptake of inositol occurred within 3 h after exposure to plant PI, while clearly detectable cytotoxicity by other biochemical methods normally is observed only after 2-3 d. It is likely that the decreased uptake of inositol by the tumor cells is either a reflection of, or a cause of, reduced intracellular turnover of endogenous PI. The present results are compatible with our previous finding that plant PI, but not animal PI, exerts cytotoxic effects. We previously proposed (Jett and Alving, 1983) that the cytotoxic effect of plant PI was the result of, or the cause of, changes in intracellular phospholipid metabolism. It now appears that decreased inositol uptake and decreased endogenous PI turnover occur quite early in the sequence of events that leads to cytotoxicity of tumor cells. Future studies will determine whether other related systems, such as prostaglandin synthesis, protein kinase activity, cyclic nucleotides, or intracellular calcium move-ment are also affected as a result of decreased PI turn-over prior to overt cytotoxicity to tumor cells.

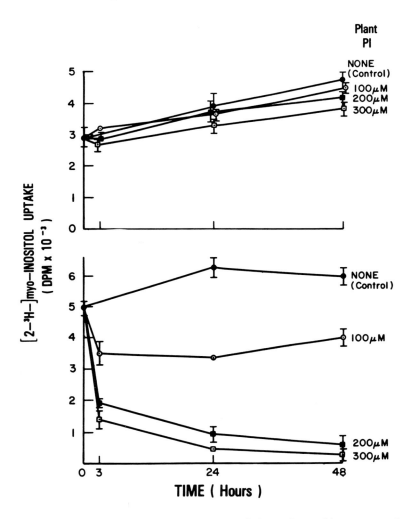

Figure 3. Effects of liposomes containing plant PI on normal and transformed cells (Top) WI-38 Normal Human Embryonic Lung Fibroblasts. The cells were incubated with plant PI liposomes, at the PI concentrations indicated, and the cells were harvested as described in the legend to Fig. 1b. (Bottom) WI-38/SV-40 transformed cells. These cells were treated exactly as described for the WI-38 normal cells. The symbols for both frames are: control without PI (●); 100 μM PI (O); 200 μM PI (■); 300 μM PI (□).

96

REFERENCES

Alving, C.R. and Swartz, G.M., Jr. (1984) In: *"Liposome Technology"*, Vol. 2 (G. Gregoriadis, ed.) pp. 55-68, CRC Press, Boca Raton.

Jett, M. and Alving, C.R. (1983) *Biochem. Biophys. Res. Commun.* *114*, 863-871.

DISCUSSION

Summarized by William W. Wells

Department of Biochemistry, Biochemistry Building,
Michigan State University, East Lansing, MI 48824

The first session of the conference was especially varied in the topics presented ranging from the mechanisms of enzymatic formation of L-myo-inositol 1-phosphate from glucose 6-phosphate to the selective cytotoxicity to tumor cells of liposomes containing PI of plant origin. Dr. Frank Eisenberg (NIH) replied to an inquiry from Dr. Wells (Michigan State Univ.) that synthetic 5-ketoglucose 6-phosphate, when added to the homogeneous L-myo-inositol 1-phosphate synthase catalyzing the formation of product from [^{14}C]glucose 6-phosphate, did not lead to dilution of the radioactivity of the product indicating that the proposed intermediate is not exchangeable and essentially has no finite existence. Dr. Eisenberg added that to actually see the proposed intermediate, the partial reactions would have to be interrupted to prevent the rapid cyclization of 5-ketoglucose 6-phosphate to inosose-2 1-phosphate. Attempts to observe cyclization products of authentic 5-ketoglucose 6-phosphate enzymatically have been unsuccessful.

In response to Dr. Bleasdale's paper, Dr. Robert Michell (Univ. Birmingham, UK) was curious about the physico-chemical differences between lung surfactants with various PG:PI ratios and asked if dietary manipulation to increase the PI-content of the surfactant of adult rats causes these animals to suffer respiratory distress syndrome? Dr. Bleasdale (Univ. Texas, Dallas) replied that when adult rats are treated in this manner PG is almost totally removed from their lung surfactant but the animals appear perfectly normal. In addition, replacement of PG in surfactant with PI did not alter pressure-volume relationships in excised lungs and did not influence a variety of parameters of surface activity that were measured using a Langmuir-Wilhelmy balance. He did not, however, rule out the possibility that PG-rich surfactant is indispensable in the perinatal period. Dr. John Fain (Brown Univ.) asked whether Dr. Bleasdale had attempted to make rabbits or rats *myo*-inositol deficient to determine the effect on the PG:PI ratio in fetal lung surfactant. Dr. Bleasdale stated that although such studies on intact animals had not been conducted, when type II pneumonocytes were cultured in a *myo*-inositol deficient medium, the cells synthesized predominantly PG and not PI. Dr. Wells commented that elevated *myo*-inositol levels had been observed in the testes of diabetic rats and this correlated well with testicular glucose 6-phosphate levels. In response, Dr. Bleasdale agreed that it would be important to examine glucose 6-phosphate levels in the placenta of diabetic animals to assess whether enhanced placental *myo*-inositol synthesis could account for the elevated fetal serum *myo*-inositol levels observed in his studies.

Dr. Douglas Greene (Univ. Pittsburg) brought up the question of possible effects of hyperglycemia on *myo*-inositol uptake by type II pneumonocytes. At concentrations of *myo*-inositol found in adult serum, the uptake of *myo*-inositol by slices of fetal rabbit lung was unaffected by glucose but at concentrations of *myo*-inositol found in fetal serum, *myo*-inositol uptake by these slices was actually stimulated by glucose. Dr. George Hauser (Harvard Univ.) noted that he had observed similar shifts between PI and PG in pinealocytes in a *myo*-inositol deficient medium despite an apparently adequate cellular *myo*-inositol level. He asked if there was evidence for specific pools of inositol for PI synthesis. Dr. Bleasdale replied that the existence of separate pools of *myo*-inositol had not been demonstrated but he pointed out that type II pneumonocytes contained *myo*-inositol at concentrations about 10 times the Km value for PI synthase and yet PI synthesis was sensitive to changes in concentration of extracellular *myo*-inositol. To add to this point, Dr. Peter Downes (ICI, UK) offered the comment that parotid cells can be made severely *myo*-inositol deficient if one stimulates with, for example, a muscarinic agonist and simultaneously blocks IP breakdown with lithium. One sees instead of an increased PG,

100

that there is a build up of CDP-DG, and *myo*-inositol levels as high as 10 mM are required to reverse this situation. Thus, the question of *myo*-inositol pools is also relevant in this system. In reply to a question from Dr. Joseph Eichberg (Univ. Houston), there were no available data on glycerol 3-phosphate levels in type II pneumonocytes to support the suggestion that this metabolite could regulate the shift to PG synthesis during lung development.

Following the paper by Dr. Holub, Dr. Lapetina (Wellcome Research Laboratories) asked whether the substitution of eicosapentaenoic acid for arachidonic acid at position *sn*-2 of phosphatidylinositol affected the action of phospholipase C. Such studies have not been conducted nor is the precise fatty acid composition of PIP or PIP_2 relative to PI known although stearoyl-arachidonoyl species do predominate in all three phosphoinositides. In response to Dr. Lapetina's question to Dr. Holub, Dr. Johnston (Univ. Texas, Dallas) stated he had examined amnion tissue and found no specificity of phospholipase C for particular molecular species of PI, though these studies did not include treatment with eicosapentaenoic acid as described by Dr. Holub. Regarding the increased arachidonoyl composition in PI of platelets of cholesterol-fed animals, Dr. Peggy Zelenka (NIH), asked if other phospholipids were similarly affected. Dr. Holub responded that more arachidonate was observed in platelets of cholesterol fed subjects simply because there was correspondingly more PI. The question was raised by Dr. Michell whether in the biosynthesis of PI, both *sn*-1 and *sn*-2 acyl groups were remodeled to which Dr. Holub replied that 1-stearoyl, 2-arachidonoyl species were synthesized from pools of PI having essentially the same composition as their corresponding CDP-DG precursor. Dr. Hanahan (Univ. Texas, San Antonio) noted that platelets have a short half-life and raised the interesting point that differences in the mean age of platelets due to possible changes in $t_{1/2}$ as a result of dietary supplementation with cod-liver oil might complicate interpretation of the observed decrease in platelet aggregation. There was no readily available information in the literature to address this question.

Following the paper by Dr. Sherman (Washington Univ.), Dr. Berridge (Univ. Cambridge, UK) reported that atropine depressed the levels of PI in insects in analogy to the observations presented by Sherman, but also it depressed the level of IP_2 and IP_3 suggesting that the polyphosphoinositide pathway is also involved here. This raised questions concerning the origin of IP_3 and 1,2 cyclic inositol phosphate. Li^+ does not appear to affect IP_3ase but at high levels may affect IP_2ase. Sherman suggested that the increased 1,2 cyclic IP may arise from PI metabolism related to Na^+ and K^+-ATPase since Simmons and Winegrad have shown a high dependence on myo-inositol and arachidonic acid for Na^+ and K^+-ATPase. Dr. Peter Downes commented that in brain slices Li^+ can completely block hydrolysis of the IP but only partially inhibits hydrolysis of the IP_2 and has little or no inhibition of IP_3 hydrolysis. It was agreed that the number of enzymes involved in the degradation of the IP_3 and IP_2 intermediates and the exact positions of the phosphomonoesterases remained to be determined. Dr. Ata Abdel-Latif (Med. Coll. Georgia) questioned Dr. Nicolas Bazan on his report that the effect of propranolol in the retina appeared to be contradictory, that is, in the presence of propranolol an increase in diacylglycerol was observed. In reply, Dr. Bazan explained that at first no increase in diacylglycerol was observed, but later an increase in sn-1-stearoyl, 2-arachidonoylglycerol did occur and arose from PIP_2. The conclusion from studies with propranolol was that multiple enzymes of phospholipid metabolism are affected simultaneously. The significance of effects of myo-inositol at 1-10 mM was questioned by Dr. Carl Alving (Walter Reed Army Inst.), however, it was Dr. Bazan's opinion that levels of 1 mM were within physiological limits and did not create an osmotic artifact.

The two shorter reports presented by Janowsky (NIH) and Jett (Walter Reed Army Inst.) generated considerable discussion. Dr. Janowsky replied to a question from Dr. Michell that additional α_1 and α_2 adrenergic antagonists had been tested in the hippocampal slice system and the results were consistent with the view that both α_1 and α_2 receptors appeared to be involved in norepinephrine stimulation. The bimodal dose-response curve for norepinephrine was offered as additional evidence that the catecholamine may act at both α_1 and α_2

receptors. Michell reported that Stevens and Logan at Birmingham (UK) have observed the accumulation of IP_2 and IP_3 in a similar hippocampal system mediated by vasopressin at V_1 receptors, thus demonstrating that in this brain region, activation of both amine and peptide receptors elicits a rise in cellular inositol phosphates. Dr. Darryle Schoepp (Univ. Kansas) commented that in cerebral cortex the noradrenergic stimulation appeared to be only α_1-mediated, and did not demonstrate a biphasic dose-response curve. Dr. Janowsky replied that the differences could very well be explained by differences in the brain regions in question. Regarding the carbachol-stimulation of IP accumulation in 6-hydroxy dopamine lesioned preparations, Dr. Barbara Talamo (Tufts Medical School) asked whether there was a shift in the sensitivity of the dose-response curve for norepinephrine. Though this was not investigated in this instance, it was examined for the medial forebrain bundle lesions and Dr. Janowsky observed an increase at all concentrations tested. Furthermore, norepinephrine could be tonically inhibiting the response to carbachol based on acute and chronic studies in the cortex though this has not been examined *in vitro*. Dr. John Yandrasitz (Univ. Pennsylvania) asked about pools of *myo*-inositol, phosphoinositides and inositol phosphates to explain the lack of effect on phosphoinositide turnover and specific radio-activity in response to elevated *myo*-inositol levels. Janowsky noted that of 10^7 dpm of [^{3}H]*myo*-inositol, only 10^3 dpm were found in inositol phosphates, suggesting a very minor incorporation.

The provocative observation was made by Jett and Alving that liposomes of plant PI and cholesterol were strongly cytotoxic to tumor but not normal cells whereas other lipids including PI of animal origin were harmless. Dr. Hokin-Neaverson questioned the purity of the plant PI; however, Dr. Jett indicated that no differences were observed between the commercial plant PI and plant PI purified by TLC. When plant PI was mixed with animal PI (in reply to a question by Dr. Bernard Agranoff, Univ. Michigan), the response was similar to that of animal PI alone, namely, no cytotoxicity. Dr. Eichberg suggested that plant PI might be a more ready substrate for a phospholipase A_2 thus producing higher levels of lysophosphatidylinositol. However, Dr. Alving did not

103

find this explanation attractive since the toxic effect of plant PI was not observed in normal cells where such phospholipase A_2 activity might also be expected to prevail.

In conclusion, the question of the existence of various metabolic pools of *myo*-inositol, *myo*-inositol-phosphates and phosphoinositides continued to thwart rapid advances in several systems reported in the session. More progress is needed in understanding the regulation of the synthesis and turnover of *myo*-inositol and phosphoinositides and of IP_3, IP_2 and IP breakdown, topics further considered in later sections of this book.

PART II

PHOSPHOINOSITIDE BIOSYNTHESIS AND DEGRADATION

PHOSPHATIDYLINOSITOL SYNTHASE

IN MAMMALIAN PANCREAS

Mabel Hokin-Neaverson and Gregory S. Parries

Departments of Physiological Chemistry and Psychiatry,
and the Wisconsin Psychiatric Institute,
University of Wisconsin, Madison, WI   53706

SUMMARY

In this paper we outline two current studies on phosphatidylinositol synthase in mammalian pancreas. Progress on the solubilization, stabilization, and purification of phosphatidylinositol synthase from dog pancreas microsomal membranes is reviewed, and the modulation of phosphatidylinositol synthase activity by Li^+-induced changes in inositol levels in dispersed mouse pancreas acinar cells is discussed.

INTRODUCTION

Phosphatidylinositol (PI) synthase (CDP-1,2-diacyl-sn-glycerol:myo-inositol 3-phosphatidyltransferase, EC 2.7.8.11), was first described more than 25 years ago by Agranoff et al. (1958) and Paulus and Kennedy (1960). It catalyzes the final step in PI biosynthesis, the reaction between CDP-diacylglycerol (CDP-DG) and myo-inositol to form PI and CMP, with a requirement for either Mn^{2+} or Mg^{2+} for activation:

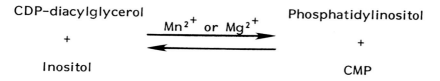

CDP-diacylglycerol   Phosphatidylinositol

$$Mn^{2+} \text{ or } Mg^{2+}$$

+   +

Inositol   CMP

The enzyme is also active in the reverse direction (Hokin-Neaverson *et al.*, 1977; Bleasdale *et al.*, 1979). In animal tissues, the enzyme is located primarily in the endoplasmic reticulum and has been largely studied in microsomal membrane fractions.

The enzyme functions to maintain normal levels of PI, and is an integral part of the well-known cycle of PI breakdown and subsequent resynthesis which occurs in many tissues in response to a wide variety of hormones, neurotransmitters, and other agents (reviewed by Michell, 1975; Irvine *et al.*, 1982). Bleasdale *et al.* (1979) have suggested a role for the reverse reaction catalyzed by PI synthase in the regulation of the utilization of CDP-DG for the synthesis of PI and phosphatidylglycerol for lung surfactant.

Our interest in PI synthase, in addition to its role in the resynthesis phase of the PI cycle, was spurred some time ago by investigations in the mammalian pancreas which suggested that hormone-stimulated PI breakdown might be catalyzed by the reverse reaction of the enzyme to give CDP-DG and inositol as the first breakdown products (Hokin-Neaverson *et al.*, 1978). However, we have since used the ability of Li^+ to inhibit *myo*-inositol 1-phosphatase (Naccarato *et al.*, 1974; Hallcher and Sherman, 1980) to determine unequivocally, that this is not the case. In the presence of Li^+, the first water-soluble product of PI breakdown is inositol 1-phosphate in mouse pancreas acinar cell dispersions stimulated with either CCK-8 or acetylcholine (ACh) (Hokin-Neaverson and Sadeghian, 1984). This confirmed that stimulated PI breakdown involves hydrolysis to diacylglycerol (DG) and inositol 1-phosphate catalyzed by phospholipase C-type enzyme, as was first proposed (Hokin, M.R. and Hokin, L.E., 1964). An increased accumulation of inositol 1-phosphate in the presence of Li^+ in insect salivary gland, parotid gland and rat brain has been reported by Berridge *et al.* (1982).

Recent work in this laboratory has dealt with two aspects of PI synthase in mammalian pancreas. We have carried out studies on the solubilization and purification of PI synthase from dog pancreas microsomes in order to obtain sufficient quantities of highly purified enzyme for detailed characterization of its molecular properties. We have also examined in the intact cell whether failure to regenerate inositol from stimulated PI breakdown, due to Li$^+$ inhibition of inositol 1-phosphatase, would compromise PI synthase activity. The purpose of this article is to outline and discuss some of this work. Both aspects of the work are currently in progress and will be published in detail elsewhere (Parries and Hokin-Neaverson, 1984).

SOLUBILIZATION, STABILIZATION, AND PARTIAL PURIFICATION OF PI SYNTHASE FROM MAMMALIAN PANCREAS

Detailed biochemical investigations of PI synthase have been severely hampered by the lack of substantial amounts of highly purified enzyme. One of our aims has been the purification of PI synthase from mammalian pancreas in order to further study its molecular properties. PI synthase in mammalian pancreas was first studied by Prottey and Hawthorne (1967) using guinea pig tissue. We have used dog pancreas as a source of enzyme, and have partially characterized enzyme activity from this tissue. Some properties of PI synthase activity in dog pancreas microsomal membranes are summarized in Table I. In general, the characteristics of PI synthase activity in dog pancreas are similar to those reported in other tissues. However, the enzyme in dog pancreas microsomes exhibits a specific activity of 30 nmol PI formed $\times$ min^{-1} $\times$ mg^{-1} protein, which is higher than the values reported in the literature for most other tissues, such that the pancreas represents a relatively enriched source of enzyme for purification.

PI synthase is a membrane-bound protein and, as is the case with other membrane proteins, its purification presents several difficulties. Since it is tightly associated with the membrane, the enzyme must be first solubilized in stable form by an appropriate detergent, and then purified by procedures in which the detergent does not interfere. Because of these difficulties, the membrane-bound enzymes of PI metabolism have been much less extensively studied than the soluble enzymes of PI metabolism.

TABLE I

Some Properties of PI Synthase Activity in Dog Pancreas Microsomes

Specific activity (V_{max}):	30 nmol PI $\times$ min^{-1} $\times$ mg^{-1} protein
pH optimum:	Forward direction, 8.5–9.0
	Reverse direction, 6.5–7.0
Apparent substrate affinities:	K_m (CDP-DG) = 18 μM
	K_m (*myo*-inositol) = 0.76 mM
Metal ion activation:	Apparent K_a (Mn^{2+}) = 42 μM
	Apparent K_a (Mg^{2+}) = 2.5 mM

The standard assay mixture (250 μl) contained 100 μM CDP-DG (derived from egg phosphatidylcholine), 1.0 mM MnCl$_2$, 5.0 mM myo-[2-^{3}H]inositol, 50 mM Tris-HCl, pH 8.5 (37°C), 100 mM KCl, and 15-20 μg microsomal protein. After a 5 min incubation at 37°C, [^{3}H]PI formation was determined essentially as described (Hokin-Neaverson et al., 1977).

Yeast PI synthase has been recently purified to near homogeneity from *Saccharomyces cerevisiae* (Fischl and Carman, 1983), but attempts to purify the enzyme from mammalian sources have been much less successful. Instability of the solubilized enzyme has been a major problem. Rao and Strickland (1974) reported a 15-fold purification of the enzyme starting from rat brain microsomal membranes. However, this partially purified enzyme was said to have the properties of a membrane fragment with a molecular weight greater than 300,000. These authors found a high degree of instability of the enzyme in response to the various treatments which they tried, and they reported a general lack of success in several attempts to further purify the enzyme.

TABLE II

Solubilization of PI Synthase From Dog Pancreas Microsomes with Various Detergents

Detergent	PI Synthase Activity	
	Uncentrifuged	105,000 x g Supernatant Fraction
	(% of control)	
None	100	0
1% n-Octyl-glucoside	95	95
1% NaCholate	60	31
0.5% NaDeoxycholate	28	21
0.5% Digitonin	14	7
1% Triton X-100	21	–

Dog pancreas microsomes were treated with various detergents in a medium which contained 10 mM Tris-HCl, pH 8.4, 10% glycerol, 1.0 mM dithiothreitol, 1.0 mM MnCl$_2$, and 0.25 mM myo-inositol. After 1 h at 4°C, PI synthase activity was measured in an assay mixture which contained 50 mM Tris-HCl, pH 8.5, 3.0 mM MnCl$_2$, 3.0 mM myo-[2-^{3}H]inositol, and 1.0 mM CDP-DG. The final concentration of detergent in the assay was 0.6% (0.3% in the case of digitonin and deoxycholate). After a 30 min incubation at 37°C, [^{3}H]PI formation was measured essentially as described (Hokin-Neaverson et al., 1977).

Takenawa and Egawa (1977) reported a 28-fold purification of PI synthase from rat liver microsomes in the presence of the nonionic detergent Triton X-100, but with a yield of only 3%. Although this enzyme preparation was thought to be nearly homogeneous as determined

111

by SDS-polyacrylamide gel electrophoresis, the final specific activity was only 11 nmol PI formed x min^{-1} x mg^{-1} protein, which is lower than the specific activity of the enzyme in intact microsomes from several tissues including our preparation of microsomes from dog pancreas. More recently, PI synthase from rat brain was solubilized with Triton X-100 and purified 93-fold from the tissue homogenate by CDP-DG-Sepharose affinity chromatography (Eichberg et al., 1983). This enzyme preparation resolved into 3 Coomassie Blue stained bands after SDS-polyacrylamide gel electrophoresis, but the overall yield of activity was only 4%.

In our efforts to purify PI synthase from dog pancreas microsomes, the first task was to obtain conditions under which the enzyme could be solubilized in a stable form. Attempts to solubilize PI synthase from mammalian pancreas with Triton X-100 or other polyoxyethylene type detergents were unsuccessful. The major difficulty was the instability of the enzyme in the presence of these detergents. Rao and Strickland (1974) and Takenawa and Egawa (1977) also reported difficulty in stabilizing the enzyme in the presence of Triton X-100. A variety of other detergents were examined for their effectiveness in the solubilization of PI synthase from dog pancreas. Of the detergents tried, a few of which are shown in Table II, the only one which both solubilized and did not inactivate the enzyme was the nonionic detergent n-octyl-glucoside (OG). This was fortunate, since OG has a number of properties which make it a detergent of choice for the study of membrane proteins. It is a chemically homogeneous compound and is relatively non-denaturing to proteins (Stubbs et al., 1976). In addition, OG has an exceptionally high critical micelle concentration (20-25 mM) which facilitates its removal from detergent-protein micelles. It has been successfully used in the solubilization of a variety of membrane-bound proteins (Baron and Thompson, 1975; Schneider et al., 1979; Eichberg et al., 1983). The use of OG will also facilitate one of our overall aims, which is to reconstitute the purified enzyme into lipid vesicles for further characterization studies. OG can be readily removed by dialysis (Helenius et al., 1979) and has been used to reconstitute a number of membrane-bound proteins (Petri and Wagner, 1979; Mimms et al., 1981).

Treatment of dog pancreas microsomal membranes with 1% OG (33 mM) results in nearly quantitative sol-ubilization of PI synthase as determined by the amount of enzyme activity in the supernatant after centrifugation at 105,000 x g for 1 h (Table II). Although some other commonly used detergents, especially sodium cholate, also resulted in solubilization of enzymatic activity, OG was the only detergent which solubilized the enzyme in stable form and in high yield.

Solubilization of PI synthase by OG is dependent on the ratio of detergent to microsomal membrane. Optimal solubilization of the enzyme occurs with 1% OG at a deter-gent to microsomal protein ratio (w/w) of 4-6 to 1. Under the assay conditions employed here, the enzyme does not exhibit activity in the presence of solubilizing levels of OG, but enzyme activity is fully restored by dilution of the OG to below its critical micelle concen-tration. This apparent decrease in enzyme activity at high OG levels may be due to a "substrate dilution" effect, whereby formation of OG-CDP-DG mixed micelles effectively reduces the amounts of CDP-DG available to micelles which contain PI synthase. The OG-solubilized enzyme is stable over time at 4°C; 97% of the enzyme activity remains after 1 d and 67% remains after 3 d. The rate of loss of PI synthase activity is approximately the same as in intact microsomes.

In finding conditions in which the OG-solubilized enzyme was most stable, recovery of enzyme activity was found to be absolutely dependent on the presence of Mn^{2+} during OG-solubilization (Table III). The loss of activity in the absence of Mn^{2+} appears to be irreversible since there is no recovery of enzyme activity when the OG-treated membranes are subsequently assayed in the pres-ence of Mn^{2+}. Interestingly, Mg^{2+} at concentrations up to 10 mM, does not substitute for Mn^{2+} in the stabilization of the enzyme even though Mg^{2+} can substitute for Mn^{2+} as a cofactor for enzymatic activity. The concentration of Mn^{2+} required to give half-maximal stabilization of the OG-solubilized enzyme is approximately 40 μM, which is similar to the concentration of Mn^{2+} which results in half-maximal activation of the enzyme (Table I).

TABLE III

Stabilization of OG-Solubilized PI Synthase by Manganese

Addition to solubilization medium	PI synthase activity	
	Intact microsomes	OG-solubilized microsomes
	(% recovered)	
No addition	100	0.5
1 mM EDTA	98	0.3
1 mM MgCl$_2$	94	0.2
1 mM MnCl$_2$	91	91

Dog pancreas microsomes were treated either without or with 1% OG in a solubilization medium which contained 10 mM Tris-HCl, pH 8.4 (4°C), 25 mM KCl, 0.5 mM dithiothreitol, and other additions as indicated above. After 1 h at 4°C, the samples were centrifuged at 105,000 × g for 1 h. PI synthase activity in the intact and OG-solubilized microsomes were determined as described in Table II.

Treatment of dog pancreas microsomes with OG results in a nearly 4-fold enrichment in PI synthase activity. Since low concentrations of detergent can permeabilize membranes, the extraction procedure included washing the membranes with 15 mM OG prior to solubilization to facilitate the removal of soluble enzymes which may have been trapped within the intravesicular spaces of the microsomes during the initial homogenization of the pancreas tissue.

The OG-solubilized enzyme was further purified by ion-exchange chromatography on DEAE-cellulose, with an overall yield of greater than 90%. PI (0.5 mM) and egg phospholipids stabilized the enzyme during the chromato-

114

graphy step. SDS-polyacrylamide gel electrophoresis of the active fraction from the ion-exchange resin indicated the presence of 2 major protein bands with M_r values of approximately 30,000 and 60,000, with a number of minor protein bands also present. Although the enzyme preparation was clearly not homogeneous, it had a specific activity of 290 nmol PI formed $\times$ min^{-1} $\times$ mg^{-1} protein, which is much higher than that reported for the enzyme purified from rat liver (Takenawa and Egawa, 1977) and similar to that purified from rat brain (Eichberg et al., 1983). Current efforts towards further purification of the dog pancreas enzyme involve affinity chromatography on CDP-DG-Sepharose.

LITHIUM-INDUCED EFFECTS ON PI SYNTHASE ACTIVITY IN DISPERSED ACINAR CELLS FROM MOUSE PANCREAS

The level of inositol in mouse pancreas tissue rises from approximately 0.5 mM to approximately 1 mM during stimulation of PI breakdown by 10 μM ACh (Hokin-Neaverson et al., 1975). This range of concentrations might be involved in the regulation of PI synthase activity in the functioning pancreas cell, since the K_m of pancreas PI synthase for inositol is approximately 0.8 mM (Table I). The increase in the level of inositol in stimulated cells can be prevented by the addition of Li$^+$ to the incubation medium to inhibit inositol 1-phosphatase (Hokin-Neaverson and Sadeghian, 1984). The half-maximal Li$^+$ concentration for this effect is 1.25 mM and, with 10 mM Li$^+$, virtually all of the water-soluble material released from PI remains as inositol 1-phosphate rather than being hydrolyzed to inositol.

We were interested to see whether these changes in the level of inositol due to failure to regenerate inositol from inositol 1-phosphate in stimulated cells exposed to Li$^+$ might modulate the activity of PI synthase. This might result in a decrease in PI synthesis and an increase in CDP-DG levels with some diversion of CDP-DG towards the synthesis of phosphatidylglycerol (PG) via the intermediate formation of phosphatidylglycerophosphate (PGP) (Fig. 1).

We examined the effects of Li$^+$ on CDP-DG, PG and PI metabolism in dispersed mouse pancreas acinar cells.

115

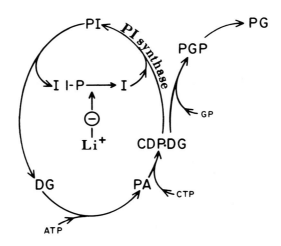

<u>Figure 1</u>. The PI cycle and the point of action of Li[+], with the alter-
nate pathway for CDP-DG utilization for PG synthesis. I, inositol; I
1-P, inositol 1-phosphate; GP, glycerophosphate.

The results from some of these experiments are summarized in Table IV. In the experiments shown, dispersed
cells were prepared and incubated essentially as described
elsewhere (Hokin-Neaverson and Sadeghian, 1984).

To measure CDP-DG levels, [5-³H]cytidine was
added to the incubation medium and [³H]CDP-DG was
extracted and assayed as described (Hokin-Neaverson *et
al.*, 1977). To measure PG and PI synthesis, [³²P]ortho-
phosphate was added to the incubation medium and
[³²P]PG and [³²P]PI were extracted and assayed as
described (Geison *et al.*, 1976). In incubations where
[³H]CDP-DG was assayed, ACh was added after 20 min.
Where [³²P]PG was assayed, ACh was added at zero time.
Where [³²P]PI was assayed, ACh was added at zero time
and atropine (1 µM) was added after 20 min; this was to
encourage the net synthesis of PI which occurs during
reversion to the unstimulated state (Hokin-Neaverson,
1974). The total incubation time in each case was 40 min.

TABLE IV

Changes in Levels of [^{3}H]CDP-DG, [^{32}P]PG and [^{32}P]PI
in Unstimulated and ACh-Stimulated Dispersed Cells from
Mouse Pancreas: Effects of Li$^+$ and Inositol

Additions	[^{3}H]CDP-DG	[^{32}P]PG	[^{32}P]PI
	(percent of control value)		
None	100	100	100
Inositol	48	61	121
Li$^+$	214	213	100
Li$^+$ + inositol	52	61	136
ACh	106	82	606
ACh + inositol	54	74	623
ACh + Li$^+$	732	618	495
ACh + Li$^+$ + inositol	144	257	610

Additions were: inositol, 5 mM; LiCl, 10 mM; acetylcholine, 10 μM, plus eserine, 0.1 mM. Values are from 2 dispersed cell preparations. Actual radioactivities per 0.5 ml cell suspension in the controls, (no additions), for [^{3}H]CDP-DG were 537 ± 24 cpm and for [^{32}P]PG and [^{32}P]PI corrected to a specific radioactivity of 10^8 cpm per μmol for medium ^{32}P were, respectively, 38 ± 3 cpm and 690 ± 10 cpm. Other conditions were as described in the text.

As shown in Table IV, Li$^+$ had a large effect on the level of [^{3}H]CDP-DG, especially in the stimulated cells. There was a 2-fold increase in unstimulated cells and a 7-fold increase in stimulated cells. At the 40 min time interval used, in the absence of Li$^+$, there was no difference between [^{3}H]CDP-DG levels in the presence or absence of ACh. The effect of Li$^+$ on the [^{3}H]CDP-DG level was completely reversed in unstimulated cells and was substantially reversed in the ACh-stimulated cells by addition of inositol to the medium, indicating that inositol was a limiting factor for utilization of CDP-DG by PI synthase in both conditions. The marked effect of Li^+ in the stimulated cells in the absence of added inositol was presumably due to an increase in the rate of synthesis of CDP-DG driven by the rise in PA level which occurs as a consequence of PI breakdown (Hokin-Neaverson, 1974), coupled with the failure to increase inositol to a level which was not limiting for PI synthase activity.

Inositol itself, in the absence of Li^+, reduced [^{3}H]CDP-DG levels by 50% and increased [^{32}P]PI by 20%, suggesting that inositol levels are somewhat limiting for PI synthase activity in these cells in the unstimulated state. The changes in levels of [^{32}P]PG were directly correlated with the changes in the levels of [^{3}H]CDP-DG, giving increased diversion of CDP-DG to the PG pathway in the presence of Li^+, and reversal by inositol. No effect of Li^+ on the [^{32}P]PI level was observed in unstimulated cells. There was a 20% decrease in [^{32}P]PI in cells which had been treated with ACh followed by atropine, and this was reversed by inositol.

The results indicate that failure, due to Li^+ inhibition of inositol 1-phosphatase, to regenerate inositol from inositol 1-phosphate derived from PI breakdown can result in levels of inositol which are limiting to PI synthase activity, particularly in stimulated cells, and this leads to a dramatic increase in CDP-DG level, with some diversion of CDP-DG to the PG pathway in stimulated cells, and to a slower rate of PI synthesis during reversion to the unstimulated state.

It should be noted that Li^+ did not affect the amount of stimulated PI breakdown in dispersed acinar cells from mouse pancreas (Hokin-Neaverson and Sadeghian, 1984); nor did Li^+, at concentrations up to 10 mM, affect PI synthase activity in pancreas microsomal membranes.

Li^+ is widely used in the treatment of psychiatric disorders and the serum concentrations of Li^+ used therapeutically are close to the half maximal concentrations of Li^+ for inhibition of inositol 1-phosphatase (Hallcher and

Sherman, 1980). The levels of Li^+ used therapeutically can be expected therefore to lead to a rise in the inositol 1-phosphate level and, in tissues in which there is no readily available alternate source of inositol, this may lead to a decrease in the level of inositol. In their pioneering work on the actions of Li^+ in the brain, Allison and his coworkers (Allison and Stewart, 1971; Allison and Blisner, 1976; Allison et al., 1976) observed reduced brain inositol levels and increased brain inositol 1-phosphate levels in vivo in Li^+-treated rats and showed that the effects were mediated by a muscarinic cholinergic mechanism. Sherman et al. (1981) showed that the inositol 1-phosphate which accumulates is derived from PI breakdown. The results we have observed in the pancreas cells suggest that the failure to regenerate inositol from PI breakdown in the presence of Li^+ in tissues in which inositol levels are limiting for PI synthase activity could lead to changes in the composition of cellular membranes with the maintenance of higher than normal steady-state levels of CDP-DG, and with increased synthesis of PG and possibly some decrease in PI synthesis. Under conditions of chronic Li^+ intake, these changes, over time, might lead to significant changes in membrane composition. Such changes may be relevant to the mechanism of action of Li^+ therapeutically, or to its side effects.

ACKNOWLEDGEMENTS

The work from this laboratory was supported by NIH grant NS-13878 and NIMH grant MH-26494.

REFERENCES

Agranoff, B.W., Bradley, R.M. and Brady, R.O (1958) J. Biol. Chem. 233, 1077-1083.

Allison, J.H. and Stewart, M.A. (1971) Nature 233, 267-268.

Allison, J.H. and Blisner, M.E. (1976) Biochem. Biophys. Res. Commun. 68, 1332-1338.

Allison, J.H., Blisner, M.E., Holland, W.H., Hipps, P.P. and Sherman W.R. (1976) Biochem. Biophys. Res. Commun. 71, 664-670.

Baron, C. and Thompson, T.E. (1975) Biochim. Biophys. Acta 382, 276-285.

Berridge, M.J., Downes, C.P. and Hanley, M.R. (1982) *Biochem. J. 206*, 587-595.

Bleasdale, J.E., Wallis, P., MacDonald, P.C. and Johnston, J.M. (1979) *Biochim. Biophys. Acta 575*, 135-147.

Eichberg, J., Bostwick, J.R. and Ghalayini, A. (1983) in: *"Neural Membranes"* (G.Y. Sun, N. Bazan, J.Y. Wu, G. Porcellati and A.Y. Sun, eds.) pp. 191-213, Humana Press, Clifton, New Jersey.

Fischl, A.S. and Carman, G.M. (1983) *J. Bacteriol. 154*, 304-311.

Geison, R.L., Banschbach, M.W., Sadeghian, K. and Hokin-Neaverson, M. (1976) *Biochem. Biophys. Res. Commun. 68*, 343-349.

Hallcher, L.M. and Sherman, W.R. (1980) *J. Biol. Chem. 255*, 10896-10901.

Helenius, A., McCaslin, D.R., Fries, E. and Tanford, C. (1979) *Methods Enzymol. 56*, 734-749.

Hokin, M.R. and Hokin, L.E. (1964) in: *"Metabolism and Physiological Significance of Lipids"* (R.M.C. Dawson and D.N. Rhodes, eds.) pp. 423-434, Wiley, New York.

Hokin-Neaverson, M. (1974) *Biochem. Biophys. Res. Commun. 58*, 763-768.

Hokin-Neaverson, M. and Sadeghian, K. (1984) *J. Biol. Chem. 259*, 4346-4352.

Hokin-Neaverson, M. Sadeghian, K., Majumder, A.L. and Eisenberg, F., Jr. (1975) *Biochem. Biophys. Res. Commun. 67*, 1537-1543.

Hokin-Neaverson, M., Sadeghian, K., Harris, D.W. and Merrin, J.S. (1977) *Biochem. Biophys. Res. Commun. 78*, 364-371.

Hokin-Neaverson, M., Sadeghian, K., Harris, D.W. and Merrin, J.S. (1978) in: *"Cyclitols and Phosphoinositides"* (W.W. Wells and F. Eisenberg, Jr., eds.) pp. 349-359, Academic Press, New York.

Irvine, R.F., Dawson, R.M.C., and Freinkel, N. (1982) in: *"Contemporary Metabolism"* (N. Freinkel, ed.) Vol. 2, pp. 301-342, Plenum Medical Book Co., New York.

Michell, R.H. (1975) *Biochim. Biophys. Acta 415*, 81-147.

Mimms, L.T., Zampighi, G., Nazaki, Y., Tanford, C. and Reynolds, J.A. (1981) *Biochemistry 20*, 833-840.

Naccarato, W.F., Ray, R.E. and Wells, W.W. (1974) *Arch. Biochem. Biophys. 164*, 194-201.

Parries, G.S. and Hokin-Neaverson, M. (1984) *Biochemistry*, in press.

Paulus, H. and Kennedy, E.P. (1960) *J. Biol. Chem.* *235*, 1303-1311.

Petri, W.A. and Wagner, R.R. (1979) *J. Biol. Chem.* *254*, 4313-4316.

Prottey, C. and Hawthorne, J.N. (1967) *Biochem. J.* *105*, 379-392.

Rao, R.H. and Strickland, K.P. (1974) *Biochim. Biophys. Acta* *348*, 306-314.

Schneider, W.J., Basu, S.K., McPhaul, M.J., Goldstein, J.L. and Brown, M.S. (1979) *Proc. Natl. Acad. Sci. USA* *76*, 5577-5581.

Sherman, W.R., Leavitt, A.L., Honchar, M.P., Hallcher, L.M. and Phillips, B.E. (1981) *J. Neurochem.* *36*, 1947-1951.

Stubbs, G.W., Smith, H.G., Jr. and Litman, B.J. (1976) *Biochim. Biophys. Acta* *425*, 46-56.

Takenawa, T. and Egawa, K. (1977) *J. Biol. Chem.* *252*, 5419-5423.

THE ENZYMOLOGY OF PHOSPHOINOSITIDE CATABOLISM, WITH PARTICULAR REFERENCE TO PHOSPHATIDYL-INOSITOL 4,5-BISPHOSPHATE PHOSPHODIESTERASE

R.F. Irvine, A.J. Letcher, D.J. Lander
and R.M.C. Dawson

Department of Biochemistry, AFRC Institute of
Animal Physiology, Babraham, Cambridge CB2 4AT, UK

ABSTRACT

The properties and control mechanisms of the enzymes involved in phosphoinositide catabolism are reviewed, in particular those of the PIP_2 phosphodiesterase of rat brain. We have previously shown (Irvine et al., 1984a) that under approximately physiological ionic conditions, this enzyme will not hydrolyze its substrate in a lipid mixture similar to that of the inner half of a plasma membrane; however, if the substrate presentation is changed, the enzyme can become very active. We have tried to characterize more precisely this change in substrate presentation by mixing the PIP_2 substrate with dipalmitoyl phosphatidylethanolamine, pig-liver phosphatidylethanolamine, lysophosphatidylcholine, dioleoylglycerol and distearoylglycerol. Although these experiments confirm the profound effect that substrate structure can have on PIP_2 phosphodiesterase, an explanation in terms of a "bilayer" versus "non-bilayer" configuration is too simplistic to explain the enzyme's preferences. Furthermore, it may be that the head-groups of phospholipids adjacent to the PIP_2 substrate can also directly affect the activity. We suggest that in vivo PIP_2 phosphodiesterase is not controlled by calcium, nor by enzyme protein configuration, but solely by the manner in which the substrate is presented to the enzyme.

INTRODUCTION

In studying phosphoinositide-metabolizing enzymes the principal objective is to shed light on the mechanisms of control of phosphoinositide turnover. We believe that the properties of the enzymes in question, as studied in the test tube, can give us insight into their control and function *in vivo*. Thus, in addition to characterizing the basic properties of some of these enzymes, we have over the last few years started to submit them to conditions which they may experience *in vivo* with respect to the ionic environment and the presentation of the substrate, and this approach has lead to some interesting and, we hope, useful ideas on how they may function *in situ*. Most of our recent work has focussed on PIP_2 phosphodiesterase and phosphomonoesterase, but the other enzymes of the phosphoinositide pathway (Fig. 1) may be subject to similar principles of control. The possible control mechanisms and basic properties of these enzymes were reviewed recently (Irvine, 1982) and firstly a brief updating and modification of that review is necessary.

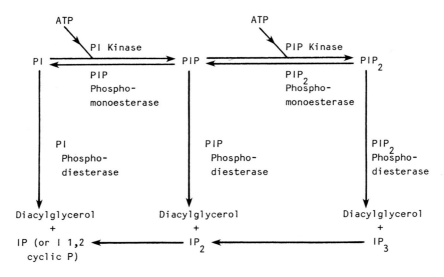

Figure 1. Pathways of phosphoinositide hydrolysis

SUBCELLULAR SITES OF

POLYPHOSPHOINOSITIDE METABOLISM IN RAT LIVER

M.A. Seyfred, C.D. Smith, L.E. Farrell and W.W. Wells

Department of Biochemistry, Michigan State University
East Lansing, MI 48824

SUMMARY

Although the polyphosphoinositides, phosphatidyl-inositol 4-phosphate (PIP) and phosphatidylinositol 4,5 bisphosphate (PIP_2), have often been proposed as important components of biological membranes, the subcellular distribution of PIP, PIP_2, and the enzymes involved in their metabolism have not been fully characterized. We have therefore prepared various subcellular organelles from rat liver and determined their PI kinase and PIP kinase activities. Poly-phosphoinositide-degrading enzymes have also been assayed using [^{32}P]PIP and [^{32}P]PIP_2 as substrates. Plasma membranes exhibit the highest specific activity for PIP kinase, but PI kinase appears to be fairly evenly distributed throughout the cell. The plasma membrane expressed the highest hydrolytic activity toward PIP_2 while similar rates of PIP hydrolysis were observed in all the fractions analyzed. Rates of incorporation of ^{32}P into PIP and PIP_2 in subcellular organelles were also deter-mined after fractionation of isolated rat hepatocytes incubated with ^{32}P$_i$. [^{32}P]PIP_2 synthesis is confined almost exclusively to the plasma membrane, while [^{32}P]PIP is found in other subcellular fractions as well. These studies suggest that the primary site of PIP_2 metabolism is at the plasma membrane, and that PIP is metabolized throughout the cell.

INTRODUCTION

PIP and PIP_2 have recently received increasing attention in many biological systems (Michell, 1975; Downes and Michell, 1982). Most studies have focused on the receptor-mediated breakdown of phosphoinositides in response to certain hormones and other factors (Farese *et al.*, 1980; Akhtar and Abdel-Latif, 1980; Weiss *et al.*, 1982; Farese *et al.*, 1982; Rhodes *et al.*, 1983; Billah and Lapetina, 1983). Other investigators have proposed roles for PIP and PIP_2 in Ca^{2+} mobilization (Michell, 1975; Hendrickson and Reinertgen, 1971; Buckley and Hawthorne, 1972), regulation of enzyme activity (Lipsky and Lietman, 1980) and membrane fluidity (Sheetz *et al.*, 1982).

The subcellular sites of PIP and PIP_2 metabolism were initially investigated by analyzing the distribution of phosphatidylinositol (PI) kinase and phosphatidylinositol 4-phosphate (PIP) kinase, the enzymes responsible for their synthesis (Michell and Hawthorne, 1966; Michell *et al.*, 1967). These studies proposed that PIP and PIP_2 metabolism is chiefly restricted to the plasma membrane. However, more recent studies, using highly purified preparations of various subcellular organelles, have revealed that PI kinase and PIP kinase are more widely distributed throughout the cell (Behar-Bannelier and Murray, 1980; Collins and Wells, 1983; Jergil and Sundler, 1983; and Smith and Wells, 1983). Turnover of PIP and PIP_2 is also dependent on the breakdown of the lipids; however, studies of the mechanisms and distribution of the enzymes responsible for these processes remain largely incomplete.

A second approach to determining subcellular sites of metabolism of PIP and phosphatidylinositol 4,5-bisphosphate (PIP_2) is to analyze the distribution of PIP and PIP_2 in a cell; however, their low levels make analysis by existing chemical means virtually impossible. Incorporation of ^{32}P from endogenously synthesized $[\gamma-^{32}P]ATP$, followed by fractionation of the cells and analysis of ^{32}P-labeled PIP and PIP_2, has given indications of the rates of PIP and PIP_2 turnover in certain systems (Kirk *et al.*, 1981; Creba *et al.*, 1983; Litosch *et al.*, 1983). In these studies, however, the subcellular sites of PIP and PIP$_2$ metabolism could not be determined due to the lack of a reliable subcellular fractionation scheme that can be applied to cell culture systems.

In the present report, we have analyzed the activities of PI kinase and PIP kinase in various subcellular fractions isolated from rat liver. The ability of these isolated organelles to hydrolyze PIP and PIP$_2$ was also investigated. In addition, we have examined the sites of PIP and PIP$_2$ turnover using suspensions of isolated rat hepatocytes. These two approaches both indicate that PIP$_2$ metabolism is most prominent in the plasma membrane, whereas PIP is metabolized throughout the cell.

EXPERIMENTAL PROCEDURES

Materials. Adult male Sprague-Dawley rats (180-250 g) were used for all studies. [^{32}P]Orthophosphate, carrier-free, was purchased from ICN Pharmaceuticals, Inc. [γ-^{32}P]ATP was prepared by the method of Glynn and Chappell (1964) as modified by Walsh *et al.* (1971). Phospholipids were obtained from Sigma or Serdary Research Laboratories. Triton X-100 was obtained from Research Products, Inc. Neomycin (Sigma) was immobilized on Glycophase-CPG beads (Pierce Chemical Co.) as described by Schacht (1981). Glass distilled organic solvents were obtained from MCB. Collagenase (E.C. 3.4.99.5) CLSII was purchased from Worthington. Penicillin-streptomycin solution (10,000 units/ml penicillin, 10 mg/ml streptomycin), MEM non-essential amino acids solution (100X), MEM amino acids solution with 100 mM L-glutamine (50X) and MEM vitamin solution (100X) were obtained from Grand Island Biological Co. Percoll was purchased from Pharmacia. All other chemicals were of reagent grade.

Isolation of Subcellular Organelles from Rat Liver. Plasma membranes were isolated according to the method of Emmelot *et al.*, 1974. Mitochondria were prepared as described by Saltiel *et al.* (1982). Microsomes were pelleted from a 15,000 x *g* supernatant fraction by centrifugation at 100,000 x *g* for 60 min, and the resulting supernatant was used as cytosol. Nuclei were isolated according to a modification (Smith and Wells, 1983) of the method of Blobel and Potter (1966). Nuclear envelopes were prepared from the purified nuclei as described by Kay *et al.* (1972). Lysosomes were isolated by centrifugation on Percoll density gradients similar to the method described by Pertoft *et al.* (1978). The gradient was fractionated and the region which had the highest level of hexosaminidase activity (density = 1.12 - 1.14) was pooled and frozen. Lysosomal membranes were pelleted by centrifugation of the thawed preparations at 100,000 × g for 60 min. All of the preparations were resuspended in 0.25 M sucrose containing 25 mM N-2-hydroxyethylpiperazine-N'-2-ethanesulfonic acid (HEPES), pH 7.5. The purity of isolated subcellular fractions was assessed by marker enzyme analysis according to Leighton *et al.* (1968). Marker enzymes included hexosaminidase (Sellinger *et al.*, 1960) for lysosomes, alkaline phosphodiesterase I (Aronson and Touster, 1974) for plasma membranes, glucose 6-phosphatase in the presence of 40 mM *L*-tartrate (Brightwell and Tappel, 1968) for microsomes and fumarase (Hill and Bradshaw, 1969) for mitochondria.

Assay of PI Kinase and PIP Kinase. Samples (typically 50 µg of protein in 15 µl) were preincubated at 30°C for 5 min before the addition of an equal volume of either PI kinase assay mixture (2 mM PI; 0.8% Triton X-100; 20 mM $MgCl_2$; 4 mM $[\gamma-^{32}P]ATP$, 2500–3000 cpm/pmol; and 50 mM HEPES, pH 7.5) or PIP kinase assay mixture (1 mM PIP; 20 $MgCl_2$; 4 mM $[\gamma-^{32}P]ATP$; and 50 mM HEPES, pH 7.5). The suspensions were incubated an additional 2 min at 30°C and the reactions were terminated by the addition of 1.5 ml of chloroform/methanol (1:2, v/v), and the phospholipids were extracted as described by Schacht (1981). The lipid residues were redissolved in 50 µl of chloroform/methanol/H_2O (75:25:2, by vol.) and analyzed by thin-layer chromatography on silica gel H plates (Supelco) developed with chloroform/methanol/20% methylamine (60:36:10, by vol.) as described by Volpi *et al.* (1983). The plates were exposed to Kodak XAR-5 X-ray film for autoradiography. Spots corresponding to PIP and PIP_2 were scraped into scintillation vials and counted with Safety Solve (Research Products, Int.) scintillation fluid. The coefficients of variation for the kinase assays were less than 5% among duplicate samples and typically less than 10% among different preparations.

Preparation of Phosphatidylinositol [4-^{32}P]Phosphate and Phosphatidylinositol [4,5-^{32}P]Bisphosphate. Erythrocyte ghosts, prepared by the method of Schneider and Kirschner (1970), were incubated with 2 mM [γ-^{32}P]ATP, 10 mM MgCl$_2$, and 50 mM HEPES, pH 7.5, for 60 min at room temperature. Phospholipids were then extracted and [^{32}P]PIP and [^{32}P]PIP$_2$ were purified by affinity chromatography on a column of immobilized neomycin as described by Schacht (1981). Analysis of the products by thin-layer chromatography, as described above, indicated radiochemical purities of greater than 95% for [^{32}P]PIP and 93% for [^{32}P]PIP$_2$. The labeled phospholipids were stored in chloroform/methanol/H$_2$O (75:25:2 by vol.) at -20°C and were reextracted (Schacht, 1981) before use.

Assay of [^{32}P]PIP and [^{32}P]PIP$_2$ Hydrolysis. Samples (typically 50 µg of protein in 15 µl) were preincubated with 20 mM EDTA, 2 mM MgCl$_2$, or 2 mM CaCl$_2$ for 5 to 10 min at 30°C. The assays were initiated by the addition of 15 µl of 0.6 mM [^{32}P]PIP or [^{32}P]PIP$_2$ in 0.8% Tween 80 containing 25 mM HEPES, pH 7.5. The reactions were terminated by the addition of 1.5 ml of chloroform/methanol (1:2, v/v), 0.5 ml of chloroform, and 0.5 ml of 2.4 N HCl. The mixtures were vortexed and then centrifuged to separate the phases. Radioactivity in the aqueous phases was determined as Cerenkov radiation and the amount of PIP or PIP$_2$ hydrolyzed was calculated. Blanks consisted of [^{32}P]PIP or [^{32}P]PIP$_2$ incubated in the absence of enzyme or stopped immediately after the addition of sample. The coefficients of variation for the hydrolysis assays were less than 3% among duplicate samples and less than 7% among different preparations.

Hepatocyte Isolation. Hepatocytes were isolated from fed rats by the method of Seglen (1976) as modified by Kurtz and Wells (1981). The procedure was further modified by supplementing the perfusion buffer with 10 mM glucose and the normal concentration of MEM non-essential and essential amino acids. Typically, 4.0 – 6.0 x 10^8 cells of 85-95% viability, as judged by Trypan blue exclusion, were obtained from a single animal. The cells were suspended at a concentration of 1.0 – 2.0 x 10^6 cells/ml in low phosphate (0.1 mM) Dulbecco's modified

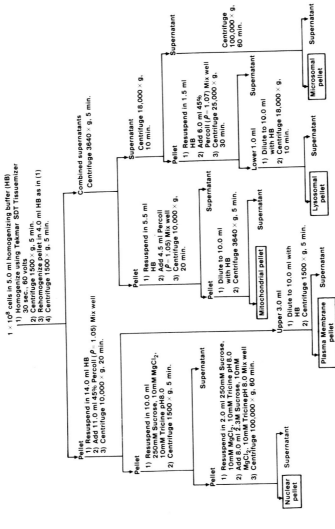

Figure 1. Flow diagram for the subcellular fractionation of rat hepatocytes. Hepatocytes were isolated and labeled with ^{32}Pi for various times as described under Experimental Procedures. The cells were transferred into 7 volumes of ice-cold HB and pelleted at 4°C. The cell pellet was gently resuspended in 5.0 ml of ice-cold HB and subjected to the above fractionation scheme at 4°C.

Eagle's medium containing 2% bovine serum albumin (Sigma, Fraction V) and 20 mM HEPES, pH 7.3. Cell suspensions of 35 ml were placed in 490 cm^2 tissue culture roller bottles which were flushed every 30 min with a 10 sec burst of 95% O_2/5% CO_2, sealed and rotated at 2 rpm at 37°C. Under these conditions, the cells remained well suspended and exhibited no loss in viability over a 3 h period.

Hepatocyte Incubations. Carrier-free ^{32}P was added to the cell suspensions at the level of 5-10 μCi/10^6 cells. At various times of incubation up to 120 min, the cells were diluted with 7 volumes of ice-cold homogenizing buffer (HB) which consisted of 250 mM sucrose, 1 mM ethylenediaminetetraacetic acid (EDTA), 10 mM NaN$_3$, 20 mM NaF, 75 mg/l phenylmethylsulfonyl fluoride (PMSF) and 10 mM N-tris[hydroxymethyl]methylglycine (Tricine), pH 8.0. The cells were pelleted at 75 x g for 5 min at 4°C. The cell pellet was resuspended in 5.0 ml of ice-cold HB.

Cell Homogenization and Subcellular Fractionation. The cell suspension was homogenized and fractionated at 4°C as described in Figure 1. After the second homogenization step, cell breakage exceeded 95% as judged by phase contrast microscopy. The stock 45% Percoll solution was prepared in HB. Resuspension of the 1500 x g pellet and subsequent "nuclear pellets" was accomplished using 3 strokes of a tight fitting glass-glass Potter-Elvehjem homogenizer. All other pellets were resuspended using 3 strokes of a Teflon-glass Potter-Elvehjem homogenizer. The final pellets were resuspended in 1.0 ml of HB. Complete fractionation of five groups of cells was accomplished in 3.5 - 4.0 h.

Phospholipid Isolation and Deacylation. The total phospholipid fraction was isolated by the method of Schacht (1981). The phospholipids were deacylated using the mild alkaline hydrolysis procedure described by Kates (1972). A minor modification was made which allowed the pH of the polar fraction to remain slightly alkaline (pH 8) by using only 0.1 ml of Dowex 50 [H$^+$]. This modification resulted in increasing the recovery of glycerophosphocholine. The polar fraction was evaporated using a N$_2$ stream at 37°C and dissolved in 100 μl of 20 mM ammonium borate, 100 mM ammonium formate, pH 9.5.

Analysis of Glycerophosphoesters by High Performance Anion Exchange Liquid Chromatography. The deacylated products were analyzed by high performance liquid chromatography (HPLC) on a 250 x 4 mm BioRad Aminex A-27 anion exchange column using a modification of the ion exchange procedure described by Dittmer and Wells (1969). A polyphasic gradient beginning with 100 mM ammonium formate, 20 mM ammonium borate, pH 9.5 and ending with 750 mM ammonium formate, 20 mM ammonium borate, pH 9.5 was established by a Beckman Model 342 Gradient Liquid Chromatograph and used to elute the glycerophosphoesters (Fig. 2A). Fractions of 0.78 ml were collected at a flow rate of 0.6 ml/min in Pyrex tubes. The fractions were either counted by Cerenkov radiation or evaporated overnight at 150°C in a vented hood. The residue which remained in the tubes after evaporation was digested with 0.3 ml of 14% H_2SO_4/12% $HClO_4$ (2:1, v/v) for 90 min at 180°C after placing glass marbles on top. The phosphate content of the fractions was determined by the method of Ames (1966) with detection of the reduced phosphomolybdate complex at 820 nm.

Determination of Cellular [γ-^{32}P]ATP Specific Radioactivity. A 3.5 ml aliquot of cell suspension was removed after various times of incubation with ^{32}P and placed in 500 µl of ice cold 5% $HClO_4$. The $HClO_4$-treated samples were homogenized using a glass-glass Potter Elvehjem homogenizer, placed on ice for 15 min and then centrifuged. The supernatant fluid was removed and neutralized with 2 M $KHCO_3$ at 4°C. The $KClO_4$ was pelleted and the supernatant fluid was removed. The specific radioactivity of [γ-^{32}P]ATP in the supernatant fraction was determined using the method of Hawkins *et al.* (1983).

Other Methods. Phospholipid concentrations were determined by total phosphate analysis by the method of Ames (1966). Protein concentrations were determined by the method of Böhlen *et al.* (1973), except that the samples were incubated with 1% sodium dodecylsulfate before derivatization with fluorescamine. Bovine serum albumin (Sigma) was used as a standard. DNA was measured by the method of Karsten and Wollenberger (1977).

RESULTS

Characterization of Polyphosphoinositide Metabolism in Isolated Organelles

Assessment of organelle purity. The isolated subcellular fractions were assayed for their content of selected marker enzymes so that the amount of cross-contamination of the fractions could be determined. The fold-purification of each organelle is indicated in Table I. Calculation of the cross-contamination of the fraction indicated that the plasma membrane preparations contained 46% microsomes and 2% mitochondria. The microsomal preparations contained 95% microsomal protein with essentially no plasma membrane contamination. The mitochondrial preparations contained 74% mitochondria with 18% and 6% contamination by microsomes and plasma membranes, respectively. The lysosomal preparations consisted of 44%

TABLE I

Assessment of Organelle Purity

Subcellular Fraction	Marker Enzyme	RSA[a]
Plasma Membrane	Alkaline phosphodiesterase I	12.1
Microsome	Glucose-6-phosphatase	4.6
Mitochondria	Fumarase	2.5
Lysosome	Hexosaminidase	46.2

Subcellular organelles were isolated from rat liver and assayed for marker enzymes as described (Experimental Procedures). Data are mean values derived from 3 preparations.
[a] RSA (Relative Specific Activity) = specific activity in purified organelle/specific activity in whole homogenate.

lysosomal protein containing 55% microsomal contamination and approximately 1% contamination by plasma membranes. Nuclear envelopes do not express an activity that can be used as a specific marker enzyme; however, the method used yields essentially pure nuclei (Smith and Wells, 1983).

Distribution of PI kinase and PIP kinase. The highest specific activity of PI kinase was associated with the microsomal preparations (Table II); however, significant levels of PI kinase activity were found in all of the organelles studied, except the mitochondria. PIP kinase was most active in the plasma membrane preparations and the other organelles studied expressed only low activity.

Hydrolysis of [^{32}P]PIP and [^{32}P]PIP$_2$ by subcellular fractions. The hydrolysis of exogenous [^{32}P]PIP and [^{32}P]PIP$_2$ by isolated organelles is described in Table III. In the presence of 10 mM EDTA, PIP was hydrolyzed at similar rates by the plasma membrane, microsomal, lysosomal membrane, and nuclear envelope preparations. This hydrolysis was in all cases inhibited by 1 mM MgCl$_2$ and 1 mM CaCl$_2$. The highest hydrolytic activity toward PIP$_2$ was expressed by the plasma membrane, although some activity was present in mitochondria, microsomes, lysosomal membranes, and cytosol. The hydrolysis of PIP$_2$ was stimulated by 1 mM MgCl$_2$, and to a lesser extent by 1 mM CaCl$_2$.

Subcellular Sites of ^{32}P Incorporation into Polyphosphoinositides in Isolated Rat Hepatocytes

Purity of subcellular fractions. Subcellular fractions were obtained by using the fractionation scheme outlined in Figure 1 and analyzed for various marker enzymes as summarized in Table IV. The fractionation process resulted in a plasma membrane fraction which consisted of 28% plasma membranes but also contained 57% microsomes. The mitochondrial fraction was composed of 60% mitochondria, 36% microsomes but only 2% plasma membranes. The lysosomal fraction contained 19% lysosomes, approximately equal amounts of mitochondria and microsomes but only 2% plasma membranes. The microsomal fraction was the purest fraction obtained and was composed of nearly 86% microsomes and 9% plasma membranes.

TABLE II

PI Kinase and PIP Kinase Activities in Rat Liver Organelles

Subcellular Fraction	PI Kinase $(nmol \times min^{-1} \times$	PIP Kinase mg^{-1} protein)
Plasma membrane	0.65	1.10
Microsome	1.45	0.12
Mitochondria	0.09	0.02
Lysosomal membrane	0.52	0.07
Nuclear envelope	0.40	0.17
Cytosol	0.45	0.22

Organelles were isolated and assayed for PI kinase in the presence of 1 mM PI and 0.4% Triton X-100, and PIP kinase in the presence of 0.5 mM PIP as described (Experimental Procedures). Data are mean values derived from 3 experiments.

HPLC analysis of deacylated phospholipids. Phospholipid standards were deacylated and subjected to anion exchange HPLC as described (Experimental Procedures). Eight glycerophosphoesters were separated including glycerophosphoinositol 4-phosphate (GPIP) and glycerophosphoinositol 4,5-bisphosphate (GPIP$_2$) (Fig. 2A). Greater than 95% recovery of deacylated phospholipids was obtained. ^{32}P-labeled phospholipids isolated from rat hepatocyte cell homogenates were similarly deacylated and chromatographed. As shown in Figure 2B, radioactivity corresponded very well with lipid phosphorus with the exception that very little radioactivity was incorporated into phosphatidylserine (PS) (GPS). Also, there was no detectable lipid phosphorus associated with PIP (GPIP) or PIP$_2$ (GPIP$_2$), however, there was substantial ^{32}P incorporation into these phospholipids.

Seyfred et al.

TABLE III

Hydrolysis pf PIP and PIP_2 by Subcellular Fractions

Subcellular Fraction	Addition	PIP	PIP_2
		(nmol hydrolyzed x min^{-1} x mg^{-1} protein)	
Plasma membrane	10 mM EDTA	1.71	0.25
	1 mM $MgCl_2$	0.23	3.12
	1 mM $CaCl_2$	0.43	0.76
Microsome	10 mM EDTA	1.06	0.12
	1 mM $MgCl_2$	0.25	0.52
	1 mM $CaCl_2$	0.34	0.27
Mitochondria	10 mM EDTA	0.31	0.00
	1 mM $MgCl_2$	0.20	1.25
	1 mM $CaCl_2$	0.21	0.24
Lysosomal membrane	10 mM EDTA	1.45	0.00
	1 mM $MgCl_2$	0.93	0.47
	1 mM $CaCl_2$	1.23	0.00
Nuclear envelope	10 mM EDTA	2.25	0.05
	1 mM $MgCl_2$	1.62	0.00
	1 mM $CaCl_2$	1.60	0.00
Cytosol	10 mM EDTA	0.78	0.00
	1 mM $MgCl_2$	0.41	0.40
	1 mM $CaCl_2$	0.15	0.41

Organelles were isolated and assayed for their ability to hydrolyze [^{32}P]PIP and [^{32}P]PIP$_2$ as described (Experimental Procedures). Data are mean values derived from 3 preparations.

TABLE IV

Assessment of Organelle Purity

Subcellular Fraction	Relative Specific Activity
Nuclei (DNA/protein, w/w)	3.1
Plasma Membrane (Alk. Phosphodiesterase I)	4.4
Mitochondria (Fumarase)	2.2
Lysosomes (Hexosaminidase)	13.3
Microsomes (Glucose-6-phosphatase)	2.8

Subcellular organelles were isolated from cultured rat hepatocytes and assayed for DNA and marker enzymes as described (Experimental Procedures). Data are mean values derived from 4 experiments.

Relative Specific Activity =

$$\frac{\text{Specific activity of enzyme or DNA/protein (w/w) in fraction}}{\text{Specific activity of enzyme or DNA/protein (w/w) in cell homogenate}}$$

Estimation of possible post-homogenization artifacts. Although the subcellular fractionation process was relatively rapid, enzyme inhibitors were present in the buffer, and the fractionation was conducted at 0-4°C, the possibility of large changes in the level of ^{32}P-labeled phospholipids after homogenization did exist. To investigate this possibility, an aliquot of cell homogenate was added to methanol/chloroform (2:1, v/v) immediately after homogenization. Another aliquot was kept on ice and methanol/chloroform (2:1, v/v) was added following the completion of the subcellular fractionation. The phospholipids were isolated, deacylated and analyzed by anion exchange HPLC as described (Experimental Procedures).

In three different cell preparations, no change in the amount of ^{32}P in PIP_2 and only a 1-2% decrease in ^{32}P-labeled PI occurred during fractionation. ^{32}P-labeled phosphatidic acid (PA) increased an average of 7% during fractionation while PIP decreased 13%.

The isolation of purified nuclei requires the use of divalent cations, usually Mg^{2+} or Ca^{2+}, to prevent aggregation of nuclei with other subcellular organelles (Tata, 1974). Therefore, a similar experiment was performed to determine if any changes in ^{32}P-labeled phospholipids occurred in the crude nuclear fraction in the presence of 10 mM $MgCl_2$. Analysis of the glycerophosphoesters showed that large changes in the amount of ^{32}P in PI and PA occurred during the isolation of nuclei in the presence of buffer containing $MgCl_2$ even at 4°C. Only small changes (ca. 5-10%) were seen in the amount of ^{32}P in PIP and PIP_2 (data not shown). Therefore, the incorporation rates of ^{32}P into phospholipids isolated from partially purified nuclei that were measured are only valid for PIP and PIP_2.

The possibility of exchange of radiolabeled phospholipids between organelles after homogenization was also investigated by mixing ^{32}P-labeled microsomes obtained using the scheme depicted in Figure 1 with non-radiolabeled rat hepatocyte cell homogenate which was then fractionated by the standard procedure. Analysis of the fractions revealed 81.5% of the radioactivity was recovered in the microsomal fraction, 9% in the plasma membrane, 8% in the mitochondrial fraction and 1.4% of the radioactivity was recovered in the lysosomal fraction.

Rate of incorporation of ^{32}P into phospholipids of various subcellular fractions. Isolated hepatocytes were incubated with $^{32}P_i$ for various times and fractionated. The phospholipids were isolated, deacylated and analyzed by anion exchange HPLC as described (Experimental Procedures). The results are shown in Figures 3A-D. The data shown are the levels of ^{32}P in the various glycerophosphoesters which are measures of the amounts of ^{32}P incorporated into the corresponding phospholipids. The rates of ^{32}P incorporation per mg protein into PI and PA were similar between the lysosomal, microsomal and plasma membrane fractions but slightly slower in the

150

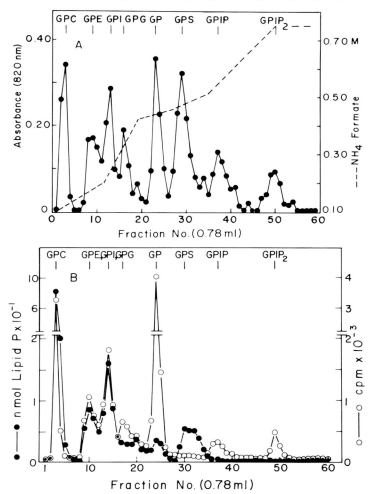

Figure 2.   Anion exchange HPLC elution profile of polar deacylated phospholipid products.   Phospholipids were isolated from hepatocytes, labeled for 60 min with $^{32}P_i$, deacylated with mild alkali and the polar products were analyzed on an Aminex A-27 exchange column as described (Experimental Procedures).   A. Standards. ●——●, absorbance at 820 nm of reduced phosphomolybdate complexes; - - -, ammonium formate concentration.   B. Glycerophosphoesters prepared from hepatocyte homogenate. 0——0, radioactivity as determined by Cerenkov radiation; ●——●, nmol lipid phosphorus.   The position of elution of polar deacylation products from phospholipid standards is indicated.   See Preface for abbreviations.

151

mitochondrial fraction. The most striking result was the rapid and extensive amount of ^{32}P incorporated into PIP and PIP$_2$ in the plasma membrane fraction compared with the other subcellular fractions. After 30 min of ^{32}P$_i$ labeling, the amount of ^{32}P incorporated per mg protein into PIP$_2$ was approximately 5 and 2.5 times greater than the amount of ^{32}P incorporated into PI and PIP, respectively, in the plasma membrane fraction. ^{32}P incorporation into PIP was nearly 2 times greater than PI after 30 min of labeling with ^{32}P$_i$ and only slightly less than the amount of ^{32}P incorporated into PA in the plasma membrane fraction.

Analysis of cell homogenate ^{32}P-labeled phospholipids illustrates that after 120 min of ^{32}P$_i$ labeling, the rate of ^{32}P incorporation into PI was still increasing. However, the incorporation of ^{32}P into PA, PIP and PIP$_2$ approached a steady state, reflecting the specific radioactivity of [γ-^{32}P]ATP which reached steady state after 60-90 min of ^{32}P labeling.

Table V illustrates the high level of ^{32}P per mg protein in PIP (GPIP) and PIP$_2$ (GPIP$_2$) in the plasma membrane compared with the other organelles after 90 min. The lysosomal fraction had 20% while the other fractions had approximately 10% of the amount of ^{32}P in PIP in the plasma membrane fraction. The amount of ^{32}P-labeled PIP$_2$ in the nuclear, lysosomal, mitochondrial and microsomal fractions was only 1-4% of the amount of ^{32}P-labeled PIP$_2$ isolated from the plasma membrane fraction. Attempts to chemically measure the mass of GPIP and GPIP$_2$ by phosphate analysis were unsuccessful because of the low quantity of PIP and PIP$_2$ in the various subcellular fractions.

DISCUSSION

In early reports on the distribution of PI kinase and PIP kinase in rat liver it was proposed that these enzymes are located in the plasma membrane. In view of a number of studies from our laboratory and others which have identified PI kinase, and in some cases PIP kinase, in highly purified preparations of other subcellular organelles, we have conducted an investigation of PIP and PIP$_2$ metabolism in rat liver organelles isolated by methods

TABLE V

Distribution of ^{32}P Incorporation into PIP and PIP$_2$
of Various Subcellular Fractions

Subcellular Fraction	GPIP	GPIP$_2$
Plasma Membrane	100	100
Nuclei	10.1 ± 2.3	4.1 ± 1.6
Mitochondria	9.3 ± 0.74	1.4 ± 0.31
Lysosomes	20.1 ± 3.1	2.8 ± 0.97
Microsomes	10.5 ± 1.2	3.3 ± 1.0

Hepatocytes were radiolabeled with ^{32}P for 90 min and processed as described (Experimental Procedures). Incorporation of ^{32}P into PIP (GPIP) and PIP$_2$ (GPIP$_2$) per mg of protein in each subcellular fraction was measured. Data are expressed as the percentage of ^{32}P incorporation/mg protein in nuclei, mitochondria, lysosomes and microsomes compared with the plasma membrane fraction and are mean values ± S.D. derived from 3 hepatocyte preparations.

which give the highest purity that can currently be obtained. Using these preparations, we have found that PI kinase activity is widely distributed throughout intracellular membranes. In addition, the cytosol contains significant levels of PI kinase activity which may act on these membranes as well. In contrast, plasma membrane PI kinase exhibits at least a 6-fold higher specific activity than any other membrane. In addition, PIP kinase activity was observed in the cytosol.

We have also used suspensions of isolated rat hepatocytes, incubated with ^{32}P$_i$, to determine where turnover of PIP and PIP$_2$ occurs in intact cells. To obtain useful information on the sites of PIP and PIP$_2$ metabolism in these hepatocytes, a fractionation procedure was developed which yields reasonably purified prepara-

153

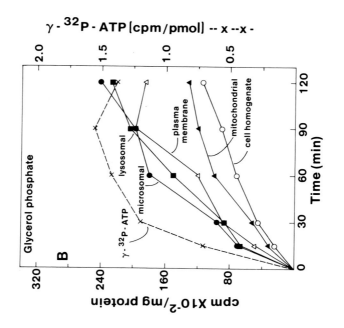

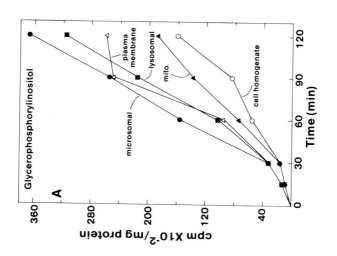

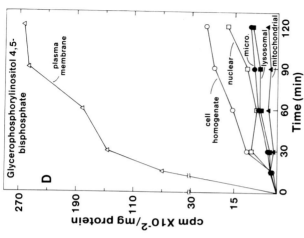

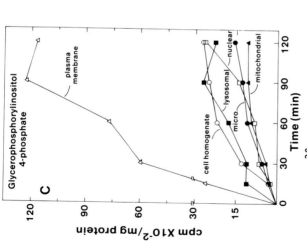

Figure 3. Time course of ^{32}P incorporation into various phospholipids isolated from subcellular fractions of rat hepatocytes. Hepatocytes were isolated and labeled with ^{32}Pi for various times as described (Experimental Procedures). The ^{32}P-labeled hepatocytes were homogenized and fractionated as described in Figure 1. The phospholipids were isolated from the various subcellular fractions, deacylated with mild alkali and the glycerophosphate esters analyzed by anion exchange HPLC as described (Experimental Procedures). A. Glycerophosphoinositol; B. Glycerol phosphate; C. Glycerophosphoinositol 4'-phosphate; D. Glycerophosphoinositol 4,5-bisphosphate. ○——○, Cell homogenate; □——□, Nuclear fraction; △——△, Plasma membrane fraction; ●——●, Microsomal fraction; ▲——▲, Mitochondrial fraction; and X——X, [γ-^{32}P]ATP. The results shown are from one of three similar hepatocyte preparations.

tions of subcellular organelles in a relatively short time and introduces minimal post-homogenization artifacts. An HPLC method for the rapid separation of the deacylation products of all the major phospholipids, including PIP and PIP_2, was developed and automated to process the numerous samples required for this investigation. The results from these phospholipid labeling studies indicate that although plasma membranes have the highest rates of ^{32}P incorporation into PIP, other subcellular membranes also contain PIP. In contrast, PIP_2 synthesis is confined almost exclusively to the plasma membrane. The two methods of analysis thus support one another since PI kinase seems to be widely distributed and PIP synthesis occurs in all membranes. Similarly, the highest PIP kinase activity is associated with the plasma membrane which is also the major site of PIP_2 synthesis. The cytosolic PIP kinase appears to be prohibited from acting on other intracellular membranes and clearly requires further study to determine its function and regulation.

Studies of the metabolism of biological components require consideration of the degradation of the components as well as their synthesis. We have accordingly also begun to characterize hepatic PIP and PIP_2 hydrolysis. Exogenous [^{32}P]PIP was hydrolyzed by all of the subcellular fractions analyzed. This hydrolysis was most active in the presence of EDTA and was in all cases inhibited by $MgCl_2$ and $CaCl_2$. [^{32}P]PIP_2 hydrolysis was stimulated by divalent metals and was most active in the plasma membranes. Further studies are required to elucidate the mechanisms of the observed PIP and PIP_2 hydrolysis and the specificities of the enzymes involved since we did not attempt to eliminate the possibility of hydrolysis of PIP and PIP_2 by non-specific phosphatases.

We have come to view PIP as a component of almost all intracellular membranes in the hepatocytes and presumably in other cells. Therefore, studies on the mechanisms of subcellular membrane function should include consideration of the effects of this phospholipid. In contrast, PIP_2 appears to be concentrated in the plasma membrane. Thus, it may be involved in specific functions such as the transduction of receptor-mediated events.

ACKNOWLEDGEMENT

This work was supported by Grant AM 32930 from the United States Public Health Service.

REFERENCES

Akhtar, R.A. and Abdel-Latif, A.A. (1980) *Biochem. J.* *182*, 783-791.

Ames, B.N. (1966) *Methods Enzymol.* *8*, 115-118.

Aronson, N.N., Jr. and Touster, O. (1974) *Methods Enzymol.* *31*, 90-102.

Behar-Bannelier, M. and Murray, R.K. (1980) *Biochem. J.* *187*, 147-156.

Billah, M.M. and Lapetina, E.G. (1983) *Proc. Natl. Acad. Sci. USA* *80*, 965-968.

Blobel, G. and Potter, V.R. (1966) *Science (Wash. D.C.)* *154*, 1662-1665.

Bohlen, P., Stein, S., Dairman, W. and Udenfriend, S. (1973) *Arch. Biochem. Biophys.* *155*, 213-220.

Brightwell, R. and Tappel, A.L. (1968) *Arch. Biochem. Biophys.* *124*, 333-343.

Buckley, J.T. and Hawthorne, J.N. (1972) *J. Biol. Chem.* *247*, 7218-7223.

Collins, C.A. and Wells, W.W. (1983) *J. Biol. Chem.* *258*, 2130-2134.

Creba, J.A., Downes, C.P., Hawkins, P.T., Brewster, G., Michell, R.H. and Kirk, C.J. (1983) *Biochem. J.* *212*, 733-747.

Dittmer, J.C. and Wells, M.A. (1969) *Methods Enzymol.* *14*, 519-523.

Downes, P. and Michell, R.H. (1982) *Cell Calcium* *3*, 467-502.

Emmelot, P., Bos, C.J., van Hoeven, R.P. and van Blitterswijk, W.J. (1974) *Methods Enzymol.* *31*, 75-90.

Farese, R.V., Saber, M.A., Vandor, S.L. and Larson, R.E. (1980) *J. Biol. Chem.* *255*, 5728-5734.

Farese, R.V., Larson, R.E. and Sabir, M.A. (1982) *J. Biol. Chem.* *257*, 4042-4045.

Glynn, I.M. and Chappell, J.B. (1964) *Biochem. J.* *90*, 147-149.

Hawkins, P.T., Michell, R.H. and Kirk, C.J. (1983) *Biochem. J.* *210*, 717–720.

Hendrickson, H.S. and Reinertgen, J.C. (1971) *Biochem. Biophys. Res. Commun.* *44*, 1258–1264.

Hill, R.L. and Bradshaw, R.A. (1969) *Methods Enzymol.* *13*, 96–99.

Jergil, B. and Sundler, R. (1983) *J. Biol. Chem.* *258*, 7968–7973.

Karsten, U. and Wollenberger, A. (1977) *Anal. Biochem.* *77*, 464–470.

Kates, M. (1972) in: *"Techniques of Lipidology: Isolation, Analysis and Identification of Lipids"* pp. 558–559, North-Holland Publishing Co., Amsterdam.

Kay, R.R., Fraser, D. and Johnston, I.R. (1972) *Eur. J. Biochem.* *30*, 145–154.

Kirk, C.J., Michell, R.H. and Hems, D.A. (1981) *Biochem. J.* *194*, 155–165.

Kurtz, J.W. and Wells, W.W. (1981) *J. Biol. Chem.* *256*, 10870–10875.

Leighton, F., Poole, B., Beaufay, H., Baudhuin, P., Coffey, J.W., Fowler, S. and DeDuve, C. (1968) *J. Cell Biol.* *37*, 482–512.

Lipsky, J.J. and Lietman, P.S. (1980) *Antimicrob. Agents Chemother.* *18*, 532–535.

Litosch, I., Lin, S-H. and Fain, J.N. (1983) *J. Biol. Chem.* *258*, 13727–13732.

Michell, R.H. (1975) *Biochim. Biophys. Acta* *415*, 81–147.

Michell, R.H. and Hawthorne, J.N. (1966) *Biochem. Biophys. Res. Commun.* *21*, 333–338.

Michell, R.H., Harwood, J.L., Coleman, R. and Hawthorne, J.N. (1967) *Biochim. Biophys. Acta* *144*, 649–658.

Pertoft, H., Warmegard, B. and Hook, M. (1978) *Biochem. J.* *174*, 309–317.

Rhodes, D., Prpic, V., Exton, J.H. and Blackmore, P.F. (1983) *J. Biol. Chem.* *258*, 2770–2773.

Saltiel, A.R., Siegel, M.I., Jacobs, S. and Cuatrecasas, P. (1982) *Proc. Natl. Acad. Sci. USA* *79*, 3513–3517.

Schacht, J. (1981) *Methods Enzymol.* *72*, 626–631.

Schneider, R.P. and Kirschner, L.B. (1970) *Biochim. Biophys. Acta* *202*, 283–294.

Seglen, P.O. (1976) *Methods Cell Biol.* *13*, 29–83.

Sellinger, O.Z., Beaufay, H., Jacques, P., Doyen, A. and DeDuve, C. (1960) *Biochem. J.* *74*, 450–456.

Sheetz, M.P., Febbroriello, P. and Koppel, D.E. (1982) *Nature (Lond.)* *296*, 91–93.

Smith, C.D. and Wells, W.W. (1983) *J. Biol. Chem.* *258*, 9360–9373.

Tata, J.R. (1974) *Methods Enzymol.* *31*, 253–262.

Volpi, M., Yassin, R., Naccache, P.H. and Sha'afi, R.I. (1983) *Biochem. Biophys. Res. Commun.* *112*, 957–964.

Walsh, D.A., Perkins, J.P., Brostrom, C.O., Ho, E.S. and Krebs, E.G. (1971) *J. Biol. Chem.* *246*, 1968–1976.

Weiss, S.J., McKinney, J.S. and Putney, J.W., Jr. (1982) *Biochem. J.* *206*, 555–560.

LOSS OF PHOSPHATIDYLINOSITOL AND GAIN IN

PHOSPHATIDATE IN NEUTROPHILS

STIMULATED WITH FMET-LEU-PHE

S. Cockcroft and D. Allan

Department of Experimental Pathology, School of Medicine,
University College, London, UK

SUMMARY

Neutrophils stimulated with fMetLeuPhe show a loss of phosphatidylinositol (PI) and a gain in phosphatidate (PA). In cells prelabeled with $^{32}P_i$, it would be expected that the newly synthesized PA would have the same specific radioactivity as cellular ATP provided that PI loss is catalyzed by phospholipase C and the resultant diacylglycerol is phosphorylated by ATP. Instead, the specific radioactivity of the newly-formed PA was found to be less than a tenth of the specific radioactivity of ATP. The source of the newly formed PA is most likely to be PI because: (1) Increase in PA mass is accompanied by an equivalent decrease in mass of PI. (2) In cells pulse-labeled with [^{3}H]glycerol, label lost from PI is recovered in PA. (3) Specific radioactivity of the new [^{32}P]PA is similar to the specific radioactivity of [^{32}P]PI. The most plausible explanation for these results is that PI is directly converted to PA by phospholipase D action. This conclusion is supported by observations that neither inositol phosphates nor diacylglycerol increase at the expense of PI or polyphosphoinositides in fMetLeuPhe-stimulated cells.

INTRODUCTION

Neutrophils form the body's first line of defence against invasion by foreign microorganisms. They migrate to sites of infection by responding to chemotactic stimuli released by bacteria. Synthetic compounds based on these chemotactic factors have been widely used in the study of neutrophil activation. The N-formyl tripeptide fMetLeuPhe (FMLP) is one of many of these analogues and most of the investigations described below were performed using this agonist. Other known soluble stimuli for these cells include complement-derived factors, C5A and C3A, and leukotriene B_4.

The chemotactic response can be repressed and the secretory response enhanced by addition of cytochalasin B to cells prior to stimulation (Bennett *et al.*, 1980). Release of β-glucuronidase can be quantitated easily and this allows the cellular response of the cell to be monitored with precision.

The binding of FMLP to its receptor is the initial step in a process which leads ultimately to the secretion of granular contents into the external medium. The intervening events that occur, such as changes in the level of cytosolic Ca^{2+} and cAMP are common to many other cells. Another common response shared by many cells is a receptor-mediated change in the metabolism of phosphatidylinositol (PI), an event observed thirty years ago by Hokin and Hokin (1953). The concept that has arisen subsequently from the results of many of the investigations of stimulated PI turnover is that stimulation of receptors that control a rise in cytosolic Ca^{2+} also causes breakdown of PI (and possibly, the polyphosphoinositides) by phospholipase C action, yielding inositol phosphates and diacylglycerol as the initial reaction products. The resultant diacylglycerol is phosphorylated by cellular ATP to form PA which is converted back to PI via CDP-diacylglycerol (Michell *et al.*, 1981). It should be stressed that the cycle of reactions suggested to explain the turnover of the headgroup of PI is based more on known *in vitro* enzymatic reactions rather than a secure knowledge that the cycle indeed functions as postulated in intact cells.

Most of the data available from the studies performed since the 1950's on a variety of cell-types including neutrophils, fits the paradigm of the PI cycle outlined above. However, the analysis of lipid metabolism in neutrophils presented here argues against the proposed involvement of phospholipase C and instead suggests that the loss of PI and gain in PA can be accounted for by phospholipase D action.

PI AND PA METABOLISM IN STIMULATED NEUTROPHILS

Stimulation of PI metabolism in neutrophils was first reported by Karnovsky *et al.* (1961) and Sastry and Hokin (1966). These studies used particulate stimuli which activate phagocytosis and demonstrated that stimulation of the neutrophil led to enhanced labeling of PI (a measure of resynthesis which is secondary to breakdown) when cells were incubated with either $^{32}P_i$ or [^{3}H]inositol. We reported that FMLP also stimulates the enhanced labeling of PI in cells prelabeled with $^{32}P_i$ or [^{3}H]-inositol (Bennett *et al.*, 1980; Cockcroft *et al.*, 1980a,b).

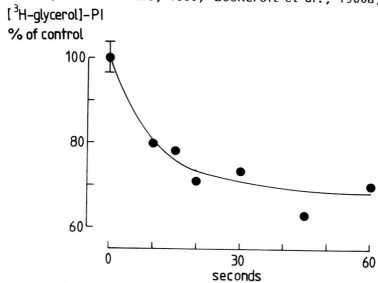

Figure 1. Time-course of PI loss from rabbit neutrophils stimulated with FMLP. Rabbit neutrophils were exposed to [^{3}H]glycerol (90 min) then chased with glycerol (10 mM) (two 20 min periods) with a wash in between. Cells were stimulated with 10 nM FMLP for the indicated times.

These initial experiments were followed by studies of PI breakdown using [^{3}H]glycerol to pulse-label the lipids (Cockcroft *et al.*, 1981). Glycerol can cross the membrane readily and thus it can be easily chased. Because it is incorporated only during *de novo* synthesis of lipids, ^{3}H incorporated into glycerolipids in a short incubation period may not reflect the mass distribution of lipids. On the other hand, it could be argued that because the lipid pool which is receptor-sensitive would have a higher rate of turnover and would therefore quickly incorporate radioisotope, this would provide a sensitive way of measuring the breakdown of this particular pool. It was found that in neutrophils that were exposed to [^{3}H]glycerol and then stimulated with FMLP, there is approximately 35% breakdown of labeled PI but only about 15-20% loss of PI phosphorus. Therefore the pool of PI that is receptor-linked is turning over faster than the bulk of the cellular PI.

Maximal PI loss is observed at 15s after FMLP addition and remains at this level for up to 60s as determined radiochemically (Fig. 1). The synthesis of PA and PI was initially studied by measuring the incorporation of ^{32}P$_i$ into PA and PI in cells incubated with ^{32}P$_i$ for one hour at which time the ATP pool had reached isotopic equilibrium. The increase in PA labeling preceded that of PI as would be expected if PI was resynthesized from PA. What is interesting to note is the time-course of PA labeling: it is slow compared to PI loss (Fig. 2). This suggested to us that since breakdown of PI was nearly complete in 10s and the formation of radiochemical PA was only complete at 5 min, this reflected the slow conversion of diacylglycerol to PA. Thus we decided to quantitate diacylglycerol at 10s, the time of near-maximal PI loss but where the radiochemical increase in PA is small. To our surprise we found that there was no increase in the level of diacylglycerol (as determined by gas chromatography) at 10s (Cockcroft, S. and Allan, D., unpublished observations). This discrepancy led us to study the mass production of PA and to compare it with the radiochemical generation of PA.

PA constitutes only about 0.3% of the total phospholipids and therefore in those experiments in which PA mass was determined it was necessary to use human

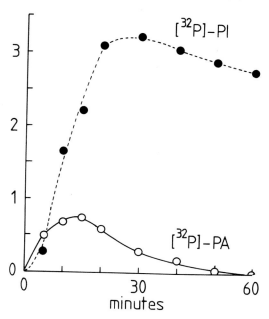

Figure 2. Time-course of incorporation of $^{32}P_i$ into PA and PI in FMLP-stimulated rabbit neutrophils. Rabbit neutrophils were preincubated with $^{32}P_i$ for 1 h prior to addition of FMLP (10 nM). The radioactivity in the control incubations increased linearly with time and this was subtracted from the stimulated values for each time point. O, PA. •, PI. (Reproduced from Cockcroft et al, 1981, Biochem. J. 200, 501-508, with permission).

instead of rabbit cells. While only $1-2 \times 10^9$ rabbit cells could be obtained, human cells were readily available in the range of $5-8 \times 10^9$ cells per day. Cells from these two sources differ in that human neutrophils require higher concentrations of FMLP for optimal secretion (10^{-7} rather than 10^{-8} M).

Secondly, the extent of secretion is different. Rabbit cells release up to 80% of β-glucuronidase whilst human cells release up to 50% with an optimum concentration of FMLP. However, the extent of changes in PI and PA metabolism are similar.

Fig. 3A shows the relationship between lysosomal enzyme release and PA formation in FMLP-stimulated human neutrophils. Net formation of PA shows a similar time-course to enzyme secretion increasing from 3.4 nmol to 21.4 nmol/2 $\times$ 10^8 cells at 20s. This level subsequently declines to 15 nmol at 2 min and this decrement can be accounted for by an increase in diacylglycerol (Table I). The initial rate of PA formation is 32 nmol $\times$ min^{-1} $\times$ 10^{-8} cells. The increase in PA formation could be accounted for by the decrease in PI mass (data not shown).

TABLE I

Effect of FMLP on Diacylglycerol Generation
in Human Neutrophils

	PA	DG	
Control	3.4	4 ± 0.7	(4)
FMLP, 10s	19	3.9 ± 0.65	(4)
FMLP, 20s	21.4	N.D.	
FMLP, 120s	15	10 ± 1	(4)

Results are presented as nmol of lipid per 2 x 10^8 neutrophils. Diacyl-glycerol (DG) was determined by gas chromatography. N.D., not determined.

Fig. 3A also shows the relationship between [^{32}P]PA and the mass of PA. The increase in the radioactivity of PA in ^{32}P-labeled neutrophils shows a quite different pattern to that of net increase in PA mass. Two distinct phases are apparent; an initial increase in labeling which levels off at 20s followed by a further increase over the remaining 2 min. The pattern observed for PA labeling

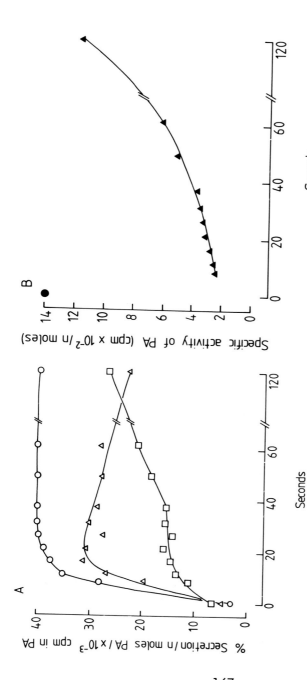

Figure 3A. Time-course of PA formation (mass and $^{32}P_i$ incorporation) and its relationship to enzyme secretion in FMLP-stimulated human neutrophils. Human neutrophils were exposed to $^{32}P_i$ for 90 min prior to FMLP (100 nM) addition. Each incubation mixture contained 2.8×10^8 cells. $\triangle$, nmol of PA; $\square$, cpm of ^{32}P PA; O, β-glucuronidase secretion.

Figure 3B. Time-course of changes in specific radio-activity of new PA in FMLP-stimulated neutrophils. The data in Fig.3A are replotted as specific radio-activity of the newly synthesized PA (cpm/nmol). Specific radioactivity of ATP in this experiment was 3641 cpm/nmol. ●, specific radioactivity in control cells. ▲, specific radioactivity in FMLP-stimulated neutrophils.

in human neutrophils is similar to the results from rabbit neutrophils (see Fig. 2 and Cockcroft et al., 1980b).

The results shown in Fig. 3B are calculated from Fig. 3A as specific radioactivity of the newly-formed [^{32}P]PA. The specific radioactivity of [^{32}P]PA in control cells was 1483 cpm/nmol, but 10s after addition of FMLP, the specific radioactivity of the newly-formed PA was 235 cpm/nmol. This then increased over the whole 2 min period. If PA formation were due to phosphorylation of diacylglycerol by cellular ATP, then the specific radio-activity of the newly formed PA should be the same as that of ATP. The specific radioactivity of ATP γ phosphate in this experiment was 3641 cpm/nmol and remained the same in both the control and in the cells stimulated with FMLP over the whole 2 min period. Thus the observed specific radioactivity of the newly-formed PA is less than a tenth of that expected if PA was formed by phosphorylation of diacylglycerol with ATP. The increase in the specific radioactivity of PA in the stimulated cells during the 2 min period probably reflects the exchange of the phosphate of the new PA with ATP. The initial specific radioactivity of the newly formed PA bears a close resemblance to the specific radioactivity of PI (114 cpm/nmol in this experiment) and is markedly different from the specific radioactivity of the other phospholipids (10 cpm/nmol for phosphatidylcholine; <1 cpm/nmol for phosphatidylethanolamine and phosphatidylserine in this experiment.

It is clear that the newly formed PA in FMLP-stimulated neutrophils cannot be derived by phosphoryla-tion of diacylglycerol with bulk cellular ATP although it is possible that some compartmentalised ATP pool of low specific radioactivity is the phosphate donor in the synthesis of the newly-formed PA. However, pools of ATP of differing specific radioactivity were not observed when neutrophils were fractionated on a discontinuous sucrose gradient which allowed the cytosol to be separated from the granules and the plasma membrane fraction (Bennett et al., 1982). A second argument that phosphorylation of diacylglycerol is unlikely to be involved concerns the initial rate of PA formation (32 nmol x min^{-1} x 10^{-8} cells). This is far higher than any observed rate of diacylglycerol kinase activity in the neutrophil (see below) or in any other cell (Call and Rubert, 1973; Allan *et al.*, 1980). The third possibility, that the newly-formed PA arises from *de novo* synthesis is also unlikely because of the observed low specific radioactivity of the newly-formed PA. A similar argument applies to a fourth alternative which is that cell activation leads to inhibition of phosphatidate phosphohydrolase which causes a redirection of glycerolipid synthesis resulting in the accumulation of PA (Brindley *et al.*, 1975).

Since the specific radioactivity of the newly formed PA is closest to the specific radioactivity of PI, it is most likely that the latter lipid is the source of PA. This proposal is supported by the redistribution of radioactivity that occurs in human neutrophils pulse-labeled with [^{3}H]glycerol and then stimulated with FMLP. Approximately 80% of the ^{3}H lost from PI could be accounted for by the increase in ^{3}H in PA (Fig. 4). Moreover, the time-course of [^{3}H]PA formation resembled the time-course of changes in the mass of PA.

The formation of diacylglycerol was also monitored in neutrophils stimulated with FMLP for either 10s or 2 min (Table I). It is clear that at 10s there was no change in diacylglycerol at a time of active PI loss. At 2 min,however, an increase in diacylglycerol is observed which accounts for the concomitant decline in PA mass.

To further discount the possibility that PI is degraded in a reaction catalyzed by phospholipase C, the generation of inositol phosphates was measured in [^{3}H]-inositol-labeled cells stimulated with FMLP. LiCl (10 mM) was present for 30 min prior to stimulation to inhibit inositol 1-phosphatase.

Neutrophils were labeled with [^{3}H]inositol for 2.5 h to incorporate ^{3}H into the inositol lipids. The radioactivity in the lipids was comprised of [^{3}H]PI (90%) and [^{3}H]polyphosphoinositides (10%). The water-soluble components from both control and stimulated neutrophils (FMLP, 1 μM for 6 min) were analyzed as described by Berridge *et al.* (1982). There was no increase in any of the inositol phosphates on addition of FMLP (Table II). Similar results were obtained when neutrophils were stimulated for 10s, 2 min or 6 min.

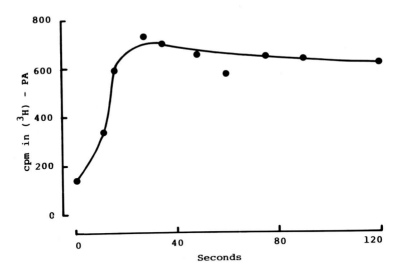

Figure 4. Time-course of FMLP-stimulated [³H]glycerol-labeled PA formation. Human neutrophils were pulse-chased with [³H]glycerol as
described in Fig. 1. Cells were stimulated with 100 nM FMLP for the
indicated times.

These results argue against any action of phospholipase C on either PI or the polyphosphoinositides in
FMLP-stimulated neutrophils. However, a neutrophil
membrane fraction enriched in plasma membranes does
contain a Ca^{2+}-activated polyphosphoinositide phosphodiesterase of the C-type (manuscript in preparation).
Addition of Ca^{2+} (0.5 mM) to [³²P]-labeled plasma membrane fraction resulted in the loss of 60% of ³²P from the
polyphosphoinositides and this was entirely accounted for
by the increase in ³²P in IP_2 and IP_3 (Fig. 5). The loss
of label from the polyphosphoinositides was accompanied
by an equivalent loss of lipid phosphorus. The other
product of polyphosphoinositide breakdown was diacylglycerol which was available for conversion into PA when
ATP was present thus demonstrating the presence of
diacylglycerol kinase [Call and Rubert, (1973) could not
demonstrate any diacylglycerol kinase activity in human
neutrophils probably because of failure to take adequate
precautions against proteolysis]. The activity of this

TABLE II

Lack of Effect of FMLP on Generation of Inositol
Phosphates in Human Neutrophils

	Control (dpm)	FMLP 10^{-6} M (dpm)
Inositol	$3,662,904 \pm 198,270$	$3,707,699 \pm 42,836$
GPI	$25,279 \pm 3,027$	$36,254 \pm 4,634$
IP	$5,888 \pm 545$	$8,277 \pm 788$
IP_2	$4,267 \pm 335$	$4,297 \pm 306$
IP_3	$1,907 \pm 76$	$2,767 \pm 63$

Human neutrophils were exposed to [^{3}H]inositol (50 µCi/ml) for 2.5 h. Cells were washed and incubated with inositol (10 mM) and LiCl (10 mM) for 30 min prior to stimulation with FMLP for 6 min (3 x 10^8 cells per incubation). Radioactivity in total phosphoinositides was 1,159,258 ± 77,145 dpm. Polyphosphoinositides accounted for approximately 10% of the total radioactivity in phosphoinositides. Data are mean values ± S.D. derived from triplicate observations.

enzyme with endogenously generated substrate was about 50 pmol of PA formed x min^{-1} x mg^{-1} protein. This is roughly equivalent to 20 pmol x min^{-1} x 10^{-8} cells calculated on the basis that 10^9 neutrophils yield 4 mg of plasma membranes. The activity of diacylglycerol kinase observed was similar to the activity seen in red blood cells using endogenous diacylglycerol (generated from the breakdown of polyphosphoinositides in Ca^{2+}-treated membranes) (unpublished observations).

Despite the presence in plasma membranes of neutrophils, of a Ca^{2+}-activated phospholipase C that is specific for polyphosphoinositides, no breakdown of the polyphosphoinositides was observed when intact neutrophils were stimulated with FMLP. Breakdown of polyphospho-

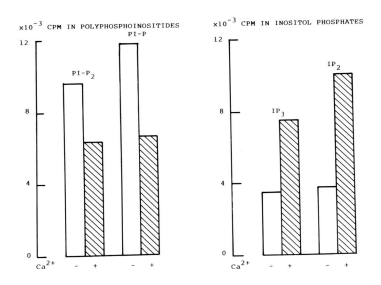

Figure 5. Loss of ^{32}P from polyphosphoinositides with a concomitant gain in ^{32}P in inositol bisphosphate and inositol trisphosphate in Ca^{2+}-treated plasma membrane fraction of human neutrophils. A plasma membrane-enriched fraction was prepared from $^{32}P_i$-labeled human neutrophils as described previously (Bennett et al., 1982). It was incubated for 10 min in the presence or absence of Ca^{2+} (0.5 mM). The decrease in ^{32}P from the polyphosphoinositides was compared with the increase in ^{32}P in IP_2 and IP_3.

inositides was measured in cells prelabeled with $^{32}P_i$ for 90 min, at which time the monoester phosphates of the polyphosphoinositides reached isotopic equilibrium with cellular [^{32}P]ATP. There was no apparent loss of ^{32}P from the polyphosphoinositides with FMLP during the course of 1 min (Fig. 6). Secretion of β-glucuronidase and an increase in PA labeling were, however, observed as usual (Fig. 6). Thus the rise in cytosolic Ca^{2+} with FMLP is clearly not sufficient to activate the polyphosphoinositide phosphodiesterase.

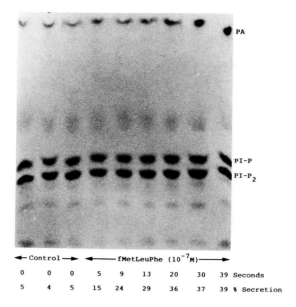

Figure 6. Effect of FMLP on polyphosphoinositides as a function of time. Human neutrophils were incubated in the presence of $^{32}P_i$ for 90 min, at which time the monoester groups of the polyphosphoinositides reached isotopic equilibrium with cellular [^{32}P]ATP. FMLP (100 nM) was added and samples were withdrawn at appropriate times. Secretion was monitored in parallel with the radioactivity in the individual phospholipids. FMLP addition stimulated an increase in the labeling of PA but no changes in either PIP or PIP$_2$ was observed.

CONCLUSIONS

From the available data on FMLP-stimulated neutrophils it can be concluded that the conversion of PI to PA is direct, with no diacylglycerol as an intermediate. This is based on the following observations: (1) The newly-formed PA has a ^{32}P-specific radioactivity too low for it to be a product of diacylglycerol phosphorylation by cytosolic ATP. This pool of PA is most likely to be derived from breakdown of an existing phospholipid. This other lipid is most likely to be PI because, (a) the specific radioactivity of PI is closest to the specific radioactivity of the newly formed PA, (b) increase in PA mass is accompanied by an equivalent decrease in mass of PI, (c) in cells pulse-labeled with [^3H]glycerol, label lost

173

from PI is recovered in PA. (2) No chemical or radio-chemical formation of diacylglycerol is observed at the time of active PI degradation. (3) No generation of inositol phosphates is observed. (4) The amount of PA formed requires the activity of diacylglycerol kinase to be 30-fold higher than any observed rate of diacylglycerol kinase (Call and Rubert, 1973; Allan *et al.*, 1980).

Based on these observations, we propose that the initial change in phospholipid metabolism in FMLP-stim-ulated neutrophils is the removal of inositol from PI to form PA. This PA then accumulates radioactivity by phosphorylation-dephosphorylation exchange. These reactions, which presumably occur at the plasma mem-brane (Bennett *et al.*, 1982), are outlined in Figure 7.

If PI is metabolized by phospholipase D, then the water-soluble product has to be inositol rather than inositol phosphates. It is therefore interesting to note that in most systems where the water-soluble products of PI hydrolysis have been studied *e.g.* pancreas (Hokin-Neaverson *et al.*, 1975), liver (Prpic *et al.*, 1982) and platelet (Bell and Majerus, 1980) it is actually a net increase in inositol rather than its phosphorylated deriva-tives which accounts for the loss from PI. The failure to form inositol phosphates has been explained by the activity of the enzyme inositol 1-phosphatase (Prpic *et al.*, 1982; Bell and Majerus, 1980). However, in the platelet where the water soluble derivatives were analyzed at 10s after cell activation, this assumption may not necessarily be correct. Only in the pancreas has it been suggested that PI may be degraded by alternative routes such as phospholipase D or the reversal of the synthetic route of PI (Hokin-Neaverson *et al.*, 1975).

The majority of data on PI metabolism in at least 30 different cell-types has been interpreted in terms of loss of PI (or polyphosphoinositides) catalyzed by phospho-lipase C. However, our data on the neutrophil do not support such a mechanism for the degradation of the inositol-containing lipids. A synthesis of the available data has to take into account not only those studies where formation of inositol phosphates has been reported but also those situations, such as the neutrophil, where it does not occur.

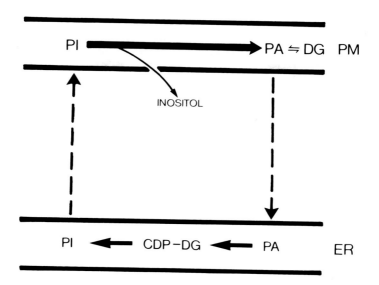

Figure 7. A proposed scheme for the interconversion of PI and PA in stimulated neutrophils. Abbreviations: DG, diacylglycerol; CDP-DG, CDP-diacylglycerol; PM, plasma membrane; ER, endoplasmic reticulum.

Our interpretation of the available data is that metabolism of PI can occur by 4 distinct routes; degradation by phospholipases A_2, C or D and utilization of PI for synthesis of polyphosphoinositides. Which pathway dominates in any one cell-type may depend on the nature of the stimulus.

The biological relevance of the conversion of PI to PA observed in neutrophils is not known. At least, its role does not lie in the generation of inositol trisphosphate or diacylglycerol, two molecules which have recently been suggested to be putative second messengers (Michell, 1983; elsewhere in this volume). One likely possibility is that the conversion of PI into PA changes a phospholipid which is inhibitory to fusion to generate another which promotes fusion (Sundler and Papahadjopoulos, 1981; Sundler et al., 1981). This may, in conjunction with other cellular events, lead to the fusion of the granule with the plasma membrane and thus to exocytosis of the granular constituents.

ACKNOWLEDGEMENTS

We thank the Medical Research Council, U.K. for their support.

REFERENCES

Allan, A., Thomas, P. and Gatt, S. (1980) *Biochem. J.* *191*, 669-672.

Bell, R.L. and Majerus, P.W. (1980) *J. Biol. Chem.* *255*, 1790-1792.

Bennett, J.P., Cockcroft, S. and Gomperts, B.D. (1980) *Biochim. Biophys. Acta* *601*, 584-591.

Bennett, J.P. Cockcroft, S. and Gomperts, B.D. (1982) *Biochem. J.* *208*, 801-808.

Berridge, M.J. Downes, C.P. and Hanley, M.R. (1982) *Biochem. J.* *206*, 587-595.

Brindley, D.N., Allan, A. and Michell, R.H. (1975) *J. Pharm. Pharmac.* *27*, 462-464.

Call, F. II and Rubert, M. (1973) *J. Lipid Res.* *14*, 466-474.

Cockcroft, S., Bennett, J.P. and Gomperts, B.D. (1980a) *FEBS Letters* *10*, 115-118.

Cockcroft, S., Bennett, J.P. and Gomperts, B.D. (1980b) *Nature (Lond)* *288*, 275-277.

Cockcroft, S., Bennett, J.P. and Gomperts, B.D. (1981) *Biochem. J.* *200*, 501-508.

Hokin, M.R. and Hokin, L.E.. (1953) *J. Biol. Chem.* *203*, 967-977.

Hokin-Neaverson, M., Sadeghian, K., Majumder, A.L. and Eisenberg, F. Jr. (1975) *Biochem. Biophys. Res. Commun.* *67*, 1537-1544.

Karnovsky, M.L. and Wallach, D.F.H. (1961) *J. Biol. Chem.* *236*, 1895-1901.

Michell, R.H. (1983) *Trends in Biochem. Sci.* *8*, 263-265.

Michell, R.H., Kirk, C.J., Jones, L.M., Downes, C.P. and Creba, J.A. (1981) *Phil. Trans. R. Soc. Lond. B.* *296*, 123-137.

Prpic, V., Blackmore, P.F. and Exton, J.H. (1982) *J. Biol. Chem.* *257*, 11327-11331.

Sastry, P.S. and Hokin, L.E. (1966) *J. Biol. Chem.* *241*, 3354-3361.

Sundler, R. and Papahadjopoulos, D. (1981) *Biochim. Biophys. Acta* *647*, 743-750.

Sundler, R., Duzgunes, N. and Papahadjopoulos, D. (1981) *Biochim. Biophys. Acta* 647, 751–758.

THE *DE NOVO* PHOSPHOLIPID SYNTHESIS EFFECT: OCCURRENCE, CHARACTERISTICS, UNDERLYING MECHANISMS AND FUNCTIONAL SIGNIFICANCE IN HORMONE ACTION AND SECRETION

Robert V. Farese

Veterans Administration Hospital, Department of
Medicine, University of South Florida College of Medicine,
Tampa, FL

SUMMARY

A number of hormones and other agents have been
found to provoke rapid, relatively large increases in
phosphatidic acid (PA), phosphatidylinositol (PI), phosphatidylinositol 4-phosphate (PIP), phosphatidylinositol
4,5-bisphosphate (PIP_2) and diacylglycerol (DG) in
certain of their target tissues. This effect is due to *de
novo* synthesis of PA and may be the consequence of
increased availability of substrate (*e.g.*, glycerol 3-P) or
increased activity of glycerol 3-P acyltransferase. This
"*de novo* effect" appears to be a post-second messenger
event and, in some tissues, may be the consequence of
increases in cAMP and/or Ca^{2+}. Increases in PA, DG,
PIP and PIP_2 (and possibly inositol-phosphates) may
serve to alter the uptake and distribution of cellular Ca^{2+}
and increase protein kinase C activity. Changes in local
concentrations of lipids and phospholipids may also provoke direct effects on membrane-bound receptors, trans-
porters and enzymes. The *de novo* phospholipid effect
may be important in the following: (a) regulation of
steroidogenesis by ACTH, luteinizing hormone, angiotensin and K^+ in adrenal and ovarian tissues; (b) insulin
action on pyruvate dehydrogenase and glucose and lipid
metabolism in adipose tissue and muscle; (c) handling of

179

Ca^{2+} and $PO_4{}^{3-}$ in parathyroid hormone-sensitive tissues; (d) light-stimulation of the retina; and (e) glucose-induced insulin secretion in pancreatic islets.

THE *DE NOVO* PHOSPHOLIPID SYNTHESIS EFFECT: OCCURRENCE AND CHARACTERISTICS

During the past several years, it has become apparent that phosphoinositide metabolism can be perturbed by a mechanism considerably different from the classical "phosphatidylinositol (PI) effect", *i.e.*, phospholipase C-mediated PI hydrolysis and resynthesis via diacylglycerol (DG) and phosphatidic acid (PA) (see Hokin and Hokin, 1953; and Michell, 1975). This newly recognized mechanism, which involves *de novo* synthesis of PA followed by rapid increases in the levels of PI, phosphatidylinositol 4-phosphate (PIP) and phosphatidylinositol 4,5-bisphosphate (PIP_2) became apparent during studies on the adrenal cortex, where it was observed that ACTH induced rapid increases in the mass of PIP and PIP_2 (Farese *et al.*, 1979 and 1980a). Subsequently, it was realized that increases in PIP and PIP_2 were preceded by increases in PI and PA (Farese *et al.*, 1980b, 1980c; see also Laychock *et al.*, 1978) and were associated with increases in other phospholipids, including phosphatidylglycerol (PG) (Farese *et al.*, 1980b and c), phosphatidylcholine (PC) and phosphatidylethanolamine (PE), (Farese *et al.*, 1982a) and DG (Farese *et al.*, 1981a). These findings [recently confirmed by Igarashi *et al.* (1983)] were interpreted to suggest that ACTH provokes a rapid increase in *de novo* PA synthesis, with a large fraction of the newly synthesized PA being channeled through the PI cycle.

Similar phospholipid effects (*i.e.*, *de novo* synthesis of PA, PI, PIP and PIP_2) were observed during the actions of: luteinizing hormone in ovarian tissues (Davis *et al.*, 1983) and testicular Leydig cells (Lowitt *et al.*, 1982); insulin in both adipose tissue (Farese *et al.*, 1982b) and cultured myocytes (Farese, unpublished observations); parathyroid hormone (PTH) in kidney cortex tubules (Farese *et al.*, 1980d, 1981b; Bidot-Lopez et, al., 1981; Meltzer *et al.*, 1982); angiotensin-II (A-II), K^+, ACTH and serotonin in rat adrenal glomerulosa tissue (Farese *et al.*, 1983a); light-induced stimulation of the

180

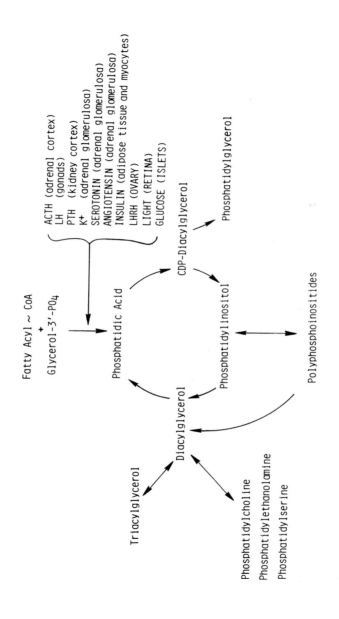

Figure 1. The de novo phospholipid effect

retina (Schmidt, 1983); and glucose-induced insulin secretion in pancreatic islets (Farese, unpublished observations)(Fig. 1). Obviously, the *de novo* phospholipid synthesis effect appears to be widespread in the actions of many hormones and other agents in their target tissues. Moreover, these effects occur rapidly (within minutes) and involve substantial increases in phospholipids and lipids, *e.g.*, approximately 40-100% for DG and less abundant, acidic phospholipids, such as PA, PI, PS, PIP and PIP_2, and 10-20% for the more abundant, basic phospholipids, PC and PE.

Increases in phospholipid levels after hormonal treatment were originally documented by measurement of phospholipid-phosphorus mass (by colorimetry) after purification by thin-layer chromatography. In the case of some hormones, *e.g.*, insulin, increases in phospholipid mass were also associated with increases in ^{32}P incorporation into phosphoinositides and other phospholipids (Stein and Hales, 1974; DeTorrontegui and Berthet, 1965; Garcia-Sainz and Fain, 1980). In rat adipose tissue, we have documented that insulin-induced increases in ^{32}P incorporation into PI, PIP, PIP_2, PS and PC + PE are not due to changes in ATP specific radioactivity (Farese *et al.*, 1984b), and increases in radiolabeling reflect increases in phospholipid mass, since degradation is not increased by insulin (unpublished).

Increased ^{32}P incorporation into phospholipids has also been observed in ACTH-treated adrenal sections (Farese *et al.*, 1980b), but in adrenal cells, incorporation is dose-dependent (Farese *et al.*, 1983b): at lower ACTH concentrations, ^{32}P incorporation increases in parallel with increases in phospholipid mass, but, at higher ACTH concentrations, incorporation decreases, despite continued increases in mass. The later decrease in specific radioactivity of phospholipids appears to be at least partly due to a concomitantly observed ACTH-dose-dependent decrease in ATP specific radioactivity (Farese *et al.*, 1983b). The latter may be related to the massive generation of cAMP, which may mobilize intracellular phosphate (perhaps via turnover of the α and β phosphates of ATP). Along these lines, Meltzer *et al.* (1982) have observed that during PTH action in dog kidney tubules, incorporation of ^{32}P into phosphoinositides increases rapidly along with phospholipid mass, but this is soon followed by a decline in radiolabeling, despite continued increases in phospholipid mass; this dissociation may also be due to cAMP accumulation. Direct addition of cAMP to incubations of adrenal sections leads to an increase in the mass of phosphoinositides, but little or no increase in ^{32}P incorporation over a 60 min period, if cAMP and $^{32}P_i$ are added simultaneously at zero time (Farese et al., 1980b, 1983b); if, however, adrenal sections are prelabeled for 2 h with $^{32}P_i$, cAMP provokes a rapid increase in [^{32}P]-phospholipids, but this is short-lived and is replaced by an inhibitory pattern for ^{32}P incorporation (see Fig. 2). This secondary decrease in incorporation seems to be best explained by a decrease in phospholipid-precursor specific radioactivity, but, in previous studies (Farese et al., 1983b) we did not observe a decrease in the specific radioactivity of total cellular ATP after cAMP treatment (this may reflect heterogeneity of ATP radiolabeling).

Because of the difficulties in interpretation of results derived from acute ^{32}P-labeling studies, we have recently employed a cell culture system (BC3H-1 myocytes) in which intracellular phosphate pools and all phospholipids are radiolabeled to constant specific radioactivity by incubation with $^{32}P_i$ for 72 h prior to treatment with insulin. Using this prelabeling approach, ^{32}P in phospholipids reflects changes in phospholipid mass, and increases due to turnover will not be discernable because demonstration of this depends upon the conversion of unlabeled to labeled phospholipids (which occurs only during acute labeling). In this cell culture system, we have observed that within 10-30 min, insulin provokes 40-60% increases in ^{32}P (reflecting mass) in PA, PI, PIP and PIP_2, 20% increases in PS, PC and PE, and 100% increases in DG. Thus, we have verified that previously observed insulin-induced increases in phospholipid mass are accurately reflected by changes in ^{32}P content.

We have also used the ^{32}P-prelabeled myocyte culture system to verify that insulin-induced increases in phospholipids were due to increased synthesis, rather than decreased degradation; thus, in pulse-chase experiments (unpublished), insulin did not decrease the degradation rates of PA, PI, PIP, PIP_2, PS, PC or PE. Moreover, our findings suggested that insulin provoked

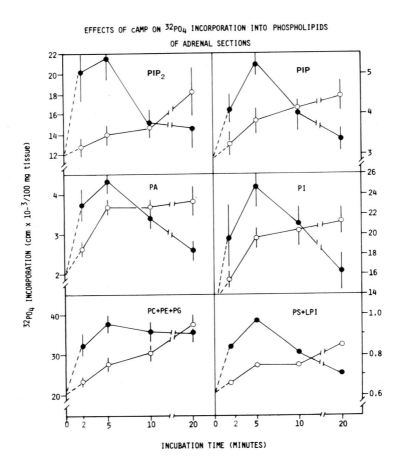

EFFECTS OF cAMP ON $^{32}PO_4$ INCORPORATION INTO PHOSPHOLIPIDS
OF ADRENAL SECTIONS

Figure 2. Adrenal sections were preincubated for 2 h with $^{32}P_i$. Cyclic AMP (5 mM) was added (closed circles), and incubations were continued for the indicated times. Control values are shown by open circles. Mean values ± S.E. of 6 determinations.

an initial rapid burst of synthesis, followed by a more moderate increase in synthesis after a new steady-state had been attained, *i.e.*, after 15-30 min (see below).

THE *DE NOVO* PHOSPHOLIPID SYNTHESIS EFFECT:
UNDERLYING MECHANISMS

The mechanisms whereby the above agents increase the synthesis of PA, PI and other phospholipids are presently uncertain. ACTH, PTH and A-II increase glycogenolysis and lipolysis, and these could increase the availability of glycerol 3-P and fatty acids for PA synthesis. In pancreatic islets, which are freely permeable to glucose, glucose may directly increase glycerol 3-P availability, and, in target tissues of insulin, this could be effected by an increase in glucose uptake. However, in myocyte cultures, we have found (unpublished) that insulin increases PA, PI, PIP and PIP_2 equally well in the absence of extracellular glucose, and must therefore operate through a mechanism independent of glucose uptake. This mechanism may involve changes in glycerol 3-phosphate acyltransferase, since insulin increases the activity of this enzyme in liver and adipose tissue (Sooranna and Saggerson, 1976; Bates *et al.*, 1977). On the other hand, we have measured glycerol 3-P acyltransferase activity in adrenal homogenates, and have found no changes after ACTH (or cycloheximide – see below) treatment *in vivo* (unpublished); perhaps ACTH increases PA synthesis simply by increasing glycogenolysis and glycolysis.

As with ACTH treatment, increases in PA, PI, PIP, PIP_2, PC and PE have also been observed during stimulation of adrenal tissue *in vitro* by cAMP, the generally accepted second messenger for ACTH (Farese, 1980a, 1980b and 1981b). Similarly, dibutyryl cAMP, like PTH, provokes increases in these phospholipids in kidney tissue (Bidot-Lopez *et al.*, 1981) [the failure of Meltzer *et al.* (1982) to observe effects of cAMP may have been due to their using 1 mM instead of 5 mM dibutyryl cAMP, which is more effective (unpublished)]. These findings suggest that the increase in the synthesis of PA and other phospholipids is a "post-second messenger event" in the actions of ACTH and PTH; this contrasts with the phosphoinositide hydrolysis effect, which may be a "pre-second messenger" event in the action of agents which mobilize Ca^{2+} via surface receptors, (see Downes and Michell, 1982). In the case of A-II, Ca^{2+} is thought to be the major second messenger, and along these lines, we

185

have found (Farese *et al.*, 1983a) that Ca^{2+} is required for A-II-induced increases in phospholipid mass in adrenal zona glomerulosa tissue, and, furthermore, that the Ca^{2+} ionophore, A23187, like A-II, increases the mass of PA, PI, PIP and PIP_2 in rabbit kidney cortex (unpublished observations). In the case of insulin, its second messenger has not been identified, but there clearly must be a signal that relays information from the insulin surface receptor to glycerol 3-P acyltransferase in the endoplasmic reticulum. Changes in cGMP and Ca^{2+} may occur during insulin action (Czech, 1977) and theoretically could be involved, but dibutyryl cGMP and A23187 do not mimic insulin effects on phospholipid metabolism in cultured myocytes (unpublished observations). Another possibility is that an increase in phosphatase activity, which may occur in the action of a putative, insulin-inducible chemical mediator (see Czech, 1981), could activate glycerol 3-P acyltransferase (see Nimmo and Houston, 1978). In addition, increases in *de novo* synthesis of PA and PI may be triggered by other undefined second messengers, and there may be initial *direct* effects of insulin and other hormones on PIP and PIP_2 metabolism (*e.g.*, a decrease in degradation or an increase in kinase-mediated synthesis) since the increases in PIP and PIP_2 occur in the plasma membrane. Along the latter lines, Jolles *et al.* (1981) have observed direct effects of ACTH on PIP_2 synthesis in a membrane/cytosol system from rat brain.

Increases in PA, PI, PIP, PIP_2, PC, PE, PS and PG, which are provoked by ACTH, LH, insulin, PTH, cAMP, and A-II have been found to be blocked by prior administration of protein synthesis inhibitors, *e.g.*, cycloheximide and puromycin (for a review, see Farese, 1983a). In addition to preventing the induction of increases in these phospholipids, addition of cycloheximide to ACTH- or insulin-pretreated adrenal or muscle tissues (respectively) results in a very rapid decrease of phospholipid mass from elevated to (but generally not below) control levels. The latter observation was originally attributed (Farese, 1980c) to a cycloheximide-induced decrease in phospholipid (PA) synthesis, coupled with an inherently fast, but unchanged, rate of phospholipid degradation. However, we have recently found in the cultured myocyte that cycloheximide markedly increases the degradation rates (as compared to the control rates which were measured in pulse-chase experiments – see above) for PA, PI, PIP, PIP_2 and PS in insulin-treated (but not control) myocytes (unpublished). We have also found that cycloheximide and puromycin cause a rapid decrease in phospholipid mass in ACTH-stimulated adrenal tissue, and this is followed by a secondary increase in ^{32}P incorporation into phospholipids in acute radiolabeling studies (unpublished). As a result, while the mass of PI and other phospholipids decreases with cycloheximide treatment, acute incorporation of ^{32}P (but not [^{3}H]gly-cerol) may paradoxically increase; these changes are similar to those observed during phospholipase C action, viz., enhanced removal of phospholipid head groups, followed by generation of DG and subsequently, by partial resynthesis and radiolabeling of phospholipids (particularly PA and PI). We therefore presently believe that protein synthesis inhibitors activate a non-specific phospholipase C (effecting hydrolysis of PA, PC and PE, as well as PI) and possibly other phospholipases and lipases (since DG may also decrease – unpublished) in hormone-stimulated (but not control) target tissues. Just how inhibition of protein synthesis and concomitant hor-mone action may be linked to phospholipid degradation is presently unknown. In any event, these problems, e.g., paradoxical changes in mass and ^{32}P-labeling, phospho-lipid degradation, and formation of potentially noxious by-products, etc., should be kept in mind when protein synthesis inhibitors are employed. For example, it is not possible to study cycloheximide-induced inhibition of the de novo phospholipid effect in acute ^{32}P-labeling studies; meaningful results will only be apparent in studies of changes in phospholipid mass or phospholipids that are prelabeled with ^{32}P to constant specific radioactivity. In addition, noxious by-products of phospholipase action (e.g., fatty acids or lysophospholipids) could be respon-sible for the partial inhibition of DG kinase which was observed in ACTH + cycloheximide-treated adrenal tissue (Farese et al., 1981a), and during inhibition of ACTH effects on steroidogenesis (Farese et al., 1980c and 1981b) or insulin effects on pyruvate dehydrogenase (see below).

Induced increases in *de novo* phospholipid synthesis have been found to require Ca^{2+}. With ACTH and cAMP in the adrenal cortex (Farese, 1981c), this requirement appears to be for intracellular, rather than extracellular, Ca^{2+}. Whether intracellular Ca^{2+} is required permissively or serves as a mediator in *de novo* phospholipid synthesis (*e.g.*, to enhance glycogenolysis and provision of glycerol 3-P) is presently unknown. It may be noted that agonist-induced increases in Ca^{2+} may also decrease PI synthesis in certain conditions (see Egawa *et al.*, 1981), and it is presently not possible to make any generalizations.

The recent finding that cycloheximide provokes a rapid increase in phospholipid degradation leads us to reconsider previous calculations of phospholipid synthesis during ACTH action in adrenal tissue (Farese *et al.*, 1980c). Those calculations were based upon the erroneous assumption that the rapid decreases in phospholipids which were observed during cycloheximide-induced reversal of ACTH effects simply reflected the usual rates of degradation, and the rates of phospholipid synthesis were clearly overestimated. In cultured myocytes (unpublished), it was possible to perform prolonged pulse-chase experiments and accurately assess degradation rates. In this system, the half-life for ^{32}P-prelabeled phospholipids was found to be less than 30 min in the presence of cycloheximide, but greater than 4 h in the absence of cycloheximide. Moreover, insulin did not influence the fractional rate of degradation of phospholipids in this tissue, but nevertheless caused PI levels to increase by 50% within 15-30 minutes and stay relatively constant thereafter. It would therefore appear that insulin provokes an initial burst of phospholipid synthesis, but after 15-30 min, the rate of synthesis decreases and only approximately one-half of the expanded PI pool is renewed every 4-5 h (the latter is derived from the assumption that PI synthesis must equal PI degradation during steady-state conditions). This rate of synthesis is much slower than earlier estimates (Farese *et al.*, 1980c), and obviously requires considerably less energy expenditure.

THE *DE NOVO* PHOSPHOLIPID SYNTHESIS EFFECT: FUNCTIONAL SIGNIFICANCE

The physiological relevance of increases in phospholipid content and changes in relative phospholipid concentrations (particularly the less abundant substances) due to stimulation of *de novo* synthesis of PA is presently uncertain. Increases in DG (now observed in insulin as well as ACTH action, and probably a ubiquitous consequence of the *de novo* effect), PS, PA and PI would be expected to directly activate protein kinase C or facilitate its activation by Ca^{2+}, as has been suggested for agents which increase phosphoinositide (either PIP_2 or PI) hydrolysis (Nishizuka and Takai, 1981 and 1983). Similarly, increased generation of DG and PI, may increase substrate availability for phospholipase A_2 and DG lipase, respectively, causing increased generation of arachidonic acid, prostaglandins, thromboxanes, and products of the lipooxygenase pathway (this also seems to be caused by phosphoinositide hydrolysis). Increases in polyphosphoinositides theoretically should alter the electrochemical and biological properties of cellular membranes, *e.g.*, increases in PIP_2 in the plasma membrane increase the binding of Mg^{2+} and/or Ca^{2+}, and may also increase lateral mobility of membrane components (Sheetz *et al.*, 1982), perhaps by opposing limitations imposed by cytoskeletal structures. Increases in natural Ca^{2+} ionophores such as PA (Putney *et al.*, 1980) or inositol-1,4,5-trisphosphate (IP_3) (Streb *et al.*, 1983), a product of the hydrolysis of PIP_2 (which theoretically may be increased when PIP_2 levels are elevated), may provoke changes in Ca^{2+} uptake from extracellular sources or release from intracellular stores. This may explain changes in Ca^{2+} observed during the actions of ACTH (Leier and Jungman, 1973) and insulin (MacDonald *et al.*, 1976a, b) Interestingly, the ACTH effect on Ca^{2+} uptake is cycloheximide-sensitive.

Turning to more specific functions, a role for phospholipid metabolism in steroidogenesis seems likely. For those agents which operate via Ca^{2+} (*e.g.*, A-II or K^+), early rapid effects of PIP_2 or PI hydrolysis may initiate or amplify increases in Ca^{2+} (Farese *et al.*, 1982). In addition, increases in the levels of phospholipids, via the *de novo* phospholipid effect (which occurs concomitantly

189

with PIP_2 and/or PI hydrolysis in adrenal glomerulosa tissue during the action of A-II and K^+ - see Farese *et al.*, 1983 and 1984) may be important for steroidogenesis *per se*. Along the latter lines, the *de novo* phospholipid effect has been observed in the action of all tested steroidogenic agents (see above), but there have been a few instances in which this effect could not be demonstrated, *e.g.*, in a small number of experiments on bovine adrenal glomerulosa cells (Elliott *et al.*, 1983) and in the gonadotropin-primed rat ovary (Tanaka and Strauss, 1982). In considering these negative experiments, it should be noted that : (a) the method for demonstrating the mass effect (by P_i content) is not particularly sensitive for small changes; (b) ^{32}P incorporation may reflect PI turnover better than increases in *de novo* synthesis in acute ^{32}P-labeling experiments; and (c) many tissues and collagenase-dispersed cell preparations are anatomically and functionally heterogeneous (and damaged), and may fail to reveal hormone-induced changes in the correct phospholipid pool of the correct cell population. The sensitivity of the *de novo* phospholipid effect (but *not* the phosphoinositide-turnover effect) to cycloheximide, and the close correlation (Farese *et al.*, 1980c and 1981c) between cycloheximide-induced inhibition of phospholipid metabolism and steroidogenesis (rather than between protein synthesis *per se* and steroidogenesis, as shown in adrenal sections *in vitro* - see Farese *et al.*, 1969 and 1981c) is also in keeping with the possibility that the *de novo* effect may be important in the activation of steroidogenesis by a variety of agents. As noted above, however, the cycloheximide effect on steroidogenesis may also be explained by a noxious effect secondary to phospholipase activation in hormone-stimulated (but not, or less so, in control) tissues.

If phospholipids are truly important in steroidogenesis, there are a number of mechanisms that could theoretically be operative, *e.g.*, direct stimulation of mitochondrial cholesterol side-chain cleavage by PIP, PIP_2, or PG-P (Farese and Sabir, 1979, 1980); protein kinase C activation and subsequent protein phosphorylation; mobilization of intracellular Ca^{2+}; and interaction of phospholipids and cholesterol-carrier proteins. In this regard, phorbol esters and sub-stimulatory concentrations of A23187, have been found to activate steroidogenesis in adrenal glomerulosa cells (Kojima et al., 1983); this could reflect participation of protein kinase C and Ca^{2+} mobilization in the activation of steroidogenesis, and this could be provoked by increases in DG, PA or IP_3 by either the de novo effect or by PI or PIP_2 hydrolysis.

Recently, a steroidogenic peptide has been isolated by Pederson and Brownie (1983) in the cytosol of ACTH-treated adrenal tissue, and cycloheximide inhibits this effect of ACTH. This peptide is hydrophobic, and, like acidic phospholipids (Farese and Sabir, 1979 and 1980), promotes the interaction of cholesterol with cytochrome-P_{450} and subsequent cleavage of the cholesterol side-chain, and may thus be important during ACTH action. Unfortunately, further information on the turnover (i.e., metabolic lability) of this steroidogenic peptide is presently not available; similarly, the reason why this peptide is not found in the mitochondrial cell fractions, where ACTH is known to promote an increase in activity for cholesterol side-chain cleavage, is not apparent. Although this peptide may ultimately be proved to be more important than phospholipids in the direct activation of steroidogenesis during ACTH action, activation of phospholipid metabolism may nevertheless precede and cause changes in the generation or availability of the steroidogenic peptide, e.g., by effecting changes in proteolysis or translocation to the cytosol; thus, phospholipid metabolism may prevail as the cycloheximide-sensitive step in the action of steroidogenic agents. In support of the latter possibility is the close correlation of changes in phospholipids during either rapid (Farese, 1980) or relatively slow (Farese et al., 1981c) reversal of ACTH effects on steroidogenesis.

As noted above, insulin-induced increases in phospholipids in both adipose tissue and muscle, and increases in glycerol 3-P acyltransferase activity in adipose tissue are of great interest, since these are the major target tissues of insulin. Activation of pyruvate dehydrogenase (PDH) in rat adipose tissue (Farese et al., 1984b) seems to be well correlated to insulin-induced increases in phospholipids (PI and PS) in dose-response and time-sequence studies, and the metabolic requirements for full activation of PDH and phospholipid metabolism seem to be similar [in adipose tissue, both require an exogenous carbohydrate source and are inhibited by cycloheximide and puromycin (see also Sica and Cuatrecasas, 1973)]. Activation of PDH, as well as glucose oxidation, has also been observed in the action of exogenously added phospholipase C on adipose tissue, and it is conceivable that increases in DG, PA and/or IP_3 may play key roles in these processes (Honeyman *et al.*, 1983). Further support for a role for phospholipids in PDH activation derives from our recent observation (unpublished) that cycloheximide provokes very rapid decreases in PDH activity, and, concomitantly, PA, PI, PIP, PS and DG levels in insulin-treated myocytes. The best correlation was between decreases in PDH, PA and DG; this suggests that activation of protein kinase C and Ca^{2+} mobilization may be important events in the activation of PDH by insulin. Alternatively, PS may directly increase PDH activity (Kiechle *et al.*, 1982) or, along with DG, PI and PA, may activate protein kinase C.

Changes in phospholipid metabolism would also appear to be a logical candidate for the increase in glucose transport observed during insulin action, since the latter has been shown to be due to the translocation of transporting units from intracellular membranes to the plasma membrane (Cushman and Wardzala *et al.*, 1980, and Suzuki and Kono, 1980). However, against this possibility is the observation that translocation of glucose transporting units to the plasma membrane was observed during combined treatment with insulin and protein synthesis inhibitors (Kono *et al.*, 1981); in addition, we have found that protein synthesis inhibitors do not block insulin effects on glucose transport or uptake in both adipose tissue and cultured myocytes (unpublished observations), despite inhibiting PDH activity and oxidation of glucose to CO_2 (unpublished). Unfortunately, there are several difficulties in these studies: (a) protein synthesis inhibitors provoke small but potentially important increases in phospholipids in control tissues (perhaps by activating glycogenolysis), and, moreover, have mild insulin-like effects (unpublished); and (b) protein synthesis inhibitors may not fully inhibit insulin-induced increases in PA, DG (due to apparent activation of phospholipase C) or other related substances, despite decreasing the levels of PI, PS and other phospholipids (see above). Obviously, further work is required to evaluate the role of phospholipid metabolism in the effect of insulin on glucose transport. Similarly, further work will be required to determine whether insulin-induced increases in phospholipids may mediate other effects of insulin, e.g.: (a) PA and/or PS have been reported to activate both PDH (see above) and low Km phosphodiesterase (Macaulay et al., 1982); and (b) internalization of the insulin receptor (Kadle et al., 1983) and amino acid transport (Elsas et al., 1968) are inhibited by protein synthesis inhibitors, and this inhibition could be due to phospholipase activation, rather than a requirement for a labile protein.

The role of phospholipid metabolism in the action of PTH is also conjectural at this time. The phosphaturic effect of PTH was found to be sensitive to cycloheximide (Farese et al., 1980d), and changes in phospholipids could therefore be important. Changes in Ca^{2+} transport in PTH-sensitive tissues (bone and kidney) may also be influenced by changes in phospholipids, in view of the observed increases in PIP, PIP_2, PA and, presumably, IP_3 and their effects on Ca^{2+} (see above).

In addition to increases in phospholipids and DG serving to activate specific enzymes, transporting units and receptors, the generalized increase in phospholipid synthesis that occurs during the actions of ACTH, PTH, insulin, angiotensin, and gonadotropins may serve to trigger the cellular hypertrophy that occurs during the actions of these tropic agents in certain of their target tissues. Increases in PC and PE, as well as other phospholipids, would ultimately lead to an expansion of intracellular membranes, and this may increase water imbibition, and perhaps lead to increased synthesis of RNA and proteins to provide a balanced increase in membrane-associated constituents. Increases in DG may also activate protein kinase C, which, in turn may activate proteins important in protein synthesis (e.g., ribosomal protein 6 – see LePeuch et al., 1983) or nucleic acid synthesis. Along these lines, it is of interest to note that the ACTH-induced increase in RNA polymerase (which is probably largely responsible for the effects of ACTH on adrenal protein synthesis) is inhibited by cycloheximide (Farese and Schnure, 1967).

The increase in *de novo* synthesis in PA, PI, PIP and PIP_2 during glucose stimulation of pancreatic islets is of great interest, since glucose metabolism *per se* by glycolysis (perhaps to generate glycerol 3-P and PA as suggested by our findings) is recognized to be exceedingly important for provoking insulin secretion (Ashcroft, 1980). While glucose also increases PI and PIP_2 hydrolysis in islets (Clements *et al.*, 1981; Laychock, 1983), these effects are Ca^{2+}-dependent, not directly related to glucose metabolism, and may be secondary events. As such they are clearly distinct from Ca^{2+}-independent PIP_2 hydrolysis, which is thought to be the primary biochemical event in receptor-mediated mobilization of Ca^{2+} (see Downes and Michell, 1982). It is intriguing to speculate that the primary effect of glucose may be to increase *de novo* synthesis of PA, PI, PIP, PIP_2 and DG. Subsequent increases in PA, DG and IP_3 may mobilize Ca^{2+} (Putney *et al.*, 1980; Streb *et al.*, 1983) and activate protein kinase C (Nishizuka and Takai, 1981) and phospholipase C as well (Dawson *et al.*, 1983), causing Ca^{2+}-dependent PIP_2 and/or PI hydrolysis, exocytosis, and insulin secretion.

CONCLUSION

The list of substances that provoke the *de novo* phospholipid synthesis effect continues to expand (see Fig. 1). Clearly, this effect is relatively common, and may, in fact, operate concurrently with PIP_2 and/or PI hydrolysis. Undoubtedly, in many cases, the occurrence of the *de novo* effect has gone unrecognized for a variety of reasons: (a) tacit assumptions that increases in ^{32}P-labeling are solely attributable to phosphoinositide hydrolysis; (b) failure of ^{32}P incorporation to reflect the *de novo* effect in cases where cAMP-generation is markedly increased; and (c) failure to apply appropriate methodology for its documentation, *i.e.*, measurement of phosphorus mass or ^{32}P-radioactivity after prelabeling of phospholipids to constant specific radioactivity. Hopefully, this recognition will improve, as it is likely that the *de novo* effect is an important intracellular effector mechanism for many hormones and other agents. Along these lines, it should be noted that the *de novo* and phosphoinositide hydrolysis effects share several important features, *e.g.*, increased generation of DG, PA,

CDP-DG, PG-P, PG and possibly inositol-phosphates (perhaps, however, in different cellular compartments), and these substances (through changes in membrane composition and function, protein kinase C activation and protein phosphorylations and ionic shifts) may be responsible for many changes in cellular functions. Major tasks for the future will be to further define the underlying mechanisms, the intracellular anatomical sites and the specific functional consequences of the *de novo* phospholipid effect.

ACKNOWLEDGEMENTS

This work was supported by funds from the National Institutes of Health (1 RO1 HL 28290-02) and the Research Service of the Veterans Administration.

REFERENCES

Ashcroft, S.J.H. (1980) *Diabetologia 18*, 5-15.

Bates, E.J., Topping, D.L., Sooranna, S.R., Saggerson, D. and Mayes, P.A. (1977) *FEBS Letters 84*, 225-228.

Bidot-Lopez, P., Farese, R.V. and Sabir, M.A. (1981) *Endocrinology 108*, 2078-2082.

Clements, R.S. Jr., Evans, M.H. and Pace, C.S. (1981) *Biochim. Biophys. Acta 647*, 1-9.

Cushman, S.W. and Wardzala, L.J. (1980) *J. Biol. Chem. 255*, 4758-4762.

Czech, M.P. (1977) *Ann. Rev. Biochem. 46*, 359-384.

Czech, M.P. (1981) *Am. J Med. 70*, 142-150.

Davis, J.S., Farese, R.V. and Clark, M.R. (1983) *Endocrinology 112*, 2212-2214.

Dawson, R.M.C., Hemington, N.L. and Irvine, R.F. (1983) *Biochem. Biophys. Res. Commun. 117*, 196-201.

DeTorrontegui, G. and Berthet, J. (1965) *Biochim. Biophys. Acta 116*, 447-481.

Downes, C.P. and Michell, R.H. (1981) *Biochem. J. 198*, 133-140.

Downes, P. and Michell, R.H. (1982) *Cell Calcium 3*, 467-502.

Egawa, K., Sacktor, B. and Takenawa, T. (1981) *Biochem. J. 194*, 129-136.

Elliott, M.E., Goodfriend, T.L. and Farese, R.V. (1983) *Life Sci. 33*, 1771-1778.

Elsas, L.J., Albrecht, I. and Rosenberg, L.E. (1968) *J. Biol. Chem.* *243*, 1846-1853.

Farese, R.V. (1983) *Endocrine Reviews* *4*, 78-95

Farese, R.V. and Sabir, M.A. (1979) *Biochim. Biophys. Acta* *575*, 299-304.

Farese, R.V. and Sabir, A.M. (1980) *Endocrinology* *106*, 1869-1880.

Farese, R.V. and Schnure, J.J. (1967) *Endocrinology* *80*, 872-882.

Farese, R.V., Linarelli, L.G., Glinsmann, W.H., Ditzion, B.R., Paul, M.I. and Pauk, G.L. (1969) *Endocrinology* *85*, 867-874.

Farese, R.V., Sabir, M.A. and Vandor, S.L. (1979) *J. Biol. Chem.* *254*, 6842-6844.

Farese, R.V., Sabir, M.A., Vandor, S.L. and Larson, R.E. (1980a) *J. Biol. Chem.* *255*, 5728-5734.

Farese, R.V., Sabir, M.A. and Larson, R.E. (1980b) *J. Biol. Chem.* *255*, 7232-7237.

Farese, R.V., Sabir, M.A. and Larson, R.E. (1980c) *Proc. Natl. Acad. Sci. USA* *77*, 7189-7193.

Farese, R.V., Bidot-Lopez, P., Sabir, A., Smith, J., Schinbeckler, B. and Larson, R. (1980d) *J. Clin. Invest.* *65*, 1523-1526.

Farese, R.V., Sabir, M.A. and Larson, R.E. (1981a) *Biochemistry* *20*, 6047-6051.

Farese, R.V., Sabir, M.A. and Larson, R.E. (1981b) *Endocrinology* *109*, 1895-1901.

Farese, R.V., Sabir, M.A. and Larson, R.E. (1981c) *Endocrinology* *109*, 1424-1427.

Farese, R.V., Sabir, M.A. and Larson, R.E. (1982a) *Biochemistry* *21*, 3318-3322.

Farese, R.V., Larson, R.E. and Sabir, M.A. (1982b) *J. Biol. Chem.* *257*, 4042-4045.

Farese, R.V., Larson, R.E., Sabir, M.A. and Gomez-Sanchez, C.E. (1983a) *Endocrinology* *113*, 1377-1386.

Farese, R.V., Sabir, M.A., Larson, R.E. and Trudeau III, W. (1983b) *Cell Calcium* *4*, 195.

Farese, R.V., Larson, R.E. and Davis, J.S. (1984a) *Endocrinology* *114*, 302-304.

Farese, R.V., Farese Jr., R.V., Sabir, M.A., Larson, R.E., Trudeau III, W.L. and Barnes, D. (1984b) *Diabetes*, in press.

Garcia-Sainz, J.A. and Fain, J.N. (1980) *Biochem. J.* *186*, 781-789.

Honeyman, T.W., Strohsnitter, W., Scheid, C.R. and Schimmel, R.J. (1983) *Biochem. J.* *212*, 489-498.

Hokin, M.R. and Hokin, L.E. (1953) *J. Biol. Chem.* *203*, 967-977.

Igarashi, Y., Cheng, B.-L., Hsu, D. and Kimura, T. (1983) *Fed. Proc.* *42*, 1920.

Jolles, J., Zwiers, H., Dekker, A., Wirtz, K.W.A. and Gispen, W.H. (1981) *Biochem. J.* *194*, 283-291.

Kadle, R., Kalter, V.G., Raizada, M.K. and Fellows, R.E. (1983) *J. Biol. Chem.* *258*, 13116-13119.

Kiechle, F.L., Strauss III, J.F., Tanaka, T. and Jarett, L. (1982) *Fed. Proc.* *41*, 1082.

Kojima, I., Lippes, H., Kojima, K. and Rasmussen, H. (1983) *Biochem. Biophys. Res. Commun.* *116*, 555-562.

Kono, T., Suzuki, K., Damsey, L.E., Robinson, F.W. and Blevins, T.L. (1981) *J. Biol. Chem.* *256*, 6400-6407.

Laychock, S.G. (1983) *Biochem. J.* *216*, 101-106.

Laychock, S.G., Shen, J.C., Carmines, E.L. and Rubin, R.P. (1978) *Biochim. Biophys. Acta* *528*, 355-363.

Leier, D.J. and Jungmann, R.A. (1973) *Biochim. Biophys. Acta* *329*, 196-210.

LePeuch, C.J., Ballester, R. and Rosen, O.M. (1983) *Proc. Natl. Acad. Sci. USA* *80*, 6858-6862.

Lowitt, S., Farese, R.V., Sabir, M.A. and Root, A.W. (1982) *Endocrinology* *111*, 1415-1417.

Macaulay, S.L., Kiechle, F.L. and Jarett, L. (1982) *Fed Proc.* *41*, 1082.

McDonald, J.M., Bruns, D.E. and Jarett, L. (1976a) *Biochem. Res. Commun.* *71*, 114-121.

McDonald, J.M., Bruns, D.E. and Jarett, L. (1976b) *Proc. Natl. Acad. Sci. USA* *73*, 1542-1546.

Meltzer, E., Weinreb, S., Bellorin-Font, E., and Hruska, K.A. (1982) *Biochim. Biophys. Acta* *712*, 258-267.

Michell, R.H. (1975) *Biochim. Biophys. Acta* *415*, 81-147.

Nimmo, H.G. and Houston, B. (1978) *Biochem. J.* *176*, 607-610.

Nishizuka, Y. (1983) *Trends Biochem. Sci.* *8*, 13-16.

Nishizuka, Y., and Takai, Y. (1981) in: *"Protein Phosphorylation"* (O. Rosen and E.G. Krebs, eds.) pp. 237-249, Cold Spring Harbor Laboratory.

Pedersen, R.C. and Brownie, A.C. (1983) *Proc. Natl. Acad. Sci. USA* *80*, 1881-1886.

Putney, J.W. Jr., Weiss, S.J., Van DeWalle, C.M. and Haddas, R.A. (1980) *Nature* *284*, 345-347.

Schmidt, S. (1983) *J. Biol. Chem.* *258*, 6863-6868.

Sheetz, M.P., Febbroriello, P. and Koppel, D.E. (1982) *Nature* *296*, 91-93.

Sica, V. and Cuatrecasas, P. (1973) *Biochemistry* *12*, 2282-2290.

Sooranna, S.R. and Saggerson, E.D. (1976) *FEBS Letters* *64*, 36-39.

Stein, J.M. and Hales, C.N. (1974) *Biochim. Biophys. Acta* *337*, 41-49.

Streb, H., Irvine, R.F., Berridge, M.J. and Schulz, I. (1983) *Nature* *306*, 67-69.

Suzuki, K. and Kono, T. (1980) *Proc. Natl. Acad. Sci. USA* *77*, 2542-2545.

Tanaka, T. and Strauss III, J.F. (1982) *Endocrinology* *110*, 1592-1598.

HYDROLYSIS OF PHOSPHATIDYLINOSITOL 4-PHOSPHATE AND PHOSPHATIDYLINOSITOL(4,5)-BISPHOSPHATE BY SPECIFIC PHOSPHOMONOESTERASES IN HUMAN ERYTHROCYTES

F.B.St.C. Palmer and S.E. Mack

Department of Biochemistry, Dalhousie University
Halifax, Nova Scotia, Canada B3H 4H7

SUMMARY

Evidence has been obtained for the sequential dephosphorylation of PtdIns(4,5)P_2 in human erythrocytes by two specific phosphatases. A Mg^{2+}-dependent phosphatase, partially-purified from cytosol, specifically removes the 5-phosphate from PtdIns(4,5)P_2 and lyso-PtdIns(4,5)P_2. A cation-independent phosphatase which only hydrolyzes PtdIns4P and lysoPtdIns4P is localized in the membrane. Maximal hydrolysis of exogenous substrates by both phosphatases requires cationic and non-ionic detergents. Under these conditions, Ca^{2+} inhibits the PtdIns(4,5)P_2 phosphatase but not the PtdIns4P phosphatase by competing with Mg^{2+}. Treating the membranes with ATP (or ADP) and a divalent cation (Ba^{2+}, Mg^{2+}, Ca^{2+}, Mn^{2+}) irreversibly inactivates the PtdIns4P phosphatase. Endogenous PtdIns4P and PtdIns(4,5)P_2 are degraded by these enzymes only after membranes are disrupted with Triton X-100. A similar differential location of the two activities was observed in other mammalian erythrocytes. Homogenates of various rat tissues also exhibit cation-independent PtdIns4P phosphatase activity which, except for brain, exceeds the Mg^{2+}-dependent PtdIns(4,5)P_2 phosphatase activity.

199

INTRODUCTION

The monoesterified phosphate groups of phosphatidylinositol 4-phosphate (PtdIns4P) and phosphatidylinositol (4,5)-bisphosphate (PtdIns(4,5)P_2) turn over rapidly via an ATP-dependent cycle (Abdel-Latif, 1983). Phosphatidylinositol (PtdIns) is phosphorylated first in the 4, then the 5 position by specific PtdIns and PtdIns4P kinases. PtdIns(4,5)P_2 is sequentially dephosphorylated in the reverse order. Crude and partially purified preparations from several tissues convert PtdIns(4,5)P_2 into PtdIns. PtdIns(4,5)P_2 and PtdIns4P appear to compete for the same phosphatase in kidney extracts (Cooper and Hawthorne, 1975) and a highly purified polyphosphoinositide phosphatase from rat brain cytosol acts equally on PtdIns4P and PtdIns(4,5)P_2 (Nijjar and Hawthorne, 1977). Few other studies have specifically assessed both enzymatic activities but similarities in the characteristics and subcellular distribution of the two phosphatase activities in several tissues suggested that a single enzyme catalyzes removal of both phosphates. However, the possibility of two distinct phosphatases must now be reconsidered since highly specific PtdIns(4,5)P_2 phosphatases, which do not catalyze the hydrolysis of PtdInsP *in vitro*, have been partially purified from human erythrocytes and protozoa (Palmer, 1981; Roach and Palmer, 1981).

RESULTS AND DISCUSSION

A number of phosphatases occurring in human erythrocyte cytosol can be separated from hemoglobin by ammonium sulphate precipitation at 0° (Roach and Palmer, 1981). A fraction obtained by DEAE-Sepharose chromatography had Mg^{2+}-dependent activity with PtdIns(4,5)P_2 and lysoPtdIns(4,5)P_2 but was inactive with PtdIns4P and structurally related compounds (Fig. 1). Inositol 1,4,5-trisphosphate was also hydrolyzed. This activity in erythrocyte cytosol has been attributed to the PtdIns(4,5)P_2 phosphatase (Downes *et al.*, 1982) or possibly to the active cytosolic 2,3-bisphosphoglycerate phosphatase. Since neither GroPIns(4,5)P_2 nor 2,3-bisphosphoglycerate were hydrolyzed, there seems to be a separate phosphatase. Inositol 1,4,5-trisphosphate phosphatase has also been described in human erythrocyte

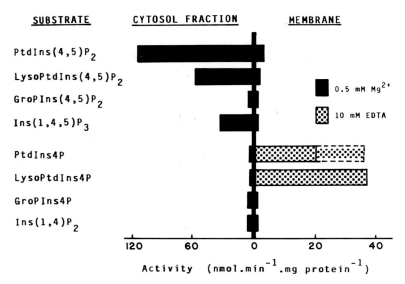

SUBSTRATE　　　　CYTOSOL FRACTION　　　　MEMBRANE

PtdIns(4,5)P$_2$

LysoPtdIns(4,5)P$_2$

GroPIns(4,5)P$_2$

Ins(1,4,5)P$_3$

PtdIns4P

LysoPtdIns4P

GroPIns4P

Ins(1,4)P$_2$

0.5 mM Mg^{2+}

10 mM EDTA

120　　60　　0　　20　　40

Activity (nmol.min^{-1}.mg protein^{-1})

Figure 1. Specificity of human erythrocyte polyphosphoinositide phos-phatases. Assays were with 1 mM substrate and 0.2% (w/v) Triton X-100 at several CTAB/substrate ratios and with either 0.5 mM Mg^{2+} or 10 mM EDTA. Maximal activity was obtained at CTAB/substrate ratios of 2.5 for cytosol PtdIns(4,5)P$_2$ phosphatase and 1.5 for membrane PtdIns4P phos-phatase. Stippled bar shows that portion of the activity observed with Mg^{2+} that was obtained in the presence of EDTA. Extension of stippled bar for PtdIns4P represents the increased activity with pre-solubilized membranes (see text).

membranes but, in our experiments, the membranes were virtually inactive. The membranes catalyzed specifically the dephosphorylation of PtdIns4P and lysoPtdIns4P and this activity was not inhibited by EDTA. The very low activity with PtdIns(4,5)P$_2$ was completely inhibited by EDTA and therefore is due to a different phosphatase. PtdIns4P hydrolysis was more dependent upon how the membranes were disrupted with nonionic detergent than was the hydrolysis of lysoPtdIns4P (Mack and Palmer, 1984). Activity comparable to that with lysoPtdIns4P was achieved only when the membranes were solubilized with Triton X-100 before adding them to the substrate/cetyltri-methylammonium bromide (CTAB)/Triton mixture in the assay. The specificities of the cytosolic and membrane-bound enzymes were tested further with a variety of phosphate esters. Neither preparation was active with

201

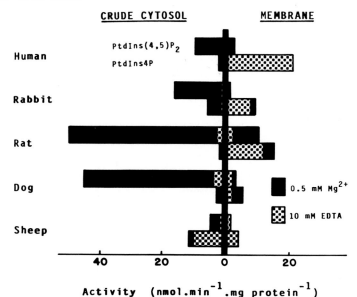

CRUDE CYTOSOL   **MEMBRANE**

Human   PtdIns(4,5)P$_2$
        PtdIns4P

Rabbit

Rat

Dog   0.5 mM Mg^{2+}

                                 10 mM EDTA

Sheep

40   20   0   20

Activity (nmol.min^{-1}.mg protein^{-1})

Figure 2. Distribution of polyphosphoinositide phosphatases in some mammalian erythrocytes. Optimal assay conditions for PtdIns(4,5)P$_2$ and PtdIns4P phosphatases were as determined with human erythrocyte preparations (see legend to Fig. 1).

phosphatidic acid, lysophosphatidic acid or a selection of nucleoside phosphates, sugar phosphates and artificial substrates for non-specific phosphatases (Roach and Palmer, 1981; Mack and Palmer, 1984).

The clearest evidence for the differential location of a Mg^{2+}-dependent PtdIns(4,5)P$_2$ phosphatase (cytosol) and a cation-independent PtdIns4P phosphatase (membrane) was obtained with human erythrocytes. In comparative studies with erythrocytes of other species (Fig. 2) only the crude cytosol (20–50% saturated $(NH_4)_2SO_4$ precipitable protein) was used since the other phosphatases in these preparations have negligible activity with polyphosphoinositides. This accounts for the lower specific activity for human erythrocyte cytosolic PtdIns(4,5)P$_2$ phosphatase than shown in Fig. 1. A comparable distribution of activities was observed in erythrocytes of rabbit and rat, the latter having very high cytosolic PtdIns(4,5)P$_2$ phosphatase activity. Dog

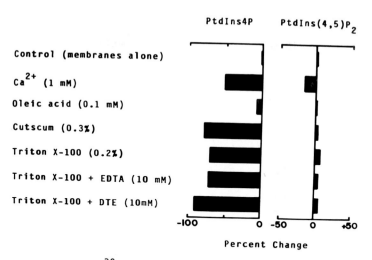

Figure 3. Loss of $[^{32}P]$polyphosphoinositides from human erythrocyte membranes. Polyphosphoinositides were radiolabeled by incubating washed erythrocytes with $^{32}P_i$ (Downes and Michell, 1981). Membranes were isolated following hemolysis in 10 mM Tris/1 mM EDTA (pH 7.2) and incubated at 37°C in 50 mM Pipes (pH 7.2). After solubilization in 1% SDS/10 mM EDTA, the phosphoinositides were separated from the total reaction mixture by thin-layer chromatography.

erythrocytes also had high cytosolic $PtdIns(4,5)P_2$ phosphatase activity but lacked appreciable PtdIns4P phosphatase activity. Sheep erythrocytes differed most from human, with very little $PtdIns(4,5)P_2$ phosphatase activity and with most of the EDTA insensitive PtdIns4P phosphatase activity in the cytosol. Nevertheless, the data support the idea of distinct $PtdIns(4,5)P_2$ and PtdIns4P phosphatases in mammalian erythrocytes.

The relevance of these *in vitro* results to *in vivo* function is as yet unclear. Conditions permitting either phosphatase to degrade its substrate in membranes without first altering membrane structure have not been found. Some characteristics of the hydrolysis of polyphosphoinositides in EDTA-washed membranes from erythrocytes that were incubated previously with $^{32}P_i$ are shown in Fig. 3. As expected, Ca^{2+} stimulated polyphosphoinositide phosphodiesterase activity resulting in loss of radioactive PtdIns4P and PtdIns(4,5)P$_2$. Non-ionic detergents having a hydrophobic aryl moiety selectively facilitate the EDTA-insensitive dephosphorylation of PtdIns4P (radioactive product identified as inorganic phosphate by ion-exchange chromatography). This occurs at concentrations of detergents which substantially disrupt the membranes but are insufficient for complete solubilization. A variety of other ions, detergents and lipids (including fatty acids and lysophospholipids) were ineffective. Radioactivity was also lost from PtdIns(4,5)P$_2$ when cytosolic proteins (source of PtdIns(4,5)P$_2$ phosphatase) and Mg^{2+} were added to the reaction mixture (data not shown).

Optimal stimulation of the PtdIns(4,5)P$_2$ phosphatase *in vitro* requires CTAB, Triton X-100 and Mg^{2+}. Under these conditions, the Mg^{2+}-requirement was that of the enzyme (Km = 12 µM) and was independent of cation binding by the substrate (Roach and Palmer, 1981). Ca^{2+} also did not interact with PtdIns(4,5)P$_2$ but inhibited the phosphatase by competing with Mg^{2+} for the enzyme (K$_i$ = 50 µM). The greater affinity of the enzyme for Mg^{2+} and the low intracellular concentration of Ca^{2+} make the *in vivo* regulation of the cytosolic PtdIns(4,5)P$_2$ phosphatase by these cations unlikely. The membrane PtdIns4P phosphatase was not affected by Ca^{2+}. However, it was inactivated by brief incubation of membranes with a divalent cation and ATP, GTP or ADP. Deoxynucleotides, phosphate esters and monovalent cations were ineffective. The mechanism of this effect is not yet clear. Phosphorylation of the enzyme seems unlikely since inactivation was not prevented by bovine heart protein kinase inhibitor or reversed by exposure to alkaline phosphatase.

Finally, to determine if similar polyphosphoinositide phosphatases might be more widely distributed, dialysed crude homogenates of rat tissues were surveyed for Mg^{2+}-dependent and cation-independent phosphatase activity with both PtdIns4P and PtdIns(4,5)P$_2$ as substrate (Fig. 4). Activities with both substrates were low in heart and skeletal muscle. All other tissues exhibited Mg^{2+}-dependent activity with PtdIns(4,5)P$_2$, this activity being very high in brain. There was also a low residual cation-independent activity. On the other

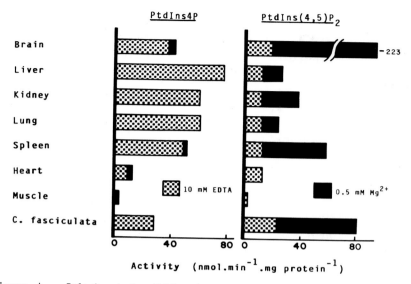

Figure 4. Polyphosphoinositide phosphatases in tissue homogenates. Assays were as described for Fig. 1 except that the CTAB/substrate ratios were as optimized for brain (0.6 and 1.2 for PtdIns4P and PtdIns(4,5)P$_2$ phosphatase respectively).

hand, there was negligible Mg^{2+}-dependent activity with PtdIns4P in any tissue; but the cation-independent PtdIns(4,5)P$_2$ phosphatase activity exceeded both the EDTA-insensitive activity with PtdIns(4,5)P$_2$ and, except for brain and spleen, the Mg^{2+}-dependent PtdIns(4,5)P$_2$ phosphatase activity. Similar data were obtained with homogenates of the protozoan *Crithidia fasciculata*. These results suggest the possibility that cation-independent PtdIns4P phosphatases contribute to the dephosphorylation of PtdIns(4,5)P$_2$ in other tissues and organisms.

ACKNOWLEDGEMENTS

This work was supported by the Medical Research Council of Canada.

Palmer and Mack

REFERENCES

Abdel-Latif, A.A. (1983) in: *"Handbook of Neurochemistry"*, Vol. 3, 2nd ed. (A. Lajtha, ed.) pp. 91–131, Plenum Press, New York.
Cooper, P.H. and Hawthorne, J.N. (1975) *Biochem. J.* *150*, 537–551.
Downes, C.P and Michell, R.H. (1981) *Biochem. J.* *198*, 133–140.
Downes, C.P., Mussat, M.C. and Michell, R.H. (1982) *Biochem. J.* *203*, 169–177.
Mack, S.E. and Palmer, F.B.St.C. (1984) *J. Lipid Res.* *25*, 75–85.
Nijjar, M.S. and Hawthorne, J.N. (1977) *Biochim. Biophys. Acta* *480*, 390–402.
Palmer, F.B.St.C. (1981) *Can. J. Biochem.* *59*, 469–476.
Roach, P.D. and Palmer, F.B.St.C. (1981) *Biochim. Biophys. Acta* *661*, 323–333.

DRUG-INDUCED MODIFICATIONS OF

PHOSPHOINOSITIDE METABOLISM

P. Marche, A. Girard and S. Koutouzov

INSERM U7, Department of Pharmacology
Hôpital Necker, 75015 Paris, France

SUMMARY

The isolated red cell ghost was used as a biomembrane model to investigate the influence of aminoglycosides upon phosphoinositide metabolism by measuring ^{32}P incorporation into the polyphosphoinositides (PIP and PIP_2). A rapid and dose-dependent decrease in $[^{32}P]PIP_2$ and an increase in $[^{32}P]PIP$ were observed. Studies with neomycin revealed that these changes could likely be ascribed to an inhibition of the PIP-kinase activity by the drug. The order of potency of various aminoglycosides for the impairment of phosphoinositide metabolism varied as a function of the cationicity of the drugs. A possible relationship between the aminoglycoside-induced nephrotoxicity and the drug-induced alteration of phosphoinositide turnover is discussed.

INTRODUCTION

In the proximal renal tubule where aminoglycosides accumulate and can exert their nephrotoxic potential, several observations suggest that phosphoinositides play an important role at the drug-binding site. Neomycin has been reported to alter phosphoinositide metabolism *in vivo* and *in vitro* and strong interactions between phosphoinositides and aminoglycosides have been described (Schibeci and Schacht, 1977; Schacht, 1978; Palmer, 1981; Sastrasinh *et al.*, 1982). In an attempt to elucidate the biochemical events that may account for the aminoglycoside-induced alterations of phosphoinositide metabolism, we investigated the influence *in vitro* of neomycin and other aminoglycosides upon the turnover of polyphosphoinositides. Membranes isolated from rat erythrocytes were chosen as the experimental model because conditions which allow the selective study of each of the enzymes involved in the ^{32}P-labeling of PIP and PIP_2 are well described in this system. Under these conditions, PIP and PIP_2 are the only phospholipids that incorporated ^{32}P (Downes and Michell, 1981; Koutouzov and Marche, 1982; Marche *et al.*, 1982). Results are discussed in terms of the mechanism of action of aminoglycosides within membranes.

METHODS

Fifteen to eighteen week-old male rats of the Wistar strain were supplied by Iffa-Credo (France) and [γ-^{32}P]-ATP was from Amersham (UK). Amikacin, dibekacin and netilmicin were generously provided by Bristol, Bristol/Bellon and Unilabo, respectively. Gentamicin and tobramycin were from Sigma. Blood sampling and isolation of ghost membranes were performed according to Koutouzov *et al.* (1982). For ^{32}P-incorporation into membrane phospholipids, 0.3 mg membrane protein were preincubated for 10 min at 37°C in 50 mM Tris-HCl buffer, pH 7.5, in the presence (or absence for controls) of the antibiotic. After the preincubation, $MgCl_2$ (5 mM) and [^{32}P]ATP Na_2 (2 mM) were added. The total assay volume was 0.5 ml. The reaction was stopped after 15 min and samples were processed for phospholipid extraction, separation and assay of ^{32}P as described previously (Marche *et al.*, 1982). Studies of the neomycin effect on Mg^{2+} (or Ca^{2+})-stimulated polyphosphoinositide breakdown were carried out on ^{32}P-prelabeled membranes as already described in Marche *et al.* (1983). Results are expressed as mean values ± S.E. of at least 4 separate experiments.

RESULTS AND DISCUSSION

The effects of neomycin on the ^{32}P-incorporation into polyphosphoinositides were first studied as a function of the drug concentration. The results presented in Figure 1A indicate that in the isolated erythrocyte membrane, neomycin considerably affects the incorporation of ^{32}P into PIP_2 and PIP, which nevertheless remain the only

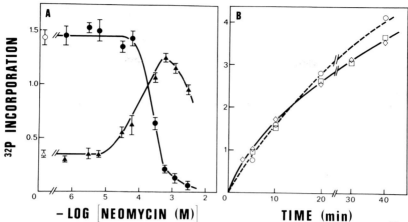

Figure 1. A. Effect of neomycin at various concentrations on the ^{32}P-labeling of PIP$_2$ (●) and PIP (▲). B. Time-course of the ^{32}P-labeling of: (O) [^{32}P]PIP$_2$ (no drug, 5 mM MgCl$_2$); (□) [^{32}P]PIP (0.3 mM neomycin, 5 mM MgCl$_2$); (◊) [^{32}P]PIP (no drug, 40 mM MgCl$_2$).

radioactive lipids. Neomycin (10^{-5} - 10^{-3} M) induced an increase in [^{32}P]PIP and a decrease in [^{32}P]PIP$_2$; it is noteworthy that the total radioactivity incorporated into both PIP$_2$ and PIP in neomycin-treated samples did not vary significantly from that incorporated into controls. At concentrations greater than 1 mM, neomycin decreased ^{32}P-labeling of both PIP and PIP$_2$, likely by decreasing PI-kinase activity. The time course of the effects of neomycin (0.3 mM) is represented in Figure 1B which clearly shows that during the first 20 min the curve depicting the ^{32}P-labeling of PIP in the presence of the drug could not be distinguished from that depicting the ^{32}P-labeling of PIP$_2$ in the absence of the drug. These curves are superimposable with that representing the ^{32}P-labeling of PIP in the absence of neomycin, when the incubation medium contained MgCl$_2$ (40 mM).

In another set of experiments, the effects of neomycin on the Mg^{2+}-activated PIP$_2$-specific phosphomonoesterase activity and on the Ca^{2+}-activated polyphosphoinositide phosphodiesterase activity, were measured. The results presented in Table I, demonstrate that neomycin has no influence on these enzyme activities. With respect to the ^{32}P-labeling of polyphosphoinositides, our findings

TABLE I

Effect of Neomycin on the Mg^{2+} (or Ca^{2+})-Stimulated
Breakdown of Membrane-Bound $[^{32}P]$Polyphosphoinositides

Addition	$[^{32}P]PIP_2$	$[^{32}P]PIP$
	(^{32}P nmol $\times$ 15 $min^{-1} \times mg^{-1}$ protein)	
None	1.32 ± 0.06	0.36 ± 0.04
Neomycin (0.3 mM)	1.33 ± 0.06	0.36 ± 0.05
$MgCl_2$ (5 mM)	0.84 ± 0.04	0.60 ± 0.01
$MgCl_2$ + neomycin	0.79 ± 0.05	0.63 ± 0.01
$CaCl_2$ (1 mM)	1.09 ± 0.04	0.21 ± 0.01
$CaCl_2$ + neomycin	1.08 ± 0.04	0.21 ± 0.02

that 0.3 mM neomycin (i) could mimic the effect of $MgCl_2$ (40 mM) and (ii) did not affect the $PIP_2 \rightarrow PIP$ conversion, suggest therefore that the drug could act by inhibiting the PIP-kinase activity. This hypothesis is supported by the fact that when membranes were phosphorylated in the presence of $MgCl_2$ (40 mM), neomycin induced a decrease in $[^{32}P]PIP_2$ without modifying $[^{32}P]PIP$ (Marche et al., 1983). This also indicates that neomycin (up to 1 mM) did not affect the PI-kinase activity. Inhibition of the PIP-kinase activity by neomycin is likely a consequence of drug-phosphoinositide interactions (Schacht, 1978; Palmer, 1981) which lower the substrate availability. A direct drug-enzyme interaction cannot, however, be ruled out.

In Table II, are represented the effects of various aminoglycosides on the ^{32}P-labeling of PIP_2 and PIP. Each of the drugs studied behaved qualitatively as neomycin, i.e., they induced a decrease in $[^{32}P]PIP_2$ and an increase in $[^{32}P]PIP$. Quantitatively, however, a statistical analysis revealed that, according to their ability to impair the metabolism of $[^{32}P]PIP$ and $[^{32}P]PIP_2$, the aminoglycosides tested can be ranked: neomycin > tobramycin > gentamicin $\sim$ dibekacin $\sim$ netilmicin > amikacin. Streptomycin has been reported to be uneffective (Marche

TABLE II

Effect of Aminoglycosides on the Incorporation of $^{32}P_i$ into PIP_2 and PIP

Addition	$[^{32}P]PIP_2$	$[^{32}P]PIP$
	(percent of control)	
amikacin	97 ± 10	114 ± 5
dibekacin	67 ± 13	193 ± 18
gentamicin	86 ± 11	161 ± 6
neomycin	39 ± 7	320 ± 32
netilmicin	75 ± 13	152 ± 10
tobramycin	71 ± 13	220 ± 22

Aminoglycosides were at a concentration of 0.6 mM.

et al., 1983). This rank order correlates roughly with the order of aminoglycosides when ranked according to their cationic charge. Nevertheless, tobramycin was significantly more potent than dibekacin, gentamicin and netilmicin even though all these compounds exhibit similar cationic characteristics. The nephrotoxic potential of aminoglycosides appears also to be correlated with their cationicity (Humes *et al.*, 1982). This potential, however, likely results from both the extent to which the drug accumulates within tubules and the intrinsic toxicity of the drug on the tubular cell itself (Soberon *et al.*, 1979). In the kidney cortex of rats, therefore, the lesser accumulation of tobramycin compared to gentamicin may explain the lower degree of nephrotoxicity of the former compound (Soberon *et al.*, 1979; Kaloyanides and Pastoriza-Munoz, 1980) although its capacity to alter phosphoinositide metabolism was greater.

The polyphosphoinositides have been described to be intimately involved in the processes of transmembrane Ca^{2+} flux (Creba *et al.*, 1983; Berridge, 1983) and aminoglycosides have been shown to be able to replace Ca^{2+} from biomembranes both *in vivo* and *in vitro*

211

Marche et al.

(Williams *et al.*, 1981; Lullman and Vollmer, 1982). It is therefore conceivable that alterations in phosphoinositide metabolism and membrane permeability and/or transport capacity by aminoglycosides play critical roles in the pathogenesis of cellular injury.

REFERENCES

Berridge, M.J. (1983) *Biochem. J. 212*, 849–858.

Creba, J.A., Downes, C.P., Hawkins, P.T., Brewster, G., Michell, R.H. and Kirk, C. (1983) *Biochem. J. 212*, 733–747.

Downes, C.P. and Michell, R.H. (1981) *Biochem. J. 198*, 133–140.

Humes, H.D., Weinberg, J.M. and Krauss, T.C. (1982) *Am. J. Kid. Dis. 2*, 5–29.

Kaloyanides, G.J. and Pastoriza-Munoz, E. (1980) *Kidney Int. 18*, 571–582.

Koutouzov, S. and Marche, P. (1982) *FEBS Letters 144*, 16–20.

Koutouzov, S., Marche, P., Cloix, J.F. and Meyer, P. (1982) *Am. J. Physiol. 243*, H590–H597.

Lullman, H. and Vollmer, B. (1982) *Biochem. Pharmacol. 31*, 3769–3773.

Marche, P., Koutouzov, S. and Meyer, P. (1982) *Biochim. Biophys. Acta 710*, 332–340.

Marche, P., Koutouzov, S. and Girard, A. (1983) *J. Pharmacol. Exp. Ther. 227*, 415–420.

Palmer, F.B.St.C. (1981) *J. Lipid Res. 22*, 1296–1300.

Sastrasinh, M., Krauss, T.C., Weinberg, J.M. and Humes, H.D. (1982) *J. Pharmacol. Exp. Ther. 222*, 350–358.

Schacht, J. (1978) *J. Lipid Res. 19*, 1063–1067.

Schibeci, A. and Schacht, J. (1977) *Biochem. Pharmacol. 26*, 1769–1774.

Soberon, L., Bowman, R.L., Pastoriza-Munoz, E. and Kaloyanides, G.J. (1979) *J. Pharmacol. Exp. Ther. 210*, 334–343.

Williams, P.D., Holohan, P.D. and Ross, C.R. (1981) *Toxic. Appl. Pharmac. 61*, 234–242.

DISCUSSION

Summarized by William R. Sherman

Department of Psychiatry, Washington University
School of Medicine, 4940 Audubon Avenue,
St. Louis, MO 63110

PHOSPHATIDYLINOSITOL SYNTHASE IN MAMMALIAN PANCREAS

M. Hokin-Neaverson and G.S. Parries

Dr. J. Eichberg (Univ. Houston) commented that, with Dr. A. Ghalayini, he had recently achieved a purification of the PI synthase to about 1000 units per milligram of protein from a brain microsomal fraction. He also pointed out that the yields of PI synthase based on whole homogenates rather than on the microsomal fraction were quite low because the latter constitutes only a fraction of total organ protein. Thus in his work on brain microsomes, comparable activities to those of Dr. Hokin-Neaverson from dog pancreas were obtained and with recoveries as high as 70%.

Dr. R.B. Moore (Univ. South Alabama) asked whether differential inhibition of the PI synthase was observed *in vitro* with δ- and γ-hexachlorocyclohexane. He remarked that, in early work *in vivo*, the δ-isomer seemed to be a general inhibitor of PI turnover but the γ-isomer inhibited only the hormonally-stimulated component. Dr. Hokin-Neaverson said that while the δ-form was 90% inhibitory the γ- and β-isomer did not inhibit, with each at a concentration of 4 mM. She pointed out that there is a large difference in the solubility of these isomers which might account for the results obtained in earlier work.

213

Dr. A.A. Abdel-Latif (Med. Coll. Georgia) asked if acute effects of lithium on the appearance of inositol phosphate were observed. He commented that in iris muscle he finds an effect only after 15 min or longer which may be a matter of insufficient radioactivity at earlier time periods. Dr. Hokin-Neaverson replied that the appearance of inositol phosphate correlated exactly with the disappearance of PI over a time course of about 1 to 15 min, and that the rate of appearance was dependent upon the concentration of ACh.

THE ENZYMOLOGY OF PHOSPHOINOSITIDE CATABOLISM, WITH PARTICULAR REFERENCE TO PHOSPHATIDYL-INOSITOL 4,5-BISPHOSPHATE PHOSPHODIESTERASE

R.F. Irvine, A.J. Letcher, D.J. Lander and R.M.C. Dawson

Dr. P.W. Majerus (Washington Univ.) commented that the two soluble phospholipase C's from ram seminal vesicles had only 0.1-1% activity when PIP_2 rather than PI was employed as substrate even with high concentrations of Ca^{2+}. He also pointed out that Dr. S.E. Rittenhouse, using endogenous substrates, had similarly found limited PIP_2 breakdown. Dr. Irvine responded that, using brain homogenate as the enzyme source, with a concentration of PI ten times that of PIP_2 and in the presence of the appropriate PE mixture, the PIP_2 is hydrolyzed at ten times the rate of the PI. He also pointed out that Low and Weglicki had found that myocardial phospholipase C's catalyze the hydrolysis of both of the polyphosphoinositides. The specificities of the enzymes are thus subject to extreme variability, one aspect of which is the composition of the lipids associated with the substrate.

Dr. C.P. Downes (ICI, UK) pointed out that a variety of different tissues with subtle and specific lipid compositional differences apparently respond identically when stimulated suggesting that control of the hydrolytic events might not after all be so dependent on lipid environment in the intact cell. Dr. Irvine said that in his view the lipid compositional changes effected a kind of substrate control (rather than enzyme control or control by calcium) and he felt that he is modeling some analogous but undefined situation in the cell.

Dr. J.N. Hawthorne (Univ. Nottingham, UK) asked how there could be any PIP_2 left in these experiments with brain homogenates. Given the very high brain phosphomonoesterase activity, the kinases, being about 1000 times less active, would not be able to keep up the synthetic pace. Dr. Irvine agreed that there had to be controls on both the kinase and monoesterase activities. It was also evident that the diesterase was controlled by the receptor, and that the kinase as well seemed to be influenced by receptor activation, as shown by PIP_2 levels rising above resting levels following stimulation. Dr. Irvine said that his studies were more directed toward finding ways to achieve high enough levels of phospholipase C activity in order to mimic the cellular response toward PIP_2 and that the kinase studies should really be performed under similar conditions, in order to be comparable.

Dr. Y. Oron (Tel Aviv Univ.) pointed out that in early studies he had found that PI kinase would only phosphorylate 1% of the endogenous membrane substrate and then stop, and that soluble PIP kinase would phosphorylate only soluble synthetic substrate and not touch membrane-bound PIP. He averred that, nevertheless, the cell functions well and so it is we who are doing something wrong. Dr. R.H. Michell (Univ. Birmingham, UK) pointed out that, in liver cells, the entire PIP and PIP_2 pool is turning over at least once every 5 min in a balanced steady-state process. The cells are thus controlling these events in a way not apparent in isolated enzyme preparations, thus it is up to us to find a way to see that the *in vitro* assays fit the cellular model, not the other way round.

Dr. J.N. Fain (Brown Univ.) asked if Dr. Irvine was proposing that the hormone-receptor interaction caused the PIP_2 to be exposed to the enzyme, and, if that is the case, how could it not be exposed to both monoesterase *and* diesterase. Why not, instead, say that the hormone activates the phosphodiesterase as has been shown by other laboratories? Dr. Irvine said that he did not mean to imply that the lipid was exposed by receptor occupancy. In his opinion adding the hormone to the membrane preparation, with resulting phospholipase C action, does not discriminate between substrate-activation and enzyme-activation.

Dr. S.E. Rittenhouse (Harvard Univ.) asked if arachidonate or arachidonate-rich PA had been examined as stimulators of phospholipase C action on PIP_2. Dr. Irvine said that 30% PA did nothing and that its effect on PI phosphodiesterase activity required too high Ca^{2+} levels but that these matters needed further study.

SUBCELLULAR SITES OF POLYPHOSPHOINOSITIDE METABOLISM IN RAT LIVER

M.A. Seyfred, C.D. Smith, L.E. Farrell and W.W. Wells

Dr. J.W. Putney (Med. Coll. Virginia) remarked that in parotid gland, both IP_2 and IP_3 appeared at similar rates, but the PIP did not decrease in level whereas PIP_2 levels fell, suggesting that there was PIP in the cell that was not hormone-sensitive. He felt that this fits with Dr. Wells' finding of wide PIP distribution in the cell while PIP_2 may be localized in the plasma membrane.

Dr. C.R. Alving (Walter Reed Army Inst.) said that, using monoclonal antibodies to PIP, he had found this lipid predominantly on the outer surface of Chinese hamster ovary cells and wondered whether the enzymes of its metabolism might also be there. Dr. Wells could not comment on the enzyme localization in the plasma membrane but thought the PIP might well be distributed on both sides. Dr. R. Sundler (Univ. Lund, Sweden) commented that Golgi was very enriched in PI kinase and wondered if Dr. Wells' hepatocyte plasma membranes might contain Golgi. Dr. Wells said that his fractionation would have removed them, although he did not specifically assay Golgi marker enzymes.

The discussion then centered on the problems of attributing cellular locale to enzymes and substrates that are so sensitive to physical state. Answering a question from Dr. M.J. Dimino (East Virginia Med. School) about whether added Triton X-100 or added PI influenced the reaction, Dr. Wells said that the rates of phosphorylation could be stimulated by both but neither was obligatory.

Dr. Michell commented that it was very difficult to attribute a V_{max} to a specific cellular element in the case of these substrates and enzymes, since the enzyme activities were so dependent on little-understood variables. Dr. Wells pointed out that added PI was used in his assays, which should give V_{max} results. Dr. Dimino commented that, in ovarian tissue Triton X-100 was necessary, but that small amounts of exogenous PI did not raise the rate, while larger amounts inhibited phosphorylation. Dr. Michell concluded the discussion by reiterating his concern over the difficulty of establishing a set of conditions for analysis that would truly reflect the activity of the intact cell, which was a recurrent point of view in this session.

LOSS OF PHOSPHATIDYLINOSITOL AND GAIN IN PHOSPHATIDATE IN NEUTROPHILS STIMULATED BY F-MET-LEU-PHE

S. Cockcroft and D. Allen

Part of the discussion over this paper concerned itself with evidence that polyphosphoinositide phospholipase C activation, with subsequent Ca^{2+} level increases, and rapid phorbol ester stimulation of secretion, all occur in neutrophils (or similar cells) just as is currently being reported in hepatocytes, pancreatic acinar cells, etc. It was suggested that the stimulated phospholipase D response might thus be an independent event. Dr. Putney referred to his work with Dr. P.P. Godfrey on prelabeled DMSO-differentiated HL-60 cells which are thought to be a model for human neutrophils. This preparation exhibits, when stimulated, a brisk increase in IP_3 and the other inositol phosphates which cannot arise by any known route other than phospholipase C action. Dr. H.M. Korchak (New York Univ.) reported that her experiments with Quin-2 showed Ca^{2+} fluxes to be similar in the neutrophil to those in other cells and that protein kinase C appears to be a cytosolic component of neutrophils as well. None of this was claimed to refute Dr. Cockcroft's evidence (which she suggested may support an alternative stimulated metabolism of phosphoinositides) but rather supported the suggestion that the matter is more complex than is accounted for by phospholipase D acting alone in response to stimuli. Dr. Majerus pointed

217

out that the low specific radioactivity of the PA in Dr. Cockcroft's experiments could result from *de novo* synthesis of PA, by acylation of glycerol 3-P of low specific radioactivity, thus leading to an unexpectedly low level of ^{32}P in the PA. He remarked that, in platelets for example, the radiolabeling of glycerol 3-P is typically much lower than that of ATP.

THE *DE NOVO* PHOSPHOLIPID SYNTHESIS EFFECT: OCCURRENCE, CHARACTERISTICS, UNDERLYING MECHANISMS AND FUNCTIONAL SIGNIFICANCE IN HORMONE ACTION AND SECRETION

R.V. Farese

Dr. Michell, referring to experiments with fat pads that were described in Dr. Farese's presentation, asked how it was possible for a two-fold increase in amounts of PA and PI to occur accompanied by only a two-fold increase in radiolabeling following stimulation. That was, he felt, far too small an increase in radiolabel to be understandable. Dr. Farese suggested that the result might be due to selectively-radiolabeled pools of lipid precursors. Dr. M.J. Berridge (Univ. Cambridge, UK) inquired about the insulin effects on myocyte phospholipid metabolism, specifically the evidence that insulin produced a large increase in diacylglycerol. He asked if Dr. Farese thought that the insulin response occurred via hydrolysis of phosphoinositides, to which Dr. Farese responded that he thought that was a real possibility. Dr. Putney commented that, to him, this seemed unlikely since, for example in the hepatocyte, stimulation of the α_1-receptor produces all of the now-recognized PI-related effects, and yet the physiological effect on the cell is exactly the opposite of the effect of insulin. That signified to him that insulin action does not involve the same kind of lipid changes that accompany the actions of α_1-agonists and muscarinic agonists.

PART III

RECEPTOR–MEDIATED ALTERATIONS IN PHOSPHOINOSITIDE METABOLISM

INOSITOL LIPID BREAKDOWN IN RECEPTOR-MEDIATED RESPONSES OF SYMPATHETIC GANGLIA AND SYMPATHETICALLY INNERVATED TISSUES

R.H. Michell*, E.A. Bone*, P. Fretten*, S. Palmer* and C.J. Kirk*, Michael R. Hanley**, Hilary Benton*** Stafford L. Lightman**** and Kathryn Todd****

*Department of Biochemistry, University of Birmingham, Birmingham B15 2TT; **Department of Biochemistry, Imperial College of Science and Technology, South Kensington, London SW7 2AZ; ***Cancer Research Campaign Cell Proliferation Unit, Royal Postgraduate Medical School, Ducane Road, London, W12 0HS; ****Medical Unit, Charing Cross and Westminster Medical School, Page Street, London, SW1P 2AP, UK

SUMMARY

Some peripheral tissues, including hepatocytes and aortic smooth muscle, possess vasopressin-responsive receptors of the V_1-type. Activation of these receptors causes phosphodiesterase-catalyzed breakdown of phosphatidylinositol 4,5-bisphosphate (PtdIns4,5P$_2$) and leads to a rise in cytosol [Ca^{2+}]. It has previously been assumed that these V_1-receptors are mainly responsive to changes in the concentration of circulating arginine-vasopressin (AVP). However, we have now identified a vasopressin-like neuropeptide (VLP) in the major noradrenergic neurones of sympathetic ganglia and in nerve fibers within sympathetically innervated tissues. It seems possible that VLP will prove to be a co-transmitter that is released with noradrenaline at sympathetic nerve terminals and varicosities, in which case VLP, rather than AVP, may be the predominant activator of V_1-receptors in vivo. In addition, AVP produces a striking stimulation of PtdIns4,5P$_2$ breakdown in isolated superior cervical sympathetic ganglia: this response is mediated by V_1-receptors and it persists when ganglia are stimulated in a medium of low Ca^{2+} concentration or in the presence of indomethacin. A smaller stimulation of inositol lipid breakdown occurs in ganglia depolarized by high $[K^+]$ (in the presence of 10 µM atropine): this response requires extracellular Ca^{2+} and is insensitive to V_1-receptor blockade.

INTRODUCTION

Arginine-vasopressin (AVP) is a neuropeptide that is synthesized in specialized neurons in the hypothalamus, from which nerve fibers run both to other brain regions and to the neurohypophysis (Sofroniew, 1983). Neurohypophysial AVP is released into the bloodstream, where it acts as a circulating hormone that inhibits urine output from the kidney: the V_2-receptors that mediate this effect do so by stimulating renal tubular adenylate cyclase (Jard, 1980). Elevation of the plasma AVP concentration may also increase blood pressure (by contracting arterial smooth muscle) and elevate the blood glucose concentration (by causing activation of hepatic glycogen phosphorylase). These latter effects are mediated by stimulation of V_1-receptors, which activate a phosphodiesterase-catalyzed hydrolysis of PtdIns4,5P$_2$ (Michell *et al.*, 1981, 1984; Kirk *et al.*, 1981; Kirk, 1982; Creba *et al.*, 1983; Thomas *et al.*, 1983) and bring about an increase in the Ca^{2+} concentration in the cytosol of stimulated cells (Kirk, 1982). It has long been clear that circulating AVP concentrations are adequate to control urine output, but under most physiological circumstances there is insufficient AVP in the blood to produce substantial effects on either blood pressure or hepatic glycogen metabolism. Thus the physiological function of V_2-receptors is understood, but there has been doubt as to whether V_1-receptors are of physiological importance except in times of stress (*e.g.* after severe hemorrhage).

Until now, work on the peripheral role(s) of receptors responsive to vasopressin has been based on the assumption that the only vasopressin or vasopressin-like material outside the CNS, other than oxytocin, is the AVP present in the bloodstream. However, our recent experiments establish that the large noradrenergic neurones of

222

sympathetic ganglia contain a vasopressin-like peptide (VLP), thus raising the possibility that the V_1- or V_2- receptors of peripheral tissues may sometimes be stimulated *in vivo* by VLP released as a co-transmitter with noradrenaline at sympathetic nerve terminals and varicosities (Hanley *et al.*, 1984). This work will be briefly summarized in the first part of this paper.

The experiments in which we discovered VLP were undertaken after we had made the unexpected observation that incubation of isolated superior cervical sympathetic ganglia with added AVP causes a striking stimulation of inositol lipid metabolism. Like the equivalent responses of hepatocytes and vascular smooth muscle, this response is mediated by V_1-receptors. These ganglia also exhibit a stimulation of inositol lipid metabolism when depolarized by incubation with high extracellular $[K^+]$. The second part of this paper describes these responses, whose function is still unknown.

A VASOPRESSIN-LIKE PEPTIDE IN THE SYMPATHETIC NERVOUS SYSTEM

Superior cervical and coeliac ganglia were extracted with acid and the extracts examined for vasopressin-like and oxytocin-like immunoreactivity by RIA using antibodies showing a high degree of discrimination between the two peptides (Lightman and Forsling, 1980). A material was found that was immunoreactive with both types of antisera (Table I). Gel filtration of the extracts revealed that this substance was substantially larger than either oxytocin or AVP (Hanley *et al.*, 1984). This material, which we refer to as vasopressin-like peptide (VLP), was present in the ganglia of both normal rats and Brattleboro rats, despite the fact that the Brattleboro animals lack hypothalamic and neurohypophysial AVP.

Staining sections of sympathetic ganglia with antibodies to either vasopressin or oxytocin revealed immunoreactivity in at least two structures. The most striking staining was throughout the cytoplasm of all large neurones within the ganglia, with the nuclei remaining unstained (Fig. 1a). There was also clear staining of an intrinsic plexus of nerve fibers in the spaces between the neurones (Hanley *et al.*, 1984). Since most of the

TABLE I

Vasopressin-like and Oxytocin-like Immunoreactivity
in Extracts of Rat Superior Cervical Sympathetic Ganglia

	Apparent peptide concentration	
	Normal rats	Brattleboro rats
	(fmol per ganglion)	
Vasopressin-like	211 ± 26 (12)	85 ± 12 (12)
Oxytocin-like	23 ± 0.8 (14)	25 ± 1.6 (13)

Extracts of ganglia were prepared, and their contents of immunoreactive peptides assayed, as described elsewhere (Hanley et al., 1984).

neurones visible in the sections appeared to be stained, it seemed likely that these were the large noradrenergic neurones of the ganglia. This was confirmed by studies in which the same sections of rat superior cervical sympathetic ganglia were first stained with an anti-vasopressin antibody, then destained and stained a second time with an antibody to dopamine β-hydroxylase: there was a close, and maybe complete, correspondence in the patterns of cells stained by the two antibodies (Hanley et al., 1984).

These studies raised the possibility that VLP, a new peptide with an immunological similarity to both vasopressin and oxytocin, might be synthesized in the major noradrenergic neurones of the sympathetic nervous system, in which case it also seemed possible that this material would be transported to the periphery and there released as a co-transmitter with noradrenaline. Staining of sympathetically innervated peripheral tissues revealed a vasopressin-positive nerve fiber pattern consistent with this view: Fig. 1b shows the vasopressin-positive innervation of an artery in the kidney. Although these studies point strongly to the possibility that VLP might be a neurally released mediator of some of the effects of

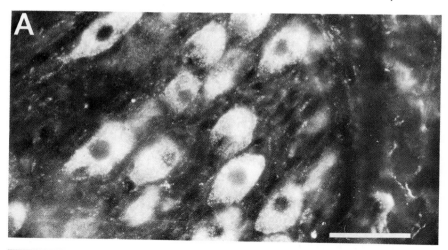

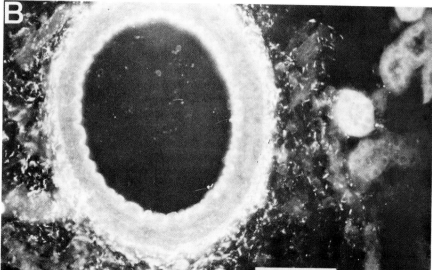

<u>Figure 1</u>. Immunofluorescence micrographs of vasopressin-like immuno-reactivity in sympathetic ganglia and peripheral nerve fibers. Scale bars indicate 100 μm. A. Section of a coeliac ganglion from a macaque monkey (<u>Macaca cynomologous</u>) stained with an antibody to arg-vaso-pressin. Note the stained neuronal cytoplasm and unstained nuclei. B. Cross-section of the renal artery of a mouse stained with the same anti-vasopressin antibody. Note the dense innervation of the outer muscle layers.

Michell et al.

sympathetic innervation, we do not yet have direct evidence either for its release at peripheral sites or for actions at these sites.

VASOPRESSIN-STIMULATED INOSITOL LIPID BREAKDOWN IN SUPERIOR CERVICAL SYMPATHETIC GANGLIA

Activation of a variety of cell-surface receptors has long been known to stimulate inositol lipid metabolism in diverse tissues, and recent work has identified phosphodiesterase-catalyzed hydrolysis of PtdIns4,5P$_2$ as the initiating step in this process (see Michell *et al.*, 1984, for a review). The most likely functions of this reaction are mobilization of Ca^{2+} in the cytosol of stimulated cells (Michell, 1975, 1982a; Downes and Michell, 1982; Streb *et al.*, 1983) and activation of protein kinase C (Nishizuka, 1983): these processes are discussed by other authors elsewhere in this volume.

Amongst the tissues in which receptors cause stimulation of inositol lipid metabolism are many regions of the nervous system, including sympathetic ganglia exposed to muscarinic cholinergic stimuli (Hokin *et al.*, 1960). The functions of this type of biochemical response to receptor activation in nervous tissue remain obscure (Hawthorne and Pickard, 1979), but it seems reasonable to guess that they may be related to generation of the same intracellular Ca^{2+} and protein phosphorylation signals as in better understood peripheral tissues (Michell, 1982b, 1982c; Downes, 1983), with these relatively slow signals having some role in modulating the responsiveness of the stimulated neurones to other inputs.

Previous studies of stimulated inositol lipid metabolism in nervous tissues have almost always employed measurements of the labeling of PtdIns, a poor assay method, for assessing responses to stimuli. A much better method is to measure the accumulation of inositol phosphates in stimulated cells, preferably in the presence of Li$^+$ (an inhibitor of inositol 1-phosphatase; Berridge *et al.*, 1982, 1983). Using this method, we confirmed that cholinergic stimulation does indeed stimulate the breakdown of inositol lipids in isolated superior cervical ganglia from rats (Table II). This response was abolished by atropine, as expected for a response mediated by muscarinic receptors.

226

TABLE II

Accumulation of Labeled Inositol Phosphates in
Rat Superior Cervical Ganglia Exposed to
Vasopressin and to Muscarinic Cholinergic Stimulation

| | Labeled inositol phosphates | | | |
	GroPIns	InsP	$InsP_2$	$InsP_3$
	(dpm per 10^3 dpm in lipids)			
Control	31.5 ± 2.5	215 ± 28	138 ± 29	9.9 ± 1.0
100 µM carba-myl-β-methyl-choline	36.1 ± 2.9	406 ± 37	188 ± 23	13.5 ± 1.1
10 µM atropine	35.3 ± 5.7	258 ± 34	123 ± 32	12.2 ± 2.4
100 µM carba-myl-β-methyl-choline and 10 µM atropine	30.2 ± 2.1	115 ± 17	74 ± 15	7.4 ± 1.1
0.32 µM vaso-pressin	24.7 ± 2.3	1303 ± 129	472 ± 52	68.3 ± 7.8

Ganglionic lipids were prelabeled with [^{3}H]inositol and the ganglia were stimulated for 60 min in the presence of 10 mM LiCl. Experimental details as in Hanley et al. (1984). Values are mean ± SEM (n for each experimental group is at least 6 ganglia).

Having established the applicability of this assay method to isolated ganglia, we then tested several stimuli known to be present in ganglia or to affect ganglionic function. Noradrenaline, LHRH and angiotensin brought about no accumulation of inositol phosphates in the ganglia, despite the fact that they are all potent stimuli for inositol lipid breakdown in appropriate target tissues (Michell, 1975, 1982a; Michell et al., 1981, 1984; Creba et al., 1983). Vasopressin has no known direct effects upon sympathetic ganglia, but it potently stimulates inositol lipid breakdown both in peripheral tissues (*e.g.* liver) and in nervous tissue (L. Stephens and S. Logan, personal communication). To our surprise, vasopressin not only stimulated inositol lipid breakdown in isolated sympathetic ganglia but also evoked a far greater response than did muscarinic cholinergic stimulation (Table II).

The magnitude of this response to vasopressin has allowed us to detect it after only a few seconds of stimulation. At these early times (15 sec and 60 sec) there was significant accumulation of inositol bisphosphate and trisphosphate in the ganglia (Hanley *et al.*, 1984), but inositol monophosphate accumulation was not detected until after longer periods of incubation. Thus it seems likely that in the vasopressin-stimulated ganglia, as in other stimulated cells, the stimulated reaction in inositol lipid metabolism is phosphodiesterase-catalyzed hydrolysis of PtdIns4,5P_2 (and possibly also of PtdIns4P); the later accumulation of inositol monophosphate probably reflects phosphatase action on inositol 1,4,5-trisphosphate (Michell *et al.*, 1984). This response persists in ganglia incubated in media of low Ca^{2+} concentration either in the presence or absence of 1 μM nifedipine.

Stimulation of inositol lipid metabolism in peripheral tissues is mediated by V_1-receptors; these receptors are sensitive to inhibition by V_1-antagonists which abolish the vasopressor effects of vasopressin, and they recognize oxytocin as an agonist of lower affinity and efficacy than AVP. A selective V_1-receptor antagonist (Kruszynski *et al.*, 1980) abolishes the stimulation of inositol lipid breakdown that is evoked in ganglia by AVP (Hanley *et al.*, 1984), and oxytocin is a poor agonist by comparison with AVP (Fig. 2). Thus it seems that the stimulation of PtdIns4,5P_2 breakdown that is evoked by AVP in isolated superior cervical sympathetic ganglia has biochemical and pharmacological characteristics that are essentially identical to those of the equivalent responses of hepatocytes (Creba *et al.*, 1983; Thomas *et al.*, 1983; Michell *et al.*, 1984), aorta smooth muscle (Takhar and Kirk, 1981) and hippocampus (L. Stephens and S. Logan, personal communication).

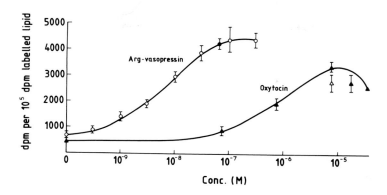

<u>Figure 2</u>. Accumulation of InsP₃ in rat superior cervical sympathetic ganglia incubated with arg-vasopressin or oxytocin for 10 min in the presence of 10 mM LiCl. Ganglia were prelabeled with [³H]inositol in the absence of any stimulus and were then stimulated with peptides for 10 min.

INOSITOL LIPID BREAKDOWN IN GANGLIA DEPOLARIZED BY INCUBATION IN THE PRESENCE OF HIGH EXTRACELLULAR [K⁺]

It was reported many years ago that depolarization of sympathetic ganglia stimulates their inositol lipid turnover, and that this response persists in ganglia deprived of their cholinergic input (Nagata *et al.*, 1973). When we found that sympathetic ganglia both responded to AVP and possessed an intrinsic plexus of nerve fibers containing VLP, we wondered whether internally released VLP might mediate the stimulation of inositol lipid metabolism that was seen in K⁺-depolarized ganglia. We therefore examined ganglia, incubated with 80 mM K⁺ in the presence of 10 μM atropine, for accumulation of inositol phosphates. We found a consistent and substantial accumulation of inositol phosphates in ganglia incubated in this medium (Table III). This response was abolished when ganglia were incubated in a medium of low Ca²⁺ concentration (not shown), a result in accord with the idea that inositol lipid breakdown can be activated by a neuro-

229

TABLE III

Accumulation of Labeled Inositol Phosphates in Rat Superior Cervical Ganglia Incubated in 80 mM K^+

	Labeled inositol phosphates
	(dpm per 10^3 dpm in lipids)
Control	8.3 ± 0.7
80 mM K^+	17.4 ± 1.0
80 mM K^+ and 1 μM V_1-antagonist	15.6 ± 1.0

Ganglionic lipids were prelabeled by incubation with [^{3}H]inositol, and ganglia were exposed to high K^+ for 60 min by substitution of some of the Na^+ of the normal medium with K^+. Incubations included 10 mM LiCl. Values given are mean ± SEM (n for each experimental group is at least 12 ganglia). Values presented are the sum of the labeling in Ins$\underline{P}$, Ins$\underline{P}_2$, Ins$\underline{P}_3$.

transmitter that is released within the ganglia by a Ca^{2+}-dependent mechanism. However, this response was not significantly inhibited by the V_1-antagonist that was effective in preventing the AVP-stimulated inositol lipid breakdown, suggesting that VLP is not the endogenously released mediator of the effect of depolarization. Attempts to identify this mediator by the use of other pharmacological antagonists (propranolol, yohimbine, prazosin, phenoxybenzamine, mepyramine, methysergide) indicated that it is unlikely to be noradrenaline, histamine or 5-hydroxytryptamine: its identity remains to be determined.

DISCUSSION

The conclusions that arise naturally from the results presented briefly in this paper fall into three groups: the presence of VLP in the sympathetic nervous system, the presence and biochemical mechanism of "vasopressin receptors" of the V_1-type in sympathetic ganglia, and the probable presence in these ganglia of an endogenously released neurotransmitter which employs the same mechanism of action as muscarinic cholinergic and V_1-receptors. Although we do not yet understand the relationship between these three characteristics of sympathetic ganglia, it seems likely that they will all ultimately prove to be related either to information transmission through the sympathetic nervous system or to the modulation of this information flow.

The Possible Significance of the Co-Existence of VLP and Noradrenaline in the Principal Neurones of Sympathetic Ganglia

Recent studies of the nervous system have revealed numerous situations in which classical amine neurotransmitters and putative peptide neurotransmitters coexist (Lundberg and Hokfelt et al., 1983). Indeed, several neuropeptides have been identified in neurones present in either mammalian or amphibian sympathetic ganglia: examples include LHRH, somatostatin, substance P, neuropeptide Y and calcitonin gene-related peptide (see Hanley et al., 1984 for references). However, most of these peptides, including neuropeptide Y, are somewhat restricted in their distributions amongst ganglionic neurones and sympathetic nerve fibers, and the only peptide with a very widespread association with dopamine-β-hydroxylase is VLP.

It is therefore of some interest to ask what might be the significance of the probable co-existence and simultaneous release of noradrenaline and VLP at sympathetic nerve terminals. Unfortunately this type of question cannot yet be answered with precision for any system in which cells respond to the simultaneous release of a classical neurotransmitter and a novel peptide. However, we do know that cells in peripheral tissues, of which hepatocytes are the best understood, possess receptors both for catecholamines and for vasopressin (or for some related peptide such as VLP). In the liver cells of adult rats the dominant influence of noradrenaline is upon α_1-receptors, and the only vasopressin receptors appear to be of the V_1-type (Williamson *et al.*, 1981; Kirk, 1982). Since α_1-adrenoceptors and V_1-vasopressin receptors share the same mechanism of action, a combination of small quantities of noradrenaline and VLP might provide an economical method for effectively activating the Ca^{2+}-signalling pathway.

In a more general sense, it may be of importance that noradrenaline and vasopressin (or VLP?) or both ligands which are capable of activating more than one type of cell-surface receptor, with different receptors activating different signalling pathways in the responsive cells: adenylyl cyclase is activated by β_1- and β_2-adrenoceptors and by V_2-vasopressin receptors and is inhibited by α_2-adrenoceptors, and PtdIns4,5P$_2$ breakdown and Ca^{2+} mobilization are responses evoked by α_1-adrenoceptors and by V_1-vasopressin receptors. Since the cells of each peripheral tissue that responds to sympathetic innervation will express a characteristic set of receptors selected from amongst these five candidates, the release of two transmitters rather than one could substantially diversify the potential response patterns of cells exposed to a mixture of catecholamines and peptides of neural and humoral origin.

To date, the only observed physiological action of the VLP present in ganglionic extracts has been its antidiuretic action when injected into ethanol-anesthetized and water-loaded rats (Hanley *et al.*, 1984), and further speculations about its possible role in the responses of peripheral tissues to activation of the sympathetic nervous system must be deferred until we know more about the actions of VLP *in vivo* and *in vitro*.

Inositol Lipid Breakdown in Ganglia Exposed to Receptor Stimuli or Depolarized by High K$^+$

The primary transmission of information through sympathetic ganglia is by rapid cholinergic transmission. This employs nicotinic cholinergic receptors, and there is no evidence that inositol lipids play any role in this process. However, it has been known for many years that the released acetylcholine also activates muscarinic receptors, with these causing a slow excitatory post-synaptic potential ("slow EPSP") which facilitates signal transmission through the ganglia. This is achieved by muscarinic cholinergic inhibition of a hyperpolarizing membrane K^+ conductance (the "M-current") (Brown, 1983). It is not known how receptor activation leads to inhibition of the M-current, but the occurrence of mus-carinically-controlled $PtdIns4,5P_2$ breakdown in the ganglia might point to a mechanism involving a rise in cytosol Ca^{2+} and/or the activation of protein kinase C (see the Introduction).

The most surprising facet of the results reported here is the fact that AVP produces a stimulation of inositol lipid breakdown in sympathetic ganglia that is much greater than the response evoked by maximal stimu-lation of muscarinic cholinergic receptors. Although there have previously been hints that AVP might exert some influence on the sympathetic nervous system (Cowley and Barber, 1983), we know of no previous studies that have reported direct effects of neurohypophysial peptide hor-mones on sympathetic ganglia. As mentioned earlier, the changes in inositol lipid metabolism seen in AVP-treated ganglia appear to be identical in type, if not in magni-tude, to the changes evoked by AVP both in peripheral tissues (hepatocytes, Creba *et al.*, 1983; arterial smooth muscle, Takhar and Kirk, 1981; renal mesangial cells, Troyer *et al.*, 1983) and in hippocampal slices (L. Stephens and S. Logan, personal communication). A wealth of evidence implicates AVP-stimulated $PtdIns4,5P_2$ breakdown in a signalling process involving Ca^{2+}, and possibly also protein kinase C, in hepatocytes, so it is tempting to ascribe a similar function to this reaction in ganglia. It will be of particular interest to analyze the action of AVP on the electrophysiological characteristics of signal transmission through ganglia.

AVP is present in the blood perfusing ganglia, and VLP exists in a plexus of nerve fibers within ganglia, so it is difficult to be sure which of these two peptides is the normal physiological stimulus to the ganglionic V_1-receptors. One clue may come from our studies of ganglia stimulated with high K^+. Despite the presence of the VLP-positive fiber plexus within the ganglia, administration of a V_1-vasopressin antagonist did not significantly reduce the inositol lipid breakdown in response to depolarization, suggesting that no agonist active at V_1-receptors was released within depolarized ganglia. Possibly, therefore, the V_1-receptors of ganglia serve to communicate changes in circulating AVP to the ganglia, somehow modulating their activity. A major value of such crosstalk between circulating hormonal AVP and a sympathetic nervous system whose transmitters include VLP might lie in the coordination of the activities of two systems employing closely related peptide signals with similar modes of action.

The data we have obtained from ganglia depolarized with high K^+ reveal that there must be at least one more signal, in addition to muscarinic cholinergic and V_1-vasopressin stimuli, that can stimulate inositol lipid breakdown in sympathetic ganglia. Moreover, this is a signal intrinsic to the ganglia and it can be released in a Ca^{2+}-dependent manner, suggesting that it is a neurotransmitter released within the ganglia. On present evidence, this transmitter appears not to be noradrenaline, histamine, 5-hydroxytryptamine or LHRH, but we still do not know its identity. Its function is unknown.

V_1-Receptors: Hormone Receptors or Neurotransmitter Receptors?

In the Introduction we alluded to the fact that circulating concentrations of AVP are usually too low to have substantial effects on peripheral tissues possessing V_1-receptors. These receptors are, however, present in concentrations adequate to provoke major effects on cell function if occupied. A possible solution to this long-standing puzzle may well lie in the discovery of VLP as a possible co-transmitter with noradrenaline in the sympathetic nervous system: peripheral V_1-receptors may normally respond mainly to VLP released locally from sympathetic nerve fibers, rather than to circulating AVP. The same pattern probably applies to cells bearing V_1-receptors in the hippocampus and maybe other brain regions: they are likely to respond to AVP, VLP or some related neuropeptide released nearby. It seems possible that sympathetic ganglia might be the only structures endowed with sufficient V_1-receptors to display an appreciable biological response at concentrations of circulating AVP close to the normal range: possibly these responses serve to effectively coordinate the actions of circulating AVP and neurally derived VLP?

ACKNOWLEDGEMENTS

Some parts of this work were supported by grants and studentships from the Medical Research Council.

REFERENCES

Berridge, M.J., Downes, C.P. and Hanley, M.R. (1982) *Biochem. J. 206*, 587–595.

Berridge, M.J., Dawson, R.M.C., Downes, C.P., Heslop, J.P. and Irvine, R.F. (1983) *Biochem. J. 212*, 473–483.

Brown, D.A. (1983) *Trends Neurosci. 6*, 302–307.

Cowley, A.W. and Barber, B.J. (1983) in: *"The Neurohypophysis: Structure, Function and Control"* (B.A. Cross and C. Leng, eds.) pp. 415–424, Elsevier, Amsterdam.

Creba, J.A., Downes, C.P., Hawkins, P.A., Brewster, G., Michell, R.H. and Kirk, C.J. (1983) *Biochem. J. 212*, 733–747.

Downes, C.P. (1983) *Trends Neurosci. 6*, 313–316.

Downes, C.P. and Michell, R.H. (1982) *Cell Calcium 3*, 467–502.

Hanley, M.R., Benton, H., Lightman, S.L., Todd, K., Bone, E.A., Fretten, P., Palmer, S., Kirk, C.J. and Michell, R.H. (1984) *Nature 309*, 258–261.

Hawthorne, J.N. and Pickard, M.R. (1979) *J. Neurochem. 32*, 5–14.

Hokin, M.R., Hokin, L.E. and Shelp, W.D. (1960) *J. Gen. Physiol. 44*, 217–226.

Jard, S. (1980) in: *"Cellular Receptors for Hormones and Neurotransmitter"* (D. Schulster and A. Levitzki, eds.) pp. 253–266, John Wiley, Chichester.

Kirk, C.J. (1982) *Cell Calcium 3*, 399–410.

Kirk, C.J., Creba, J.A., Downes, C.P. and Michell, R.H. (1981) *Biochem. Soc. Trans. 9*, 377–379.

Kruszynski, M., Lammek, B., Manning, M., Seto, J., Haldar, J. and Sawyer, W.H. (1980) *J. Med. Chem. 23*, 364–368.

Lightman, S.L. and Forsling, M. (1980) *Clin. Endocrinol.* *12*, 39-46.

Lundberg, J.M. and Hokfelt, T. (1983) *Trends Neurosci.* *6*, 325-333.

Michell, R.H. (1975) *Biochim. Biophys. Acta* *415*, 81-147.

Michell, R.H. (1982a) *Cell Calcium* *3*, 285-294.

Michell, R.H. (1982b) *Neurosci. Res. Prog. Bull.* *20*, 338-350.

Michell, R.H. (1982c) in: *"Phospholipids in the Nervous System"*, Vol. 1, Metabolism (L. Horrocks, G.B. Ansell and G. Porcellati, eds.) pp. 315-325, Raven Press, New York.

Michell, R.H., Kirk, C.J., Jones, L.M., Downes, C.P. and Creba, J.A. (1981) *Phil. Trans. Roy. Soc. Lond.* *B 296*, 123-137.

Michell, R.H., Hawkins, P.T., Palmer, S. and Kirk, C.J. (1984) in: *"Calcium Regulation in Biological Systems"* (M. Endo, ed.) Takeda Symposium on Bioscience, in press.

Nagata, Y., Mikoshiba, K. and Tsukada, Y. (1973) *Brain Res.* *56*, 259-269.

Nishizuka, Y. (1983) *Trends Biochem. Sci.* *8*, 13-16.

Sofroniew, M.W. (1983) *Trends Neurosci.* *6*, 467-472.

Streb, H., Berridge, M.J., Irvine, R.F. and Schulz, I. (1983) *Nature* *306*, 67-69.

Takhar, A.P. and Kirk, C.J. (1981) *Biochem. J.* *194*, 167-172.

Thomas, A.P., Marks, J.S., Coll, K.E. and Williamson, J.R. (1983) *J. Biol. Chem.* *258*, 5716-5725.

Troyer, D.A., Kreisberg, J.I., Schwertz, D.W. and Venkatachalam, M.A. (1983) *Fed. Proc.* *42*, 1259.

Williamson, J.R., Cooper, R.H. and Hoek, J.B. (1981) *Biochim. Biophys. Acta* *639*, 243-295.

HORMONE EFFECTS ON PHOSPHOINOSITIDE
METABOLISM IN LIVER

J.H. Exton, V. Prpic, R. Charest, D. Rhodes,
and P.F. Blackmore

Howard Hughes Medical Institute, Laboratories for the
Studies of Metabolic Disorders and Department of
Physiology, Vanderbilt University School of Medicine,
Nashville, TN 37232

SUMMARY

Vasopressin, epinephrine and angiotensin II markedly
stimulated the breakdown of phosphatidylinositol and
release of myo-inositol in isolated rat liver cells, but
these changes were not significant before 2 min. It is
therefore concluded that the degradation of phosphatidylinositol is too slow to account for the increase in cytosolic
Ca^{2+} and activation of phosphorylase caused by these
agonists, which occur within 2 s. Breakdown of PIP_2 and
release of IP_3 induced by vasopressin and epinephrine
occurred more rapidly than the breakdown of phosphatidylinositol, being significant at 3 s with 10^{-7} M vasopressin
and 10^{-5} M epinephrine. The concentration-dependence
curves for the effects of vasopressin or epinephrine on
PIP_2 breakdown or IP_3 accumulation were not well
correlated with those on cytosolic Ca^{2+} or phosphorylase
a. Furthermore, addition of Li^+ ions enhanced, after
30-60 s, the effects of low concentrations of vasopressin
and epinephrine on IP_3 levels, but not on cytosolic Ca^{2+}
or phosphorylase a. These findings may be reconciled
with the hypothesis that PIP_2 breakdown and IP_3 release
are integrally involved in the effects of vasopressin and
epinephrine on cytosolic Ca^{2+}, if it is postulated that an
extremely small breakdown of the polyphosphoinositide is
sufficient to alter cellular Ca^{2+} fluxes. In addition, the
failure of Li^+ ions to modify the effects of the agonists on

cytosolic Ca^{2+} or phosphorylase could be explained if only those changes in IP_3 that occur within 30-60 s of hormone treatment control cell Ca^{2+}. It is concluded that although much more experimentation is required, the bulk of existing evidence supports the hypothesis that PIP_2 breakdown plays a primary role in the actions of Ca^{2+}-dependent hormones in liver.

INTRODUCTION

There is much evidence that vasopressin and α-adrenergic agonists such as epinephrine and nor-epinephrine exert their effects in liver and certain other tissues through Ca^{2+}-dependent mechanisms (Exton, 1981; Williamson *et al.*, 1981). Recently it has also been demonstrated that these agents rapidly increase the concentration of Ca^{2+} in the cytosol of liver cells (Charest *et al.*, 1983). However, the molecular mechanisms coupling the vasopressin and α-adrenergic receptors to the cellular Ca^{2+} changes responsible for the rise in cytosolic Ca^{2+} remain very obscure.

It is generally agreed that the Ca^{2+}-dependent hormones promote the release of Ca^{2+} from intracellular sites (Exton, 1981; Williamson *et al.*, 1981), although some argument remains concerning the anatomical location of these sites. There is also evidence for a change in the flux of Ca^{2+} across the plasma membrane (Assima-copoulos-Jeannet *et al.*, 1977; Barritt *et al.*, 1981a) which has been interpreted as an increase in Ca^{2+} entry due to the opening of Ca^{2+} channels, or an inhibition of Ca^{2+} efflux due to decreased activity of the plasma membrane Ca^{2+} pump. An inhibition by Ca^{2+}-dependent hormones of Ca^{2+} transport in rat liver plasma membrane vesicles has recently been demonstrated (Prpic *et al.*, 1984), but the time course of the effect is relatively slow indicating that it is involved in the maintenance rather than the generation of the increase in cytosolic Ca^{2+}.

It has been suggested that Ca^{2+}-dependent hormones exert their effects on cellular Ca^{2+} fluxes through changes in phosphoinositides (Michell, 1975; Michell *et al.*, 1981). Initially, it was proposed that the breakdown of phosphatidylinositol was the important link (Michell, 1975, 1979), but later it was postulated that the degradation of PIP_2 was the primary factor (Kirk *et al.*, 1981; Michell *et al.*, 1981). However, many aspects of the proposal require further exploration and this is the subject of this article.

RESULTS

Hormonal Stimulation of Phosphatidylinositol Breakdown in Hepatocytes

The effects of Ca^{2+}-dependent hormones on the breakdown of phosphatidylinositol in liver were assessed by measuring the release of *myo*-inositol since isotopic studies indicated that the accumulation of inositol phosphates due to the breakdown of the phospholipid was negligible in comparison to the accumulation of free inositol (Prpic *et al.*, 1982). *myo*-Inositol release from rat hepatocytes was markedly increased by vasopressin and epinephrine after a lag-time of about 2 min (Fig. 1). A similar pattern of [³H]*myo*-inositol release was observed if the phosphatidylinositol of the hepatocytes was radiolabeled by intraperitoneal injection of the rats with [³H]*myo*-inositol 18-20 h prior to cell preparation (Prpic *et al.*, 1982). Angiotensin II, epinephrine and ATP also promoted the release of [³H]*myo*-inositol from [³H]phosphatidylinositol but the ionophore A23187 and glucagon were ineffective (Prpic *et al.*, 1982). Calcium depletion of the hepatocytes using the chelator EGTA abolished the effects of the hormones on phosphatidylinositol breakdown (Prpic *et al.*, 1982). However, readdition of Ca^{2+} to the EGTA-treated cells fully restored the response (Prpic *et al.*, 1982) indicating that the chelator was not acting by damaging the cells, in contrast to what has been surmised in the literature (Litosch *et al.*, 1983).

The slowness of the phosphatidylinositol response (Fig. 1) relative to the changes in Ca^{2+} and phosphorylase caused by the Ca^{2+}-dependent hormones (Charest *et al.*, 1983; Blackmore *et al.*, 1983) rendered it unlikely that this response is primarily involved in the generation of these changes. In addition, the Ca^{2+} dependence of the phosphatidylinositol breakdown raised the possibility that it was secondary to the increase in Ca^{2+}. However, it must be recognized that Ca^{2+} dependence may reflect a Ca^{2+} requirement of one of the reac-

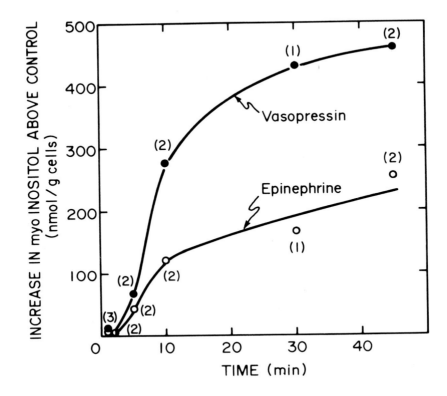

Figure 1. Release of myo-inositol from hepatocytes incubated with 10^{-8} M vasopressin or 10^{-5} M epinephrine. myo-Inositol was measured by gas-liquid chromatography. From Prpic *et al*. (1982) with permission.

tions involved in the response and does not necessarily mean that the response is mediated by an increase in Ca^{2+}. In support of this argument is the observation that the increase in cytosolic Ca^{2+} caused by A23187 or glucagon does not stimulate phosphatidylinositol break-down (Prpic *et al*., 1982).

240

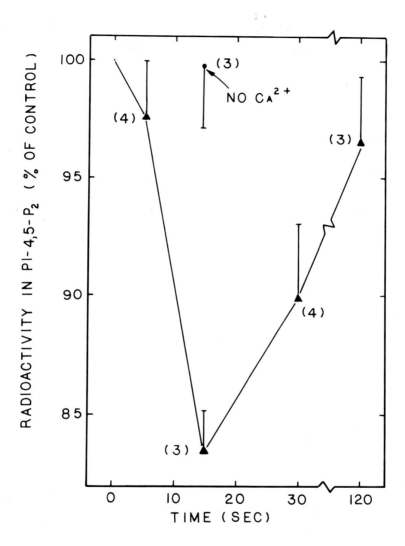

<u>Figure   2</u>.   Time   course   of   PIP$_2$ breakdown   induced by vasopressin in hepatocytes. Hepatocytes prelabeled with [^{3}H]<u>myo</u>-inositol were incubated with 10^{-8} M vasopressin and radioactivity in PIP$_2$ was   determined   at   the   times   shown.   From Rhodes <u>et</u> <u>al</u>. (1983) with permission.

Hormonal Stimulation of PIP_2
Breakdown in Hepatocytes

Kirk *et al.* (1981) were the first to report that vasopressin rapidly stimulated PIP_2 breakdown in rat hepatocytes. We have confirmed this (Rhodes *et al.*, 1983) using hepatocytes prelabeled with [^{3}H]*myo*-inositol (Fig. 2) as have other groups using cells radiolabeled in several ways (Thomas *et al.*, 1983; Litosch *et al.*, 1983). The effect is detectable at 5 s, maximal at 15-30 s, and greatly diminished at 2 min (see also Kirk *et al.*, 1981; Creba *et al.*, 1983; Thomas *et al.*, 1983; Litosch *et al.*, 1983). This is clearly much faster than the phosphatidylinositol response reported by most workers (Kirk *et al.*, 1981; Prpic *et al.*, 1982; Thomas *et al.*, 1983; *cf.* Litosch *et al.*, 1983). It is also of greater relative magnitude *i.e.*, 20-50% decrease from control at 30 s (Rhodes *et al.*, 1983; Thomas *et al.*, 1983; Litosch *et al.*, 1983) compared with 0-20% for phosphatidylinositol (Kirk *et al.*, 1981; Prpic *et al.*, 1982; Thomas *et al.*, 1983; Litosch *et al.*, 1983).

However, a question remains as to whether or not the PIP_2 response can be elicited by physiological concentrations of Ca^{2+}-dependent hormones. We reported that significant changes in this phospholipid were not detectable at 15 s with a vasopressin concentration of 10^{-9} M, whereas this concentration of the hormone produced maximal changes in cell Ca^{2+} and activation of phosphorylase (Fig. 3). Furthermore, 10^{-7} M epinephrine, which markedly affected cellular Ca^{2+} fluxes (Blackmore *et al.*, 1978; Charest *et al.*, 1984) and fully activated phosphorylase in hepatocytes at 15 s (see Fig. 5), did not produce detectable changes in PIP_2 levels at this time. Creba *et al.* (1983) also found large differences between the dose response curves for the effects of vasopressin or angiotensin II on PIP_2 breakdown and phosphorylase activation in liver cells. On the other hand, Thomas *et al.* (1983) found that the dose response curve for the effect of vasopressin on PIP_2 breakdown was well correlated with that on phosphorylase.

These findings raise questions about the hypothesis that PIP_2 breakdown is involved in the coupling of the receptors for vasopressin or epinephrine to the observed changes in Ca^{2+} fluxes.

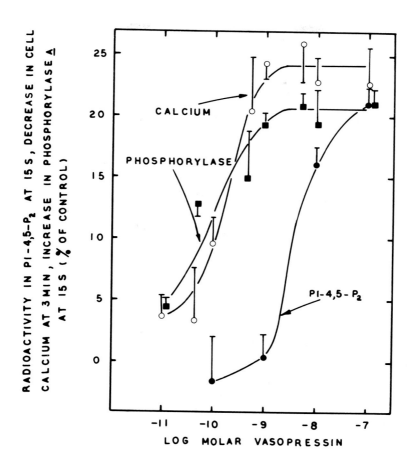

<u>Figure 3</u>. Dose response curves for the effects of vasopressin on phosphorylase $\underline{a}$, cell Ca^{2+} and PIP_2 in hepatocytes. Hepatocytes prelabeled with [^{3}H]<u>myo</u>-inositol were incubated with the concentrations of vasopressin shown. [^{3}H]PIP_2 and phosphorylase $\underline{a}$ were measured at 15 s and cell Ca^{2+} at 3 min after exposure to vasopressin. From Rhodes <u>et al</u>. (1983) with permission.

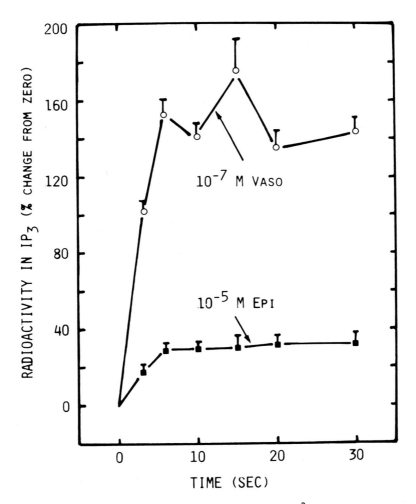

<u>Figure 4</u>. Time course of the accumulation of [^{3}H]IP$_3$ in hepatocytes prelabeled by incubation for 90 min with [^{3}H]<u>myo</u>-inositol and then exposed to 10^{-7} M vasopressin or 10^{-5} M epinephrine. From Charest <u>et al</u>. (1984) with permission.

Hormonal Stimulation of the Release of Inositol Phosphates in Hepatocytes

A more sensitive way of measuring the breakdown of a compound is to measure the release of products, especially when the compound is rapidly resynthesized $i.e.$ turns over rapidly. This approach was therefore taken with PIP_2 which appears to be rapidly resynthesized (Fig. 2) and whose breakdown products are 1,2-diacylglycerol and IP_3. Figure 4 shows the time course of the effects of vasopressin and epinephrine at maximally effective concentrations on the release of radiolabeled IP_3 from liver cells previously incubated for 90 min with [^{3}H]myo-inositol to radiolabel the phosphoinositides. It is seen that vasopressin is more effective than epinephrine in stimulating the formation of IP_3, but that the effects of both agents are significant at 3 s. It is possible that the effect occurs as fast as that (2 s) on cytosolic Ca^{2+} and phosphorylase (Charest et $al.$, 1983; Charest et $al.$, 1984), but sampling difficulties render this point uncertain. More detailed studies of the early time course with lower concentrations of agonists show a transient rapid phase of IP_3 release followed by a slower sustained phase (data not shown).

Figure 5 shows the concentration-dependence curves for the stimulation of IP_3 release by vasopressin and epinephrine at 15 s. For comparison the changes in cytosolic Ca^{2+} and phosphorylase at this time are shown. As in the case of PIP_2 breakdown, higher doses of vasopressin and epinephrine are required for IP_3 release than for increasing cytosolic Ca^{2+} and phosphorylase a. As expected from Figure 4, there is a poorer correlation between the effects of epinephrine than vasopressin on IP_3 formation and the other parameters at 15 s. In order for these data to be reconcilable with the hypothesis that IP_3 is integrally involved in the coupling of receptors for the Ca^{2+}-dependent hormones to the cellular Ca^{2+} changes induced by these agents, it is necessary to postulate that only a 10 or 40% increase in IP_3 is capable of maximally increasing cytosolic Ca^{2+} and phosphorylase (Fig. 5).

Further experiments to explore the role of IP_3 were performed using the monovalent cation Li^+. Addition of this ion to hepatocytes produces a marked enhancement of

245

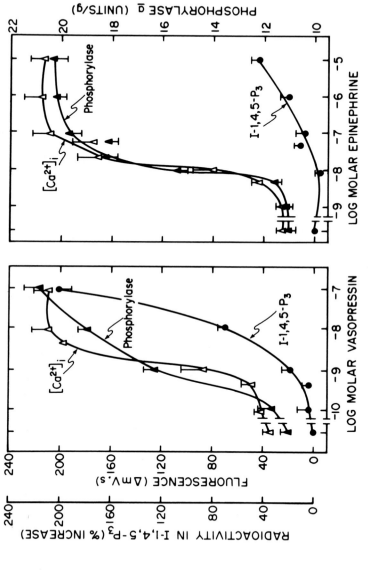

Figure 5. Dose response curves for the effects of vasopressin and epinephrine on phosphorylase $\underline{a}$, cytosolic Ca^{2+} and $[^3H]IP_3$ in hepatocytes prelabeled with $[^3H]myo$-inositol. Hepatocytes prelabeled as described in Fig. 3 were incubated with vasopressin and epinephrine at the concentrations shown and the three parameters were measured after 15 s. From Charest $\underline{et\ al}$. (1984) with permission.

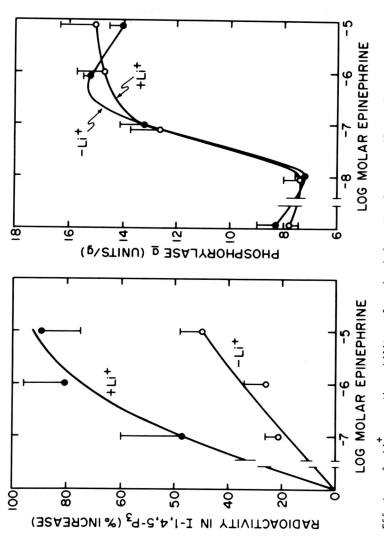

<u>Figure 6.</u> Effects of Li$^+$ on the ability of epinephrine to increase IP$_3$ and phosphorylase $\underline{a}$ in hepatocytes. Hepatocytes were prelabeled as described in Fig. 3 and incubated with epinephrine at the concentrations shown with or without 10 mM LiCl. [^{3}H]Inositol trisphosphate and phosphorylase $\underline{a}$ were measured after 5 min.

247

Exton et al.

the effects of epinephrine, vasopressin and angiotensin II on inositol trisphosphate (*e.g.*, Fig. 6), although the effect is not evident until after 30-60 s (data not shown). However, Li^+ addition does not alter the effects of these agents on cytosolic Ca^{2+} or phosphorylase with respect to either their time course or dose response. These findings are not easily explicable if IP_3 is the intracellular messenger involved, unless further postulations are made (see Discussion).

DISCUSSION

PIP_2 Breakdown and Phosphorylase Activation

The data presented show that a supraphysiological concentration of vasopressin (10^{-7} M) stimulates a transient PIP_2 breakdown and a prolonged IP_3 accumulation in hepatocytes which begins as early as 3 s. A very high concentration of epinephrine (10^{-5} M) produces a smaller accumulation of IP_3 which is also detectable at 3 s. However, the breakdown of PIP_2 or the release of IP_3 induced by these agonists may be slower than the 2 s required for the increases in cytosolic Ca^{2+} and phosphorylase *a*. Furthermore, 100-fold lower concentrations of the hormones barely elicit the IP_3 response, although they increase cytosolic Ca^{2+} and phosphorylase *a* maximally or half-maximally. There is also an apparent discrepancy between the increase in IP_3 associated with maximal activation of phosphorylase by vasopressin compared with that when epinephrine is the agonist (Figure 5). For these findings to fit with the hypothesis that PIP_2 breakdown is responsible for the alterations in cell Ca^{2+} and phosphorylase caused by vasopressin and epinephrine in liver, it must be postulated that small changes in the breakdown of this phospholipid are sufficient to elicit the responses.

Kirk *et al.* (1981) and Creba *et al.* (1983) also pointed out a discrepancy between the vasopressin concentrations required to stimulate PIP_2 breakdown and activate phosphorylase in hepatocytes. By comparison with data from Cantau *et al.* (1980), Creba *et al.* (1983) suggested that the dose response curves for vasopressin binding to hepatocytes and for stimulation of PIP_2 breakdown were highly correlated, and that a very small fractional occupancy of vasopressin receptors and resulting breakdown of PIP_2 were sufficient to cause Ca^{2+} mobilization.

Thomas et al. (1983) also studied the concentration dependencies of the vasopressin effects on phosphorylase activation and PIP_2 breakdown in hepatocytes. In contrast to our work (Rhodes et al., 1983) and that of Creba et al. (1983), these workers reported a good correlation between the phospholipid change and the increase in phosphorylase a, although some key points in the dose response curves were lacking and the parameters were measured at different times (30 s and 90 s).

Consideration of IP_3 as a Possible Second Messenger

The time course of IP_3 accumulation in response to vasopressin in hepatocytes is initially similar to that of PIP_2 breakdown (Rhodes et al., 1983; Charest et al., 1984). The dose response curves for the effects of vasopressin on these two changes are also very similar (Rhodes et al., 1983; Charest et al., 1984). These considerations support the conclusion that the inositol trisphosphate which accumulates is the breakdown product of PIP_2, namely IP_3.

It has been suggested that this compound might act as a second messenger for Ca^{2+}-dependent hormones in their target cells. Although this hypothesis is attractive, it has to be reconciled with certain data. For example, Li^+ ions markedly enhanced the accumulation of IP_3 observed at 30 s or longer with low concentrations of epinephrine (Fig. 6) or vasopressin (Charest et al., 1984), but did not alter the effects of these agonists on cytosolic Ca^{2+} or phosphorylase at any time. Lithium apparently acts by inhibiting the phosphatase acting on IP_3 and would have been expected to enhance the effects of submaximally effective concentrations of the agonists on Ca^{2+} mobilization. On the other hand, Li^+ was not effective before 30-60 s and it is possible that IP_3 acts only during the initial phase of hormone action (perhaps intracellular Ca^{2+} release) which is later superseded by another phase (perhaps inhibition of the plasma membrane Ca^{2+} pump). Accordingly, changes in

the level of IP_3 occurring beyond 60 s may not be
relevant to the control of cytosolic Ca^{2+} by Ca^{2+}-
dependent hormones. Alternatively, it can be postulated
that the IP_3 which is initially released is from a plasma
membrane pool of PIP_2 which controls cell Ca^{2+} fluxes,
whereas the IP_3 which is later released is from a larger
pool which doesn't affect cell Ca^{2+}. The data with epinephrine and the biphasic time course of IP_3 accumulation
observed with submaximally effective concentrations of
agonists would be consistent with the idea of more than
one pool. In support of the hypothesis that IP_3 plays an
important role in the initial changes in cell Ca^{2+}, one can
also cite recent reports that preparations of IP_3 promote
the release of Ca^{2+} from permeabilized pancreatic acinar
or hepatic cells (Streb *et al.*, 1983; Joseph *et al.*, 1984;
Burgess *et al.*, 1984) or liver fractions enriched in
mitochondria or microsomes (P. Thiyagarajah, R. Charest
and P.F. Blackmore, unpublished observations).

Speculation on the Role of Phosphoinositide Changes
in the Actions of Calcium-Dependent Hormones in Liver

Clearly, the hypothesis that IP_3 acts as a second
messenger for Ca^{2+}-dependent hormones in liver and that
PIP_2 breakdown plays a role in the control of liver cell
Ca^{2+} requires more support. In particular, information is
needed on the following key points: (1) the mechanisms
coupling the breakdown of PIP_2 with receptor occupancy
in the plasma membrane, (2) the coupling of this breakdown with the cellular Ca^{2+} changes responsible for the
rise in cytosolic Ca^{2+}. With respect to coupling to the
receptors, it needs to be convincingly demonstrated that
breakdown of PIP_2 occurs in the plasma membrane *in vitro*
upon addition of agonists. Concerning the linkage of
PIP_2 degradation with cellular Ca^{2+} fluxes, the hypothesis
that this breakdown opens Ca^{2+} "gates" in the plasma
membrane remains to be demonstrated. As discussed
extensively elsewhere (Blackmore *et al.*, 1982; Charest *et
al.*, 1984), there is much evidence that Ca^{2+} influx is not
responsible for the increase in cytosolic Ca^{2+} that occurs
during the first minute of exposure of hepatocytes to
vasopressin or epinephrine. On the other hand, there is
evidence that Ca^{2+} flux across the plasma membrane is
altered at later times due to inhibition of the Ca^{2+} pump
(Prpic *et al.*, 1984; Charest *et al.*, 1984). With respect

to the mobilization of intracellular Ca^{2+}, evidence is accumulating to support the idea that IP_3, one of the primary breakdown products of PIP_2, acts as a second messenger which releases Ca^{2+} from intracellular sites (Streb *et al.*, 1983). Concerning the other primary product, namely 1,2-diacylglycerol, there is no evidence that this alters cell Ca^{2+} distribution, and it is also now thought that its metabolite, phosphatidic acid, is unlikely to act as a biological ionophore (Barritt *et al.*, 1981b; Holmes and Yoss, 1983).

A possible role for the 1,2-diacylglycerol generated in liver cells by Ca^{2+}-dependent agents such as vasopressin (Thomas *et al.*, 1983) is activation of the Ca^{2+}-phospholipid-dependent protein kinase (protein kinase C) discovered by Nishizuka and associates (Takai *et al.*, 1979). This enzyme is present in liver and is stimulated by 1,2-diacylglycerol (Kishimoto *et al.*, 1980). Certain phorbol esters can substitute for diacylglycerol and activate the enzyme (Castagna *et al.*, 1982). Such phorbol esters added to hepatocytes do not alter cytosolic Ca^{2+} or activate phosphorylase, although they inhibit the effects of α_1-adrenergic agonists on these parameters (R. Charest and P.F. Blackmore, unpublished observations). They increase the phosphorylation of three unknown cytosolic proteins which are also phosphorylated in response to vasopressin and α-adrenergic agonists (in addition to other proteins), but not in response to the ionophore A23187 (Garrison, 1983). Thus phosphoinositide breakdown induced by Ca^{2+}-dependent hormones in the liver may not only be related to the initial mobilization of Ca^{2+} from intracellular stores, but may also serve to activate protein kinase C which phosphorylates specific proteins. The identity of these proteins as well as the functional significance of their phosphorylation remain to be defined.

REFERENCES

Assimacopoulos-Jeannet, F.D., Blackmore, P.F. and Exton, J.H. (1977) *J. Biol. Chem.* *252*, 2662-2669.
Barritt, G.J., Dalton, K.A. and Whiting, J.A. (1981a) *FEBS Lett.* *125*, 137-140.
Barritt, G.J., Parker, J.C. and Wadsworth, J.C. (1981b) *J. Physiol. (London)* *312*, 29-55.

Blackmore, P.F., Brumley, F.T., Marks, J.L. and Exton, J.H. (1978) *J. Biol. Chem.* *253*, 4851-4858.

Blackmore, P.F., Hughes, B.P., Shuman, E.A. and Exton, J.H. (1982) *J. Biol. Chem.* *257*, 190-197.

Blackmore, P.F., Hughes, B.P., Charest, R., Shuman, E.A. and Exton, J.H. (1983) *J. Biol. Chem.* *258*, 10488-10494.

Burgess, G.M., Godfrey, P.P., McKinney, J.S., Berridge, M.J., Irvine, R.F. and Putney, J.W., Jr. (1984) *Nature 309*, 63-66.

Cantau, B., Keppens, S., de Wulf, H. and Jard, S. (1980) *J. Receptor Res.* *1*, 137-168.

Castagna, M., Takai, Y., Kaibuchi, K., Sano, K., Kikkawa, U. and Nishizuka, Y. (1982) *J. Biol. Chem.* *257*, 7847-7851.

Charest, R., Blackmore, P.F., Berthon, B. and Exton, J.H. (1983) *J. Biol. Chem.* *258*, 8769-8773.

Charest, R., Prpic, V., Exton, J.H. and Blackmore, P.F. (1984) *J. Biol. Chem.*, in press.

Creba, J.A., Downes, C.P., Hawkins, P.T., Brewster, G., Michell, R.H. and Kirk, C.J. (1983) *Biochem. J.* *212*, 733-747.

Exton, J.H. (1981) *Mol. Cell. Endocrinol.* *23*, 233-264.

Garrison, J.C. (1983) in: *"Isolation, Characterization and Use of Hepatocytes"* (R.A. Harris and N.W. Cornell, eds.) pp. 551-559, Elsevier/North-Holland Biomedical Press, Amsterdam, The Netherlands.

Holmes, R.P. and Yoss, N.L. (1983) *Nature 305*, 637-638.

Joseph, S.K., Thomas, A.P., Williams, R.J., Irvine, R.F. and Williamson, J.R. (1984) *J. Biol. Chem. 259*, 3077-3081.

Kirk, C.J., Creba, J.A., Downes, C.P. and Michell, R.H. (1981) *Biochem. Soc. Trans.* *9*, 377-379.

Kishimoto, A., Takai, Y., Mori, T., Kikkawa, U. and Nishizuka, Y. (1980) *J. Biol. Chem.* *255*, 2273-2276.

Litosch, I., Lin, S.-H. and Fain, J.N. (1983) *J. Biol. Chem.* *258*, 13727-13732.

Michell, R.H. (1975) *Biochim. Biophys. Acta 415*, 81-147.

Michell, R.H. (1979) *Trends Biochem. Sci.* *4*, 128-131.

Michell, R.H., Kirk, C.J., Jones, L.M., Downes, C.P. and Creba, J.A. (1981) *Phil. Trans. Royal Soc. Lond. B 296*, 123-127.

Prpic, V., Blackmore, P.F. and Exton, J.H. (1982) *J. Biol. Chem.* *257*, 11323-11331.

Prpic, V., Green, K.C., Blackmore, P.F. and Exton, J.H. (1984) *J. Biol. Chem.* *259*, 1382-1385.

Rhodes, D., Prpic, V., Exton, J.H. and Blackmore, P.F. (1983) *J. Biol. Chem.* *258*, 2770-2773.

Streb, H., Irvine, R.F., Berridge, M.J. and Schulz, I. (1983) *Nature* *306*, 67-69.

Takai, Y., Kishimoto, A., Iwasa, Y., Kawahara, Y., Mori, T. and Nishizuka, Y. (1979) *J. Biol. Chem.* *254*, 3692-3695.

Thomas, A.P., Marks, J.S., Coll, K.E. and Williamson, J.R. (1983) *J. Biol. Chem.* *258*, 5716-5725.

Williamson, J.R., Cooper, R.H. and Hoek, J.B. (1981) *Biochim. Biophys. Acta* *639*, 243-295.

RELATIONSHIP BETWEEN VASOPRESSIN ACTIVATION OF RAT HEPATOCYTE GLYCOGEN PHOSPHORYLASE, INHIBITION OF Ca^{2+}-Mg^{2+}-ATPase AND PHOSPHOINOSITIDE BREAKDOWN

J.N. Fain, S.-H. Lin, I. Litosch, and M. Wallace

Section of Biochemistry, Brown University
Providence, Rhode Island   02912

SUMMARY

Rat hepatocyte glycogen phosphorylase is activated by vasopressin, α_1-adrenergic agonists and an active phorbol ester in the presence of 1 μM A23187. Alpha$_1$-adrenergic agonists and vasopressin stimulate a rapid degradation of phosphatidylinositol and phosphatidylinositol 4,5-bisphosphate as well as increase the conversion of phosphatidylinositol to phosphatidylinositol 4,5-bisphosphate in rat hepatocytes. Phosphoinositide breakdown due to vasopressin was maximal at 30 s and unaffected by the presence or absence of Ca^{2+} in the incubation medium. These effects were seen in hepatocytes pre-labeled by incubation *in vitro* with ^{32}P or [^{3}H]inositol as well as by injection of [^{3}H]inositol into rats 18 h prior to sacrifice. The site of phosphatidylinositol breakdown due to vasopressin appeared to be the plasma membrane since addition of vasopressin to isolated rat liver plasma membranes increased phosphatidylinositol breakdown. Vasopressin addition to rat hepatocytes also resulted in an immediate inhibition of the Ca^{2+}-Mg^{2+}-ATPase activity of the plasma membranes. The data indicate that breakdown of phosphatidylinositol and phosphatidylinositol 4,5-bisphosphate are associated with receptor activation, but their relationship to the inhibition of the Ca^{2+}-Mg^{2+}-ATPase remains to be established.

255

ACTIVATION OF GLYCOGEN PHOSPHORYLASE BY Ca^{2+}

This article reviews our studies on the regulation of phosphoinositide turnover in rat hepatocytes by vasopressin and α_1-catecholamines as well as their effects on glycogen phosphorylase and Ca^{2+}-Mg^{2+}-ATPase activity. Initially it was thought that all effects of hormones on glycogen phosphorylase in liver were due to cyclic AMP. Sherline *et al.* (1972) first found that phenylephrine (a catecholamine which predominantly stimulates α_1-adrenoceptors) increased glycogen phosphorylase without affecting cyclic AMP in perfused rat livers. Furthermore, while epinephrine increased cyclic AMP accumulation and glycogen phosphorylase activity, the addition of an α-adrenergic antagonist (phentolamine) blocked the rise in cyclic AMP but not that of phosphorylase. Sherline *et al.* (1972) suggested that the activation of glycogen phosphorylase due to α-adrenergic stimulation was secondary to vasoconstriction which elevated AMP, the only other compound known at that time to regulate glycogen phosphorylase. However, Tolbert *et al.* (1973) found similar increases due to α-adrenergic activation of gluconeogenesis in rat hepatocytes that were unrelated to changes in total cyclic AMP. The use of rat parenchymal cells (hepatocytes) ruled out effects via vasoconstriction.

Subsequent studies have confirmed these findings and it is currently thought that an elevation in cytosolic Ca^{2+} rather than cyclic AMP is responsible for the increase in gluconeogenesis and glycogen phosphorylase seen in rat hepatocytes after addition of α-adrenergic agonists, vasopressin or angiotensin. The alpha effects of catecholamines can be divided into α_1 involving elevations in cytosolic Ca^{2+} and α_2 involving inhibition of adenylyl cyclase (Fain and Garcia-Sainz, 1980). The receptors for alpha catecholamines in rat liver are predominantly of the α_1 type as are the effects of catecholamines on glycogenolysis (Hoffman *et al.*, 1980) and gluconeogenesis (Kneer *et al.*, 1979).

The effects of various hormones on rat hepatocytes are summarized in Table I. Recently Shukla *et al.* (1983) found that platelet activating factor (1-O-alkyl-2-acetyl-*sn*-glycero-3-phosphocholine) increased glycogen phosphorylase activity. Fain *et al.* (1984) found that the

TABLE I

Agent	Elevations in Cyclic AMP	Effects on Glycogen Phosphorylase	Effect on Phosphatidyl-inositol Turnover
A23187	No	Activates in presence of extracellular Ca^{2+}	None
4β-phorbol 12β-myristate 13α-acetate	No	Activates in presence of submaximal A23187	None
Vasopressin (V-1) or Angiotensin	No	Activates	Increase
Catecholamines (α₁)	No	Activates glycogen phosphorylase in absence of extracellular Ca^{2+}	Increase
Platelet Activating Factor	No	Activates	Increase
Glucagon and Catecholamines (β)	Yes	Activates	None
Catecholamines (α₂)		Inhibition of adenylyl cyclase not seen in rat hepatocytes	
Vasopressin (V-2)		Stimulation of adenylyl cyclase not seen in rat hepatocytes	

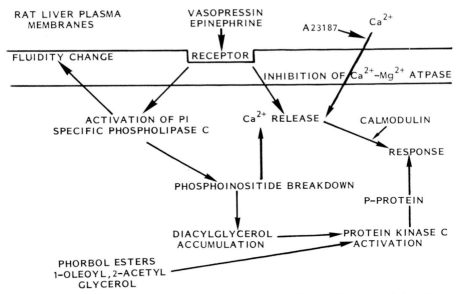

<u>Figure 1</u>. A model for vasopressin and epinephrine action on rat hepato-cytes. Glycogen phosphorylase (the response) is activated by Ca^{2+} in the presence of diacylglycerol. Active phorbol esters and 1-oleoyl-2-acetyl glycerol mimic the action of diacylglycerol on protein kinase C activation. In addition to glycogen phosphorylase activation there is an increase in membrane fluidity and an inhibition of Ca^{2+}-Mg^{2+} ATPase activity. It is unclear what is responsible for these changes but the primary effect of the hormones is probably activation of a phospholipase C which degrades membrane phosphoinositides (PI). Concurrently there is a release of bound Ca^{2+} from intracellular stores owing to the increase of inositol trisphosphate resulting from breakdown of PIP$_2$.

active phorbol ester in the presence of 1 μM A23187 also increased glycogen phosphorylase. It is also of interest that no effects of inhibition of adenylyl cyclase by catecholamines or stimulation by vasopressin have been observed in rat hepatocytes (Table I).

A cyclic AMP-independent mechanism for regulation of glycogen phosphorylase involving diacylglycerol and Ca^{2+} is depicted in Fig. 1. Vasopressin, catecholamines acting on α_1-receptors (α_1-catecholamines) and platelet activating factor elevate cytosolic Ca^{2+} and increase diacylglycerol secondarily to phosphoinositide breakdown. One function for phosphoinositide breakdown is to provide

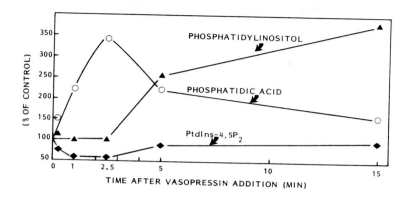

Figure 2. Effect of vasopressin on net uptake of ^{32}P into phosphatidic acid, phosphatidylinositol, and phosphatidylinositol 4,5-bisphosphate. Rat hepatocytes (4-5 x 10^6 cells/ml) were incubated with or without vasopressin (20 milliunits/ml) for the indicated times in phosphate-free Krebs-Ringer bicarbonate buffer containing 5 μCi of ^{32}P/ml. Uptake of ^{32}P into phospholipids in the presence of vasopressin is expressed relative to control hepatocytes incubated for the same length of time. Each time point had a control incubation. Hepatocytes were incubated with ^{32}P for 5-7 min prior to the start of the experiment. The amount of label present in the different phospholipids at the start of the experiment and the end of the experiment was as follows in controls: phosphatidic acid, 1580 and 5345 cpm; phosphatidylinositol, 685 and 1825 cpm; PIP$_2$, 1880 and 2720 cpm. Reproduced with permission from Litosch et al. (1983).

diacylglycerol that activates protein kinase C (Takai *et al.*, 1984). In the presence of low concentrations of A23187 (1 μM), an increase in glycogen phosphorylase activity can be seen in rat hepatocytes due to an active phorbol ester that may involve protein kinase C activation (Fain *et al.*, 1984). An unresolved issue is whether phosphoinositide breakdown has a causal relationship to the release of intracellular stores of bound Ca^{2+} in hepatocytes.

STIMULATION OF PHOSPHATIDYLINOSITOL SYNTHESIS
BY VASOPRESSIN

Long before the possibility of a separate second messenger for epinephrine action on hepatocytes was recognized, De Torrontegui and Berthet (1966) found that epinephrine, but not glucagon, increased the uptake of ^{32}P into all phospholipids by rat liver slices. Subsequently, Kirk et al. (1977, 1981) and Tolbert et al. (1980) found that epinephrine or vasopressin increased uptake of ^{32}P into phosphatidylinositol and phosphatidic acid, but not into other phospholipids. These effects are illustrated in Fig. 2 using hepatocytes which had been incubated with ^{32}P for only 5-7 min prior to the addition of vasopressin. There was an early rise in uptake of label into phosphatidic acid followed by a rise in phosphatidylinositol. There was no effect of the hormone on uptake of label into phosphatidylinositol 4,5-bisphosphate. Under similar conditions, there was no increase due to vasopressin of [^3H]inositol incorporation into phosphatidylinositol (Tolbert et al., 1980).

The relationship between vasopressin or α_1-catecholamine stimulation of ^{32}P uptake into phosphatidylinositol and phosphoinositide breakdown, as well as the rise in cytosolic Ca^{2+} is unknown at the moment. Putney et al. (1980) suggested that phosphatidic acid might serve as a Ca^{2+} ionophore and mediate vasopressin elevation of cytosolic Ca^{2+}. Probably the rise in phosphatidic acid formation is too slow to account for the rise in cytosolic Ca^{2+} (Thomas et al., 1983; Blackmore et al., 1983; Joseph and Williamson, 1983). However, inositol trisphosphate derived from breakdown of phosphatidylinositol 4,5-bisphosphate rather than phosphatidic acid is the current candidate for the intracellular Ca^{2+}-releasing agent (Streb et al., 1983).

It has been assumed that the increase in phosphatidic acid and phosphatidylinositol formation is secondary to an increased availability of diacylglycerol derived from phosphoinositide breakdown. However, this is possibly an over-simplification and other factors may be involved. Ca^{2+} is a known inhibitor of the enzymes involved in the formation of both phospholipids and a decrease in Ca^{2+} at their site of synthesis should relieve an inhibitory constraint on their formation. This has been suggested by Fain (1982). Probably the site for resynthesis of phosphatidic acid and phosphatidylinositol is the plasma membrane since we find CDP-DG:inositol phosphatidyltransferase activity in plasma membrane vesicles that is inhibited by 10 to 100 nM free Ca^{2+} (Wallace and Fain, 1984, unpublished studies).

PHOSPHOINOSITIDE BREAKDOWN

Most recent work has focused on phosphoinositide breakdown in rat hepatocytes. Lin and Fain (1981) and Litosch *et al*. (1983) utilized hepatocytes in which phosphatidylinositol was labeled by injection of [^{3}H]inositol 18 h prior to the isolation of hepatocytes. The addition of vasopressin to these hepatocytes induced a rapid breakdown of phosphoinositides (Fig. 3). There was a marked decrease in labeled phosphatidylinositol after only 7.5 s and the maximal effect was seen after only 30 s incubation with vasopressin. The amount of [^{3}H]inositol present in hepatocyte phosphatidylinositol 4,5-bisphosphate 18 h after its administration was less than 2% of the total radiolabel present in phosphoinositides (Litosch *et al.*, 1983).

Lin and Fain (1981) found a selective loss of [^{3}H]phosphatidylinositol from plasma membrane-rich vesicles isolated after 5 min incubation of hepatocytes with vasopressin. However, it was necessary to homogenize the hepatocytes at 5°C and use conditions which minimized activation of lysosomal phospholipases as well as redistribution of phosphatidylinositol by transfer proteins.

The next step was to demonstrate that the direct addition of vasopressin or epinephrine to isolated rat liver plasma membrane preparations increased phosphoinositide breakdown. Wallace *et al*. (1982) found that vasopressin alone gave a 7% breakdown of total phosphatidylinositol (based on phosphate analysis) while deoxycholate gave a 21% breakdown and the combination produced a 32% breakdown which was significant (Table II). This effect was probably not through activation of phospholipase C by an elevation of media Ca^{2+} by vasopressin since all studies were performed in buffer containing 0.5 mM EGTA.

261

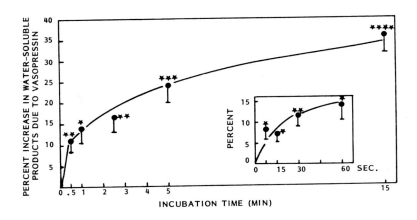

Figure 3. Vasopressin-stimulated appearance of water-soluble tritium label from in vivo [³H]inositol-labeled hepatocytes. Hepatocytes, labeled in vivo with [³H]inositol, were incubated with or without vaso-pressin for the indicated times. Results are shown as the increment in water-soluble label due to vasopressin from four separate hepatocyte preparations. The amount of water-soluble label in control hepatocytes was 840 cpm at the start and 900 cpm at 15 min. * p < 0.025; ** p < 0.01; *** p < 0.005; **** p < 0.001. Reproduced with permission from Litosch et al. (1983).

Subsequently, Fain *et al*. (1983b) found a similar effect on breakdown of labeled phosphatidylinositol in hepatic plasma membranes obtained from rats injected 18 h previously with [³H]inositol. There was a 12% decrease in labeled phosphatidylinositol in membranes incubated with 50 mU/ml of vasopressin in buffer containing 1 mM $CaCl_2$ and 17% breakdown in Ca^{2+}-free buffer containing 0.5 mM EGTA (Table III). Wallace *et al*. (1983) also found an 8% breakdown of total rat hepatic plasma mem-brane phosphatidylinositol due to norepinephrine.

TABLE II

Effects of Vasopressin and Deoxycholate on Phospholipids of Rat Liver Plasma Membranes.

	% change from basal due to added agents		
	0.5 mg/ml Deoxycholate	50 mU/ml Vasopressin	Deoxycholate + Vasopressin
Phosphatidylcholine	+ 8 ± 5	− 1 ± 3	+ 5 ± 2
Phosphatidylethanolamine	+ 4 ± 2	− 2 ± 3	+ 2 ± 3
Phosphatidylserine	+ 1 ± 5	− 3 ± 6	+ 3 ± 5
Phosphatidylinositol	−21 ± 8 [*]	− 7 ± 6	−32 ± 8 [+]

For each experiment, the entire plasma membrane preparation from 40 g of liver (4 rat livers) was resuspended in 3-4 ml of 20 mM N-2-hydroxyethyl-piperazine-N'-2-ethanesulfonic acid buffer, pH 7.4, containing 100 mM KCl, 1 mM $MgCl_2$ and 0.5 mM EGTA. This suspension was diluted 1:1 in buffer with or without deoxycholate and/or vasopressin for incubation at 37°C over 30 min in a final volume of 0.5 ml (2 mg protein/ml). All experiments were done with freshly prepared membranes incubated for 30 min at 37°C. Phospholipids were extracted, chromatographed, and individual phospholipids were assayed for inorganic phosphorus. Results are expressed as percent change from control ± S.E. for 10 experiments each with duplicate samples for all conditions. Reproduced with permission from Wallace et al. (1982).

[*] Significantly lower than control p<0.05, Wilcoxon rank sum analysis.
[+] Significantly lower than either control or deoxycholate alone, p< 0.05, Wilcoxon rank sum analysis.

Fain et al.

TABLE III

Stimulation of Labeled Phosphatidylinositol Degradation by Vasopressin or Catecholamine Addition to Hepatocyte Membranes.

Additions	Plus Ca^{2+}	Ca^{2+}-Free
	% Change due to added hormones	
Vasopressin 10 mU/ml	$- 7 \pm 1$	$- 3 \pm 2$
Vasopressin 50 mU/ml	-12 ± 2	-17 ± 5
Epinephrine 10 μM	-12 ± 3	0 ± 3
Epinephrine 50 μM	-10 ± 3	-10 ± 5

Livers were removed from inositol-deficient rats 18 h after the injection of 50 μCi of myo-[2-^{3}H]inositol. The plasma membranes were incubated for 20 min in buffer containing 100 mM KCl, 20 mM HEPES (pH = 7.4), 10 mM $MgCl_2$, 0.2 mM ATP, 1 mM creatine phosphate plus creatine phosphokinase and 0.5 mM EGTA (for Ca^{2+}-free) or with 1 mM $CaCl_2$ (for Ca^{2+}). The values are the means ± S.E. of 3 paired experiments. [^{3}H]Phosphoinositides added to each tube contained 1650 cpm. Reproduced with permission from Fain et al. (1983b).

The direct effect of norepinephrine on phosphatidylinositol degradation in rat liver plasma membranes was independently reported by Harrington and Eichberg (1983). Their results were similar in that added Ca^{2+} was not required but they found effects only in the presence of cytosol. The localization and nature of the enzyme activated by hormone remain to be elucidated.

ROLE OF Ca^{2+} IN PHOSPHOINOSITIDE BREAKDOWN

Studies in which extracellular Ca^{2+} was depleted by chelation with EGTA demonstrated that a stimulation of glycogen phosphorylase due to vasopressin was abolished while incorporation of ^{32}P into phosphatidylinositol was not abolished (Tolbert et al., 1980). Similar results have been seen by others with respect to stimulation by vasopressin of breakdown of both phosphatidylinositol and phosphatidylinositol 4,5-bisphosphate (Fain et al., 1983b; Billah and Michell, 1979; Kirk et al., 1978). In contrast, Prpic et al. (1982) and Rhodes et al. (1983) observed that prolonged chelation by isolating, washing and then incubating hepatocytes in Ca^{2+}-free buffer containing 1 mM EGTA abolished all effects of vasopressin. These results indicate that after brief chelation of extracellular Ca^{2+}, vasopressin is unable to activate glycogen phosphorylase, but is able to stimulate phosphoinositide turnover which suggests that elevation of Ca^{2+} by vasopressin is not responsible for phosphoinositide turnover.

There are differences between vasopressin and α_1-catecholamine action that remain to be explained. Both hormones appear to elevate cytosolic Ca^{2+} and increase phosphoinositide turnover. However, the activation of glycogen phosphorylase by α_1-catecholamines is not abolished by brief chelation of extracellular Ca^{2+} under conditions in which that of vasopressin is abolished (Fain, 1978; Fain et al., 1983b).

There is no evidence that A23187, even at a concentration of 10 μM, increases phosphoinositide turnover (Litosch et al., 1983; Michell et al., 1981). Takenawa et al. (1982) did find an increase in diacylglycerol accumulation in hepatocytes incubated with A23187 but this was apparently the result of non-specific stimulation of triacylglycerol and phospholipid breakdown. This may account for the ability of 10 μM A23187 to activate glycogen phosphorylase since it both elevates cytosolic Ca^{2+} and increases diacylglycerol production. The data shown in Table IV indicate that at low levels of A23187 (1 μM) there is little activation of glycogen phosphorylase, but if combined with 4β-phorbol 12β-myristate 13α-acetate (PMA) there was a 5-fold greater effect than seen with either agent alone (Fain et al., 1984). These data support the

265

TABLE IV

Stimulation of Rat Hepatocyte Glycogen Phosphorylase by Phorbol Ester in the Presence of A23187

Additions	% increase in glycogen phosphorylase
1 μM A23187	4.7 ± 1.5*
1.6 μM, 4α-phorbol 12β-myristate 13α-acetate	10.6 ± 3.0**
A23187 + phorbol ester	48.0 ± 6.7***

Rat hepatocytes (5.0×10^6/tube) were incubated for 15 min prior to the addition of A23187 or 4α-phorbol 12β-myristate 13α-acetate. The reactions were stopped 2 min after the addition of these agents. The values are the means ± S.E. of 16 paired experimental replications. Reproduced with permission from Fain et al. (1984).

* Significant with $p < 0.01$ by t test.
** Significant with $p < 0.005$ by t test.
*** Significant with $p < 0.001$ by t test.

model depicted in Fig. 1 where diacylglycerol and Ca^{2+} are considered as the twin arms or messengers which synergistically activate glycogen phosphorylase. Similar synergistic effects of PMA and A23187 have been seen with respect to insulin release by rat pancreatic islets (Zawalich et al., 1983), aldosterone secretion in isolated adrenal glomerulosa cells (Kojima et al., 1983), platelet secretion (Rink et al., 1983; Yamanishi et al., 1983) and DNA synthesis by lymphocytes (Mastro and Smith, 1983).

PHOSPHATIDYLINOSITOL 4,5-BISPHOSPHATE BREAKDOWN

Michell *et al.* (1981) first suggested that breakdown of phosphatidylinositol 4,5-bisphosphate by vasopressin in rat hepatocytes is the primary event. All loss of phosphatidylinositol was postulated to be a consequence of phosphorylation to phosphatidylinositol 4-phosphate and then to phosphatidylinositol 4,5-bisphosphate. More recently, Rink *et al.* (1983) suggested that inositol trisphosphate which is the product derived by phosphodiesteratic cleavage of phosphatidylinositol 4,5-bisphosphate (*i.e.*, by a phospholipase C type enzyme) is the second messenger which releases Ca^{2+} from intracellular stores.

The question arises as to why vasopressin or α_1-catecholamines increase phosphatidylinositol breakdown rather than that of phosphatidylinositol 4,5-bisphosphate (PIP_2) in isolated membranes. To date, no one has successfully prepared plasma membranes from rat liver such that they could demonstrate hormone-activated loss of PIP_2. This could be due to rapid metabolism of the compound during isolation. The breakdown of phosphatidylinositol does not seem to require phosphorylation to phosphatidylinositol 4,5-bisphosphate in isolated plasma membranes (Fain *et al.*, 1983b). Our working hypothesis is that vasopressin activates a phospholipase C which degrades any phosphoinositide present in or on the inner surface of the plasma membrane.

Possibly, when cells are incubated for short periods of time with ^{32}P or [^3H]inositol there is preferential labeling of the polyphosphoinositides because of their more rapid turnover. This is illustrated in the studies shown in Fig. 2 utilizing hepatocytes incubated for only 5-7 min with ^{32}P prior to the addition of vasopressin. Under these conditions there was 2-3 times as much label in PIP_2 as in phosphatidylinositol. There was a 50% drop of label in PIP_2 which was maximal 60 s after addition of hormone and by 5 min had returned to control values. In contrast, no net loss of label from phosphatidylinositol was seen and by 5 min there was a net accumulation.

Rather different results were noted if hepatocytes were incubated with [^3H]inositol for 60 min prior to the

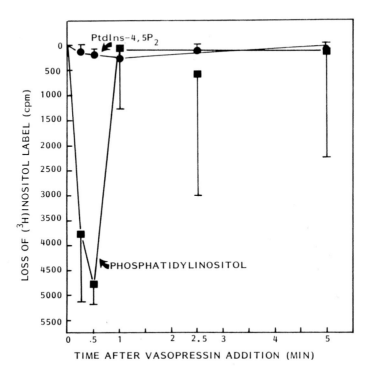

Figure 4. Comparison of changes in [^{3}H]inositol-labeled phospho-inositides based on total changes in radioactivity. Rat hepatocytes labeled for 60 min *in vitro* with [^{3}H]inositol were incubated with or without vasopressin (20 milliunits/ml) for the indicated times. Results are the mean ± S.E. for five hepatocyte preparations plotted as changes in total radioactivity. The change in phosphatidylinositol 4-phosphate at 30 s was 28 cpm. The cpm present in the phosphoinositides at the start of the incubation were 20,150 in phosphatidylinositol, 390 in PIP, and 480 in PIP$_2$. Reproduced with permission from Litosch *et al.* (1983).

addition of vasopressin (Fig. 4). Under these conditions the label present in phosphatidylinositol was 42 times greater than that in PIP$_2$. The addition of vasopressin resulted in a marked drop in label present in phospha-tidylinositol which was near maximal at 15 s, maximal at 30 s, and back to control values by 1 min. The net change in labeled PIP$_2$ was much less.

268

TABLE V

Effect of Stimulation with Vasopressin for 30 Seconds
on Hepatocyte Phospholipids.

Phospholipid	Control	+ Vasopressin
	nmol	
Phosphatidylinositol (20)	54 ± 4	47 ± 3
Phosphatidylinositol 4-phosphate (5)	6.2 ± 1.5	4.9 ± 0.8
Phosphatidylinositol 4,5-bisphosphate (10)	2.3 ± 0.4	3.2 ± 0.5*
Phosphatidylcholine (17)	157 ± 8	150 ± 9
Phosphatidylethanolamine (15)	109 ± 10	104 ± 9
Phosphatidic Acid (7)	9 ± 1	12 ± 1

Hepatocytes, (4 x 10^6 cells/ml) were incubated with or without 20 munits/ml vasopressin for 30 s. Phospholipids were separated by two-dimensional chromatography and the lipid content determined by phosphorus analysis. Results are the mean ± S.E. of N determinations indicated in parenthesis. The increase in PIP_2 due to vasopressin was significant with a $p < 0.01$ by paired comparisons and the increase in phosphatidic acid due to vasopressin was also significant ($p < 0.05$). Reproduced with permission from Litosch et al (1983).

A third type of response to vasopressin is noted if instead of measuring changes in [^{3}H]inositol or ^{32}P label the total content of phosphoinositides was examined (Table V). Thirty seconds after the addition of vasopressin there was a 39% increase in total PIP_2 and a 15% decrease in total phosphatidylinositol content.

These results suggest that vasopressin increases the total formation of PIP_2 as well as its degradation and at 30 s the stimulation of formation is greater than of degradation. One problem is that none of the studies with ^{32}P-labeling has shown an increase in labeled PIP_2 formation at early time points indicating that the rise in total PIP_2 arises from a pool of ATP which equilibrates poorly with rapidly labeled ^{32}P-ATP.

EFFECTS OF VASOPRESSIN AND CATECHOLAMINES ON MEMBRANE FLUIDITY AND Ca^{2+}-Mg^{2+}-ATPase

In addition to breakdown of phosphatidylinositol, the addition of vasopressin or α_1-catecholamine agonists to isolated rat liver plasma membranes results in the release of labeled $^{45}Ca^{2+}$ and increased fluidity (Burgess et al., 1983). A fourth effect of vasopressin and α_1-catechol-amine agonists on hepatocyte plasma membranes is an inhibition of the Ca^{2+}-Mg^{2+}-ATPase activity (Lin et al., 1983). After incubation of rat hepatocytes with vaso-pressin, the activity of Ca^{2+}-Mg^{2+}-ATPase in plasma membranes was decreased by 15-30%. The effect of vasopressin on the activity of this enzyme did not require the presence of extracellular calcium and maximal inhibi-tion was seen after only 15 s exposure to vasopressin. The concentration of vasopressin needed for half-maximal inhibition of this enzyme in hepatocytes was approximately 6 nM (Lin et al., 1983). These findings indicate that the high affinity Ca^{2+}-Mg^{2+}-ATPase of hepatocyte plasma membranes could be involved in the elevation of cyto-plasmic calcium due to vasopressin and α_1-catecholamine agonists.

Possibly the hormone-receptor complex causes the release of Ca^{2+} bound to the receptor or closely associ-ated proteins and this increases the access of phospha-tidylinositol or phosphatidylinositol 4,5-bisphosphate to a phospholipase C enzyme that liberates diacylglycerol and inositol phosphates. Alternatively, the hormone-receptor complex could activate a membrane-bound phospholipase C resulting in breakdown of the membrane-bound phospho-inositides that results in liberation of bound intracellular stores of Ca^{2+} as well as increased influx of Ca^{2+}.

PHOSPHOINOSITIDE TURNOVER IN
BLOWFLY SALIVARY GLANDS

In blowfly salivary glands, breakdown of phospho-inositides has been linked to the receptor mechanism involved in elevation of intracellular Ca^{2+} (Fain and Berridge, 1979a,b; Berridge and Fain, 1979). Both [^{3}H]inositol and ^{32}P are selectively incorporated into a hormone-sensitive pool of phosphoinositides that constitutes less than 5% of the total cellular phosphoinositide (Fain and Berridge, 1979a,b). 5-Hydroxytryptamine (5-HT) stimulates a 20% breakdown of labeled phosphatidylinositol and polyphosphoinositides within 1 min (Litosch et al., 1984) that is associated with a transient formation of diacylglycerol (Litosch et al., 1982). Inositol phosphates accumulate, indicating that hormone-stimulated phosphoinositide breakdown is mediated through a phospholipase C activity. There is a 5-fold greater loss of phosphatidylinositol than PIP_2 during cellular activation, indicating that phosphatidylinositol is the major phosphoinositide degraded (Litosch et al., 1984). Elevation of intracellular Ca^{2+} by the ionophore, A23187, does not stimulate breakdown of phosphatidylinositol and PIP_2 through a phospholipase C, but does increase conversion of PIP_2 to phosphatidylinositol via phosphomonoesterase. Under Ca^{2+}-free conditions 5-HT–stimulated loss of phosphatidylinositol is unaffected, however, the initial loss of PIP_2 is followed by its resynthesis of PIP_2. These results indicate that some conversion of phosphatidylinositol to PIP_2 occurs during hormone action. The major pathway for phosphatidylinositol and PIP_2 degradation, however, is via phospholipase C.

Prolonged stimulation of salivary glands with 5-HT depletes the hormone-regulated pool of phosphoinositides and this results in an inability of hormone to elevate intracellular Ca^{2+} (Fain and Berridge, 1979a,b). Restoration of hormone-stimulated Ca^{2+} mobilization is dependent on the extent of resynthesis of phosphoinositides. At low medium inositol (3–30 µM), there is a partial recovery of the Ca^{2+} gating response. Under these conditions, PIP_2 is preferentially resynthesized from phosphatidylinositol (Sadler et al., 1984). Phosphatidylinositol constitutes less than 20% of the labeled pool. Increasing the medium

271

inositol to 300 µM results in a full restoration of the Ca^{2+} gating response and a proportionately greater synthesis of phosphatidylinositol than PIP_2. In fully recovered glands, phosphatidylinositol constitutes the major phosphoinositide (Sadler *et al.*, 1984). These results demonstrate that restoration of both phosphatidylinositol and PIP_2 levels is necessary for full recovery of Ca^{2+} gating.

In cell-free systems obtained from prelabeled salivary glands, 5-HT stimulates a dose-dependent, rapid breakdown of phosphatidylinositol observed in the presence of 0.5 mM EGTA and in the absence of added Ca^{2+} (Fain *et al.*, 1983a). These findings demonstrate that in blowfly salivary glands, 5-HT stimulates a rapid breakdown of phosphoinositides that is coupled to the ability to elevate intracellular Ca^{2+}. The direct activation of phosphatidylinositol breakdown in a cell-free system links phosphatidylinositol breakdown to a primary receptor event.

REFERENCES

Berridge, M. and Fain, J.N. (1979) *Biochem. J.* *178*, 59–69.

Billah, M.M. and Michell, R.H. (1979) *Biochem. J.* *182*, 661–668.

Blackmore, P.E., Hughes, B.P., Charest, R., Shuman, E.A. and Exton, J.H. (1983) *J. Biol. Chem.* *258*, 10488–10494.

Burgess, G.M., Giraud, F., Poggioli, J. and Claret, M. (1983) *Biochim. Biophys. Acta* *731*, 387–396.

De Torrontegui, G. and Berthet, J. (1966) *Biochim. Biophys. Acta* *116*, 467–476.

Fain, J.N. (1978) in: *"Receptors and Recognition Series 6A"* (P. Cuatrecasas and M.F. Greaves, eds.) pp. 1–62, Chapman and Hall, London.

Fain, J.N. (1982) *Horizons in Biochemistry and Biophysics* *6*, 237–276.

Fain, J.N. and Berridge, M.J. (1979a) *Biochem. J.* *178*, 45–58.

Fain, J.N. and Berridge, M.J. (1979b) *Biochem. J.* *180*, 655–661.

Fain, J.N. and Garcia-Sainz, J.A. (1980) *Life Sci.* *26*, 1183–1194.

Fain, J.N., Lin, S.-H., Litosch, I. and Wallace, M. (1983a) *Life Sci.* *32*, 2055-2068.

Fain, J.N., Lin, S.-H., Randazzo, P., Robinson, S. and Wallace, M. (1983b) in: *"Isolation, Characterization and Use of Hepatocytes"* (R.A. Harris and N.W. Cornell, eds.) pp. 411-418, Elsevier Pub. Co., Amsterdam.

Fain, J.N., Li, S.-Y., Litosch, I. and Wallace, M. (1984) *Biochem. Biophys. Res. Commun. 119*, 88-94.

Harrington, C.A. and Eichberg, J. (1983) *J. Biol. Chem. 258*, 2087-2090.

Hoffman, B.B., Michell, T., Kilpatrick, M.D., Lefkowitz, R.J., Tolbert, M.E.M., Gilman, H. and Fain, J.N. (1980) *Proc. Natl. Acad. Sci. USA 77*, 4569-4573.

Joseph, S.K. and Williamson, J.R. (1983) *J. Biol. Chem. 258*, 10425-10432.

Kirk, C.J., Verrinder, T.R. and Hems, D.A. (1977) FEBS Lett. 83, 267-271.

Kirk, C.J., Verrinder, T.R. and Hems, D.A. (1978) *Biochem. Soc. Trans. 6*, 1031-1033.

Kirk, C.J., Michell, R.H. and Hems, D.A. (1981) *Biochem. J. 184*, 155-165.

Kojima, I., Lippes, H., Kojima, K. and Rasmussen, H. (1983) *Biochem. Biophys. Res. Commun. 116*, 555-562.

Kneer, N.M., Wagner, M.J. and Lardy, H.A. (1979) *J. Biol. Chem. 254*, 12160-12168.

Lin, S.-H. and Fain, J.N. (1981) *Life Sci. 18*, 1905-1912.

Lin, S.-H., Wallace, M.A. and Fain, J.N. (1983) *Endocrinology 116*, 2268-2275.

Litosch, I., Saito, Y. and Fain, J.N. (1982) *Am. J. Physiol. 243*, C222-226.

Litosch, I., Lin, S.-H. and Fain, J.N. (1983) *J. Biol. Chem. 258*, 13727-13732.

Litosch, I., Lee, H.S. and Fain, J.N. (1984) *Am. J. Physiol. 246*, C141-C147.

Mastro, A.M. and Smith, M.C. (1983) *J. Cell. Physiol. 116*, 51-56.

Michell, R.H., Kirk, C.J., Jones, L.M., Downes, C.P. and Creba, J.A. (1981) *Phil. Trans. R. Soc. Lond. B 296*, 123-137.

Prpic, V., Blackmore, P.F. and Exton, J.H. (1982) *J. Biol. Chem. 257*, 11323-11331.

Putney, J.W., Jr., Weiss, S.J., Van de Walle, C. and Haddas, R.A. (1980) *Nature 285*, 345-347.

273

Rhodes, D., Prpic, V., Exton, J.H. and Blackmore, P.F. (1983) *J. Biol. Chem.* *258*, 2770-2773.

Rink, T.J., Sanchez, A. and Hallam, T.J. (1983) *Nature* *305*, 317-319.

Sadler, K., Litosch, I. and Fain, J.N. (1984) *Biochem. J.* (in press).

Sherline, P., Lynch, A. and Glinsmann, W.H. (1972) *Endocrinology* *91*, 680-690.

Shukla, S.D., Buxton, D.H., Olson, M.S. and Hanahan, D.J. (1983) *J. Biol. Chem.* *258*, 10212-10214.

Streb, H., Irvine, R.F., Berridge, M.J. and Schulz, I. (1983) *Nature* *306*, 67-69.

Takai, Y., Kikkawa, U., Kaibuchi, K. and Nishizuka, Y. (1984) *Adv. Cyclic Nucleotide Res.* (in press).

Takenawa, T., Homma, Y. and Nagai, Y. (1982) *Biochem. Pharmacol.* *31*, 2663-2667.

Thomas, A.P., Marks, J.S., Coll, K.E. and Williamson, J.R. (1983) *J. Biol. Chem.* *258*, 5716-5725.

Tolbert, M.E.M., Butcher, F.R. and Fain, J.N. (1973) *J. Biol. Chem.* *248*, 5686-5692.

Tolbert, M.E.M., White, A.C., Aspry, K., Cutts, J. and Fain, J.N. (1980) *J. Biol. Chem.* *255*, 1938-1944.

Wallace, M.A., Randazzo, P., Li, S.-Y. and Fain, J.N. (1982) *Endocrinology* *111*, 341-343.

Wallace, M.A., Poggioli, J., Giraud, F. and Claret, M. (1983) *FEBS Lett.* *156*, 239-243.

Williamson, J.R., Cooper, R.H. and Hoek, J.B. (1981) *Biochim. Biophys. Acta* *639*, 243-295.

Yamanishi, J., Takai, Y., Kaibuchi, K., Sano, K., Castagna, M. and Nishizuka, Y. (1983) *Biochem. Biophys. Res. Commun.* *112*, 778-786.

Zawalich, W., Brown, C. and Rasmussen, H. (1983) *Biochem. Biophys. Res. Commun.* *117*, 448-455.

POLYPHOSPHOINOSITIDES AND MUSCARINIC CHOLINERGIC AND α_1-ADRENERGIC RECEPTORS IN THE IRIS SMOOTH MUSCLE

Ata A. Abdel-Latif, Jack P. Smith and Rashid A. Akhtar

Department of Cell and Molecular Biology
Medical College of Georgia, Augusta, GA   30912

SUMMARY

An attempt was made in this brief review first to recount some of our early studies which culminated in the characterization of the polyphosphoinositide (phosphatidylinositol 4-monophosphate, PIP and phosphatidylinositol 4,5-bisphosphate, PIP_2) effect in the iris smooth muscle, and second to present more recent data on the rapid breakdown of ^{32}P-prelabeled polyphosphoinositide (PPI) and release of [^3H]*myo*-inositol phosphates by carbachol (CCh) in this tissue. The PPI effect is defined as the agonist-stimulated breakdown of PIP_2 into diacylglycerol, measured as labeled phosphatidate, and inositol trisphosphate (IP_3). These early findings included: (1) the demonstration of an agonist-stimulated breakdown of PIP_2 which occurred at relatively short time intervals (2.5 – 10 min), when compared to the phosphatidylinositol (PI) effect reported in a variety of tissues; (2) the demonstration of the PPI effect *in vivo*, in response to electrical stimulation of the sympathetic nerve of the eye; (3) the demonstration, through pharmacologic and adrenergic denervation supersensitivity studies that PIP_2 breakdown is linked to muscarinic cholinergic and α_1-adrenergic receptors; (4) the demonstration that phosphodiesteratic cleavage of PIP_2 into diacylglycerol and IP_3, by PIP_2 phosphodiesterase, is the molecular mechanism underlying the PPI effect; (5) the demonstration of some requirement for Ca^{2+}, derived mainly from studies on the inhibitory effects of EGTA and Ca^{2+} ionophore A23187 on this phenomenon; however, the recent finding that the cationophore-stimulated breakdown of PIP_2 is blocked by prazosin leads us now to conclude that while the PPI effect in the iris needs some Ca^{2+}, it is not regulated by intracellular Ca^{2+}; (6) the demonstration of a close correlation between agonist-stimulated PIP_2 break-down and agonist-induced muscle contraction, which led us to suggest that the agonist-stimulated PIP_2 breakdown is an early event in the pathway which leads from recep-tor activation to muscle response. Data are also pre-sented which demonstrate that in the iris, the breakdown of labeled PIP_2 and release of IP_3 by CCh occur within 15 s; in contrast the release of IP occurred at longer time intervals (>1 min). Thus after incubation for 15 s with CCh there was 48% loss of ^{32}P radioactivity from PIP_2 in ^{32}P-labeled iris, 81% increase in IP_3 release and no change in the release of IP in iris prelabeled with [^3H]inositol. These data suggest that agonist-stimulated PIP_2 breakdown is probably involved in the mechanism of both the phasic (fast) and tonic (slow) components of the contractile response. Neither 2-deoxyglucose nor Li^+, when added for short time intervals (10 min), had any influence on the PPI effect. In accord with our previous studies we conclude that the phosphodiesteratic cleavage of PIP_2 is an early (initial) event in the pathway which leads from activation of Ca^{2+}-mobilizing receptors to muscle response.

INTRODUCTION

It is now a decade since we first reported the agonist-stimulated turnover of phosphatidylinositol (PI) in the rabbit iris smooth muscle, and as in other tissues this phenomenon in the iris is mediated through muscarinic cholinergic and α_1-adrenergic receptors (Abdel-Latif, 1974; Abdel-Latif et al., 1976; Akhtar and Abdel-Latif, 1983). In October of 1975, while the senior author was on a sabbatical at the University of Nottingham, England, we discovered that in iris muscle which was prelabeled with ^{32}P or [^3H]myo-inositol, acetylcholine (ACh) stimulated the breakdown of phosphatidylinositol 4,5-bis-phosphate (PIP_2), and this was accompanied by an increase in the ^{32}P labeling of phosphatidic acid (PA) and phosphatidylinositol (PI) (Abdel-Latif et al., 1977).

Muscarinic and α_1-Adrenergic Stimulated PIP$_2$ Breakdown

Furthermore, we showed that in this tissue the stimulated breakdown of PIP$_2$ is linked to the activation of muscarinic cholinergic (Abdel-Latif *et al.*, 1977) and α-adrenergic receptors (Abdel-Latif *et al.*, 1978). This phenomenon will be referred to here as the "PPI effect". Our original studies on the PPI effect were carried out at time intervals which ranged from 2.5 – 15 min (Abdel-Latif and Akhtar, 1976) and incubations were conducted either under breakdown or under incorporation conditions (Abdel-Latif *et al.*, 1977). It must be emphasized here that in contrast to hepatocytes and other isolated cells, where receptor desensitization seems to occur within seconds after hormonal stimulation, agonist-induced smooth muscle contraction can be maintained for several minutes (Rosenberger and Triggle, 1979). Since these original observations we have investigated extensively both the molecular mechanism and physiological significance of this PPI effect in the iris (for reviews, see Abdel-Latif *et al.*, 1978; Abdel-Latif and Akhtar, 1982; Abdel-Latif, 1983a). Prior to our studies on the iris, Durell *et al.* (1968) noted a possible increased production of inositol 1,4-bisphosphate(IP_2), but not in that of inositol 1,4,5-trisphosphate (IP_3), under conditions of stimulation of brain homogenate with ACh. At this symposium we shall: (a) briefly give an overview of the PPI effect in the iris; and (b) briefly summarize our more recent studies on the rapid breakdown of PIP$_2$ and release of inositol phosphates by carbachol (CCh) in this tissue.

OVERVIEW OF THE PPI EFFECT IN THE IRIS

The Rabbit Iris-Ciliary Body

In rabbit it is difficult to separate the iris from the ciliary body, and thus in our studies we routinely employ the whole iris-ciliary body, which is an instant slice. In mammals, the iris contains two sets of smooth muscles, the sphincter and dilator, and the ciliary body contains the ciliary smooth muscle (for review see Abdel-Latif, 1983b). The sphincter and ciliary muscles are supplied by cholinergic nerve fibers from the oculomotor nerve and the dilator muscle is supplied by adrenergic nerve fibers from the superior cervical sympathetic ganglion. The iris is involved in regulating the amount of light reaching the

277

TABLE I

Time Course of the Effect of ACh on PPI of ^{32}P-Prelabeled Rabbit Iris Muscle

Time of incubation (min)	No. of experiments	^{32}P Radioactivity (cpm/100)			
		PA	PI	PIP	PIP$_2$
0	4	76	63	51	508
2.5	2	72 (101)*	71 (77)	56 (60)	497 (477)
5	2	54 (111)	87 (101)	58 (61)	492 (402)
10	2	43 (75)	117 (142)	59 (60)	259 (159)
15	2	42 (57)	105 (126)	50 (54)	180 (123)

Pairs of irises were prelabeled with ^{32}P by incubation at 37°C in a medium containing 25 μCi of ^{32}P and 10 mM D-glucose in a final volume of 1 ml for 30 min. The ^{32}P-labeled irises were washed four times with excess non-radioactive cold medium, then incubated (of the pair, one was used as control and the other as experimental) in 1 ml non-radioactive medium containing 10 mM D-glucose and 10 mM 2-deoxyglucose. * ACh + serine (0.05 mM each) were added. Taken from Abdel-Latif and Akhtar (1976) with permission.

retina and the ciliary body is involved in aqueous humor formation. Binding studies revealed the presence of muscarinic and α-adrenergic receptors in the iris (Taft *et al.*, 1980).

PIP_2 Breakdown is Linked to Muscarinic and α_1-Adrenergic Receptors in the Iris

When ACh was added to ^{32}P-labeled iris (Table I) there was a significant loss of ^{32}P from PIP_2 and an increase in the labeling of PA and PI at all time intervals investigated (2.5 - 15 min). Thus after 10 min of incubation there was a 39% increase in PIP_2 breakdown and a 74% increase in PA labeling. The ACh-stimulated breakdown of PIP_2 was blocked by atropine, but not by D-tubocurarine (Table II), thus demonstrating for the first time that PIP_2 breakdown is linked to receptor activation (for reviews, see Hawthorne, 1983; Fisher *et al.*, 1984). The PPI effect was observed in the presence and absence of 2-deoxyglucose (*e.g.*, see Table IV, in Abdel-Latif *et al.*, 1977) and in the present study 2-deoxyglucose was found to have no effect on the CCh-stimulated release of inositol phosphates in the iris (see Table VIII below). These findings answer the criticism made recently by Michell and his colleagues (Downes and Michell, 1982; Michell, 1982) against the use of 2-deoxyglucose and glucose for short time intervals in our early studies (Tables I and II). 2-Deoxyglucose was previously used by others (Durell *et al.*, 1968; Schacht and Agranoff, 1974) in their studies on the PI effect in brain, and in more recent studies Kerr *et al.* (1983) administered 2-deoxyglucose as a supplement for fasting for detection of hypoglycemia in children. A particular objective of the employment of 2-deoxyglucose in these early studies was to press the point that our experiments were being carried out under breakdown conditions. Furthermore, this phenomenon was demonstrated *in vivo* in the iris (Abdel-Latif *et al.*, 1978). Thus significant loss of ^{32}P from [^{32}P]PIP_2 was observed in this tissue in response to electrical stimulation of the sympathetic nerve of the eye.

TABLE II

Effects of Atropine and D-Tubocurarine on ACh-Stimulated Breakdown of PIP_2 in Rabbit Iris Muscle Prelabeled with $^{32}P_i$

Additions	Effects of antagonists on ACh-stimulated breakdown of PIP_2 (% of control)		
	PA	PI	PIP_2
ACh	135	138	72
ACh + atropine	104	99	119
ACh + D-tubocurarine	152	143	77

Conditions of incubation were the same as described for Table I, except that the ^{32}P-prelabeled muscle was first incubated in the presence or absence of the antagonist (0.1 mM) for 5 min; then ACh and eserine (0.05 mM each) were added as indicated and the incubation was continued for another 10 min. Taken from Abdel-Latif et al. (1977) with permission.

In conclusion, in the above early studies we have conclusively demonstrated an agonist-stimulated breakdown of PIP_2, which is rapid, when compared to time course studies of the PI effect in a variety of tissues (30 – 120 min), and is linked to muscarinic and α_1-adrenergic receptor activation.

Effects of EGTA and Ca^{2+}

When we initiated these studies in 1977 there were few studies, which were mostly negative, on the requirement for Ca^{2+} in the PI effect (for summary see Abdel-Latif, 1983a). We made the assumption that the PPI effect occurs at the plasma membrane of the cell and thus we asked the question whether Ca^{2+} could be involved in the pathway which leads from receptor activation to PIP$_2$ breakdown and muscle response (Akhtar and Abdel-Latif, 1978a). In Ca^{2+}-free medium ACh stimulated appreciably the breakdown of PIP$_2$ and ^{32}P-labeling of PA, and this was blocked by EGTA (Table III). More recent studies on CCh-stimulated release of inositol phosphates in the iris confirmed these findings (data not shown).

Our studies on the requirement for Ca^{2+} can be summarized as follows: (1) the PPI effect can be demonstrated in a Ca^{2+}-free medium; (2) the PPI effect is inhibited by EGTA, and this inhibition is reversed by Ca^{2+}. Smooth muscle contraction is also inhibited by EGTA (Gabella, 1975); (3) PIP$_2$ breakdown is induced by ionophore A23187. The cationophore also induces smooth muscle contraction (Rosenberger and Triggle, 1979; Warenycia and Vohra, 1983); (4) iris microsomal PIP$_2$ phosphodiesterase is stimulated by low concentrations of Ca^{2+} and is inhibited by EGTA (Akhtar and Abdel-Latif, 1980); (5) addition of ACh or norepinephrine (NE) to iris muscle incubated in a ^{45}Ca^{2+}-containing medium brings about an increase in cellular ^{45}Ca^{2+} in the tissue (Akhtar and Abdel-Latif, 1979). Inclusion of EGTA prevented the reduction of PPI labeling induced by ACh and ionophore A23187 in synaptosomes (Fisher and Agranoff, 1981), abolished the vasopressin-stimulated PIP$_2$ breakdown in hepatocytes (Rhodes et al., 1983) and inhibited the glucose-stimulated breakdown of PPI in isolated islets of Langerhans (Laychock, 1983; Best and Malaisse, 1983). From our calcium studies on the iris we suggested, (a) that this cation could be involved in coupling receptor binding of agonist to activation of PIP$_2$ phosphodiesterase and (b) that Ca^{2+} mobilization (i.e. Ca^{2+} influx) from extracellular sites could follow the breakdown of the PPI (Abdel-Latif and Akhtar, 1982). However, in view of more recent findings in our laboratory we believe that while there could be some Ca^{2+} needed for activation of the phosphodiesterase, agonist-stimulated PIP$_2$ breakdown in the iris is not regulated by intracellular Ca^{2+}. This conclusion is based on the following reasoning: Recently Warenycia and Vohra (1983), working with vas deferens, found that ionophore A23187-induced muscle contraction is blocked by phentolamine. More recent studies on the iris indicate that prazosin, an α_1 blocker, inhibits both NE- and cationo-

TABLE III

Effects of EGTA and Ca^{2+} on ACh-Stimulated Breakdown of PIP_2 and Labeling of PA in Rabbit Iris Muscle Labeled with ^{32}P *in vitro*

Additions			^{32}P-Radioactivity in phosphoinositides (% of control)		
Ca^{2+} (1.25 mM)	ACh (0.05 mM)	EGTA (0.25 mM)	PA	PI	PIP_2
−	−	+	89	101	102
−	+	−	156	117	84
−	+	+	107	111	98
+	+	−	198	117	71
+	+	+	211	112	70

Irises, in pairs, were incubated for 30 min in a Ca^{2+}-free medium which contained 30 µCi ^{32}P/ml. At the end of incubation the irises were washed in non-radioactive Ca^{2+}-free medium that contained 0.25 mM EGTA. The irises were then incubated singly for 10 min in the presence or absence of Ca^{2+}, ACh and/or EGTA as shown in the table. Phospholipids were extracted and separated by means of two-dimensional TLC. Taken from Akhtar and Abdel-Latif (1978a) with permission.

phore-induced release of IP_3 (Akhtar and Abdel-Latif, 1984). This suggests that the cationophore-stimulated PIP_2 breakdown we reported previously (Akhtar and Abdel-Latif, 1978a, 1980) is probably due to the release of NE by the cationophore. Furthermore, it is interesting to note that the Ca^{2+} requirement for the PPI effect in the iris, synaptosomes, hepatocytes, pancreatic islets and several other tissues was demonstrated by using EGTA to chelate Ca^{2+}. The precise mechanisms for the action of EGTA in calcium metabolism are still unclear. EGTA could interact directly with the phosphodiesterase and/or its substrate, PIP_2; or it could interact with the agonist binding sites, as has recently been shown for EDTA (Hulme et al., 1983). Thus while there could be some need for Ca^{2+} in the PPI effect in the iris, we have no experimental evidence to show that this phenomenon is Ca^{2+} regulated. In addition, since only PIP_2 and PA turnover, but not that of PI, was influenced by Ca^{2+} these studies led us to suggest that PIP_2 phosphodiesterase is involved in the PPI effect. Unfortunately, our studies on the requirement for Ca^{2+} in this phenomenon led some investigators to propose that stimulated PIP_2 breakdown is secondary to the agonist-stimulated PI breakdown and Ca^{2+} gating (Michell, 1979). The use of the term increase in "intracellular Ca^{2+}" in our original studies (Akhtar and Abdel-Latif, 1978a) was probably misleading, since we were actually talking about release of this cation at the plasma membrane to activate the PIP_2 phosphodiesterase.

PIP_2 and Contraction Response in Normal and Sympathetically Denervated Iris

Further support for our hypothesis that receptor activation leads to the phosphodiesteratic cleavage of PIP_2 and that this PPI effect occurs early in the pathway which leads from receptor activation to muscle response came from our studies on agonist-stimulated PIP_2 breakdown and contraction responses in normal and denervated irises (Table IV). Among the more interesting findings in these studies were the following: (1) a good correlation was found between the ED_{50} values determined by the PIP_2 and contraction responses; furthermore, for the PIP_2 and contraction responses the dissociation constants for the receptor-antagonist complex (KB), determined with

TABLE IV

Relative Potencies (ED_{50}) of Cholinergic and α-Adrenergic Agonists for Normal and Sympathetically Denervated Irises Obtained for the PIP_2 and Contraction Responses

Iris	Agonist	ED_{50}		Muscle contraction
		PPI response		
		PIP_2 breakdown	PA labeling	
Normal[1]	ACh	6.0×10^{-6}	1.2×10^{-6}	7.6×10^{-6}
Normal[2]	NE	1.0×10^{-5}	1.2×10^{-5}	9.6×10^{-6}
Sympathetically[2] denervated	NE	1.6×10^{-6}	1.9×10^{-6}	3.7×10^{-7}

[1] From Grimes et al., 1979.
[2] From Abdel-Latif et al., 1975, 1979.

Muscarinic and α_1-Adrenergic Stimulated PIP_2 Breakdown

ACh as agonist and atropine as antagonist were 1.7×10^{-10} M and 1.14×10^{-10} M respectively (Grimes et al., 1979); (2) the ED_{50} values for both the PIP_2 and contraction responses decreased significantly for the denervated iris; (3) good correlation was obtained between the ED_{50} values for the PIP_2 and PA, but not PI, responses. Firstly, these studies demonstrated a close correlation between the PIP_2 and PA, but not PI, responses in both normal and denervated iris, and this again suggested to us that PIP_2 phosphodiesterase is involved in the PPI effect. Secondly, they showed a relationship between α_1-adrenergic stimulation of PIP_2 breakdown, receptor supersensitivity and muscle contraction.

Phosphodiesteratic Cleavage of PIP_2 is the Underlying Mechanism for the PPI Effect in the Iris

The studies on: (a) the Ca^{2+} requirement for the PPI effect; (b) the properties of PIP_2 phosphodiesterase and (c) the relative potencies (ED_{50}) of agonists stimulating PIP_2 breakdown in normal and denervated iris led us to conclude that activation of muscarinic cholinergic and α_1-adrenergic receptors must lead to the phosphodiesteratic cleavage of PIP_2 into DG, measured as PA, and IP_3 (eqn 1, from Akhtar and Abdel-Latif, 1980).

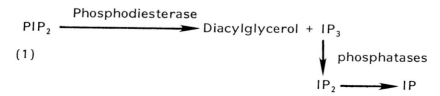

$$PIP_2 \xrightarrow{\text{Phosphodiesterase}} \text{Diacylglycerol} + IP_3$$

(1)

$$IP_3 \xrightarrow{\text{phosphatases}}$$

$$IP_2 \longrightarrow IP$$

Since efforts to monitor the PPI effect in the iris by measuring the production of diacylglycerol (DG) were unsuccessful, we decided to measure the release of the water-soluble inositol phosphates employing low-voltage paper electrophoresis. Previously the release of inositol phosphates was monitored by paper electrophoresis in brain homogenates (Durell et al., 1968) and erythrocytes (Allan and Michell, 1978) and by anion-exchange chromatography in synaptosomes (Griffin and Hawthorne, 1978). Addition of ACh to iris muscle prelabeled with [^{3}H]myo-inositol provoked a loss of radioactivity from PIP_2 and an

TABLE V

Effect of ACh on the Radioactivity of PIP_2 and Water-Soluble Inositol Phosphates in Iris Muscle Prelabeled with [^3H]*myo*-Inositol *In Vitro*

	^3H Incorporated (dpm)					
	PI	PIP	PIP_2	IP	IP_2	IP_3
Control	24315	3899	10207	1560	141	287
ACh (50 μM)	30201	3502	7353	2315	142	382

Paired irises were incubated at 37°C in a medium containing 10 μCi of [^3H]myo-inositol in a final volume of 1 ml for 60 min. The labeled irises were washed with non-radioactive medium, then incubated singly (of the pair, one was used as a control and the other as experimental) in 1 ml medium for 15 min in the presence or absence of ACh. The phospholipids were analyzed by two-dimensional TLC and the water-soluble inositol phosphates were analyzed by low-voltage paper electrophoresis. Taken from Akhtar and Abdel-Latif (1980) with permission.

increase in that of IP_3 (Table V), and this stimulation was blocked by atropine, but not by D-tubocurarine (Akhtar and Abdel-Latif, 1980). It was concluded from this study that activation of muscarinic cholinergic receptors leads to the phosphodiesteratic cleavage of PIP_2 into DG and IP_3 (eqn 1).

Effects of ACTH on PPI Metabolism and Protein Phosphorylation in the Iris

Employing iris microsomal fractions, it was observed that the ^{32}P-labeling of PIP_2 and various phosphoproteins was rapid (30 - 60 s) and as in brain (Jolles *et al.*, 1980), it was affected by ACTH (Akhtar *et al.*, 1983). Addition of $ACTH_{1-24}$ stimulated labeling of PIP_2 and inhibited that of various proteins, including one resembling the B50 protein. The neuropeptide effects were found to be independent of Ca^{2+}.

RAPID BREAKDOWN OF PPI AND RELEASE OF INOSITOL PHOSPHATES BY CARBACHOL IN THE IRIS

In the past year interest in the PPI effect was rekindled by its demonstration in a variety of tissues (Table VI), and by the finding that it occurs rather rapidly (15 – 60 s), when compared to the PI effect (2 – 5 min). Another interesting observation was the finding that Li^+, which inhibits IP-phosphatase (Hallcher and Sherman, 1980) can be used in these studies to amplify the agonist-dependent accumulation of IP (Berridge et al., 1982). The water-soluble inositol phosphates, the products of phosphoinositide breakdown, were isolated by means of anion-exchange chromatography. In view of these findings we decided to re-investigate the effect of CCh on PIP_2 breakdown in the iris at shorter time intervals (15 – 60 s).

Rapid Breakdown of PPI by Carbachol in ^{32}P-labeled Iris

As can be seen from Figure 1 the loss of ^{32}P from PIP_2 in ^{32}P-labeled iris occurred rapidly at all time intervals investigated (15 – 300 s). Thus after 15 s of incubation about 48% of the ^{32}P radioactivity in PIP_2 was lost as the result of stimulation by CCh.

Rapid Release of [^3H]Myo-inositol Phosphates by Carbachol in Iris Prelabeled with [^3H]Myo-inositol

As can be seen from Table VII, CCh induced a rapid release of IP_2 and IP_3, but not of IP, at all time intervals investigated (15 – 60 s). Thus, after 15 s of incubation with CCh the increase in the release of IP_2 and IP_3 was 37 and 81%, respectively; in contrast no change was observed in that of IP. These data are in agreement with our previous conclusion that the PPI effect occurs rapidly in the iris (Abdel-Latif et al., 1977). It also suggests that the phosphodiesteratic cleavage of PPI but not PI, is the initial reaction in the receptor-stimulated inositol lipid metabolism. It has recently been suggested that the disappearance of PI may be secondary to the PPI effect (Creba et al., 1983; Berridge, 1983; for a different view, see Litosch et al., 1983). In our earlier studies an attempt was made to treat the PI and PPI effects as two separate phenomena (Abdel-Latif et al., 1977; see also, Hawthorne, 1982). We have no experimental evidence to

TABLE VI

Receptors Coupled to PIP_2 Breakdown in Target Tissues

Tissue	Receptor	References
Iris smooth muscle	Muscarinic[a]	Abdel-Latif et al. 1977
Iris smooth muscle	α_1-Adrenergic[a] (in vitro and in vivo)	Abdel-Latif et al. 1978
Iris smooth muscle	Muscarinic[b]	Akhtar and Abdel-Latif, 1980
Brain synaptosomes	Muscarinic[a]	Fisher and Agranoff, 1981
Hepatocytes	Vasopressin[a]	Kirk et al., 1981; Creba et al., 1983; Thomas et al., 1983; Rhodes et al., 1983; Litosch et al., 1983
Parotic acinar cells	Muscarinic[a]	Weiss et al., 1982
Platelets	Thrombin[a,b]	Agranoff et al., 1983; Billah and Lapetina, 1982
Parotid gland	Substance P[b] Muscarinic[b]	Berridge et al., 1983
Blowfly salivary gland	$5-HT_1$[b]	Berridge, 1983
Pancreatic acini	Muscarinic[a] Caerulein[a]	Putney et al., 1983
Pancreatic islets	Glucose[a]	Laychock, 1983
Pituitary cells	TRH[a,b]	Rebecchi and Gershengorn, 1983; Martin, 1983

Method of assay [a] Loss of ^{32}P from PIP_2; [b] Release of IP_3

Muscarinic and α₁-Adrenergic Stimulated PIP₂ Breakdown

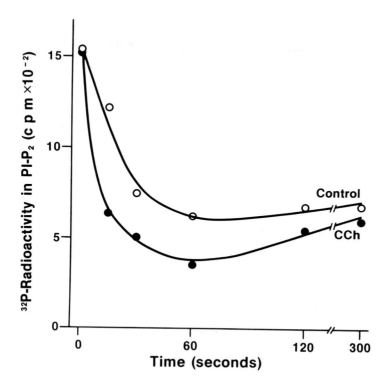

Figure 1. Time course for CCh-stimulated breakdown of PIP₂ in ³²P-labeled iris muscle. Pairs of irises were prelabeled with ³²P by incubation in 1 ml of Krebs-Ringer (pH 7.4) that contained 10 mM glucose and 25 μCi of ³²P for 1 h at 37°C. The ³²P-labeled irises were then incubated in non-radioactive Krebs-Ringer for 1 h, then incubated (of the pair, one was used as control and the other as experimental) in 1 ml of non-radioactive medium containing 10 mM D-glucose in the absence and presence of CCh (50 μM) for various time intervals. Phospholipids were extracted and [³²P]PIP₂ was separated by one-dimensional TLC as previously described (Akhtar et al., 1983). Each point represents an average of three determinations.

show a disappearance of PI in the iris. This is complicated by the fact that the release of IP in this tissue is about 8 times as high as that of IP₃ (Table VII), and by the recent demonstration of an agonist-stimulated breakdown on PI in plasma membranes of rat liver (Lin and Fain, 1981; Harrington and Eichberg, 1983).

TABLE VII

Time Course of the Effect of CCh on the Rapid Release of
[^{3}H]Inositol Phosphates in Rabbit Iris Muscle

Time of incubation (seconds)	Inositol Phosphates (cpm)								
	IP			IP$_2$			IP$_3$		
	Control	Exp.*	% of control	Control	Exp.*	% of control	Control	Exp.*	% of Control
15	13502	13406	99	1729	2376	137	1790	3241	181
30	11044	11728	106	1470	1792	122	1820	2920	160
60	10692	11484	107	1652	1998	121	1999	3284	164

In this experiment, pairs of irises were incubated in 1 ml of Krebs-Ringer bicarbonate buffer (pH 7.4) containing 10 mM D-glucose, 1.6 mM cytidine and 5 μM [^{3}H]myo-inositol (7.3 μCi/ml) for 90 min, at 37°C. CCh was then added to one of the pair and incubation continued for various short time intervals as indicated. The incubations were terminated with 10% (w/v) TCA. [^{3}H]Inositol phosphates were extracted and applied to AG 1 × 4 (formate form) columns. Free [^{3}H]myo-inositol was eluted with water. This was followed by elution of the columns with linear 0-1 M ammonium formate gradient in 0.1 N formic acid (26 ml); this step eluted glycerophosphoinositol, IP and IP$_2$. IP$_3$ was eluted by passing 15 ml of 1 M ammonium formate/0.1 N formic acid. The values are means of two different experiments.
* In the presence of CCh (50 μM).

Effects of Lithium and 2-Deoxyglucose on Carbachol-Stimulated Release of Inositol Phosphates

Exposure of the iris to 2-deoxyglucose for short time intervals had no effect on the CCh-induced release of inositol phosphates (Table VIII). Similarly, addition of Li$^+$ for 10 min had no effect on the PPI effect. However, when the iris was incubated in a medium containing [^{3}H]myo-inositol and Li$^+$ (10 mM) for longer time intervals (>30 min) there was an increase in the accumulation of IP, and this was accompanied by an increase in incorporation of the isotope into the phosphoinositides (data not shown).

CONCLUSIONS AND REMARKS

An attempt was made in this brief review to recount some of our early studies which culminated in the characterization of the PPI effect in the iris muscle. Stimulated PIP$_2$ breakdown is now believed to be controlled by all of those receptors that control a rise in Ca^{2+}, which then acts to provoke physiological responses. Calcium-mobilizing receptors, which have been shown to control PIP$_2$ breakdown, include muscarinic cholinergic, α_1-adrenergic, 5-hydroxytryptamine (5-HT$_1$), histamine (H$_1$), substance P, ACTH, thrombin and platelet activating factor. In addition, data were presented in which the CCh-stimulated [^{32}P]PIP$_2$ breakdown and CCh-stimulated release of [^{3}H]myo-inositol phosphates were monitored by TLC and anion-exchange chromatography, respectively. As we have concluded previously (Abdel-Latif et al., 1977) the PPI effect occurs rapidly in the iris (15 s in the present study), when compared to the PI effect which occurs at longer time intervals (>1 min). This could suggest that phosphodiesteratic cleavage of PIP$_2$, but not PI, is the initial reaction in the receptor-stimulated inositol lipid metabolism. Furthermore, it is in agreement with the concept that the PPI effect is a plasma membrane event which is linked to the activation of calcium-mobilizing receptors, and may play an important role in transducing the muscarinic and α_1-adrenergic signals. The mechanism of coupling between receptor activation and PIP$_2$ breakdown remains to be established. Based on studies with EGTA and ionophore A23187 we suggested previously that Ca^{2+} may be involved in coupling receptor binding of

TABLE VIII

Effects of Lithium Ion and 2-Deoxyglucose on CCh-Stimulated Release of Inositol Phosphates

Additions	Effects of pharmacological agents on release of inositol phosphates (% of control)		
	IP	IP_2	IP_3
CCh	183	179	197
CCH + Li^+	167	172	192
CCh + Li^+ + atropine	104	111	107
CCh + 2-deoxyglucose	179	168	188

In this experiment single irises were preincubated in Krebs-Ringer medium containing [^{3}H]myo-inositol for 85 min, then Li^+ (10 mM), atropine (10 µM) or 2-deoxyglucose (10 mM) were added as indicated and incubation continued for 5 min. CCh (50 µM) was then added to one of the pairs and incubation continued for 10 min. The inositol phosphates were extracted and separated by means of anion-exchange chromatography as discussed in Table VII above. The data are averages of three different experiments.

agonist to activation of PIP_2 phosphodiesterase at the plasma membrane of the iris (Akhtar and Abdel-Latif, 1978a, 1980). In view of the present finding that the cationophore-induced PIP_2 breakdown is secondary to the release of NE by the ionophore, and the fact that EGTA could inhibit the PPI effect either by directly interacting with the phosphodiesterase or the receptor protein, we conclude that, although there may be some need for Ca^{2+} at the enzyme-substrate level, this phenomenon is not regulated by intracellular Ca^{2+} in the iris muscle. The role of Ca^{2+} in the phosphoinositide effects is controversial at the present time (e.g. see Hawthorne, 1983).

Muscarinic and α_1-Adrenergic Stimulated PIP$_2$ Breakdown

The iris is a complex structure, and although extra-cellular Ca^{2+} and Na^+ are probably needed for smooth muscle contraction (Rosenberger and Triggle, 1979) and for PIP$_2$ breakdown (Akhtar and Abdel-Latif, 1982) we believe that the precise role for these cations in the PPI effect in the iris remains to be determined.

Our current thoughts on the role of stimulated PIP$_2$ breakdown in mediating Ca^{2+}-mobilizing receptors function in the iris are summarized in Figure 2. Activation of the Ca^{2+}-mobilizing receptors (R) leads to the breakdown of PIP$_2$ into DG and the water-soluble IP$_3$. In the iris PIP$_2$-PDE is both soluble and membranous (Akhtar and Abdel-Latif, 1978b). The fate and probable functions of DG and IP$_3$ could be as follows: (a) DG is phosphorylated into PA, via the plasma membrane DG kinase. Since an exchange protein for PA has not yet been identified (Somerharju *et al.*, 1983) we can assume that PA is metabolized by PA phosphohydrolase into DG and P$_i$. In the iris PA phosphohydrolase is both soluble and membranous (Abdel-Latif and Smith, 1984). (b) DG is metabolized by DG lipase and monoacylglycerol lipase to liberate arachidonic acid for prostaglandin (PG) biosynthesis (Okazaki *et al.*, 1981). Ca^{2+}-ionophore increased significantly PGE$_2$ release in the iris (Yousufzai and Abdel-Latif, 1983). Since PGE$_2$ has been shown to stimulate smooth muscle contraction, the action of the cationophore on the PPI effect could also be secondary to the release of PGE$_2$ by the ionophore. (c) DG could activate a protein kinase C-phospholipid complex (Nishizuka, 1983). (d) IP$_3$, the other product of PIP$_2$ breakdown, is metabolized by specific phosphatases to release free *myo*-inositol, which is reutilized in PIP$_2$ synthesis. A preliminary report indicates that IP$_3$ can increase Ca^{2+} efflux from a non-mitochondrial intracellular store in pancreatic acinar cells (Berridge, 1983; Streb *et al.*, 1983). (e) The structural and conformational changes which may result from the release of IP$_3$ from PIP$_2$ could cause some permeability changes at the plasma membrane. In smooth muscle, responses elicited by Ca^{2+}-mobilizing receptors have phasic ($\sim$30 s) and tonic (>1 min) components (Rosenberger and Triggle, 1979). It is possible that the stimulated release of IP$_3$ observed in the iris could be involved in both the phasic and tonic components of the contractile response.

293

Abdel-Latif et al.

Extracellular

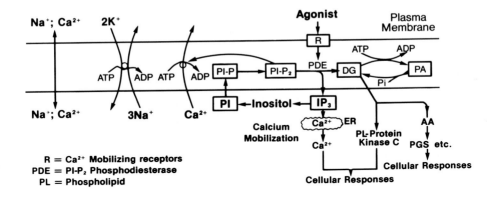

Figure 2. Scheme showing PIP_2 breakdown mediating the step between activation of Ca^{2+}-mobilizing receptors and cellular responses.

Activation of Ca^{2+}-mobilizing receptors also leads to depletion of PIP_2 at the plasma membrane. This could inhibit the Ca^{2+}-pump mechanism and consequently increase intracellular Ca^{2+} concentration. The removal of Ca^{2+} from cells is accomplished by a plasma membrane $Ca^{2+}-Mg^{2+}$-ATPase and there is some evidence suggesting links between this enzyme and PIP_2 (Buckley and Hawthorne, 1971; Peniston, 1982; Koutouzov et al., 1982; Volpi et al., 1983). Thus Peniston (1982) reported stimulation by PIP_2 of a $Ca^{2+}-Mg^{2+}$-ATPase from rat brain synaptosomal plasma membranes. Maximum stimulation was obtained with 2 µM PIP_2. The activation was greater than that produced by saturating levels of calmodulin. We believe that a possible link between Ca^{2+} pumping and Na^+ pumping and PIP_2 at the plasma membrane is worthy of investigation in more detail.

ACKNOWLEDGEMENTS

This work was supported by USPHS Grants EY04171 and EY04387 from the National Eye Institute. This is contribution number 0822 from the Department of Cell and Molecular Biology, Medical College of Georgia. We are grateful to Thomas Hadden and Joseph Latif for technical assistance.

REFERENCES

Abdel-Latif, A.A. (1974) *Life Sci. 15*, 961-973.
Abdel-Latif, A.A. (1983a) in: *"Handbook of Neurochemistry"*, Vol. 3, 2nd Ed., (A. Lajtha, ed.) pp. 91-131, Plenum Publishing Corp., New York.
Abdel-Latif, A.A. (1983b) in: *"Biochemistry of the Eye"* (R.E. Anderson, ed.) pp. 48-78, Am. Acad. Ophthalmol., San Francisco, California.
Abdel-Latif, A.A. and Akhtar, R.A. (1976) *Biochem. Soc. Trans. 4*, 317-321.
Abdel-Latif, A.A. and Akhtar, R.A. (1982) in: *"Phospholipids in the Nervous System, Vol. I: Metabolism"* (L.A. Horrocks, G.B. Ansell and G. Porcellati, eds.) pp. 251-264, Raven Press, New York.
Abdel-Latif, A.A. and Smith, J.P. (1984) *Canad. J. Biochem. Cell. Biol. 62*, 170-177.
Abdel-Latif, A.A., Green, K., Matheny, J.L., McPherson, J.C. and Smith, J.P. (1975) *Life Sci. 17*, 1821-1828.
Abdel-Latif, A.A., Owen, M.P. and Matheny, J.L. (1976) *Biochem. Pharmacol. 25*, 461-469.
Abdel-Latif, A.A., Akhtar, R.A. and Hawthorne, J.N. (1977) *Biochem. J., 162*, 61-73.
Abdel-Latif, A.A. Akhtar, R.A. and Smith, J.P. (1978a), in: *"Cyclitols and Phosphoinositides"* (Wells, W.W. and Eisenberg, F., Jr., eds.) pp. 121-143, Academic Press, New York.
Abdel-Latif, A.A., Green, K., Smith, J.P., McPherson, J.C. and Matheny, J.L. (1978b) *J. Neurochem. 30*, 517-525.
Abdel-Latif, A.A., Green, K. and Smith, J.P. (1979) *J. Neurochem. 32*, 225-228.
Agranoff, B.W., Murthy, P. and Seguin, E.B. (1983) *J. Biol. Chem. 258*, 2076-2078.
Akhtar, R.A. and Abdel-Latif, A.A. (1978a) *J. Pharmacol. Exp. Ther. 204*, 655-668.
Akhtar, R.A. and Abdel-Latif, A.A. (1978b) *Biochim. Biophys. Acta 527*, 159-170.
Akhtar, R.A. and Abdel-Latif, A.A. (1979) *Gen. Pharmacol. 10*, 445-450.

Akhtar, R.A. and Abdel-Latif, A.A. (1980) *Biochem. J.* *192*, 783-791.

Akhtar, R.A. and Abdel-Latif, A.A. (1982) *J. Neurochem.* *39*, 1374-1380.

Akhtar, R.A. and Abdel-Latif, A.A. (1983) *Exp. Eye Res.* *36*, 103-112.

Akhtar, R.H. and Abdel-Latif, A.A. (1984) *Biochem. J.* *224*, 291-300.

Akhtar, R.A., Taft, W.C. and Abdel-Latif, A.A. (1983) *J. Neurochem.* *41*, 1460-1468.

Allan, D. and Michell, R.H. (1978) *Biochim. Biophys. Acta* *508*, 277-286.

Berridge, M.J. (1983) *Biochem. J.* *212*, 849-858.

Berridge, M.J., Downes, C.P. and Hanley, M.R. (1982) *Biochem. J.* *206*, 587-595.

Berridge, M.J., Dawson, R.M.C., Downes, C.P., Heslop, J.P. and Irvine, R.F. (1983) *Biochem. J.* *212*, 473-482.

Best, L. and Malaisse, W.J. (1983) *Biochem. Biophys. Res. Commun.* *116*, 9-16.

Billah, M.M. and Lapetina, E.G. (1982) *J. Biol. Chem.* *257*, 12705-12708.

Buckley, J.T. and Hawthorne, J.N. (1972) *J. Biol. Chem.* *247*, 7218-7223.

Creba, J.A., Downes, C.P., Hawkins, P.T., Brewster, G., Michell, R.H. and Kirk, C.J. (1983) *Biochem. J.* *212*, 733-747.

Downes, P. and Michell, R.H. (1982) *Cell Calcium* *4*, 467-502.

Durell, J., Sodd, M.A. and Friedel, R.O. (1968) *Life Sci.* *7*, 363-368.

Fisher, S.K. and Agranoff, B.W. (1981) *J. Neurochem.* *37*, 968-977.

Fisher, S.K., Van Rooijen, L.A.A. and Agranoff, B.W. (1984) *Trends Biochem. Sci.* *9*, 53-56.

Gabella, G. (1975) *J. Physiol.* *249*, 28P-29P.

Griffin, H.D. and Hawthorne, J.N. (1978) *Biochem. J.* *176*, 541-552.

Grimes, M.J., Abdel-Latif, A.A. and Carrier, G.O. (1979) *Biochem. Pharmacol.* *28*, 3213-3219.

Hallcher, L.M. and Sherman, W.R. (1980) *J. Biol. Chem.* *255*, 10896-10901.

Harrington, C.A. and Eichberg, J. (1983) *J. Biol. Chem.* *258*, 2087-2090.

Hawthorne, J.N. (1982) *Nature* *295*, 281-282.

Hawthorne, J.N. (1983) *Biosci. Reports* *3*, 887-904.
Hulme, E.C., Berrie, C.P., Birdsall, N.J.M., Jameson, M. and Stockton, J.M. (1983) *Eur. J. Pharmacol.* *94*, 59-72.
Jolles, J., Zwiers, H., Dekker, A., Wirtz, K.W.A. and Gispen, W.H. (1981) *Biochem. J.* *194*, 283-291.
Kerr, D.S., Hansen, I.L. and Levy, M.M. (1983) *Metabolism*, *32*, 951-959.
Kirk, C.J., Creba, J.A., Downes, C.P. and Michell, R.H. (1981) *Biochem. Soc. Trans.* *9*, 377-379.
Koutouzov, S., Marche, P., Cloix, J.-F. and Myer, P. (1982) *Am. J. Physiol.* *11*, H590-H597.
Laychock, S.G. (1983) *Biochem. J.* *216*, 101-106.
Lin, S.-H. and Fain, J.N. (1981) *Life Sci.* *29*, 1905-1912.
Litosch, I., Lin, S. and Fain, J.N. (1983) *J. Biol. Chem.* *258*, 13727-13732.
Martin, T.F.J. (1983) *J. Biol. Chem.* *258*, 14816-14822.
Michell, R.H. (1979) *Trends Biochem. Sci.* *4*, 128-131.
Michell, R.H. (1982) in: "*Phospholipids in the Nervous System, Vol. 1: Metabolism*" (Horrocks, L.A., Ansell, G.B. and Porcellati, G., eds.) pp. 315-325, Raven Press, New York.
Nishizuka, Y. (1983) *Trends Biochem. Sci* *8*, 13-16.
Okazaki, T., Sagawa, N., Okita, J.R., Bleasdale, J.E., MacDonald, P.C. and Johnston, J.M. (1981) *J. Biol. Chem.* *256*, 7316-7321.
Peniston, J.T. (1982) *Ann. N.Y. Acad. Sci.* *402*, 296-303.
Putney, J.W., Burgess, G.M., Halenda, S.P., McKinney, J.S. and Rubin, R.P. (1983) *Biochem. J.* *212*, 483-488.
Rebecchi, M. and Gershengorn, M.C. (1983) *Biochem. J.* *216*, 287-294.
Rhodes, D., Prpic, V., Exton, J.H. and Blackmore, P.F. (1983) *J. Biol. Chem.* *258*, 2770-2773.
Rosenberger, L.B. and Triggle, D.J. (1979) *Canad. J. Physiol. Pharmacol.* *57*, 348-358.
Schacht, J. and Agranoff, B.W. (1974) *J. Biol. Chem.* *249*, 1551-1557.
Somerharju, P., van Paridon, P. and Wirtz, K.W.A. (1983) *Biochim. Biophys. Acta* *731*, 186-195.
Streb, H., Irvine, R.F., Berridge, M.J. and Schulz, I. (1983) *Nature* *306*, 67-69.

Abdel-Latif et al.

Taft, W.C., Abdel-Latif, A.A. and Akhtar, R.A. (1980) *Biochem. Pharmacol.* *29*, 2713-2720.
Thomas, A.P., Marks, J.S., Coll, K.E. and Williamson, J.R. (1983) *J. Biol. Chem.* *258*, 5716-5725.
Volpi, M., Yassin, R., Naccache, P.H. and Sha'afi, R.I. (1983) *Biochem. Biophys. Res. Commun.* *112*, 957-964.
Warenycia, M.W. and Vohra, M.M. (1983) *Canad. J. Physiol. Pharmacol.* *61*, 97-101.
Weiss, S.J., McKinney, J.S. and Putney, J.W. (1982) *Biochem. J.* *206*, 555-560.
Yousufzai, S.Y.K. and Abdel-Latif, A.A. (1983) *Exp. Eye Res.* *37*, 279-292.

THE PHOSPHATIDYLINOSITOL EFFECT IN THE RETINA:

ROLE OF LIGHT AND NEUROTRANSMITTERS

Robert E. Anderson and Joe G. Hollyfield

Cullen Eye Institute and Program in Neuroscience
Baylor College of Medicine, Houston, Texas 77030

SUMMARY

We are studying the phosphatidylinositol (PI) effect in retinas from *Xenopus laevis*. Light stimulates the incorporation of labeled inositol and phosphate, but not of glycerol, into phosphoinositides. Light and electron microscope autoradiography following incubation with [2-^{3}H]inositol demonstrates the increased labeling to be localized exclusively to the horizontal cells of the outer plexiform layer. Acetylcholine stimulates the incorporation of [2-^{3}H]inositol into PI in horizontal cells in the dark; atropine blocks this response. Glycine inhibits the light response, and strychnine reverses this effect. GABA, the neurotransmitter of *Xenopus* horizontal cells, dopamine, and norepinephrine have no effects on the incorporation of [2-^{3}H]inositol into phosphoinositides. The synthesis of phosphoinositides in *Xenopus* retinal horizontal cells appears to be regulated by at least three extracellular effectors: light and acetylcholine, both of which are stimulatory, and glycine, which is inhibitory. Under physiological conditions, the relative activity of each of these inputs precisely controls the rate of phosphoinositide turnover in this cell.

INTRODUCTION

A variety of stimuli have been shown to produce the phosphatidylinositol (PI) effect in many tissues (Michell, 1975; Abdel-Latif, 1983). We have observed that light will cause this effect in retinas of the African clawed toad, *Xenopus laevis*. The PI effect was identified biochemically by the selective incorporation of phosphate and inositol, but not of glycerol and other lipid precursors, into phosphoinositides and phosphatidic acid (PA) (Anderson and Hollyfield, 1981; Anderson *et al.*, 1983). The effect was localized by autoradiography to a single retinal neuron, the horizontal cell (Anderson and Hollyfield, 1981; Anderson *et al.*, 1983). It was also shown that certain putative neurotransmitter substances in the retina have a profound effect on the PI response in this tissue (Anderson and Hollyfield, 1984).

BIOCHEMICAL STUDIES OF THE PI RESPONSE IN THE RETINA

Retinas were dissected from dark-adapted *Xenopus laevis* under either dim red light or an infra-red image converter. In a typical experiment, groups of at least eight retinas each were incubated at 21°C for two hours in a Ringer's-bicarbonate-pyruvate medium that contained a radioactive lipid precursor. Retinas that were to remain in the dark were placed in flasks that had been double-wrapped with black electrical tape. Flasks containing light- or dark-incubated retinas were placed in the same Dubnoff metabolic incubator under a 15 watt incandescent light approximately 50 cm above the surface. Incubations were terminated by the addition of ice-cold trichloroacetic acid (final concentration 10%) and lipids were extracted with chloroform:methanol:HCl (100:100:6). After the usual washing procedures, an aliquot of the lipid extract was removed for phosphorus assay and for the determination of radioactivity in lipids. Another aliquot was applied to a thin-layer chromatoplate for determination of radioactivity in individual phospholipid classes.

Retinas incubated with [2-^{3}H]*myo*-inositol or ^{32}PO$_4$ incorporated significantly greater amounts of radioactivity into lipid in the light compared to those incubated in total

TABLE I

Effect of Light on the Incorporation of Radiolabeled Precursors into Lipids of Retinas From *Xenopus Laevis*[a]

Precursor	% Increase in light	P-Value	
[2-^{3}H]inositol	26	<0.05	(18)[b]
	21	<0.05	(18)[b]
	23	<0.05	(16)
	34	<0.001	(27)
	33	<0.025	(18)
^{32}PO$_4$	18	<0.025	(18)
	85	<0.001	(18)
	21	<0.025	(18)
[2-^{3}H]glycerol	-9	>0.4	(18)
	19	>0.2	(17)
	9	>0.3	(18)
	2	>0.9	(18)
	1	>0.9	(6)
[3-^{3}H]serine	11	>0.2	(18)
	19	>0.3	(18)
Methyl[^{3}H]choline	2	>0.4	(18)
	0	—	(17)
[1-^{3}H]ethanolamine	-11	>0.2	(18)
	-12	>0.1	(18)

[a] Data from Anderson et al. (1983).
[b] Statistical significance determined from raw data expressed as DPM/μg lipid phosphorus. Values in parentheses are the number of independent determinations.

darkness (Table I). Most of the radioactive inositol was incorporated into PI (83-86%), with the remainder being distributed in phosphatidylinositol 4-phosphate (PIP, 2-4%) and phosphatidylinositol 4,5-bisphosphate (PIP$_2$, 12-13%). Most of the $^{32}PO_4$ was incorporated into polyphosphoinositides, PIP$_2$ having 59-64% and PIP having 13-15%. PI had 17-18% of the radioactivity while PA had 6-8%. Radiolabeling of other phospholipids was negligible compared to that in these four lipid classes.

The effect of light on phospholipid metabolism was limited to those phospholipid classes involved in the classical PI response. There was no stimulation of incorporation of radioactive serine, ethanolamine, or choline into their respective phospholipid classes (Table I). Likewise, the PI response in *Xenopus* retina was not a consequence of increased *de novo* biosynthesis of these lipids, since there was no effect of light on incorporation of radioactive glycerol into any of the lipids involved in the PI response.

CELLULAR LOCALIZATION OF THE PI RESPONSE IN THE RETINA

The specific neuron(s) involved in the PI response was identified in retinas incubated with [2-^{3}H]inositol in light or darkness, as described above. The retinas were fixed for autoradiography by our modification (Anderson *et al.*, 1983) of the procedure of Gould and Dawson (1976), which utilizes dehydration techniques that do not remove radiolabeled lipids from the tissue. In our hands, less than 5% of labeled phosphoinositides are removed during the procedure. At the light microscope level, silver grains representing radioactive phosphoinositides can be seen across the entire retinal expanse (Fig. 1). However, it is quite clear that those retinas incubated in the light (Fig. 1A) have a greater number of silver grains in a specific area of the retina, the outer plexiform layer. This layer contains synapses of photoreceptor cells and bipolar cells, and cell bodies of the horizontal cells. The light-enhanced incorporation of radioactive inositol into phosphoinositides in this cell layer is depicted quantitatively in Figure 2. The number of silver grains was counted across the entire retinal expanse and normalized to regional areas. There is a

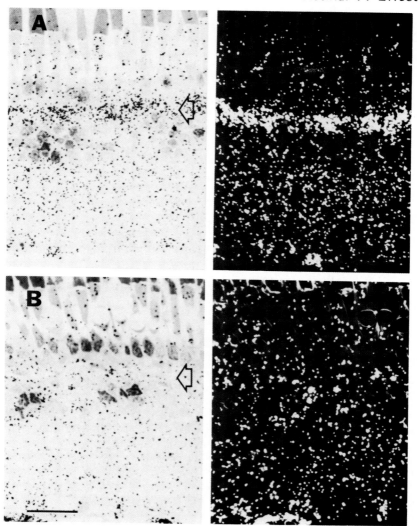

Figure 1. Transmission (left) and dark field (right) light microscope autoradiographs of a retina incubated for 2 h in the light in [2-³H]inositol and processed by the procedure designed to retain tissue lipids. In Fig. 1A, the arrows point to the region of the outer plexiform layer where light stimulated the incorporation of label into lipid. The retina shown in Fig. 1B was treated the same as the one in Fig. 1A, except that it was incubated in the dark. The bar in Fig. 1B represents 20 µm. Reproduced with permission from Anderson and Hollyfield (1981).

303

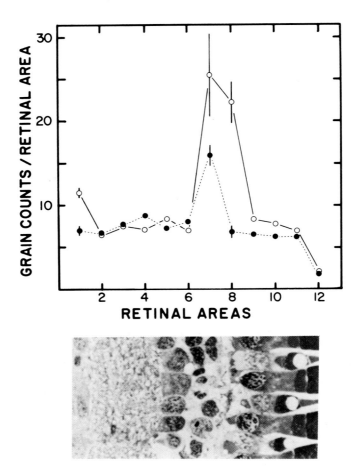

Figure 2. Regional distribution of grain counts across the retina. Plotted are the absolute grain counts per 105 μm² on the ordinate versus the region of the retina from where the counts were obtained, for retinas incubated in the light (O) or in darkness (●). The underlying micrograph aids in the specific identification of each area. Reproduced with permission from Anderson and Hollyfield (1981).

significant increase in labeling only over the outer plexiform layer. Electron microscope autoradiography showed that the label in this plexiform region was confined almost exclusively to horizontal cells (Fig. 3).

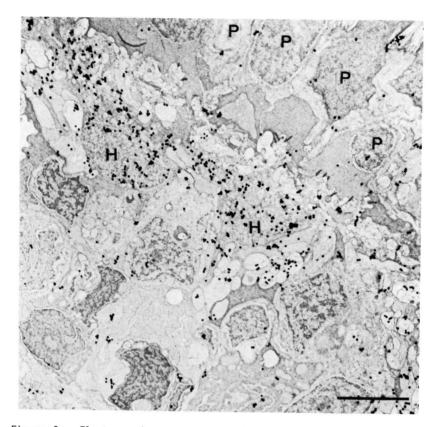

Figure 3. Electron microscope autoradiograph of the retina shown in Fig. 1A. Note the uniform distribution of the radiolabeled phosphoinositides over the horizontal (H) cells, and only sparse radiolabeling in photoreceptor (P) cells. Bar = 10 μm. Reproduced with permission from Anderson et al. (1983).

There is sparse labeling over adjacent photoreceptor cells.

EFFECT OF DIVALENT CATIONS ON THE PI RESPONSE

Since the action of light on the retina is to cause photoisomerization of the 11-*cis*-retinaldehyde to all-*trans*-retinaldehyde in visual pigments of photoreceptor cell outer segment membranes, and since the PI response

305

TABLE II

Effect of Divalent Cations on the Incorporation of Radiolabeled Precursors into Lipids of Retinas From *Xenopus Laevis*[a,b]

Cation	Precursor	% Increase Over Control	P-Value	
Mg^{2+}	[2-^{3}H]inositol	89	<0.01	(18)
	[2-^{3}H]inositol	31	<0.01	(18)
	[2-^{3}H]inositol	110	<0.001	(16)
	[2-^{3}H]inositol	174	<0.001	(18)
Mg^{2+}	$^{32}PO_4$	31	<0.01	(18)
Mg^{2+}	[3-^{3}H]serine	10	>0.2	(18)
Mg^{2+}	[2-^{3}H]glycerol	15	>0.2	(18)
Co^{2+}	[2-^{3}H]inositol	73	<0.005	(12)
Mn^{2+}	[2-^{3}H]inositol	460	<0.001	(11)
Ba^{2+}	[2-^{3}H]inositol	-35	>0.1	(12)
Ca^{2+}	[2-^{3}H]inositol	-29	>0.1	(12)

[a] Data from Anderson et al. (1983).
[b] Chloride salts, 20 mM of each divalent cation, replaced 30 mM NaCl in the Ringer's-bicarbonate-pyruvate medium. All incubations were carried out for 2 h at 21°C, in the dark.

is in horizontal cells, a second order neuron at least one synapse removed from the photoreceptor cell, the light-stimulated labeling of PI in the horizontal cell must be the result of intercellular communication between photo-receptor and horizontal cells. Photoreceptor cells of vertebrate retinas are depolarized in the dark, and hyperpolarize in response to photon capture (Penn and Hagins, 1969; Hagins *et al.*, 1970). Thus, in light,

306

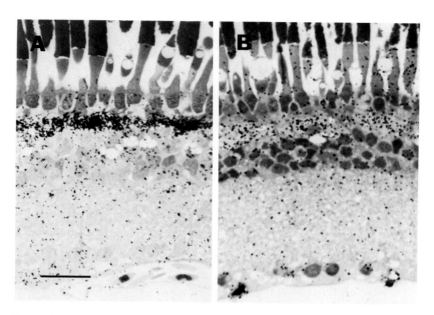

<u>Figure 4</u>. Light microscope autoradiographs of retinas incubated for 2 h in the dark with [2-^{3}H]inositol in Ringer's containing 20 mM Mg^{2+} (A) or in normal Ringer's (B, 2 mM Mg^{2+}). Note the heavy radiolabeling over the outer plexiform layer in Fig. 4A, indicative of Mg^{2+}-enhanced incorporation of [2-^{3}H]inositol into lipid. The bar in Fig. 4A represents 20 μm.

neurotransmitter release from photoreceptor cells is decreased (Trifonov, 1968; Schacher *et al.*, 1974). It is possible then that the absence, rather than the presence, of neurotransmitter might be the stimulus for the PI effect in horizontal cells. To test this possibility, dark-adapted *Xenopus* retinas were incubated in the dark in normal Ringer's or in this medium enriched in certain divalent cations chosen for their ability to inhibit (Mg^{2+}, Co^{2+}, and Mn^{2+}) or potentiate (Ba^{2+} or Ca^{2+}) neurotransmitter release from synaptic terminals (Miledi, 1973). The rationale for these experiments is that inhibition of neurotransmitter release from photoreceptor cells should mimic the effect of light. Data presented in Table II show that 20 mM Mg^{2+}, Co^{2+}, and Mn^{2+} stimulate the incorporation of [2-^{3}H]inositol into phospholipids of *Xenopus* retina. The incorporation of ^{32}PO$_4$ was also stimulated by Mg^{2+}, but [3-^{3}H]serine and [2-^{3}H]glycerol was not. On the other hand, Ba^{2+} and Ca^{2+} did not stimulate [2-^{3}H]inositol incorporation into phospho-inositides. Thus, under conditions where neurotrans-mitter release from synaptic terminals is inhibited, the PI response in this retina is stimulated. That the response is specific for PI is demonstrated by the studies showing that the incorporation of [2-^{3}H]inositol and $^{32}PO_4$, but not of [3-^{3}H]serine and [2-^{3}H]glycerol, is enhanced by the divalent cations (Table II). Also, light microscope autoradiographs of retinas incubated for two hours in the dark with [2-^{3}H]inositol in the presence or absence of 20 mM Mg^{2+} show a selective radiolabeling of cells in the outer plexiform layer (Fig. 4). This pattern is identical to that obtained when retinas are incubated in light or darkness in normal Ringer's (compare with Fig. 1). Quantitation of the grain counts across the entire retinal expanse showed that there was an increase in radiolabeling over the outer plexiform layer. Electron microscope autoradiography revealed that the increased radiolabeling was in the horizontal cells, which is similar to that seen in light (see Fig. 3).

The results of these experiments suggest that the decrease in neurotransmitter release from photoreceptor cells may be the signal for the PI response in the hori-zontal cells. However, this interpretation must be made with some caution since the divalent cations affect transmitter release from every retinal neuron and may have direct effects on lipid metabolism as well. While it is evident that the effect is specific for PI metabolism, we nevertheless sought other ways of demonstrating the role of neurotransmitters in the control of the PI response in the horizontal cell.

EFFECTS OF NEUROTRANSMITTERS ON THE PI RESPONSE

Several putative retinal neurotransmitters were tested for their effects on the PI response in *Xenopus* retina. In a typical experiment, dark-adapted retinas were incubated in the light or in darkness in the pre-sence or absence of neurotransmitter. In some studies, retinas were preincubated with specific antagonists prior to the addition of neurotransmitter. Incubation with [2-^{3}H]inositol and glycine (100 µM) for 2 hours resulted

TABLE III

Effects of Neurotransmitters on the Incorporation
of [2-³H]Inositol into Retinal Phosphoinositides[a]

Experiment 1	DPM/µg lipid-P
Light (L)	2,746 ± 911 (98)[b]
Dark (D)	1,985 ± 581 (71)
Light + Glycine (LG)	2,169 ± 660 (17)
Dark + Glycine (DG)	2,077 ± 680 (12)

L vs D, p <0.001; L vs LG, p<0.02

Experiment 2	
Light + Strychnine (LS)	2,145 ± 315 (12)
Dark + Strychnine (DS)	1,527 ± 377 (12)
Light + Strychnine + Glycine (LSG)	2,127 ± 302 (11)
Dark + Strychnine + Glycine (DSG)	1,673 ± 439 (12)

LS vs DS, p <0.001; LS vs LSG, p >0.5

Experiment 3	
Dark (D)	3,210 ± 742 (21)
Dark + Acetylcholine (DACh)	4,050 ± 825 (21)
Dark + Acetylcholine + Atropine (DAChAt)	3,180 ± 582 (12)

D, DAChAt vs DACh, p <0.005

[a] Data from Anderson and Hollyfield (1984).
[b] DPM/µg total lipid phosphorus ± standard deviation. Value in parentheses is the number of independent determinations.

in an inhibition of the light response by this transmitter candidate (Table III). Preincubation with strychnine (20 μM), an antagonist of glycine, reversed the glycine inhibition of the light response (Table III). Thus, glycine acts either directly or indirectly on horizontal cells to abolish the light-stimulated radiolabeling of phosphoinositides. That this effect is direct is suggested from the studies of Rayborn et al. (1981) which demonstrated a glycinergic interplexiform cell that makes direct synaptic contact onto horizontal cells in the Xenopus retina and releases glycine upon K$^+$ depolarization. It is not known if glycine is released in light or in darkness. However, if glycine is released in the dark, then incubation of dark-adapted retinas with strychnine should result in [2-^{3}H]inositol incorporation values similar to those observed by light stimulation. This is not the case (Table III), so it seems most likely that glycine is released in response to light, and acts directly on the horizontal cell to control the magnitude of the PI response.

Acetylcholine (50 μM) stimulated [2-^{3}H]inositol incorporation in the dark into phosphoinositides of Xenopus retinas (Table III). This stimulation was reversed by atropine (20 μM), indicating a muscarinic, receptor-mediated effect. Light microscope autoradiography and quantitation of grain count distribution revealed that the stimulated PI radiolabeling was confined to horizontal cells. An interesting finding was that atropine did not abolish the light-stimulated PI labeling in horizontal cells.

Several neurotransmitters that had no effect on the PI response in Xenopus retinas are GABA, dopamine, and norepinephrine. GABA is the neurotransmitter of horizontal cells in Xenopus retina (Hollyfield et al., 1979), and dopamine is the neurotransmitter of a class of interplexiform cells in carp retina that synapses onto horizontal cells (Dowling and Ehinger, 1978).

DISCUSSION

The physiological function of light-stimulated PI turnover in the horizontal cells is unknown. In other systems, PI turnover is most often associated with stimulus-coupled secretion events. This may be true for the horizontal cell, but to date the only known stimulus-mediated secretion event is that of GABA, the neurotransmitter of the horizontal cell, and this occurs in the dark when the cell is depolarized. The stimulated PI response occurs when the cell is hyperpolarized. Thus, some other role for stimulated PI turnover in this cell's function must be sought. In the carp retina, it has been shown that dopaminergic interplexiform cells synapse directly onto H1 horizontal cells (Dowling and Ehinger, 1978). In whole retina (Watling *et al.*, 1983) and in isolated horizontal cell preparations (van Buskirk and Dowling, 1981), dopamine stimulates adenylyl cyclase activity in these GABAergic neurons. Like our studies on light-stimulated PI metabolism, the physiological function of the activated adenylyl cyclase is not known.

There appear to be at least three inputs into horizontal cells of *Xenopus* retinas that affect phosphoinositide metabolism: 1) stimulation via a noncholinergic mechanism, 2) stimulation via a cholinergic mechanism, and 3) inhibition via a glycinergic mechanism. The functional consequences of neurotransmitter control of phosphoinositide metabolism remain to be elucidated.

ACKNOWLEDGEMENTS

The technical assistance of Paula A. Kelleher, Maureen B. Maude, Mary E. Rayborn and Janis Rosenthal is gratefully acknowledged. This research was supported in part by grants from the National Eye Institute, the National Retinitis Pigmentosa Foundation (Baltimore), the Retina Research Foundation (Houston) and Research to Prevent Blindness, Inc. Joe G. Hollyfield is an Olga K. Wiess Scholar, and Robert E. Anderson is a Dolly Green Scholar of Research to Prevent Blindness, Inc.

REFERENCES

Abdel-Latif, A.A. (1983) in: "*Handbook of Neurochemistry*" (A. Lajtha, ed.) Vol. 3, pp. 91–131, Plenum Publ. Corp., New York.
Anderson, R.E. and Hollyfield, J.G. (1981) *Biochim. Biophys. Acta 665*, 619–622.

311

Anderson, R.E. and Hollyfield, J.G. (1984) Submitted for publication.

Anderson, R.E., Maude, M.B., Kelleher, P.A., Rayborn, M.E. and Hollyfield, J.G. (1983) *J. Neurochem. 41,* 764-771.

Dowling, J.E. and Ehinger, B. (1978) *Proc. R. Soc. London (Biol.) B 210,* 7-26.

Gould, R.M. and Dawson, R.M.C. (1976) *J. Cell Biol. 68,* 480-496.

Hagins, W.A., Penn, R.D. and Yoshikami, S. (1979) *Biophys. J. 10,* 380-412.

Hollyfield, J.G., Rayborn, M.E., Sarthy, P.V. and Lam, D.M.-K. (1979) *J. Comp. Neurol. 188,* 587-598.

Michell, R.H. (1975) *Biochim. Biophys. Acta 415,* 81-147.

Miledi, R. (1973) *Proc. R. Soc. London (Biol.) 183,* 421-425.

Penn, R.D. and Hagins, W.A. (1969) *Nature 223,* 201-205.

Rayborn, M.E., Sarthy, P.V., Lam, D. M.-K. and Hollyfield, J.G. (1981) *J. Comp. Neurol. 195,* 585-593.

Schacher, S.M., Holtzman, E. and Hood, D.C. (1974) *Nature 249,* 261-263.

Trifonov, Y.A. (1968) *Biofizika 13,* 809-817.

Watling, K.J., Dowling, J.E. and Iverson, L.L. (1979) *Nature 281,* 578-580.

van Buskirk, R. and Dowling, J.E. (1981) *Proc. Natl. Acad. Sci. 78,* 7825-7829.

Ca^{2+}-MOBILIZING AGONISTS STIMULATE A
POLYPHOSPHOINOSITIDE-SPECIFIC PHOSPHOLIPASE C
IN RAT PAROTID GLAND

C. Peter Downes

ICI Pharmaceuticals Division, Bioscience Dept. II,
Alderley Park, Macclesfield, Cheshire SK10 4TG, U.K.

SUMMARY

Activation of Ca^{2+}-mobilizing receptors leads to the disappearance of inositol phospholipids from stimulated cells. Although enhanced phospholipase C activity appears to account for this effect in a variety of tissues, whether phosphatidylinositol (PtdIns) or the polyphosphoinositides (PPI) are the primary substrates for the enzyme(s) remains a controversial issue. Brief exposure of [^{3}H]inositol-labeled parotid gland slices to Ca^{2+}-mobilizing agonists (carbachol, substance P or phenylephrine) caused large increases in the levels of [^{3}H]-inositol 1-phosphate (Ins1P), [^{3}H]inositol 1,4-bisphosphate (Ins1,4P$_2$) and [^{3}H]inositol 1,4,5-trisphosphate (Ins1,4,5P$_3$). Using carbachol as the stimulus the large increases in labeled Ins1,4P$_2$ and Ins1,4,5P$_3$ were much greater than the modest decreases in the [^{3}H]PPI fractions. This demonstrates that labeled PPI are continuously hydrolyzed and replaced from the labeled PtdIns pool. When atropine was added to carbachol-treated slices the levels of [^{3}H]inositol phosphates rapidly returned to control values presumably via the activity of inositol phosphomonoesterases. The rate of disappearance of Ins1,4P$_2$ under these conditions suggests that Ins1P is formed from inositol polyphosphates and not from PtdIns. Thus, in rat parotid glands, Ca^{2+}-mobilizing agonists stimulate a phospholipase C whose primary substrates are PtdIns4P and/or PtdIns4,5P$_2$.

INTRODUCTION

Rat parotid glands possess receptors for Substance P, α_1 and muscarinic ligands. These distinct receptor populations share a common mechanism of action that involves an increase in the cytosol Ca^{2+} concentration (Putney, 1979). Activation of these Ca^{2+}-mobilizing receptors results in hydrolysis of inositol phospholipids which may be an essential reaction coupling receptor occupation to the increase in cytosol Ca^{2+} concentration (Michell, 1975; Michell et al., 1981). The changes in phospholipid metabolism were previously thought to involve a PtdIns-specific phospholipase C (Jones and Michell, 1974; Irvine et al., 1982), but more recent work points to the involvement of the PPI (Weiss et al., 1982).

In order to resolve the question of the substrate specificity of the enzyme(s) activated by Ca^{2+}-mobilizing agonists we initiated studies of the inositol phosphates formed following stimulation of rat parotid gland slices. Our findings, which have been published in detail else-where (Berridge et al., 1983; Downes and Wusteman, 1983), are briefly summarized in this communication.

METHODS

Parotid gland slices were prepared as described previously (Hanley et al., 1980), labeled with [^{3}H]inositol for 90 min and then washed extensively with modified Krebs-Ringer-bicarbonate buffer containing 10 mM unlabeled inositol. All incubations were stopped by adding trichloroacetic acid (TCA). Analysis of labeled compounds in the acid-soluble fraction was by a combination of anion exchange chromatography, high-voltage electrophoresis and paper chromatography (Berridge et al., 1983), but for routine, quantitative analyses the labeled compounds were separated on small anion exchange columns (Dowex-1, 100-200 mesh; formate form). Phospholipids in the TCA-insoluble fraction were extracted, deacylated and analyzed on the same anion-exchange columns, as described previously (Creba et al., 1983). Radioactivity was determined by liquid scintillation counting.

RESULTS AND DISCUSSION

The acid soluble fractions obtained as described above could be resolved into five labeled components which had the chromatographic and electrophoretic properties expected for free-inositol, glycerophosphoinositol (GPI), Ins1P, Ins1,4P$_2$ and Ins1,4,5P$_3$. Ca^{2+}-mobilizing agonists had little effect on the labeled GPI fraction and so these data have been omitted from the Tables. The deacylated phospholipid fraction contained three labeled compounds which behaved like the deacylation products of PtdIns, PtdIns4P and PtdIns4,5P$_2$ on Dowex-1 (formate) columns.

Effects of Ca^{2+}-mobilizing Agonists

When a labeled parotid gland slice preparation was exposed to the Ca^{2+}-mobilizing agonists carbachol, phenylephrine or substance P for 10 min there were large increases in the levels of [^{3}H]Ins1P, [^{3}H]Ins1,4P$_2$ and [^{3}H]Ins1,4,5P$_3$ (Table I). The effect was greatest for carbachol which also caused an increase in the [^{3}H]inositol fraction. Isoprenaline which acts on adenylyl cyclase linked β-receptors was ineffective (not shown in Table I, but see Berridge et al., 1982). This demonstrates that the response involves phospholipase C attack upon the polyphosphoinositides which are the only known sources of Ins1,4P$_2$ and Ins1,4,5P$_3$. However, these results, taken alone, would suggest that direct hydrolysis of PtdIns might also occur.

The response was studied in more detail using the muscarinic agonist, carbachol, as the stimulus. The results in Table II are from an experiment in which labeled parotid gland slices were incubated for 20 min without any additions, with carbachol present for the last 10 min only, or with carbachol for the first 10 min and an excess of the muscarinic antagonist, atropine, added for the final 10 min to switch off effectively the previously activated muscarinic receptors. As before, carbachol treatment induced a large rise in the levels of the [^{3}H]-inositol phosphates. After the subsequent incubation with atropine, however, these levels dropped sharply and there was a concomitant increase in the [^{3}H]inositol

315

TABLE I

The Effects of Ca^{2+}-Mobilizing Agonists on the Levels of [^{3}H]Inositol-Labeled
Water-Soluble Compounds

Drug Additions	Inositol	Radioactivity in Inositol Metabolites		
		InsIP	InsI,4P$_2$	InsI,4,5P$_3$
		DPM		
None	104,800 ± 18,400	630 ± 80	250 ± 24	800 ± 30
Carbachol (10^{-3} M)	157,200 ± 26,800	9,450 ± 900*	11,050 ± 980*	5,650 ± 310*
Carbachol (10^{-3} M) + atropine (10^{-5} M)	97,780 ± 8,330	590 ± 90	390 ± 80	950 ± 70
Phenylephrine (10^{-4} M)	113,100 ± 6,000	1,440 ± 90*	1,390 ± 70*	2,330 ± 120*
Substance P (2 × 10^{-6} M)	106,000 ± 10,200	1,190 ± 80*	1,360 ± 120*	1,830 ± 120*

Prelabeled parotid gland slices were incubated with the agonists and antagonists shown for 10 min at 37°C. Radioactivity in TCA-soluble components was determined as described in Methods. *Significantly different from control, $p<0.05$. Reproduced, with permission, from Berridge et al. (1983).

fraction. This indicates that parotid glands possess a complement of enzymes capable of hydrolyzing each of the inositol phosphates. The likely purpose of this sequence of reactions is to regenerate free inositol since this is an essential requirement for PtdIns resynthesis. Furthermore, it seems likely that measurement of the levels of [^{3}H]inositol phosphates after 10 min of carbachol treatment greatly underestimates their true rates of formation since they are continually being hydrolyzed to free inositol.

Comparison of the carbachol-induced changes in [^{3}H]inositol phosphates with the changes in the labeled phospholipid fraction reveals another interesting feature of the response. Carbachol treatment causes a 50% drop in the [^{3}H]PtdIns4,5P$_2$ fraction and a smaller, but significant fall in the level of [^{3}H]PtdIns4P. However, these changes in the levels of labeled phospholipids cannot account for the much larger accumulation of label in [^{3}H]Ins1,4P$_2$ and [^{3}H]Ins1,4,5P$_3$. This suggests that the whole PPI pool must be hydrolyzed to inositol phosphates and replenished from the labeled PtdIns pool many times during the course of even a relatively brief period of stimulation. Thus breakdown of the relatively small PPI pool can account for the disappearance of a substantial fraction of the much larger PtdIns pool.

Is PtdIns Hydrolyzed Directly in Stimulated Parotid Glands?

A series of observations, described in detail in a previous paper (Downes and Wusteman, 1983) strongly suggests that receptor activation stimulates a PPI-specific phospholipase C. These results are summarized below.

Very brief stimulation of parotid gland slices with carbachol leads to a rapid increase in the levels of [^{3}H]Ins1,4P$_2$ and [^{3}H]Ins1,4,5P$_3$ (detectable within 5 s), but there is a distinct time-lag for the formation of [^{3}H]Ins1P.

When the glands are exposed to carbachol followed 15 min later by an excess of atropine, the rate of hydrolysis of [^{3}H]Ins1,4P$_2$ observed can fully account for the formation of [^{3}H]Ins1P. Thus Ins1P could be formed

317

TABLE II

Effects of Carbachol and Subsequent Atropine Treatment on [^{3}H]Inositol-Labeled Compounds

Labeled Compound	Radioactivity (DPM)		
	Control	Carbachol	Carbachol/Atropine
Inositol	$101,100 \pm 5,550$	$124,300 \pm 13,160$	$136,200 \pm 2,960*$
InsIP	$1,030 \pm 63$	$3,550 \pm 170*$	$2,290 \pm 140**$
InsI,4P$_2$	600 ± 11	$7,820 \pm 390*$	$2,690 \pm 170**$
InsI,4,5P$_3$	500 ± 24	$3,620 \pm 124*$	$1,160 \pm 73**$
PtdIns	$134,900 \pm 4,260$	$117,500 \pm 11,260$	$115,000 \pm 4,040$
PtdIns4P	$3,160 \pm 130$	$2,300 \pm 200*$	$2,930 \pm 270$
PtdIns4,5P$_2$	$2,020 \pm 90$	$910 \pm 41*$	$1,620 \pm 70**$

Prelabeled parotid gland slices were incubated for 20 min without any additions (control), with carbachol (10^{-3} M) present for the last 10 min, or with carbachol present for 20 min with atropine (10^{-5} M) added after 10 min. *Significantly different from control, $p < 0.05$; ** significantly different from carbachol, $p < 0.05$. Reproduced, with permission, from Downes and Wusteman (1983).

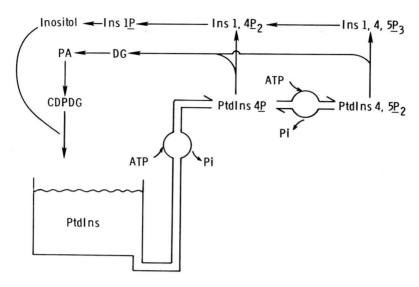

Figure 1. Receptor-stimulated inositol lipid metabolism in rat parotid glands. Activation of Ca^{2+}-mobilizing receptors stimulates a polyphosphoinositide-specific phospholipase C leading to the formation of diacylglycerol (DG) and Ins1,4,5P_3 plus Ins1,4P_2. These inositol phosphates are metabolized to free inositol for PtdIns resynthesis via Ins1P. PtdIns acts as a large reservoir and continually replenishes the PPI pools via PtdIns and PtdIns4P kinases. Abbreviations not given in the text: PA, phosphatidic acid; CDPDG, CDP-diacylglycerol; P_i, inorganic phosphate.

by hydrolysis of inositol polyphosphates and not directly from PtdIns.

Treatment of the labeled glands with the mitochondrial uncoupler 2,4-dinitrophenol, dramatically reduces the carbachol-induced formation of [³H]Ins1P, [³H]Ins1,4P_2 and [³H]Ins1,4,5P_3. This result suggests that the formation of all three inositol phosphates requires an energy-dependent step. This is easy to explain for the inositol polyphosphates, but not so for Ins1P if it is formed directly via PtdIns hydrolysis. This observation strongly supports the notion that Ins1P is formed indirectly via the inositol polyphosphates.

CONCLUSIONS

The proposed mechanism underlying agonist-stimulated metabolism of inositol phospholipids in rat parotid glands is depicted in Figure 1. Ca^{2+}-mobilizing agonists stimulate a PPI-specific phospholipase C with the formation of diacylglycerol and inositol polyphosphates. Inositol phosphates are themselves hydrolyzed in a sequence of reactions that presumably yields free inositol for lipid resynthesis. PtdIns serves as a large reservoir for the replenishment of the polyphosphoinositides. The rate of formation of potentially important intracellular signal molecules such as diacylglycerol and $Ins1,4,5P_3$ is therefore dependent upon the activity of the phospholipase C and upon the rate at which the PPIs can be resynthesized via PtdIns and PtdIns4P kinases.

REFERENCES

Berridge, M.J., Downes, C.P. and Hanley, M.R. (1982) *Biochem. J. 206,* 587-595.

Berridge, M.J., Dawson, R.M.C., Downes, C.P., Heslop, J.P. and Irvine, R.F. (1983) *Biochem. J. 212,* 473-482.

Creba, J.A., Downes, C.P., Hawkins, P.T., Brewster, G., Michell, R.H. and Kirk, C.J. (1983) *Biochem. J. 212,* 733-747.

Downes, C.P. and Wusteman, M.M. (1983) *Biochem. J. 216,* 633-640.

Hanley, M.R., Lee, C.M., Jones, L.M. and Michell, R.H. (1980) *Mol. Pharmacol. 18,* 78-83.

Irvine, R.F., Dawson, R.M.C. and Freinkel, N. (1982) in: *"Contemporary Metabolism"* (N. Freinkel, ed.) Vol. 2, pp. 301-322, Plenum Press, New York.

Jones, L.M. and Michell, R.H. (1974) *Biochem. J. 142,* 583-590.

Michell, R.H. (1975) *Biochim. Biophys. Acta 415,* 81-147.

Michell, R.H., Kirk, C.J., Jones, L.M., Downes, C.P. and Creba, J.A. (1981) *Phil. Trans. R. Soc. Ser. B. 296,* 123-137.

Putney, J.W., Jr. (1979) *Pharmacol. Rev. 30,* 209-245.

Weiss, S.J., McKinney, J.S. and Putney, J.W., Jr. (1982) *Biochem. J. 206,* 555-560.

PHOSPHATIDYLINOSITOL DEGRADATION IS DIRECTLY

PROPORTIONAL TO THE RATE OF CELL DIVISION IN

EMBRYONIC CHICKEN LENS EPITHELIA

Peggy S. Zelenka and Ngoc-Diep Vu

Laboratory of Molecular and Developmental Biology,
National Eye Institute, NIH, Bethesda, MD 20205

SUMMARY

This study probes the extent of coupling between cell division and phosphatidylinositol (PtdIns) degradation in embryonic chicken lens epithelial cells. Between 6 and 19 days of development, the rate of cell division in the central region of the lens epithelium declines gradually. By measuring the rate of PtdIns degradation in cells of this region in chicken embryos aged 6-, 9-, 12- and 19-days, we have found that the rate of PtdIns degradation is directly proportional to the rate of cell division throughout this developmental period. Furthermore, culturing explants of the central region of lens epithelia from 19-day-old embryos in the presence of fetal calf serum produces a 3 to 3.5-fold increase in both the rate of cell division and the rate of PtdIns degradation. These findings demonstrate that PtdIns degradation and cell division are tightly coupled in this tissue, and raise the possibility that PtdIns metabolism influences the regulation of the cell cycle.

INTRODUCTION

A correlation between rapid cell division and rapid PtdIns degradation has been observed in a number of cell types (Michell, 1982). In most cases, however, the cells have been observed only in one of two states: dividing or non-dividing. In contrast, cells of the central region of the embryonic chicken lens epithelium can be studied at various rates of division by taking advantage of the gradual decline in the rate of mitosis that normally occurs during development (Persons and Modak, 1970). This study compares the rate of PtdIns degradation and the rate of cell division in central lens epithelia from chicken embryos 6-, 9-, 12- and 19-days-old.

MATERIALS AND METHODS

Lens epithelia of embryonic chickens were isolated by microdissection. At each age a square explant was cut from the center of the epithelium to correspond to the "central" region described by Persons and Modak (1970). Explants were pulse-labeled for 2 h with 15.6 μCi/ml [^{3}H]inositol (12.5 Ci/mmol, NEN). Tissues were harvested at various times during a subsequent chase period and phospholipids were extracted with 1:1 $CHCl_3$:CH_3OH as previously described (Vu et al., 1983). PtdIns was isolated by thin-layer chromatography on silica gel HL plates (Analtech) using $CHCl_3$:CH_3OH:CH_3COOH:0.9% NaCl, 100:50:16:8, as the solvent system (Anderson et al., 1969). Gammexane (hexachlorocyclohexane) was purchased from Sigma, and DNA measurements were made using the fluorescence assay of Kissane and Robins (1958).

RESULTS

Cultured explants of central lens epithelia from chicken embryos aged 6-, 9-, 12- and 19-days were pulse-labeled with [^{3}H]inositol and harvested at various times during a subsequent 6 h chase period. The rate of loss of [^{3}H]PtdIns from the explants at each age is shown in Figure 1. Older embryos showed progressively slower loss of [^{3}H]PtdIns, until at 19-days, no degradation was detected even during a 24 h chase period (see Fig. 3). The slight increase in labeling of PtdIns in explants from 19-day-old embryos seen during the first 6 h of the chase period was completely eliminated by gammexane, an inhibitor of PtdIns synthesis; however, even in the presence of gammexane, there was no detectable loss of [^{3}H]PtdIns (Fig. 1). Gammexane had no effect on the loss of [^{3}H]PtdIns in explants from younger embryos (data not shown).

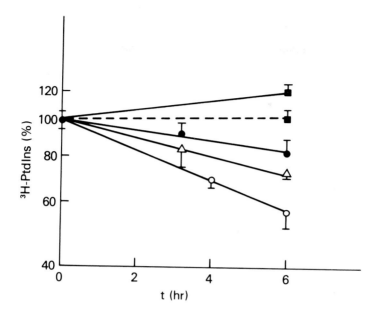

<u>Figure 1</u>. The percentage loss of [³H]phosphatidylinositol ([³H]PtdIns) per dish of 6 lens epithelial explants during a 6 h chase period. DNA per dish was constant throughout the experiment. Data are plotted as averages + SEM for embryos aged 6 days (O), 9 days (Δ), 12 days (●), and 19 days (■). The broken line represents data obtained in the presence of 0.5 mM gammexane. Straight lines were fitted to the data by least-squares analysis.

Persons and Modak (1970) have determined the percentage of dividing cells in the central region of the developing embryonic chicken lens epithelium by pulse-labeling cultured lenses with [³H]thymidine and counting the labeled nuclei after autoradiography. Comparing these values for the percentage of dividing cells with our values for the rates of PtdIns degradation (the slopes of the lines in Fig. 1), we found that both sets of data fell along the same smooth curve (Fig. 2). This indicates that the rate of cell division is directly proportional to the rate of PtdIns degradation throughout this developmental period.

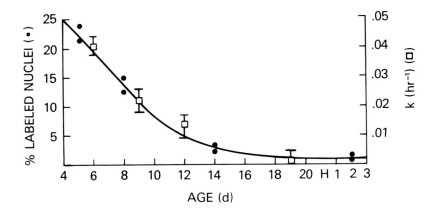

Figure 2. The percentage of labeled nuclei in the central region of the embryonic chicken lens epithelium (Persons and Modak, 1970) and the decay constant, k, (log $2/t_{1/2}$) representing the rate of degradation of phosphatidylinositol in cells of this region as a function of developmental age. "H" indicates the day of hatching. Values of k were obtained from the slopes of the lines in Figure 1, and are plotted ± the estimated error determined by the curve-fitting process.

We also tested whether stimulation of the rate of cell division would produce a proportional increase in the rate of PtdIns degradation. Explants of central lens epithelia of 19-day-old chicken embryos undergo a 3 to 3.5-fold increase in the rate of cell division when cultured in the presence of 15% fetal calf serum (Piatigorsky and Rothschild, 1972). Under these conditions the rate of PtdIns degradation also increased by a factor of 3 to 3.5 (Fig. 3), thus maintaining the previously observed proportionality between cell division and PtdIns degradation.

DISCUSSION

The above findings demonstrate that there is an exact, quantitative relationship between the rate of PtdIns degradation and the rate of cell division in the developing chicken lens epithelium. This correlation is supported by the observation that PtdIns degradation virtually ceases when the lens epithelial cells stop dividing and differenti-

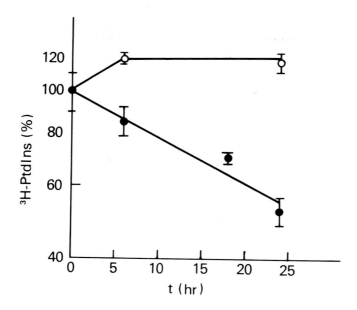

<u>Figure 3</u>. The percentage loss of [³H]phosphatidylinositol ([³H]PtdIns) per ng DNA during a 24 h chase period in lens epithelial explants of 19-day-old chicken embryos. Tissues were cultured in unsupplemented Ham's F-10 medium (O) or pre-cultured for 24 h in Ham's F-10 medium + 15% fetal calf serum (●). Data are plotted as averages + SEM.

ate to form lens fiber cells (Zelenka, 1980; Vu *et al.*, 1983). In the central region of the lens epithelium, where the cells do not normally differentiate into lens fibers *in vivo*, the rate of PtdIns degradation remains directly proportional to the percentage of dividing cells as this percentage decreases from about 22% at 5 days of development to about 1% 2 days after hatching. The decrease in the percentage of dividing cells could be caused by a general increase in the cell cycle time, or by a decrease in the number of cycling cells. In 5-day-old embryos the generation time of lens epithelial cells is 8-10 h (Zwann and Pearce, 1971). Such short generation times frequently occur in embryonic cells but increase during development as the G_1 portion of the cell cycle lengthens (Prescott, 1982). If this is the case in the

325

embryonic chicken lens epithelium, the observed correlation of PtdIns degradation with the percentage of
labeled nuclei may reflect an underlying correlation with
the length of the cell cycle.

Three mechanisms have been suggested to explain
the physiological effects associated with PtdIns degradation: (1) the diacylglycerol produced by hydrolysis of
PtdIns and its phosphorylated derivatives may activate a
calcium-dependent protein kinase (protein kinase C)
(Takai et al., 1979); (2) PtdIns degradation may lead to
an increase in cytosolic calcium levels (Michell, 1975); and
(3) PtdIns turnover may be coupled to the release of
arachidonic acid which is further metabolized to such
biologically active substances as prostaglandins and
leukotrienes (Berridge, 1981). Each of these three
mechanisms has been implicated in the regulation of cell
division. Phorbol esters, which directly stimulate protein
kinase C (Castagna et al., 1982), also stimulate division
in certain cell types (Abb et al., 1979). Calcium-
deprived cells become arrested at the transition from G_1
to S phase of the cell cycle (Gillies, 1982). In cultured
hepatocytes the effect of calcium deprivation can be
overcome by adding arachidonic acid to the culture
medium (Boyntan and Whitfield, 1980). Such observations, together with the correlation between the rate of
PtdIns degradation and the rate of cell division in
embryonic chicken lens epithelia, suggest that PtdIns
metabolism may play a role in the regulation of the cell
cycle.

ACKNOWLEDGEMENTS

We thank Dr. J. Piatigorsky for critically reading
this manuscript, and Ms. Dawn Sickles for typing.

REFERENCES

Abb, J., Bayliss, G.J. and Deinhart, F. (1979) J.
Immunol. 122, 1639-1642.
Anderson, R.E., Maude, M.B. and Feldman, G.L. (1969)
Biochim. Biophys. Acta 198, 345-353.
Berridge, M.J. (1981) Mol. Cell. Endocrinol. 24,
115-140.
Boynton, A.L. and Whitfield, J.F. (1980) Exptl. Cell
Res. 129, 474-478.

Castagna, M., Takai, Y., Kaibuchi, K., Sano, K., Kikkawa, U. and Nishizuka, Y. (1982) *J. Biol. Chem.* *257*, 7847–7851.

Gillies, R.S. (1982) *Trends Biochem. Sci.* *7*, 233–235.

Kissane, J.M. and Robins, E. (1958) *J. Biol. Chem.* *233*, 184–188.

Michell, R.H. (1982) *Cell Calcium* *3*, 429–440.

Persons, B.J. and Modak, S.P. (1970) *Exptl. Eye Res.* *9*, 144–151.

Piatigorsky, J. and Rothschild, S. (1972) *Develop. Biol.* *28*, 382–389.

Prescott, D.M. (1982) *Ann. N.Y. Acad. Sci.* *397*, 101–109.

Takai, Y., Kishimoto, A., Kikkawa, U., Mori, T. and Nishizuka, Y. (1979) *Biochem. Biophys. Res. Commun.* *91*, 1218–1224.

Vu, N.D., Chepko, G. and Zelenka, P. (1983) *Biochim. Biophys. Acta* *750*, 105–111.

Zelenka, P.S. (1980) *J. Biol. Chem.* *255*, 1296–1300.

Zwann, J. and Pearce, T.L. (1971) *Develop. Biol.* *25*, 96–118.

DISCUSSION

Summarized by Ronald P. Rubin

Department of Pharmacology, Medical College of Virginia
Richmond, Virginia 23298

Discussion of the papers in this session focused on the breakdown of phosphoinositides and the resulting formation of inositol phosphates. While the papers demonstrated that receptor agonists induce breakdown of PIP_2 in a variety of cell types, employing diverse stimuli, the mechanism by which the inositol phosphates are formed was a matter of some debate. In response to Dr. Irene Litosch's (Brown University) comment that one really cannot exclude the possibility that PI breakdown occurs in parallel with PIP_2 hydrolysis, Dr. Michell acknowledged that IP may be derived from PI breakdown directly. However, he felt that the law of parsimony militates against a situation where three lipids are being broken down in parallel. Dr. Michell felt that the simplest explanation and the one consistent with the presently available data - but certainly not the only one - is that PIP_2 is broken down to IP_3, which is then sequentially degraded by phosphatases to inositol.

Another point raised by Drs. Nicholas Bazan and Michael Berridge related to the high K^+ concentrations and the prolonged duration of stimulation employed by Dr. Michell in neural tissue. This prompted Dr. Michell to express the opinion that his findings probably reflected responses not of a rapid regulatory pathway, but rather a more gentle, modulatory one. Both Drs. Abdel-Latif and Michell agreed that the PI effect produced by K^+ depolarization had nothing to do with activation of voltage-sensitive calcium channels, but was merely a reflection of K^+-induced neurotransmitter release.

329

The findings reported by Dr. John Exton indicating that phosphoinositide breakdown was not involved in the effects of vasopressin and alpha-agonists on calcium mobilization in the liver was challenged by Dr. James Putney, who noted that, in liver, IP_3 has been shown to have biological activity and is formed during hormonal stimulation. However, Dr. Putney acknowledged the existence of the problem of precisely correlating tissue levels of IP_3 with the functional response. In response to Dr. Thomas Martin's (University of Wisconsin) question as to whether the effects of hormones on calcium pumping were mimicked either by agents that translocate calcium or alter diacylglycerol-dependent protein phosphorylation, Dr. Exton noted that ionophores abolish the ability of membrane vesicles to accumulate calcium, although he had not tested whether mitochondrial inhibitors, such as FCCP, would produce hormone-like effects on Ca-ATPase. In response to a question raised by Dr. Pedro Cuatrecasas (Wellcome Research Laboratories), Dr. Exton indicated that IP_2 and IP_3 do not influence calcium metabolism of isolated mitochondria, although he further stated that negative results might be accounted for by the absence of critical co-factors in the media.

In relation to Dr. John Fain's paper, Dr. Michell raised the question as to how, after steady state labeling of inositol phospholipids, stimulation could bring about a decreased labeling of PI with an accompanying increase in total concentration. Dr. Fain suggested that ^{32}P labeling may not be equilibrating with all pools. Dr. Downes later reaffirmed the possibility that there may be multiple pools of phosphoinositide labeling at different rates. Dr. Michell countered with the proposal that the ^{32}P labeling may be comprised of some compound in addition to $[^{32}P]$-PIP_2. The findings of Dr. Fain that phorbol esters stimulate glycogenolysis raised some questions, although Dr. Fain agreed that the action of phorbol ester is not certain to be on protein kinase C. Dr. Putney suggested that protein kinase C may not be involved in the stimulus-response pathway in hepatocytes, since an increase in $[Ca_i]$ can fully activate phosphorylase kinase, which is apparently not the case in platelets. With regard to the phorbol esters, Dr. John Williamson supported the statement of Dr. Exton by noting that phorbol esters have no effect on free calcium levels in liver cells as monitored by

330

Quin-2 fluorescence. However, he also warned that one must be certain that the vehicle e.g. ethanol or DMSO, is not responsible for any effects observed. In response to the question raised by Dr. Mark Seyfred (Michigan State University), Dr. Fain offered the suggestion that the enzymes for PI breakdown and resynthesis may be localized in the plasma membrane, as well as in the endoplasmic reticulum. He further hypothesized that resynthesis of PI is the result of calcium release from cellular membranes, which relieves the inhibitory constraint on the enzymes involved in PI synthesis.

Following Dr. Abdel-Latif's paper, Dr. Berridge acknowledged the pioneer work of Dr. Abdel-Latif on the agonist-dependent breakdown of polyphosphoinositides in 1977. He also reiterated the important fact that the observed calcium-dependency merely reflected inhibition of release of an endogenous neurotransmitter. Dr. Berridge speculated that in smooth muscle the primary phase of contraction, which is independent of extracellular calcium, may be caused by internal calcium release induced by IP_3. The secondary, tonic phase could be explained on the basis of IP_3 short-circuiting internal calcium pools, thereby placing the emphasis for calcium signalling on the plasma membrane, whereby the calcium sequestering system is toned down to allow the accumulation of cellular calcium. This hypothesis would be consistent with the findings of Drs. Fain and Exton, who demonstrated some relationship between phosphoinositide metabolism and calcium pumping activity. Dr. Abdel-Latif agreed that the calcium pump, as well as the sodium pump, was involved in phosphoinositide breakdown.

The ability of light to stimulate phosphoinositide metabolism in the retina at earlier response times was questioned by Dr. Mark Dibner (DuPont); it was suggested by Dr. Robert Anderson that this event could reflect adaptation to light, rather than a more rapid functional response of the retina. Dr. Anderson acknowledged that his system has a high signal:noise ratio and that only a small percentage of cells responded. Dr. Michell addressed the issue of Mn stimulation of radiolabeled inositol incorporation into PI, but Dr. Anderson indicated that Mn, while stimulating the exchange reaction, caused the total disruption of photoreceptor cells

331

after 4 h. Dr. Michell offered the suggestion that the actions of Mg^{2+} were likely to be exerted through some inhibitory mechanism, even though it mimics the effects of light. He proposed that Mg^{2+}, like light, decreases the inhibitory input of a neurotransmitter whose function is to decrease PI turnover. This would imply that retinal stimuli may function in a manner opposite from that observed in other model systems. Dr. Anderson supported this proposal by noting that light hyperpolarizes retinal cells.

The short papers by Drs. Peter Downes and Peggy Zelenka also prompted fruitful discussion. Dr. Abdel-Latif addressed the question as to whether lower concentrations of agonist exhibit differential effects on the formation of one or another of the inositol phosphates. Dr. Abdel-Latif indicated that when a lesser carbachol concentration is employed in smooth muscle, only IP_3 formation is observed. Dr. Downes speculated that the enzymes involved in PIP_2 breakdown are probably carefully controlled in perhaps the same way that nucleotide phosphodiesterase is regulated.

Discussion then focused on the possibility that more than one mechanism may exist for phosphoinositide breakdown. Dr. Marvin Gershengorn (NYU) stated that in GH_3 pituitary cells, calcium ionophores stimulate the hydrolysis of PIP, but not PIP_2. He suggested that the stimulation of these cells by TRH may cause the breakdown of PIP_2 causing calcium release, which in turn brings about the hydrolysis of PIP. However, Dr. Putney responded to this proposal by noting that in parotid cells the dose-response curves for formation of each of the inositol phosphates were superimposable and all were calcium-independent. Dr. Yoram Oron (Tel Aviv University) postulated the existence of a second mechanism in cell-free systems either because the polyphosphoinositides had disappeared from such a preparation and/or because this system enables the enzyme to come into contact with PI. To add to the complexities, Dr. Downes noted that the present techniques do not separate or distinguish different isomers of inositol phosphates. More sophisticated procedures are needed to ascertain the true chemical identity of the inositol phosphates that are being measured. Dr. Downes further suggested that the energy-dependent steps (kinases) in the synthesis of phosphoinositides would allow a greater degree of control of the levels of these potent putative mediators. So, by simply altering the activity of a single kinase or phosphatase, inositol phosphate levels can be drastically modified. Dr. Robert Farese supported the notion of the separate mechanisms for the breakdown of PIP_2 and PI from data derived from his studies on exocrine pancreas.

In response to the statement that people had been considering PI metabolism in lectin-treated lymphocytes and calcium mobilization since the 1960's and 70's, Dr. Peggy Zelenka stated that her system enabled her to study PI breakdown in cells which are in the complete range of division states. In response to a question regarding the effect of serum, she noted that in epithelial cells taken from explants after 15 days, stimulation with serum increases the rate of cell division, while PI metabolism becomes comparable to that of other phospholipids. These data fortify the hypothesis that PI metabolism may play a regulatory role in the cell cycle.

In conclusion, while this session established that inositol phosphate formation is enhanced by receptor agonists in several different systems, the following questions were still left unanswered: (a) What is the true chemical identity of the inositol phosphates being measured? (b) What is the biochemical mechanism of inositol phosphate formation? (c) Are the levels of inositol phosphates produced temporally and quantitatively appropriate for playing a second messenger role in the mobilization of cellular calcium? In other words, as stated by Dr. Fain, it is still not clear what the "Hokin phenomenon" truly represents.

PART IV

PHOSPHOINOSITIDES, CALCIUM AND PROTEIN PHOSPHORYLATION

MESSAGES OF THE PHOSPHOINOSITIDE EFFECT

J. W. Putney, Jr., G. M. Burgess, P.P. Godfrey
and D.L. Aub

Medical College of Virginia-VCU, Dept. of Pharmacology
Richmond, Virginia 23298

SUMMARY

In rat parotid acinar cells, Ca-mobilizing agonists cause a fall in cellular content of inositol lipids and an increase in diacylglycerol (DG) and soluble inositol phosphates (IPs). Kinetic analysis of the formation of IPs in parotid cells stimulated with methacholine indicates that the polyphosphoinositides, but not phosphatidylinositol, are degraded by phospholipase C. There is evidence for roles in cell activation for both of the initial products of this reaction, namely DG and the IPs. As DG would be expected to act through protein kinase C, this possibility was examined indirectly by using phorbol esters which are known to activate the enzyme. The results suggest a role for DG and protein kinase C in protein secretion, but not in the K^+ release response of the parotid. Inositol 1,4,5-trisphosphate (IP_3), the initial product of phosphatidylinositol 4,5-bisphosphate hydrolysis, was examined for effects on intracellular Ca^{2+} metabolism by using hepatocytes made permeable with saponin. Micromolar concentrations of IP_3 caused a rapid release of Ca^{2+} from an ATP-dependent pool believed to be endoplasmic reticulum. This result suggests that IP_3 may function as a second messenger coupling receptor activation to internal Ca^{2+} release.

INTRODUCTION

The study of agonist-induced alterations in phosphoinositide metabolism was initiated some thirty years ago when Hokin and Hokin (1953,1954) found that cholinergic stimulation of exocrine pancreas increased the incorporation of radioactive inorganic phosphate into phosphatidylinositol (PI) and phosphatidic acid (PA). Subsequent research on this phenomenon has dealt with two general questions: (1) What are the biochemical pathways involved in the phosphoinositide effect? and (2) What is the significance of this effect for the physiology of the cell? Some recent findings relevant to these basic questions are summarized here.

PATHWAYS OF THE PHOSPHOINOSITIDE EFFECT

Based on studies of the avian salt gland, Hokin and Hokin (1964) suggested that the acetylcholine-stimulated PI labelling which was observed experimentally was actually secondary to stimulation of PI breakdown by a phosphodiesterase or phospholipase C. Then, in 1975, Michell proposed that this might be the case for activation of inositol lipid turnover by Ca-mobilizing hormones in a wide range of tissues. In addition to PI, most eukaryotic cells synthesize small quantities of two phosphorylated derivatives of PI, specifically phosphatidylinositol 4-phosphate (PIP) and phosphatidylinositol 4,5-bisphosphate (PIP_2), which are formed by the sequential addition of phosphates to the 4 and 5 positions of the inositol ring by specific kinases (Michell, 1975). The action of these kinases and specific phosphomonoesterases hold the cellular levels of the three inositol lipids in rapid equilibrium.

Abdel-Latif *et al.* (1977) examined the effects of acetylcholine on ^{32}P-labeled PIP_2 and PIP in rabbit iris smooth muscle and found a substantial decrease in labeled PIP_2 due to muscarinic receptor activation. A close association between the breakdown of PI and the polyphosphoinositides was suggested by the studies of Kirk *et al.* (1981). These investigators examined the effects of vasopressin and other Ca-mobilizing hormones on radiolabeled PIP_2 in rat hepatocytes under conditions similar to those used previously for PI studies. They found that there was a parallel agonist-induced breakdown of PIP_2 and PIP which, as had been shown previously for PI, was found to be relatively resistant to Ca-depletion. Further,the Ca-ionophore A23187 in concentrations capable of fully activating phosphorylase (a Ca-mediated response) did not induce polyphosphoinositide breakdown. Kirk *et al.* (1981) therefore suggested that the three inositol-containing lipids were broken down in a single inositol lipid cycle which was more complex than originally envisioned. Because the initial rate of PIP_2 breakdown was more rapid than that previously estimated for PI breakdown, they suggested that the initial receptor-regulated event might be the phosphodiesteratic cleavage of PIP_2 rather than PI as was originally proposed.

When inositol lipids are degraded by phospholipase C, water soluble inositol phosphates are liberated which are characteristic of the specific lipids from which they are derived. Thus, PI hydrolysis would produce inositol phosphate (IP), while breakdown of PIP or PIP_2 would produce inositol bisphosphate (IP_2) or inositol trisphosphate (IP_3), respectively. By examining the water soluble inositol phosphates formed in stimulated cells it should be possible to obtain information on the specific inositol lipids degraded.

The rat parotid gland provides a useful preparation for such studies; viable, homogeneous cell suspensions can be prepared which respond to muscarinic or other receptor stimuli with a rapid breakdown of both PI (Jones and Michell, 1974) and PIP_2 (Weiss *et al.*, 1982) which is completely Ca-independent. When parotid cells prelabeled with [^{3}H]inositol were stimulated by the muscarinic agonist methacholine, all three [^{3}H]inositol phosphates were formed (Aub and Putney, 1984; Fig. 1). In these experiments, 10 mM LiCl was present which, in the parotid gland, specifically blocks the degradation of IP to free inositol (Berridge *et al.*, 1982). As shown in Fig. 1, this results in a linear increase in cellular IP, while IP_2 and IP_3 levels reach a plateau value after about 20 min. At 40 min, atropine was added to the preparation thus blocking the muscarinic receptor activation. Over the next 20 min, IP_3 and IP_2 levels decline, while the IP level remains essentially constant, or increases somewhat, due to the blockade of IP phosphatase by LiCl.

The formation of IP_3 establishes that phosphodiesteratic breakdown of PIP_2 does occur. The interpretation of the data for IP_2 and IP are less straightforward however,

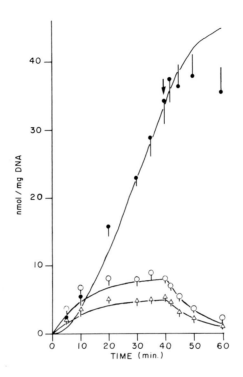

Figure 1. Accumulation of [³H]inositol phosphates in parotid acinar cells. Parotid lipids were prelabeled with [³H]inositol and cellular formation of [³H]inositol phosphates measured as previously described (Berridge et al., 1983; Aub and Putney, 1984). Methacholine (10^{-4} M) was added at t = 0, and atropine (10^{-4} M) at t = 40 min. Basal (unstimulated) values have been subtracted. Assumptions involved in converting radioactivity to nmol, and other details are described in Aub and Putney (1984) from which this is taken, with permission. (●) IP; (O) IP_2; (△) IP_3.

since IP_2 could come from breakdown of either PIP or IP_3 and similarly, IP could be derived either from PI or IP_2. Therefore, by using techniques described in detail elsewhere (Aub and Putney, 1984), these data were analyzed with a computer-assisted kinetic modeling technique so that the steady state (at t = 40 min) metabolic flux through IP_3 and IP_2 and the rate of formation of IP could be calculated. The resulting estimates for turnover of IP_3 and IP_2 were 0.43 and 1.18 nmol $\times$ mg^{-1} DNA $\times$ min^{-1}, respectively. This indicates that at steady-state IP_2 formation cannot be adequately accounted for by IP_3 breakdown and it is necessary to propose that IP_2 may also be formed by direct breakdown of PIP. Formation of IP was estimated to be 1.08 nmol $\times$ mg^{-1}DNA $\times$ min^{-1}; accordingly, IP formation can be more than adequately accounted for by the breakdown of IP_2 suggesting that direct phosphodiesteratic breakdown of PI probably does not occur in the parotid gland. Similar conclusions were reached in a recent report from Downes and Wusteman (1983). In support of this view, when the formation of inositol phosphates was examined during the first 60 sec following methacholine stimulation, cellular IP_3 and IP_2 increased immediately and in parallel while IP was not significantly increased during this interval (Aub and Putney, 1984).

A model for inositol lipid cycling which is consistent with these observations and the previously published findings of others is shown in Fig. 2. The initial reaction that occurs on stimulation is believed to be the breakdown of polyphosphoinositides to diglyceride (DG) and inositol phosphates. The DG can be rapidly phosphorylated by DG kinase to phosphatidic acid which presumably is transported to the endoplasmic reticulum where PI is synthesized. Free inositol for the synthesis of PI can be recycled from the inositol phosphates by specific phosphatases. PI is then presumably transported to the plasmalemma where the polyphosphoinositides are resynthesized by the sequential actions of PI kinase and PIP kinase.

MESSENGER FUNCTIONS OF THE PRODUCTS OF INOSITOL LIPID BREAKDOWN

The initial products of polyphosphoinositide breakdown are DG and inositol phosphates. Nishizuka and his coworkers have described a protein kinase (termed C-kinase) which in the presence of phospholipid and Ca^{2+} is activated by DG (Nishizuka, 1983). Thus DG could act as a messenger leading to the phosphorylation of specific

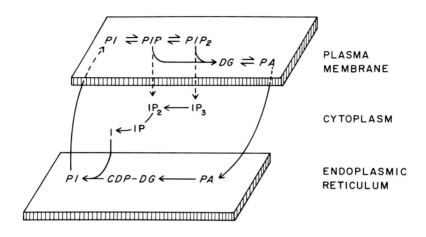

<u>Figure 2</u>. Proposed pathways of the phosphoinositide cycle in rat parotid gland. The initial reactions are believed to be the hydrolysis of PIP and PIP_2 to IP_2 and IP_3, respectively, and DG. Further details of the scheme are described in the text.

proteins and ultimately to the expression or modification of the appropriate cellular response. Investigations into the relative contribution of the C-kinase pathway in cellular activation have been greatly facilitated by the finding that certain phorbol diesters can substitute for DG in activating the enzyme in intact cells or *in vitro* (Castagna *et al.*, 1982). For example, recent studies on the actions of phorbol dibutyrate in the rat parotid gland suggest that the activation of protein secretion by cholinergic agonists may involve the C-kinase (Putney *et al.*, 1984). A limited protein secretion could be induced by phorbol dibutyrate or by the Ca-ionphore, ionomycin. When these two agents were applied in combination, their effects on protein secretion were more than additive. This synergistic action resulting presumably from separate activation of the C-kinase and Ca^{2+} pathways is similar to that seen previously with Ca ionophores and exogenous

DG in platelets (Nishizuka, 1983). Stimulation of cholinergic-muscarinic receptors in the parotid gland also causes an increased membrane permeability to K^+ which apparently results from activation of Ca-dependent K^+ channels (Putney, 1983). However, phorbol dibutyrate does not induce an increase in K^+ permeability, nor does the drug potentiate the response to ionomycin (Putney et al., 1984). This finding confirms that the interaction of phorbol dibutyrate (presumably acting on C-kinase) and Ca^{2+} on protein secretion occurs at steps subsequent to Ca^{2+}-mobilization and also suggests that the C-kinase may be involved in certain, but not all response pathways which Ca-mobilizing receptors activate.

In addition to the production of DG, polyphospho-inositide breakdown results in the liberation of water soluble inositol phosphates. Berridge (1983) has suggested that one or more of the inositol phosphates could fulfill a messenger role serving to transmit information to internal organelles. One process which such a messenger could regulate is the release of Ca^{2+} from internal organelles (Berridge, 1983). In most cases, Ca-mobilizing agonists cause an increase in Ca entry from the extra-cellular space as well as a release of Ca to the cytosol from internal stores (Putney et al., 1981). On the basis of the relative Ca-independence of stimulated phospho-inositide turnover and other points of circumstantial evidence, Michell has suggested that inositol lipid break-down could serve to couple receptor activation to cellular Ca mobilization (Michell, 1975, 1979). The suggestion that one of the inositol phosphates could act to release internal Ca provides a possible mechanism by which this coupling could occur.

The most reasonable test of this hypothesis would be to apply a purified inositol phosphate to Ca-loaded organelles (such as endoplasmic reticulum or mitochondria) under reasonably physiological conditions. Because of their highly polar nature inositol phosphates would not be expected to enter cells readily and indeed, they have no discernible effects when applied extracellularly to intact cells. To examine their actions on Ca-sequestering organelles in situ, it is therefore necessary to use pro-cedures to permeabilize specifically the plasmalemma of cells to permit access of applied inositol phosphates to the intracellular milieu. Streb *et al.*, (1983) utilized low Ca^{2+} solutions to prepare leaky acinar cells from exocrine pancreas. In the exocrine pancreas, activation of Ca-mobilizing receptors causes breakdown of polyphosphoinositides (Putney *et al.*, 1983), accumulation of IP_3, IP_2 and IP (Rubin *et al.*, 1984), and intracellular Ca^{2+} release (Stolze and Schulz, 1980). When applied to the leaky acinar cells, IP_3 (specifically, inositol 1,4,5-tris-phosphate) caused release of sequestered Ca from a pool insensitive to inhibitors of mitochondrial Ca uptake (Streb *et al.*, 1983). This finding, together with the demonstration that receptor activation leads to the accumulation of IP_3 in pancreatic acini (Rubin *et al.*, 1984), constitutes strong support for the suggested messenger role of IP_3 for this system.

Another system in which receptor activation causes considerable mobilization of internal Ca stores is the liver (Exton, 1980; Williamson *et al.*, 1981). Again, a second messenger has been proposed to release internal Ca, and again the Ca-mobilizing receptors are linked to poly-phosphoinositide hydrolysis. A number of investigators have attempted to localize the hormone-sensitive Ca pool in the liver by using subcellular fractionation techniques. While it is generally agreed that such techniques invariably show a loss of Ca (or radio-calcium) from a mitochondria-enriched fraction, there is some disagreement as to whether the mitochondria or elements of the endoplasmic reticulum are the primary source of activator Ca in intact cells (Poggioli *et al.*, 1980; Burgess *et al.*, 1983).

In recent studies, the metabolism of ^{45}Ca in saponin-permeabilized guinea pig hepatocytes was examined. By using both phosphorylase-activity and the Ca-indicator, quin-2 (Burgess *et al.*, 1983), the free cytosolic calcium concentration ($[Ca^{2+}]_i$) in unstimulated hepatocytes was estimated to be about 200 nM (Burgess *et al.*, 1983). When the plasma membranes of hepatocytes were made permeable with saponin, the cells accumulated ^{45}Ca from media containing Ca^{2+} buffered with EGTA to 200 nM. The accumulation of Ca by the permeable cells required the presence of ATP and was inhibited by a Ca-ionophore, but not by the mitochondrial poisons, dinitrophenol and oligomycin (Burgess *et al.*, 1983).

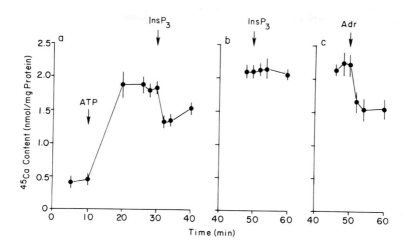

<u>Figure 3</u>.   Release of sequestered Ca by IP$_3$ (InsP$_3$ in the Fig.) in guinea-pig hepatocytes.   (a) Hepatocytes made permeable with saponin sequester ^{45}Ca on addition of ATP into non-mitochondrial pools, presumably elements of the endoplasmic reticulum. The addition of 5 μM IP$_3$ at t = 30 min causes the rapid release of about 0.5 nmol/mg protein of the sequestered ^{45}Ca; (b) 5 μM IP$_3$ does not cause ^{45}Ca release from intact hepatocytes (not saponin-treated);   (c) In intact hepatocytes, adrenaline (adr) releases a quantity of ^{45}Ca similar to that released by IP$_3$ from permeable cells. For further details, see the text and Burgess <u>et al</u>. (1984) from which this is taken, with permission.

This same preparation was used to examine the possible Ca-releasing activity of inositol phosphates (Burgess *et al.,* 1984).   As shown in Fig. 3a, IP$_3$ (specifically, inositol 1,4,5-trisphosphate) caused a rapid release of about 25% or 0.5 nmol/mg protein of the sequestered ^{45}Ca.   Inositol 1,4-bisphosphate, inositol 1-phosphate, inositol 1,2-cyclic phosphate and inositol were also tested and found to be inactive (not shown). Maximal release was obtained with 1-5 μM IP$_3$ and

Putney et al.

half-maximal release occurred with 0.2 μM IP_3. IP_3 did not cause release from intact cells (Fig. 3b); adrenaline (Fig. 3c) when applied to intact cells released a quantity of Ca similar to that released by IP_3 from permeable cells. As the Ca released by IP_3 came from a pool insensitive to mitochondrial poisons, this indicates that the IP_3-sensitive Ca pool and presumably the hormone-sensitive pool is primarily non-mitochondrial, and probably consists of one or more components of the endoplasmic reticulum. When $[Ca^{2+}]$ in the media surrounding the permeabilized hepatocytes was elevated to levels where substantial uptake of Ca by mitochondria occurred, IP_3 did not cause any Ca release from this pool but again was capable of releasing Ca from non-mitochondrial sources.

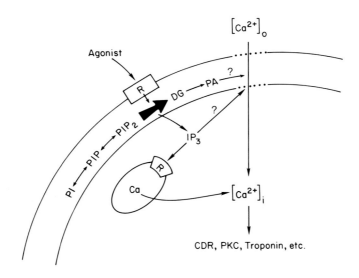

<u>Figure 4</u>. Relationship between phosphoinositide breakdown and Ca-signaling. Receptor (R)-activated hydrolysis of PIP_2 releases IP_3 which acts on another specific receptor (R) on the endoplasmic reticulum to release internal Ca. Surface membrane receptor activation also causes an increased rate of Ca influx. This could be controlled either by PA or IP_3. The subsequent increase in $[Ca^{2+}]_i$ may then regulate a variety of Ca-sensitive effectors such as calmodulin (CDR), protein kinase C (PKC), troponin, etc.

These results, when taken together with the finding that IP_3 levels are rapidly increased in intact hepatocytes treated with Ca-mobilizing hormones (Burgess *et al.*, 1984), extend to the liver the findings which Streb *et al.* (1983) obtained using pancreatic acinar cells, and suggest that this may be a general mechanism for mobilization of intracellular Ca by hormones. These findings for the guinea pig liver were recently confirmed for the case of the rat liver by Joseph *et al.* (1984).

Figure 4 illustrates a working hypothesis for the role of inositol lipid turnover in cellular Ca-mobilization. IP_3 derived from the phosphodiesteratic hydrolysis of PIP_2 causes a release of sequestered Ca^{2+} from a component of the endoplasmic reticulum. IP_3 may act by binding to a specific receptor which controls the permeability of the endoplasmic reticulum membrane to Ca^{2+}. Previously it has been suggested that the increased Ca entry across the plasma membrane which is also receptor-regulated may be mediated by phosphatidic acid (Salmon and Honeyman, 1980; Putney *et al.*, 1980). However, if IP_3 can increase the permeability of the endoplasmic reticulum for Ca^{2+}, the possibility exists that it could exert control on the plasma membrane as well. At present there are no experimental data available to distinguish between these possibilities.

ACKNOWLEDGEMENTS

Some of the work described in this chapter was supported by grants from the NIH, Nos. DE-05764 and AM-32823.

REFERENCES

Abdel-Latif, A.A., Akhtar, R.A. and Hawthorne, J.N. (1977) *Biochem. J. 162*, 61-73.

Aub, D.L. and Putney, J.W., Jr. (1984) *Life Sci. 34*, 1347-1355.

Berridge, M.J. (1983) *Biochem. J. 212*, 849-858.

Berridge, M.J., Downes, C.P. and Hanley, M.R. (1982) *Biochem. J. 206*, 587-595.

Berridge, M.J., Dawson, R.M.C., Downes, C.P., Heslop, J.P. and Irvine, R.F. (1983) *Biochem. J. 212*, 473-482.

Putney et al.

Burgess, G.M., McKinney, J.S., Fabiato, A., Leslie, B.A. and Putney, J.W., Jr. (1983) *J. Biol. Chem.* *258,* 15336–15345.

Burgess, G.M., Godfrey, P.P., McKinney, J.S., Berridge, M.J., Irvine, R.F. and Putney, J.W., Jr. (1984) *Nature 309,* 63–66.

Castagna, M., Takai, Y., Kaibuchi, K., Sano, K., Kikkawa, U. and Nishizuka, Y. (1982) *J. Biol. Chem.* *257,* 7847–7851.

Downes, C.P. and Wusteman, M.M. (1983) *Biochem. J.* *216,* 633–640.

Exton, J.H. (1980) *Am. J. Physiol. 238,* E3–E12.

Hokin, M.R. and Hokin, L.E. (1953) *J. Biol. Chem. 203,* 967–977.

Hokin, M.R. and Hokin, L.E. (1954) *J. Biol. Chem. 209,* 549–558.

Hokin, M.R. and Hokin, L.E. (1964) In: *"Metabolism and Physiological Significance of Lipids"* (R.M.C. Dawson and D.N. Rhodes, eds.) pp. 423–434, John Wiley and Sons, London.

Jones, L.M. and Michell, R.H. (1974) *Biochem. J. 142,* 583–590.

Joseph, S.K., Thomas, A.P., Williams, R.J., Irvine, R.F. and Williamson, J.R. (1984) *J. Biol. Chem. 259,* 3077–3081 .

Kirk, C.J., Creba, J.A., Downes, C.P. and Michell, R.H. (1983) *Biochem. Soc. Trans. 9,* 377–379.

Michell, R.H. (1975) *Biochim. Biophys. Acta 415,* 81–147.

Michell, R.H. (1979) *Trends Biochem. Sci. 4,* 128–131.

Nishizuka, Y. (1983) *Phil. Trans. R. Soc. Lond. B 302,* 101–112.

Poggioli, J., Berthon, B. and Claret, M. (1980) *FEBS Lett. 115,* 243–246.

Putney, J.W., Jr. (1983) *Cell Calcium, 4,* 439–450.

Putney, J.W., Jr., Weiss, S.J., VanDeWalle, C.M. and Haddas, R.A. (1980) *Nature 284,* 345–347.

Putney, J.W., Jr., Poggioli, J. and Weiss, S.J. (1981) *Phil. Trans. R. Soc. Lond. B 296,* 37–45.

Putney, J.W., Jr., Burgess, G.M., Halenda, S.P., McKinney, J.S. and Rubin, R.P. (1983) *Biochem. J.* *212,* 483–488.

Putney, J.W., Jr., McKinney, J.S., Leslie, B.A., and Aub, D.L. (1984) *Mol. Pharmacol.* (in press).

Rubin, R.P., Godfrey, P.P., Chapman, D.A. and Putney, J.W., Jr. (1984) *Biochem. J. 219,* 655–659.

Salmon, D.M. and Honeyman, T.W. (1980) *Nature 284,* 344-345.

Stolze, H. and Schulz, I. (1980) *Am. J. Physiol. 238,* G338-G348.

Streb, H., Irvine, R.F., Berridge, M.J. and Schulz, I. (1983) *Nature 306,* 67-68.

Weiss, S.J., McKinney, J.S. and Putney, J.W., Jr. (1982) *Biochem. J. 206,* 555-560.

Williamson, J.R., Cooper, R.H. and Hoek, J.B. (1981) *Biochim. Biophys. Acta 639,* 243-295.

INOSITOL TRISPHOSPHATE

AND CALCIUM MOBILIZATION

Michael J. Berridge[*] and Robin F. Irvine[+]

[*]A.F.R.C. Unit of Insect Neurophysiology and Pharmacology
Department of Zoology, University of Cambridge,
Downing Street, Cambridge CB2 3EJ and
[+]A.F.R.C. Institute of Animal Physiology,
Babraham, UK CB2 4AT

SUMMARY

Receptors that mobilize calcium stimulate the hydrolysis of phosphoinositides. Addition of 5-HT to the fly salivary gland causes the hydrolysis of phosphatidylinositol 4,5-bisphosphate (PIP_2). The resulting increases in inositol 1,4,5-trisphosphate (IP_3) and inositol 1,4-bisphosphate (IP_2) that begin to accumulate with no apparent delay is fast enough to account for the calcium-dependent component of the physiological response. In contrast, the levels of inositol 1-phosphate (IP) and free inositol begin to rise with lag periods of 20 and 60 seconds respectively. This analysis of how fast these inositol phosphates accumulate following stimulation is consistent with the idea that the primary action of hormones is to hydrolyse PIP_2 to diacylglycerol and IP_3. The idea that IP_3 may function as a second messenger to mobilize intracellular calcium has been supported by observations on pancreatic and liver cells with 'leaky' plasma membranes. Addition of IP_3 in the micromolar range induces a release of calcium from an internal membrane pool that is most likely located in the endoplasmic reticulum. The effect of IP_3 appears to be specific in that there was no release in response to IP_2, IP or inositol 1,2-cyclic phosphate. It is proposed that the IP_3 might be the long sought-after second messenger that functions to mobilize intracellular calcium.

351

INTRODUCTION

Hormones and neurotransmitters transmit information across plasma membranes using sophisticated transducing mechanisms. In those cases where information is converted into second messengers such as cyclic AMP and calcium, the transduction processes seem to depend not only upon the arrival of an external signal, but also on internal ligands (usually a phosphorylated derivative) that are cleaved during information processing to create the internal signals. Receptors that are linked to adenylate cyclase use ATP whereas the calcium-mobilizing receptors use one of the inositol lipids (Fig. 1). In the former case, an adenylate cyclase converts ATP into cyclic AMP that diffuses into the cytoplasm to activate various cellular processes. In a similar way, calcium-mobilizing receptors use phosphatidylinositol 4,5-bisphosphate (PIP_2) as a substrate that is cleaved into diacylglycerol (DG) and inositol 1,4,5-trisphosphate (IP_3), both of which may function as second messengers. One of the products (DG) remains within the plane of the membrane where it may contribute to cell activation by stimulating C-kinase to phosphorylate specific cellular proteins (Kaibuchi *et al.*, 1982; Nishizuka, 1983). The water-soluble product IP_3 is free to diffuse into the cytoplasm where it may function as a second messenger to release calcium from intracellular stores (Berridge, 1983, 1984; Streb *et al.*, 1983; Burgess *et al.*, 1984). A fascinating aspect of this receptor mechanism, therefore, is that information transduction represents a bifurcating pathway in that the hydrolysis of PIP_2 gives rise to two functional second messengers (Fig.1). The way in which these two second messengers interact with each other will be discussed later but first we must consider the evidence that IP_3 is a second messenger for calcium mobilization.

RELATION BETWEEN PHOSPHOINOSITIDES AND CALCIUM SIGNALLING

The first indication that inositol lipid metabolism might be an integral part of certain receptor mechanisms became apparent during studies on the pancreas by Hokin and Hokin (1953) and on synaptosomes by Durell *et al.* (1969). Michell (1975) subsequently noticed that all receptors which used these lipids were also those that

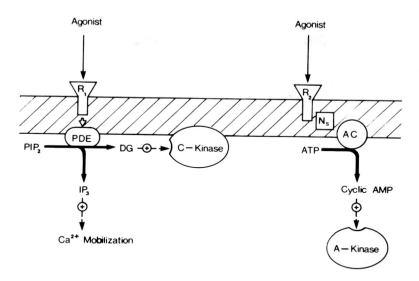

Figure 1. The role of phosphorylated precursors in the action of receptors that mobilize calcium (R_1) or generate cyclic AMP (R_2). Agonists activate R_1 receptors to stimulate the hydrolysis of phosphatidylinositol 4,5 bisphosphate (PIP_2) to diacylglycerol (DG) and inositol trisphosphate (IP_3) by a phosphodiesterase (PDE). Activation of R_2 receptors stimulates adenylate cyclase (AC) using a nucleotide-binding protein (N_s).

acted through calcium, and he proposed that this receptor-mediated hydrolysis of the phosphoinositides might be responsible for mobilizing calcium. A necessary prerequisite for such a mechanism to operate is that cleavage of the lipid should occur at resting levels of intracellular calcium (Michell and Kirk, 1981; Michell *et al.*, 1981). While there are a few examples where the breakdown of phosphoinositides is sensitive to calcium, (Cockcroft, 1982) there are many cell types where the receptor mechanism is relatively insensitive to changes in intracellular calcium (Berridge, 1981; 1982; Michell and Kirk, 1981; Michell *et al.*, 1981). When studied *in vitro*, the PIP_2 phosphodiesterase was shown to have a calcium sensitivity that enables it to hydrolyze its substrate at resting levels of calcium (Irvine *et al.*, 1984). This enzyme study also suggests that it is unlikely that the enzyme is stimulated by increases in intracellular calcium within its normal physiological range. In order to obtain more direct evidence for a causal link between inositol lipids and calcium mobilization, Fain and Berridge (1979b) studied the effect of reducing the level of these lipids in the membrane. If the salivary gland of the blowfly is hyperstimulated with 5-HT there is a massive breakdown of the phosphoinositides with a concomitant calcium-dependent inhibition of synthesis (Berridge and Fain, 1979; Fain and Berridge, 1979a). These two processes that both conspire to lower the level of inositol lipids result in a drastic reduction in calcium signalling (Fain and Berridge, 1979b). The latter could be restored, however, if such depleted glands were provided with inositol and so allowed to resynthesize phosphatidyl-inositol. The fact that the calcium signalling system in the fly gland was so sensitive to changes in the availability of the inositol lipids seemed to confirm that the latter were essential for signal transduction at calcium-mobilizing receptors. Just how these inositol lipids might function to mobilize calcium became apparent when more detailed information began to emerge suggesting that the polyphosphoinositides rather that phosphatidylinositol were the immediate substrates hydrolyzed on receptor activation.

CALCIUM-MOBILIZING RECEPTORS HYDROLYSE
PIP AND PIP$_2$ INSTEAD OF PI

Phosphatidylinositol (PI) exists in a dynamic equilibrium with phosphatidylinositol 4-phosphate (PIP) and phosphatidylinositol 4,5-bisphosphate (PIP$_2$) (Downes and Michell, 1982). Although the latter two lipids are usually present in cells as minor constituents, they are extremely important with regard to signalling because they are the substrate (in particular PIP$_2$) that is cleaved as part of the receptor mechanism. There is a PIP$_2$ phosphodiester-ase that is capable of hydrolysing PIP$_2$ to DG with the release of IP$_3$ (Thompson and Dawson, 1964; Irvine *et al.*, 1984). The first indication that these polyphos-phoinositides might play a role in receptor-mediated events was provided by Durell *et al.* (1968) who showed that acetylcholine stimulated the formation of inositol bisphosphate (IP$_2$) in brain synaptosomes. The fact that acetylcholine decreased the ^{32}P-labeling of PIP and PIP$_2$

in synaptosomes also suggested a role for the polyphos-phoinositides (Schacht and Agranoff, 1972). A study on rabbit iris smooth muscle provided the first convincing evidence that an external agonist, in this case acetyl-choline, could bring about a rapid breakdown of the polyphosphoinositides (Abdel-Latif et al., 1977) resulting in the release of IP_3 (Akhtar and Abdel-Latif, 1980). However, this breakdown of PIP_2 appeared to be calcium-dependent (Akhtar and Abdel-Latif, 1978; 1980) and was thus thought to be associated with the action of calcium on either the control of sodium and potassium permeability (Akhtar and Abdel-Latif, 1978) or on exocytosis in the case of nerve cells (Griffin and Hawthorne, 1978).

Interest in the possible involvement of the poly-phosphoinositides in calcium signalling began with the observation that vasopressin caused a very rapid dis-appearance of ^{32}P-labeled PIP_2 in liver cells through a mechanism that was largely calcium-independent (Kirk et al., 1981; Michell et al., 1981; Creba et al., 1983). It was argued that the disappearance of PIP_2 was not due to the action of a phosphomonoesterase because the levels of PIP and PI did not rise but actually declined (Michell et al., 1981; Creba et al., 1983). These observations led to the proposal that the PIP_2 hydrolysed by the receptor mechanism was rapidly replaced by phosphorylation of PI to PIP and then to PIP_2 (Michell et al., 1981; Creba et al., 1983). The fact that PIP_2 is rapidly hydrolysed by calcium-mobilizing receptors has now been confirmed in a number of different cell types including liver (Rhodes et al., 1983; Thomas et al., 1983), parotid gland (Weiss et al., 1982), blood platelets (Billah and Lapetina, 1982; Agranoff et al., 1983; Mauco et al., 1983), pancreas, (Putney et al., 1983), cloned rat pituitary cells (Rebecchi and Gershengorn 1983) and blowfly salivary gland (Berridge, 1983).

In order to obtain more information about which of these phosphoinositides was being degraded, it was decided to investigate the rate of formation of the water-soluble inositol phosphates. A number of studies have already described the existence of inositol monophosphate (IP), IP_2 and IP_3 in various cells (Allen and Michell, 1978; Griffin and Hawthorne, 1978; Akhtar and Abdel-Latif, 1980; Agranoff, et al., 1983; Berridge, et

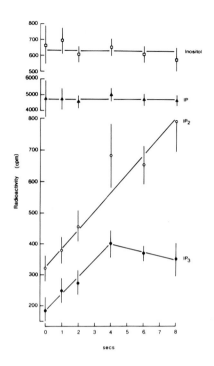

Figure 2. Changes in inositol phosphate levels in the insect salivary gland following stimulation with 10 μM 5-hydroxytryptamine (from Berridge et al., 1984 with permission).

al., 1982; Berridge, *et al.*, 1983, 1984; Rebecchi *et al.*, 1983; Rebecchi and Gerschengorn, 1983). A kinetic analysis of the formation of inositol phosphates following stimulation of the insect salivary gland with 5-HT has shown that the earliest products formed are IP_3 and IP_2 while IP and free inositol appear much more slowly (Berridge 1983; Berridge, *et al.*, 1984). Over the 8 second stimulation period shown in Figure 2, there were no changes in the levels of IP or free inositol indicating that PI is clearly not an immediate substrate for the receptor mechanism. When studied over longer incubation periods, there was an increase in IP (lag period 20 seconds) and free inositol (lag period 60 seconds) (Berridge 1983; Berridge *et al.*, 1984). Similar changes in inositol phosphates have been reported in pituitary cells (Rebecchi and Gerschengorn, 1983) and in parotid (Downes and Wusteman, 1983) immediately following stimulation. These studies on the formation of the inositol phosphates are consistent with the metabolic pathway shown in Figure 3. The rapid appearance of IP_3 indicates that a primary action of calcium-mobilizing receptors is to cleave PIP_2 as depicted in Figures 1 and 3. The rapid appearance of IP_2 is probably derived from IP_3 as shown in Figure 3 but a contribution from the breakdown of PIP (not shown in Figure 3) cannot be excluded.

In order to obtain more direct evidence that PI is not a primary substrate for the receptor, the effect of removing the polyphosphoinositides from the membrane was investigated (Berridge *et al*. 1984). Lowering the concentration of ATP with 2,4-dinitrophenol caused a large decline in the level of the polyphosphoinositides whereas there was no change in PI. When 5-HT was added, there was no release of inositol and no accumulation of inositol phosphate indicating that the receptor mechanism cannot degrade PI. On removal of DNP, the level of ATP and the polyphosphoinositides began to rise and this coincided with the appearance of inositol phosphates and free inositol. This study supports the idea that before PI can be used by the receptor mechanism it must first of all be converted into PIP_2 that is cleaved into DG and IP_3 (see also Downes and Wusteman, 1983). In this context, it is interesting to note that, out of a variety of physiological responses of the blood platelet investigated by Holmsen *et al*. (1982), PI disappearance was among the most sensitive to the depletion of cellular ATP. Since the IP_3 is produced so rapidly upon stimulation, it has been suggested that IP_3 may function as a second messenger to mobilize intracellular calcium (Berridge, 1983, 1984).

INOSITOL TRISPHOSPHATE AND CALCIUM MOBILIZATION

Any attempt to link the hydrolysis of inositol lipids to calcium signalling is complicated by the fact that many cells use both intracellular and extracellular sources of calcium. A good example of the use of intracellular calcium is found in liver where calcium-mobilizing agonists

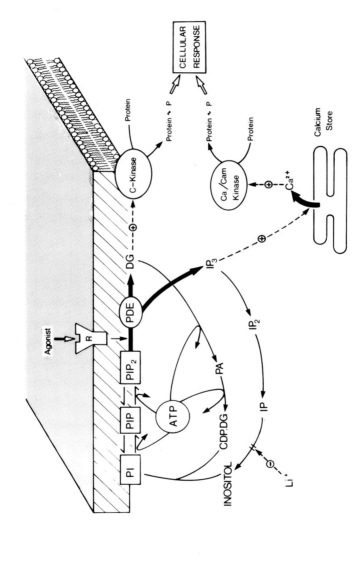

Figure 3. Relationship of agonist-dependent phosphoinositide metabolism to the bifurcating signal pathways based on diacylglycerol (DG) and calcium. In response to an agonist the receptor (R) initiates the hydrolysis of PIP_2 by a phosphodiesterase (PDE) to produce the two second messengers DG and IP_3. The former activates C-kinase whereas IP_3 acts to mobilize calcium from intracellular stores.

stimulate a massive mobilization of intracellular calcium which is then extruded from the cell resulting in a marked elevation in the concentration of extracellular calcium (Morgan *et al.*, 1982). This agonist-dependent mobilization of internal calcium is found in many other cells such as pancreas (Schulz and Stolze, 1980), smooth muscle (Triggle, 1977), parotid (Putney, 1977) pituitary (Ronning *et al.*, 1982; Thaw *et al.*, 1982) and 3T3 cells (Lopez-Rivas and Rozengurt, 1983). The IP_3 that is formed by such calcium-mobilizing receptors is an obvious candidate for the missing link between such surface receptors and the internal calcium stores (Berridge, 1983). An obvious question that arises is whether the formation of IP_3 is fast enough to account for this proposed calcium-mobilizing function. It is evident from Figure 2 that the 5-HT-dependent increase in IP_3 in the insect salivary gland is very rapid with no apparent lag period because the early time points extrapolate back to the zero time point (Fig. 2). In contrast, the 5-HT-dependent change in transepithelial potential, which is a good indicator of a change in the intracellular level of calcium, has a lag period of approximately I second (Berridge, *et al.*, 1984). It would seem, therefore, that the increase in IP_3 is fast enough to account for calcium-dependent events in the insect salivary gland.

In order to provide more direct evidence for a calcium-mobilizing role for IP_3, various permeabilized cell preparations have been investigated. Streb and Schulz (1983) have described a permeabilized pancreatic cell preparation that can respond to carbachol with a release of calcium stored in an intracellular compartment that is most likely the endoplasmic reticulum. A similar release of calcium was observed when such permeabilized pancreatic cells were treated with IP_3 (Fig.4) (Streb *et al.*, 1983). Carbachol and IP_3 appeared to be acting on the same calcium store. The effect of IP_3 was also specific since there was no release with free inositol, inositol 1,2-cyclic phosphate, IP or IP_2. The release of calcium was not maintained and the subsequent re-uptake of calcium can probably be explained by a rapid degradation of IP_3. A similar mobilization of internal calcium has been observed when the effects of IP_3 were studied on permeabilized liver cell preparations (Burgess *et al.*, 1984; Joseph *et al.*, 1984). IP_3 will also release calcium

359

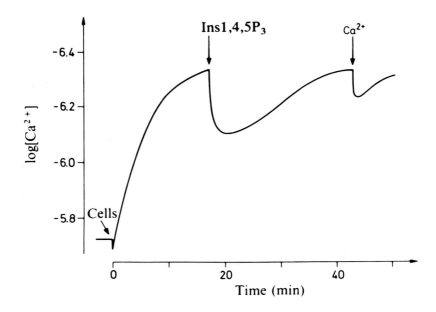

<u>Figure</u> 4. Release of intracellular calcium from 'leaky' pancreatic acinar cells by IP_3. When cells that have permeable plasma membranes were added to a low calcium medium, they gradually reduced the extra-cellular calcium concentration to approximately 0.4 µM at which point the addition of IP_3 (2.5 µM) caused a large mobilization of calcium (from Streb <u>et al</u>., 1983 with permission).

from microsomes prepared from liver (Dawson and Irvine, 1984) or from insulinoma cells (Prentki *et al.*, 1984). Addition of IP_3 (5 µM) to liver cells that had been incubated in ^{45}Ca caused a net release of 0.5 nmol/mg protein of cell-associated ^{45}Ca within 2 minutes (Burgess *et al.*, 1984). Again the effect was found to be specific in that IP_2, inositol 1,2-cyclic phosphate and inositol were ineffective. This IP_3-dependent mobilization of calcium appears to be fairly sensitive in that the EC_{50} values for this effect were 0.7 µM in the pancreas (Streb *et al.*,

1983) and 0.1-0.22 µM in liver (Burgess *et al.*, 1984; Joseph *et al.*, 1984). All the experiments carried out so far have been designed to test the possible role of IP_3 as a second messenger for calcium mobilization, but the possibility remains that it may also function to regulate calcium entry across the plasma membrane.

Most of the emphasis so far has concentrated on the idea that IP_3 functions solely as the putative second messenger responsible for mediating the action of calcium-mobilizing receptors. It might be relevant, however, to think of IP_3 as a regulator of intracellular calcium even when the cell is at rest. In other words, the level of intracellular calcium at any moment in time might be set by the level of IP_3 which acts by adjusting the calcium buffering capacity of the intracellular organelles. During stimulation, therefore, IP_3 not only contributes to the initiation of the calcium signal by releasing calcium from intracellular organelles but it may contribute to a maintenance of the calcium signal by virtue of the fact that it has effectively 'short-circuited' the internal membrane systems that normally act to sequester calcium (*i.e.*, it has reduced their capacity to buffer intracellular calcium). Since these internal organelles have a finite store of calcium, their ability to maintain an elevated intracellular level of calcium is limited and a continuation of the calcium signal will then depend more and more upon the entry of external calcium. Just such a mechanism has been proposed to account for the calcium transients (measured using aequorin) recorded in vascular smooth muscle in response to angiotensin or phenylephrine (Morgan and Morgan, 1983). It fact, it may not always be necessary for the agonist to increase the influx of external calcium because the rate of calcium entry at rest may be sufficient to maintain an elevated level of intracellular calcium once the internal buffering system has been reduced by IP_3. In other words, it is possible that in a tissue where the internal calcium stores are relatively small, a rise in intracellular calcium may be almost entirely dependent on external calcium, yet in fact be caused by IP_3 acting on an internal receptor rather than by an increase in calcium entry across the plasma membrane. In some tissues, a change in plasma membrane permeability has been documented, however, and there clearly is a need to learn more about the regulation of calcium uptake from outside the cell.

HYDROLYSIS OF PIP$_2$
A BIFURCATING TRANSDUCTION MECHANISM

A key event in the operation of calcium-mobilizing receptors is the hydrolysis of PIP$_2$ to form DG and IP$_3$ (Figs. 1 and 3). This primary biochemical event represents a bifurcation in the flow of information because the external signal is split into two second messengers. It is interesting to compare some of the properties of these two putative internal signals with that of the classical second messenger cyclic AMP. As mentioned at the beginning, they are both formed by the cleavage of a phosphorylated precursor as the result of a transduction mechanism located in the plasma membrane. The intracellular levels of these second messengers are determined by the balance between their rates of formation and removal. Agonists act by stimulating their rate of formation (Fig. 1). As with cyclic AMP, there are pathways for rapidly degrading both IP$_3$ and DG. When 5-HT is withdrawn from the insect salivary gland, there is a very rapid decline in the intracellular level of IP$_3$ (Berridge 1983), which is rapidly degraded to IP$_2$ by an IP$_3$ phosphatase (Fig. 3) (Berridge *et al.*, 1983). The other second messenger is rapidly removed either by being converted to phosphatidic acid by a DG kinase or it is hydrolysed to monoacylglycerol (MG) with the release of arachidonic acid by a DG lipase. This release of arachidonic acid represents yet another branch point of the signal cascade generated by these multifunctional receptors (Irvine and Dawson, 1980; Berridge, 1981, 1982).

It is the bifurcation of the signal pathway following cleavage of PIP$_2$ that is the most important aspect of this receptor mechanism. DG functions to activate C-kinase whereas the IP$_3$ appears to mobilize calcium. Both signal pathways contribute to the final response but they seem to use separate mechanisms because they can act synergistically with each other. The first example of such synergism was described in blood platelets where a threshold concentration of calcium ionophore was found to induce maximal secretion when combined with a low dose of phorbol ester (Kaibuchi et al., 1982). This increase in secretion induced by activating the C-kinase pathway with phorbol esters can occur without any change in the intracellular level of calcium (Rink et al., 1983). Sha'afi et al. (1983) have also noted that stimulation of the C-kinase pathway with a phorbol ester can activate neutrophils without an apparent rise in the intracellular level of free calcium. Studies on permeabilized platelets have also revealed that activation of C-kinase can potentiate the effect of low levels of calcium on the secretion of 5-HT (Knight and Scrutton, 1984). When the calcium concentration was reduced below the resting level of 10^{-7} M, there was no effect of activating the C-kinase pathway. Under normal conditions, therefore, cell activation probably depends upon the calcium and DG pathways acting in a concerted and synergistic manner. Just how these two pathways interact is still somewhat of a mystery. On the basis of experiments on 'leaky' adrenal cells, Knight and Baker (1983) have suggested that DG may act by adjusting the affinity of the exocytotic mechanism for calcium. Such a mechanism could account for the observation that the phorbol ester, 3,4-phorbol 12-myristate 13-acetate (TPA) can stimulate secretion from blood platelets without any change in the resting level of calcium (Rink et al., 1983).

Another way of attempting to identify the molecular mechanism underlying these synergistic interactions is to identify the target proteins that are phosphorylated during the action of DG and calcium. When blood platelets are stimulated with thrombin, there is an increase in the phosphorylation of a 40 K and a 20 K protein. The 40 K protein is primarily a substrate for C-kinase whereas the 20 K protein, which has been identified as the myosin light chain, is sensitive to both C-kinase and the Ca^{2+}/calmodulin-dependent protein kinase (Naka et al., 1983). The precise function of these two separate phosphorylations must await further studies to determine the role of myosin light chains in platelet activation. The two signal pathways that bifurcated at the level of the receptor converge at the final effector site in that they focus upon a common substrate. A challenge for the future will be to identify the molecular details of how these two signal pathways cooperate with each other to regulate cellular activity.

REFERENCES

Abdel-Latif, A.A., Akhtar, R.A. and Hawthorne, J.N. (1977) *Biochem. J. 162*, 61–73.

Agranoff, B.W., Murthy, P. and Seguin, E.B. (1983) *J. Biol. Chem. 258*, 2076–2078.

Akhtar, R.A. and Abdel-Latif, A.A. (1978) *J. Pharm. Exptl. Therap. 204*, 655–668.

Akhtar, R.A. and Abdel-Latif, A.A. (1980) *Biochem. J. 192*, 783–791.

Allan, D. and Michell, R.H. (1978) *Biochim. Biophys. Acta 508*, 277–286.

Berridge, M.J. (1981) *Molec. Cell. Endocrinol. 24*, 115–140.

Berridge, M.J. (1982) in: *Calcium and Cell Function*. Vol. 3., (W.Y. Cheung, ed.) pp. 1–36, Academic Press, New York.

Berridge, M.J. (1983) *Biochem. J. 212*, 849–858.

Berridge, M.J. (1984) *Biochem. J. 220*, 345–360.

Berridge, M.J. and Fain, J.N. (1979) *Biochem. J. 178*, 59–69.

Berridge, M.J., Downes, C.P. and Hanley, M.R. (1982) *Biochem. J. 206*, 587–595.

Berridge, M.J., Dawson, R.M.C., Downes, C.P., Heslop, J.P. and Irvine, R.F. (1983) *Biochem J. 212*, 473–482.

Berridge, M.J., Buchan, P.B. and Heslop, J.P. (1984) *Molec. Cell. Endocrinol.* (in preparation).

Billah, M.M. and Lapetina, E.G. (1982) *J. Biol. Chem. 257*, 12705–12708.

Burgess, G.M., Godfrey, P.P., McKinney, J.S., Berridge, M.J., Irvine, R.F. and Putney, J.W. (1984) *Nature 309*, 63–66..

Cockcroft, S. (1982) *Cell Calcium 3*, 337–349.

Creba, J.A., Downes, C.P., Hawkins, P.T., Brewset, G., Michell, R.H. and Kirk. C.J. (1983) *Biochem. J. 212*, 733–747.

Dawson, A.P. and Irvine, R.F. (1984) *Biochem. Biophys. Res. Commun. 120*, 858–864.

Downes, C.P. and Michell, R.H. (1982) *Cell Calcium, 3*, 467–502.

Downes, C.P. and Wusteman, M.M. (1983) *Biochem. J. 216*, 633–640.

Durell, J., Sodd, M.A. and Friedel, R.O. (1968) *Life Sci. 7*, 363–368.

Durell, J., Garland, J.T. and Friedel, R.O. (1969) *Science 165,* 862–866.

Fain, J.N. and Berridge, M.J. (1979a) *Biochem. J. 178,* 45–58.

Fain, J.N. and Berridge, M.J. (1979b) *Biochem. J. 180,* 655–661.

Hokin, M.R. and Hokin, L.E. (1953) *J. Biol. Chem. 203,* 967–977.

Holmsen, H., Kaplan, K.L. and Dangelmaier, C.A. (1982) *Biochem. J. 208,* 9–18.

Irvine, R.F. and Dawson, R.M.C. (1980) *Biochem. Soc. Trans. 8,* 376–377.

Irvine, R.F., Letcher, A.J. and Dawson, R.M.C. (1984) *Biochem. J.* (in press).

Joseph, S.K., Thomas, A.P., Williams, R.J., Irvine, R.F. and Williamson, J.R. (1984) *J. Biol. Chem. 259,* 3077–3081.

Kaibuchi, K., Sano, K., Hoshijima, M., Takai, Y. and Nishizuka, Y. (1982) *Cell Calcium 3,* 323–335.

Knight, D.E. and Baker, P.F. (1983) *FEBS Letters 160,* 98–100.

Knight, D.E. and Scrutton, M.C. (1984) *Nature 309, 66–68.*

Lopez-Rivas, A. and Rozengurt, E. (1983) *Biochem. Biophys. Res. Commun. 114,* 240–247.

Mauco, G., Chap, H. and Douste-Blazy, L. (1983) *FEBS Letters 153,* 361–365.

Michell, R.H. (1975) *Biochim. Biophys. Acta 215,* 81–147.

Michell, R.H. and Kirk, C.J. (1981) *Trends Pharmacol. Sci. 2,* 86–89.

Michell, R.H., Kirk, C.J., Jones, L.M., Downes, C.P. and Creba, J.A. (1981) *Phil. Trans. Roy. Soc. Ser. B. 296,* 123–137.

Morgan, J.P. and Morgan, K.G. (1982) *Pflugers Archiv. 395,* 75–77.

Morgan, N.G., Shuman, E.A., Exton, J.H. and Blackmore, P.F. (1982) *J. Biol. Chem. 257,* 13907–13910.

Naka, M., Nichikawa, M., Adelstein, R.S. and Hidaka, H. (1983) *Nature 306,* 490–492.

Nishizuka, Y. (1983) *Trends Biochem. Sci. 8,* 13–16.

Prentki, M., Biden, T.J., Janjic, D., Irvine, R.F., Berridge, M.J. and Wollheim, C.B. (1984) *Nature 309,* 562–564.

Putney, J.W. (1977) *J. Physiol. 268*, 139-149.

Putney, J.W., Burgess, G.M., Halenda, S.P., McKinney, J.S. and Rubin, R.P. (1983) *Biochem. J. 212*, 483-488.

Rebecchi, M.J. and Gerschengorn, M.C. (1983) *Biochem. J. 216*, 287-294.

Rebecchi, M.J., Kolesnick, R.N. and Gershengorn, M.C. (1983) *J. Biol. Chem. 258*, 227-234.

Rhodes, D., Prpic, V., Exton, J.H. and Blackmore, P.F. (1983) *J. Biol. Chem. 258*, 2770-2773.

Rink, T.J., Sanchez, A. and Hallam, T.J. (1983) *Nature 305*, 317-319.

Ronning, S.A., Heatley, G.A. and Martin, T.F.J. (1982) *Proc. Natl. Acad. Sci. U.S.A. 79*, 6294-6298.

Schulz, I. and Stolze, H.H. (1980) *Ann. Rev. Physiol. 42*, 127-156.

Sha'afi, R.I., White, J.R., Molski, T.F.P., Shefcyk, J., Volpi, M., Naccache, P.H. and Feinstein, M.B. (1983) *Biochem. Biophys. Res. Commun. 114*, 638-645.

Streb, H. and Schulz, I. (1983) *Am. J. Physiol. 245*, G347-G357.

Streb, H., Irvine, R.F., Berridge, M.J. and Schulz, I. (1983) *Nature 306*, 67-69.

Thaw, C., Wittlin, S.D. and Gerschengorn, M.C. (1982) *Endocrinology 111*, 2138-2140.

Thomas, A.P., Marks, J.S., Coll, K.E. and Williamson, J.R. (1983) *J. Biol. Chem. 258*, 5716-5725.

Thompson, W. and Dawson. R.M.C. (1964) *Biochem. J. 91*, 233-236.

Triggle, D.J. (1977) *Neurotransmitter-Receptor Interactions*. Academic Press, London.

Weiss, S.J., McKinney, J.S. and Putney, J.W. (1982) *Biochem. J. 206*, 555-560.

CALCIUM-PHOSPHATIDYLINOSITOL INTERACTIONS IN SECRETORY CELLS AND THE ROLE OF ARACHIDONIC ACID

Ronald P. Rubin

Department of Pharmacology, Medical College of Virginia
Richmond, VA 23298

SUMMARY

Since secretagogues express their actions by increasing Ca^{2+} availability and promoting membrane phospholipid turnover, the fundamental action of Ca^{2+} in stimulus-secretion coupling may be uncovered by elucidating its interactions with membrane phospholipids. The turnover of arachidonoyl phosphatidylinositol (PI) triggered by the Ca^{2+}-dependent activation of phospholipase A_2 and followed by the reincorporation of fatty acid into lysophospholipids, appears to be one feature that secretory systems have in common. Stimulated arachidonic acid turnover in position 2 of PI has now been demonstrated in a number of secretory systems, including adrenal cortex, rabbit peritoneal neutrophils, and exocrine pancreas. The relevance of this reaction to secretory phenomena is indicated by its requirement for Ca^{2+}, its rapid onset, and dose-response curves that parallel those of the secretory response. One or another of the products of this deacylation-reacylation cycle, including arachidonic acid metabolites and/or lysophospholipids, may participate in the cellular processes that accompany the export of secretory product from the cell. Products of the activation of phospholipase C, including diacylglycerol and/or inositol phosphates, may also serve as second messengers. Insight into the nature of the interactions between phospholipases A_2 and C with regard to their putative physiological roles should shed significant light on the vital control mechanisms that regulate the secretory process.

INTRODUCTION

The term stimulus-secretion coupling as originally defined expresses a sequence of events initiated by a stimulus combining with a membrane receptor, which leads to an increase in ionic Ca^{2+} in the cell cytosol. The elevated level of ionized Ca^{2+} then serves as a progenitor of intracellular signals leading to exocytotic secretion (Douglas, 1968). The pivotal role of calcium that is linked to exocytosis represents a fusion-fission response involving the interaction of the plasma membrane and the secretory granule membrane. The strong association between membrane fusion phenomena and the secretory response avers that the fundamental action of calcium in secretion may be revealed by a better understanding of membrane fusion phenomena. Indeed, the fusion of model membranes primarily involves the interaction of calcium and acidic phospholipids, including phosphatidylinositol (PI) (Papahadjopoulos et al., 1978). Particular attention has focused on PI with regard to its participation in events associated with receptor-agonist-mediated secretion (Laychock and Putney, 1982). The turnover of the polar head group and of arachidonic acid in position 2 of PI is catalyzed by phospholipases C and A_2, respectively; both of these reactions have been implicated in events associated with the secretory response.

Activation of a wide variety of secretory organs involves stimulation of a Ca^{2+}-dependent phospholipase A_2 that liberates arachidonic acid (Laychock and Putney, 1982) (Fig. 1). The concept that the availability of free calcium is a pivotal step in the generation of free fatty acid from phospholipid is supported by the finding that the calcium ionophore A23187 stimulates phospholipase A_2 activity in several secretory organs, including the blood platelet, adrenal cortex, neutrophil, and exocrine pancreas (Feinstein et al., 1977; Schrey and Rubin, 1979; Rubin et al., 1981a,b; Halenda and Rubin, 1982). The idea that phospholipase A_2 may regulate secretory events is appealing not only because this enzyme absolutely requires Ca^{2+} for activity, but also because of the multiple ways in which the actions of the enzyme, by regulating arachidonic acid metabolism, can be envisioned as linking membrane events and subsequent intracellular reactions associated with the secretory process.

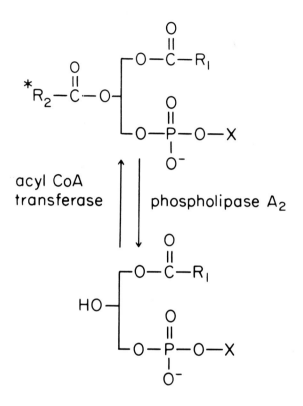

* arachidonate

Figure 1. The deacylation-reacylation cycle initiated by the Ca^{2+}-dependent activation of phospholipase A_2 causes the release of arachidonic acid from position-2 of phosphatidylinositol. The rapid reincorporation of free fatty acid into lysophosphatidylinositol is catalyzed by acylCoA transferase.

Arachidonic acid metabolites formed by way of the cyclooxygenase and lipoxygenase pathways may serve as intracellular messengers, perhaps by acting as calcium ionophores or regulating cyclic nucleotide metabolism (Laychock and Putney, 1982). Another product of phospholipase activation, the lysophospholipids, are of particular interest in this context since they possess fusogenic properties (Lucy, 1978), modulate adenylate and guanylate cyclase activities (Zwiller et al., 1976; Houslay and Palmer, 1979), and potentiate the secretory response (Martin and Lagunoff, 1979). Viewed from another perspective, the rapid reacylation of the lysophospholipid brings about the partitioning of free fatty acids into different domains of membrane phospholipids that may selectively perturb particular regions of the membrane (Fig. 1). At the same time, reacylation prevents the potentially cytotoxic lysophospholipid from rising to levels that are detrimental to the cell. The existence of phospholipases and acylating enzymes in the cell membrane may thus be important for regulating membrane functions associated with the secretory response. The characterization of this biochemical sequence in secretory cells and its possible relation to the secretory process is the subject of this chapter.

ADRENAL CORTEX

Our early studies related to phospholipids were carried out on isolated feline adrenocortical cells where calcium is an obligatory requirement for ACTH-induced steroid production and release. The presence of a Ca^{2+}-dependent phospholipase A_2, localized to the surface of adrenocortical cells, was established (Schrey et al., 1980). This phospholipase was activated by ACTH to promote the release of arachidonic acid from phospholipids and thereby stimulate prostaglandin formation. Further investigations into the fate of radiolabeled arachidonic acid described a rapid and specific turnover of this unsaturated fatty acid within a PI pool that was independent of changes in de novo synthesis and characterized as a Ca^{2+}-dependent hydrolysis of PI, followed by a rapid, selective reacylation of lysoPI (Schrey and Rubin, 1979). Adrenocortical cells incubated with [^{14}C]arachidonic acid plus ACTH or ionophore A23187 exhibited up to a 10-fold increase in the incorporation of

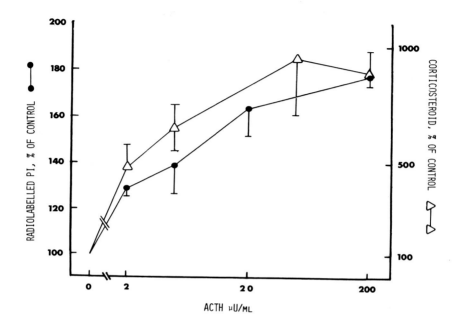

<u>Figure 2</u>. Dose-response curves of phosphatidylinositol labeling and cortisol secretion induced by ACTH. Isolated feline adrenocortical cells were incubated for 2 min with [^{14}C]arachidonic acid plus increasing doses of ACTH. (For further details see Schrey and Rubin, 1979).

label within 1-2 min. The dose-response paralleled that of cortisol secretion (Fig. 2), and other non-steroidogenic peptide hormones failed to stimulate PI labeling. Moreover, agents known to suppress phospholipase A$_2$ activity, such as p-bromophenacylbromide and chlorpromazine, were effective inhibitors of ACTH- and ionophore A23187-induced arachidonoyl PI turnover and cortisol secretion.

The pattern of distribution of radiolabeled arachidonic acid within individual phospholipids was similar to the relative concentrations of endogenous phospholipids, with the notable exception of PI, which had a specific radioactivity almost 3-fold higher than that of any other phospholipid (Laychock et al., 1978). Moreover, PI labeling with other unsaturated and monoenoic species (palmitic, stearic, and oleic acids) was relatively low, with ACTH having no effect on PI labeling by these precursors. The selective incorporation of arachidonic acid into PI may relate to a preference of adrenocortical phospholipase A_2 for this particular phospholipid to promote release of arachidonate from PI or to the known avidity of lysoPI for this fatty acid. Relevant to the latter possibility, the PI of adrenal membranes is highly enriched in arachidonic acid (Sun, 1979), and acyltransferase exhibits a marked preference for arachidonoyl-CoA when lysoPI is the acceptor substrate (Holub and Kuksis, 1978).

NEUTROPHIL

A similar deacylation-reacylation reaction involving arachidonoyl PI also is operative in the rabbit peritoneal neutrophil when challenged with secretagogues. In fact, a close correlation exists in the neutrophil between the ability of the synthetic peptide formyl-methionyl-leucyl-phenylalanine (f-Met-Leu-Phe) to enhance arachidonic acid incorporation into PI and to promote lysosomal enzyme secretion in that both processes increase in parallel with respect to time and peptide concentration (Rubin et al., 1981a). A pivotal role for calcium in this sequence was revealed by the findings that both responses are almost completely dependent upon the presence of extracellular calcium and that the calcium ionophore A23187 mimics the actions of f-Met-Leu-Phe on PI labeling and enzyme secretion (Rubin et al., 1981a,b). f-Met-Leu-Phe acts on surface membrane receptors of the neutrophil to promote calcium influx and mobilize cellular Ca^{2+} (Naccache et al., 1977; Molski et al., 1983). Thus, in accordance with these observations the hypothesis was formulated that f-Met-Leu-Phe activates a membrane-bound phospholipase A_2 by mobilizing calcium; the resulting deacylation of PI would precede the acylation of lysoPI by arachidonic acid via an acyltransferase-mediated reaction (Rubin et al., 1981a). Both components of this postulated sequence, a granulocytic phospholipase A_2 (Franson and Waite, 1979) and an acyl-CoA:lysophospholipid acyltransferase (Elsbach *et al.*, 1972) have been identified in rabbit neutrophils.

In addition to the apparent association between the turnover of arachidonoyl PI and the secretory response, existing evidence in the rabbit neutrophil also favors a key role for stimulated breakdown of PI to diacylglycerol via a phospholipase C-mediated reaction and its sub- sequent resynthesis via phosphatidic acid (PI cycle) (Cockcroft *et al.*, 1982). This stimulus-mediated change in PI metabolism is temporally correlated with secretion and is mediated by a rise in cytosolic calcium. It has been suggested that the PI cycle acts as a transmembrane signal during neutrophil activation, perhaps to control the release of arachidonic acid (Cockcroft, 1982).

More recent work in our laboratory has added to our fund of knowledge in this area by demonstrating the stimulation of arachidonoyl PI turnover is also a mode of expression of cell membrane receptors that mediate the stimulant actions of phorbol esters. These tumor promo- tors are capable of stimulating phospholipase A_2 to release free arachidonic acid and promote the formation of arachi- donate metabolites in secretory cells, including rabbit neutrophils (Hirata *et al.*, 1979; Naor and Catt, 1981). In the neutrophil, the partial refractoriness to calcium deprivation of phorbol dibutyrate (PDBu)-induced arachi- donoyl PI turnover implicates calcium in the mode of action of phorbol esters (Kramer and Rubin, unpublished observations). Indeed, PDBu causes a modest accumu- lation of cellular ^{45}Ca beginning after 2 min; this time course parallels that of arachidonoyl PI turnover and enzyme release (Becker *et al.*, 1981; Kramer and Rubin, unpublished observations).

In this context, the recently described association between the PDBu receptor and Ca^{2+}-activated phospho- lipid-dependent protein kinase (protein kinase C) (Castagna *et al.*, 1982; Niedel *et al.*, 1983) raises the question as to whether PDBu regulation of phospholipase A_2 activity may be somehow linked to protein kinase C-phosphorylation reactions. Protein kinase C is activated by diacylglycerol, one of the products of the PI cycle (Kishimoto *et al.*, 1980). PDBu, in contrast to f-Met-Leu-Phe, is unable to activate the PI cycle in rabbit neutrophils (Fig. 3). However, since the phorbol ester receptor appears to be protein kinase C, PDBu may

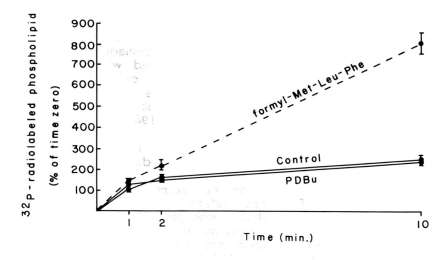

<u>Figure 3</u>. Comparative effects of phorbol dibutyrate and f-Met-Leu-Phe on incorporation of [^{32}P]orthophosphate into neutrophil phosphatidyl-inositol. Rabbit neutrophils preincubated for 30 min with ^{32}P (50 µCi/ml) were exposed to either 2 µM phorbol dibutyrate or 10 nM f-Met-Leu-Phe. Phospholipid labeling is expressed as a percent of incorporation at time zero.

directly stimulate phospholipase A_2 by mimicking the effects of PI breakdown via phosphorylation. On the other hand, PDBu may augment phospholipase A_2 activity simply by mobilizing cellular calcium. Our preliminary findings that PDBu, in contrast to calcium, is unable to activate phospholipase A_2 in preparations of sonicated neutrophils, suggests that an alteration in calcium availability constitutes an early effect of PDBu on cell surface receptors.

Thus, calcium mobilization and turnover of arachi-donoyl PI, appear to be critical links in the pathway that culminate in evoked enzyme secretion by the neutrophil.

374

Indeed, there is particularly strong evidence that arachidonic metabolites serve as mediators of the secretory process in the neutrophil, where lipoxygenase products appear to predominate quantitatively and functionally relative to cyclooxygenase products. The fact that exogenous arachidonic acid and/or its lipoxygenase metabolites mimic many of the stimulatory effects on neutrophils of f-Met-Leu-Phe (Sha'afi and Naccache, 1981) reinforces the view that stimulation of phospholipase A_2 and the resulting metabolism of arachidonic acid are a pivotal component of the mechanisms that regulate lysosomal enzyme release, as well as chemotaxis and aggregation.

EXOCRINE PANCREAS

Recent data derived from studies on exocrine pancreas complement our previous findings in that phospholipase A_2-mediated arachidonic acid release from PI also appears to be a crucial component of the activation process in the exocrine pancreas (Marshall et al., 1980, 1981; Halenda and Rubin, 1982; Rubin et al., 1982). A secretagogue-induced activation of arachidonoyl PI turnover has been demonstrated in isolated rat pancreatic acini (Halenda and Rubin, 1982); also, in mouse and rat pancreas, prelabeled with [^{14}C]arachidonic acid, a stimulus-induced loss of radiolabel from PI can be shown, as well as the formation of prostaglandins (Marshall et al., 1980, 1981; Halenda and Rubin, 1982; Rubin et al., 1982). There exists in the rat pancreas a primary and a zymogen-derived form of phospholipase A_2 (Brockerhoff and Jensen, 1974; Durand et al., 1980), as well as an active, membrane-bound acyl-CoA:1-acyl-sn-glycero-3-phospholipid acyltransferase, which preferentially incorporates arachidonic acid into lysoPI (Rubin, 1983) (Fig. 4).

The relative selectivity for the unsaturated arachidonic acid as the acyl donor over the saturated stearic acid (Rubin, 1983) is in harmony with findings that pancreatic secretagogues are unable to promote the incorporation of stearate into acinar phospholipids (Halenda and Rubin, 1982). The selectivity of the arachidonoyl PI turnover observed during the secretory response of the exocrine pancreas, and perhaps of other secretory cells (Halenda and Rubin, 1982; Rubin, 1982), may be ascribed

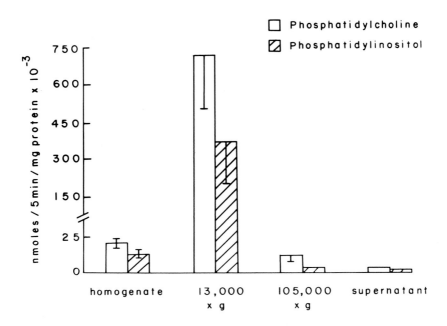

Figure 4. Distribution of acylCoA acyltransferase activity in isolated rat pancreatic acini. Subcellular fractions of homogenate were incubated for 5 min with [^{14}C]arachidonic acid plus either lysophosphatidylinositol or lysophosphatidylcholine. Each vertical bar represents the amount of corresponding phospholipid formed (n = 4).

both to the preference of a phospholipase A_2 for PI (Marshall *et al.*, 1981) and the avidity of lysoPI for arachidonic acid (Rubin, 1983). The physiological implication of these findings awaits studies to define the cellular location of pancreatic acyltransferase.

In the exocrine pancreas, secretagogues also promote the turnover of the polar head group of PI as monitored by ^{32}P and/or radiolabeled inositol, as first shown by Hokin and Hokin (1955). Later work indicated that this receptor-mediated event was associated with a decrease in PI levels and a rise in phosphatidic acid levels (Hokin-

Neaverson, 1977; Farese *et al.*, 1980) and was thought to play a key role in the reactions that couple the receptor responses to Ca^{2+} gating (Michell, 1975). Analysis of the interaction of the phospholipase C-mediated PI cycle and turnover of arachidonic acid was investigated in isolated rat pancreatic acini. The action of calcium-mobilizing receptor agonists such as carbachol was expressed by the coincident turnover of arachidonoyl PI and inositol phosphate moieties of PI; whereas, mobilization of Ca^{2+} produced by the calcium ionophore ionomycin predominantly stimulated only the deacylation-reacylation of PI (Halenda and Rubin, 1982; Putney *et al.*, 1983). Moreover, concentrations of ionomycin that elicited greater secretory responses than carbachol were less effective in stimulating the turnover of inositol phosphate and elevating phosphatidate levels. Thus, while Ca^{2+} mobilization alone appears to be an adequate stimulus for phospholipase A_2 activation and secretion, a complex series of coordinated events triggered by receptor activation may be required to stimulate phospholipase C (Berridge, 1981).

More recent work has revealed that the degree of carbachol-induced breakdown of PIP_2 and the resulting formation of IP_3 (Putney *et al.*, 1983) provides a more accurate assessment of phospholipase C activity than does PI turnover. In fact, the observed PI cycle during stimulation by receptor agonists is deemed only a secondary response to replenish the reservoir of polyphosphoinositides (Berridge, 1983). The breakdown of PIP_2 is not mimicked by ionomycin and occurs independently of agonist-induced Ca^{2+} mobilization, indicating that this breakdown occurs prior to Ca^{2+} mobilization in the stimulus-secretion pathway. In fact, recent findings suggest that one of the products of this reaction, IP_3, acts like carbachol to elicit calcium release from endoplasmic reticulum of saponin-permeabilized pancreatic acini (Streb *et al.*, 1983). Such data suggest that these water-soluble metabolites may serve as messengers linking receptor activation to calcium mobilization. It, therefore, seems clear that two separate pathways co-exist for PI degradation in the exocrine pancreas, involving phospholipases A_2 and C. Additional insight into the interactions between these two phospholipases should greatly enhance our understanding of the molecular mechanisms involved in Ca^{2+}-dependent pancreatic secretion.

TABLE I

Effect of Receptor Agonists on Levels of Inositol
Metabolites in Isolated Pancreatic Acini

Addition	Radioactivity in Inositol Metabolites (dpm)		
	Inositol phosphate	Inositol 1,4- bisphosphate	Inositol 1,4,5- trisphosphate
None	70 ± 8	10 ± 3	8 ± 3
Carbachol (10 μM)	625 ± 10	235 ± 50	315 ± 68
Cerulein (0.1 μM)	258 ± 45	68 ± 13	110 ± 28

Pancreatic acini prepared from a single rat were incubated for 60 min with 18 μCi [^{3}H]inositol. The acini were then washed and resuspended in medium containing 10 mM lithium chloride to inhibit inositol 1-phosphatase activity. Analysis of [^{3}H]inositol-labeled water soluble compounds was carried out by anion exchange chromatography on Dowex-1 formate columns (Berridge, 1983). All values are means (± S.E.) obtained from 4 different preparations.

SUMMARY AND PERSPECTIVE

In conclusion, the elucidation of a biochemical event which can be closely correlated with evoked secretion is of paramount importance in defining the role of calcium in stimulus-secretion coupling. The turnover of arachidonoyl PI appears to be one feature that secretory systems have in common, regardless of whether secretagogues trigger calcium influx or mobilize cellular calcium (Fig. 5). While arachidonic acid turnover has now been demonstrated in several diverse secretory systems (Naor and Catt, 1981; Laychock and Putney, 1982; Rubin, 1982), the question as to whether this phenomenon is a

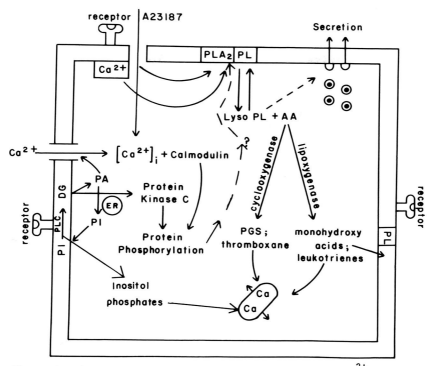

Figure 5. A hypothetical model depicting the role of Ca^{2+} and phosphatidylinositol turnover in the activation of secretion. Receptor agonist- or Ca^{2+} ionophore-induced stimulation causes ionic Ca^{2+} to accumulate within the cytosol by either entering from the extracellular fluid or being released from membrane-binding sites. Ca^{2+} mobilization activates membranous phospholipase A_2 to release arachidonic acid from phosphatidylinositol which serves as a substrate for synthesizing cyclooxygenase and lipoxygenase products. Arachidonic acid metabolites may release Ca^{2+} from intracellular organelles or be incorporated into membrane phospholipids to modify subsequent responses to stimuli (Siegel et al., 1982). The lysophospholipid (lysoPL) may also participate in the activation of the secretory process by promoting membrane fusion (exocytosis). However, a rapid reacylation of the lysoPL prevents accumulation of these potentially cytotoxic substances. In addition to Ca^{2+} mobilization, receptor-linked activation of Ca^{2+}-phospholipid-dependent protein kinase (protein kinase C) via phospholipase C-mediated inositol phospholipid breakdown to diacylglycerol (DG) may be a synergistic step in the secretory response. Alternatively, phosphatidic acid (PA) or inositol phosphates produced by this reaction may serve to mobilize cellular Ca^{2+}.

necessary prelude to secretion should be a subject for future study.

Also, the nature of the interaction between phospholipases A_2 and C with regard to their putative physiological roles in secretion remains to be elucidated. Thus, the phospholipase C-mediated turnover of polyphosphoinositides and the generation of diacylglycerol may be an important step in stimulating Ca^{2+}-phospholipid-dependent protein kinase C, which in turn may stimulate phospholipase A_2 activity. Alternatively, calcium mobilization induced either by the primary stimulus or by second messengers (e.g., inositol 1,4,5-trisphosphate) may directly activate phospholipase A_2 (Fig. 5). The resulting formation of arachidonic acid metabolites may, as in platelets, serve to activate the PI cycle (Siess et al., 1983). Whatever the complex but coordinated series of interactions proves to be, there are now sufficient grounds for believing that calcium acts in concert with messengers spawned by phospholipase activation to somehow promote the export of secretory product from the cell. The current attention rendered to phospholipase-induced phospholipid turnover as a vital component of cellular activity is likely to continue to generate new paradigms related to calcium-phospholipid interactions. Elucidation of such interactions should provide additional insight into the factors that control the secretory process.

ACKNOWLEDGEMENTS

I should like to acknowledge the invaluable contributions of my former colleagues: Stephen Halenda, Caroline Kramer, Suzanne Laychock, and Michael Schrey, who planned and carried out many of the experiments cited in this chapter. These studies were supported in part by USPHS Grant AM 28029 from the NIAMDD.

REFERENCES

Becker, E.L., Showell, H.J., Naccache, P.H. and Sha'afi, R.I. (1981) in: *"Biochemistry of the Acute Allergic Reactions"* (E.L. Becker, A.S. Simon and K.F. Austen, eds.) pp. 257-278, Alan R. Liss, New York.

Berridge, M.J. (1981) *Mol. Cell. Endocrinol.* 24, 115-140.

Berridge, M.J. (1983) *Biochem. J.* 212, 849-858.

Brockerhoff, H. and Jensen, R.G. (1974) *"Lipolytic Enzymes"*, pp. 194-265, Academic Press, New York.

Castagna, M., Takai, Y., Kaibuchi, K., Sano, K., Kikkawa, U. and Nishizuka, Y. (1982) *J. Biol. Chem.* 257, 7847-7851.

Cockcroft, S. (1982) *Cell Calcium* 3, 337-349.

Cockcroft, S., Bennett, J.P. and Gomperts, B.D. (1982) *Biochem. J.* 200, 501-508.

Douglas, W.W. (1968) *Brit. J. Pharmacol.* 34, 451-474.

Durand, S., Clemente, F. and Douste-Blazy, L. (1980) *Biomedicine* 33, 19-21.

Elsbach, P., Patriarca, P., Pettis, P., Stossel, T.P., Mason, R.J. and Vaughan, M. (1972) *J. Clin. Invest.* 51, 1910-1914.

Farese, R.V., Larson, R.E. and Sabir, M.A. (1980) *Biochim. Biophys. Acta* 633, 479-484.

Feinstein, M.B., Becker, E.L. and Fraser, C. (1977) *Prostaglandins* 14, 1075-1093.

Franson, R. and Waite, M. (1978) *Biochemistry* 17, 4029-4033.

Halenda, S.P. and Rubin, R.P. (1982) *Biochem. J.* 208, 713-721.

Hirata, F., Corcoran, B.A., Venkatsubramian, K., Shiffman, E. and Axelrod, J. (1979) *Proc. Natl. Acad. Sci. USA* 76, 2640-2643.

Hokin, L.E. and Hokin, M.R. (1955) *Biochim. Biophys. Acta* 18, 102-110.

Hokin-Neaverson, M. (1977) in: *"Function and Biosynthesis of Lipids"* (N.G. Bazan, R.R. Brenner and N.M. Giusto, eds.) pp. 429-446, Plenum Press, New York.

Holub, B.J. and Kuksis, A. (1978) *Adv. Lipid Res.* 16, 1-125.

Houslay, M.D. and Palmer, R.W. (1979) *Biochem. J.* 178, 217-221.

Kishimoto, A., Takai, Y., Mori, T., Kikkawa, U. and Nishizuka, Y. (1980) *J. Biol. Chem.* *255*, 2273-2276.

Laychock, S.G. (1983) *Diabetes* *32*, 6-13.

Laychock, S.G. and Putney, J.W., Jr. (1982) in: *"Cellular Regulation of Secretion and Release"* (P.M. Conn, ed.) pp. 53-105, Academic Press, New York.

Laychock, S.G., Shen, J.C., Carmines, E.L. and Rubin, R.P. (1978) *Biochim. Biophys. Acta* *528*, 355-363.

Lucy, J.A. (1978) in: *"Membrane Fusion"* (G. Poste and G.L. Nicholson, eds.) pp. 268-304, North Holland Publishing Company, Amsterdam.

Marshall, P.J., Dixon, J.F. and Hokin, L.E. (1980) *Proc. Natl. Acad. Sci. USA* *77*, 3293-3296.

Marshall, P.J., Boatman, D.E. and Hokin, L.E. (1981) *J. Biol. Chem.* *256*, 844-847.

Martin, T.W. and Lagunoff, D. (1979) *Nature* *279*, 250-252.

Michell, R.H. (1975) *Biochim. Biophys. Acta* *415*, 81-147.

Molski, T.F.P., Naccache, P.H. and Sha'afi, R.I. (1983) *Cell Calcium* *4*, 57-68.

Naccache, P.H., Showell, H.J., Becker, E.L. and Sha'afi, R.I. (1977) *J. Cell Biol.* *73*, 428-444.

Naor, Z. and Catt, K.J. (1981) *J. Biol. Chem.* *256*, 2226-2229.

Niedel, J.E., Kuhn, L.J. and Vandenbark, G.R. (1983) *Proc. Natl. Acad. Sci. USA* *80*, 36-40.

Papahadjopoulos, D., Portis, A. and Pangborn, W. (1978) *Ann. N.Y. Acad. Sci.* *308*, 50-66.

Putney, J.W., Jr., Burgess, G.M., Halenda, S.P., McKinney, J.S. and Rubin, R.P. (1983) *Biochem. J.* *212*, 483-488.

Rubin, R.P. (1983) *Biochem. Biophys. Res. Commun.* *112*, 502-507.

Rubin, R.P., Sink, L.E. and Freer, R.J. (1981a) *Mol. Pharmacol.* *19*, 31-37.

Rubin, R.P., Sink, L.E. and Freer, R.J. (1981b) *Biochem. J.* *194*, 497-505.

Rubin, R.P., Kelly, K.L., Halenda, S.P. and Laychock, S.G. (1982) *Prostaglandins* *24*, 179-193.

Schrey, M.P. and Rubin, R.P. (1979) *J. Biol. Chem.* *256*, 11234-11241.

Schrey, M.P., Franson, R.C. and Rubin, R.P. (1980) *Cell Calcium* *1*, 91-104.

Sha'afi, R.I. and Naccache, P.H. (1981) *Adv. Inflammation Res. 2*, 115–148.

Siegel, M.I., McConnell, R.T., Bonser, R.W. and Cuatrecasas, P. (1982) *Biochem. Biophys. Res. Commun. 104*, 874–881.

Siess, W., Siegel, F.L. and Lapetina, E.G. (1983) *J. Biol. Chem. 258*, 11236–11242.

Streb, H., Irvine, R.F., Berridge, M.J., and Schulz, I. (1983) *Nature 306*, 67–68.

Sun, G.Y. (1979) *Lipids 14*, 918–924.

Takai, Y., Minakuchi, R., Kikkawa, U., Sano, K., Kaibuchi, K., Yu, B., Matsubara, T. and Nishizuka, Y. (1982) *Prog. Brain Res. 56*, 287–301.

Zwiller, J., Ciesielski-Treska, J. and Mandel, P. (1976) *FEBS Lett. 69*, 286–290.

CALCIUM AND INOSITOL PHOSPHOLIPID DEGRADATION IN SIGNAL TRANSDUCTION

K. Kaibuchi, M. Sawamura, Y. Katakami, U. Kikkawa, Y. Takai, and Y. Nishizuka

Department of Biochemistry, Kobe University School of Medicine, Kobe 650 and Department of Cell Biology, National Institute for Basic Biology, Okazaki 444, Japan

SUMMARY

Signal-induced inositol phospholipid breakdown appears to be a sign for the transmembrane control of cellular functions and proliferation through activation of protein kinase C. This enzyme requires Ca^{2+} and phospholipid. Diacylglycerol, which is derived from the phospholipid breakdown, dramatically increases the affinity of protein kinase C for Ca^{2+}, and thereby renders this enzyme fully active without Ca^{2+} mobilization. A series of experiments with exogenous diacylglycerol and Ca^{2+} ionophore (A23187) suggests that the receptor-linked activation of protein kinase C and mobilization of Ca^{2+} are both essential and synergistically effective in eliciting full physiological cellular responses. In many, but not all, tissues such as platelets, mast cells, neutrophils and lymphocytes, stimulation of the receptors that produce cyclic AMP inhibits inositol phospholipid breakdown and blocks the activation of protein kinase C. Tumor-promoting phorbol esters, when intercalated into cell membranes, may substitute for diacylglycerol and permanently activate protein kinase C irrespective of the feedback control by cyclic AMP.

INTRODUCTION

Protein kinase C has been found first as a proteolytically activated enzyme in many tissues (Inoue *et al.*, 1977) and later shown to be activated by reversible association with membrane phospholipid (Takai *et al.*, 1979a). This protein kinase appears to be firmly linked to the signal-induced breakdown of inositol phospholipids, since diacylglycerol, one of the earliest products, activates this protein kinase by facilitating its association with membranes at a physiological concentration of Ca^{2+} (Takai *et al.*, 1979b; Kishimoto *et al.*, 1980; Kaibuchi *et al.*, 1981). Although there are some exceptions such as bovine adrenal medullary cells (Swilem *et al.*, 1983), it is generally accepted that stimulation of the receptors relating to inositol phospholipid breakdown immediately mobilizes Ca^{2+}, which appears to mediate many of the subsequent physiological responses (Michell *et al.*, 1981; Berridge, 1981). This article will briefly describe that either the signal-induced activation of protein kinase C or the mobilization of Ca^{2+} alone is not sufficient for com-

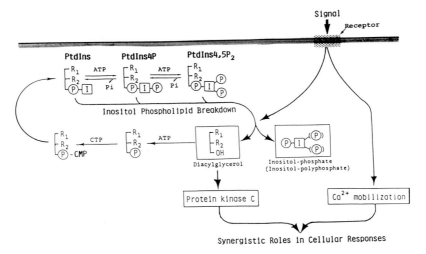

Figure 1. Inositol phospholipid turnover and signal transduction. PtdIns, phosphatidylinositol; PtdIns4P, phosphatidylinositol-4-phosphate; PtdIns4,5P$_2$, phosphatidylinositol-4,5-bisphosphate; R$_1$ and R$_2$, fatty acyl groups; I, inositol; and P, phosphoryl group.

plete signal transduction, but their synergistic actions are necessary for eliciting full cellular responses as schematically outlined in Fig. 1. Under appropriate conditions, it is possible to induce protein kinase C activation and Ca^{2+} mobilization independently by using exogenous addition of synthetic diacylglycerol or tumor-promoting phorbol ester for the former and Ca^{2+} ionophore such as A23187 for the latter (Kaibuchi et al., 1982a, 1983; Yamanishi et al., 1983). Protein kinase C is a prime target for the actions of tumor promoters as described earlier (Castagna et al., 1982).

DIRECT ACTIVATION OF PROTEIN KINASE C

In intact cell systems synthetic diacylglycerols such as 1-oleoyl-2-acetyl-glycerol or 1-acetyl-2-oleoyl-glycerol activate protein kinase C directly without interaction with cell surface receptors, although diacylglycerols having two long fatty acyl moieties such as diolein are poor activators for the enzyme, presumably due to their inability to be intercalated into the membrane (Kaibuchi et al., 1981a, 1983). To demonstrate such direct activation of protein kinase C in intact cells, platelets are mainly used as a test system, since the activation of this enzyme can be assayed by measuring the phosphorylation of its specific protein substrate having a molecular weight of approximately 40,000 (40K protein) (Sano et al., 1983).

In the experiment given in Fig. 2, it is shown that, when 1-oleoyl-2-acetyl-glycerol is suspended in 1% dimethylsulfoxide and added to intact platelets directly, 40K protein is rapidly and heavily phosphorylated just as it is after stimulation by natural extracellular messengers such as thrombin. Under these conditions neither the breakdown of inositol phospholipids nor the transient formation of endogenous diacylglycerol is observed. There is no sign of damage to cell membranes. This diacylglycerol does not appear to interact with the receptor of platelet-activating factor, since this factor induces rapid breakdown of inositol phospholipids with the concomitant formation of diacylglycerol as observed for thrombin (Ieyasu et al., 1982). Fingerprint analysis of the tryptic phosphopeptides of the 40K protein that is labeled in platelets stimulated by the synthetic diacyl-glycerol confirms that protein kinase C is indeed respons-

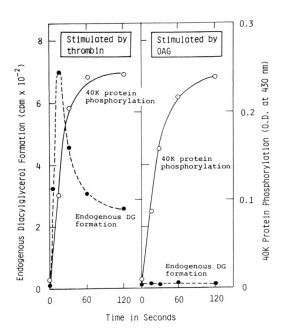

Figure 2. Phosphorylation of 40K protein and formation of endogenous diacylglycerol in human platelets stimulated by synthetic diacylglycerol and thrombin. Human platelets were labeled with either [^{3}H]arachidonate or $^{32}P_i$, and then stimulated by 1-oleoyl-2-acetyl-glycerol or by thrombin at 37°C for various periods of time as indicated. Detailed conditions are described elsewhere (Kaibuchi et al., 1983). OAG, 1-oleoyl-2-acetyl-glycerol.

ible for the phosphorylation reaction. During this stimulation of platelets the synthetic diacylglycerol is rapidly metabolized *in situ* to its corresponding phosphatidic acid, that is 1-oleoyl-2-acetyl-3-phosphoglycerol, presumably by the action of diacylglycerol kinase (Kaibuchi et al., 1983). Based on these observations, it may be concluded that under appropriate conditions exogenously added diacylglycerol can induce protein kinase C activation directly without interaction with cell surface receptors.

Tumor-promoting phorbol esters such as 12-O-tetra-decanoyl-phorbol-13-acetate (TPA) are intercalated into the membrane, substitute for diacylglycerol, and activate protein kinase C at extremely low concentrations (Castagna *et al.*, 1982). Kinetic analysis indicates that, like diacylglycerol, the tumor promoters dramatically increase the affinity of this enzyme for Ca^{2+} to the 10^{-7} M range, resulting in its full activation without detectable mobilization of this divalent cation (Yamanishi *et al.*, 1983).

SYNERGISM WITH CALCIUM FOR EXOCYTOTIC RESPONSE

Platelets are particularly suited to demonstrate the synergistic roles of protein kinase C activation and Ca^{2+} mobilization in release reactions. When platelets are stimulated by natural extracellular messengers such as thrombin, myosin light chain having a molecular weight of 20,000 (20K protein) is also phosphorylated concomitantly (Lyons *et al.*, 1975; Kawahara *et al.*, 1980; Sano *et al.*, 1983). This reaction is catalyzed by calmodulin-dependent myosin light chain kinase and is absolutely dependent on Ca^{2+} mobilization (Yagi *et al.*, 1978). If, however, platelets are stimulated by synthetic diacylglycerol, only 40K protein is phosphorylated to an extent similar to that

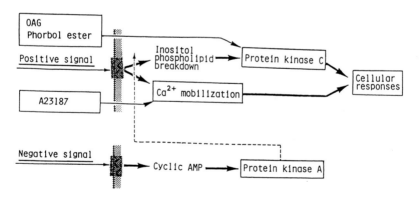

Figure 3. Synergistic actions of protein kinase C activation and Ca^{2+} mobilization for eliciting full cellular responses. In many, not all, tissues the receptor-linked cascade caused by a "positive signal" is blocked by a "negative signal" which increases cyclic AMP. In some tissues these two receptors do not appear to interact with each other but to function independently (Kaibuchi et al., 1982b). OAG, 1-oleoyl-2-acetyl-glycerol.

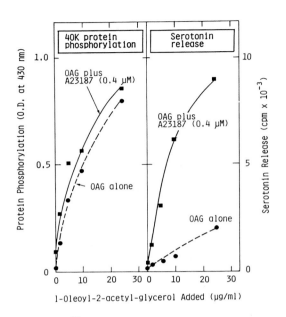

<u>Figure 4</u>. Effects of Ca^{2+} ionophore and synthetic diacylglycerol on 40K protein phosphorylation and serotonin release. Human platelets were labeled with either $^{32}P_i$ or [^{14}C]serotonin, and then stimulated by various concentrations of 1-oleoyl-2-acetyl-glycerol in the presence or absence of A23187 as indicated. Detailed conditions are described elsewhere (Kaibuchi <u>et al</u>., 1983). OAG, 1-oleoyl-2-acetyl-glycerol.

induced by thrombin. The 20K protein is not phosphorylated to a measurable extent, suggesting that Ca^{2+} is not mobilized under these conditions. This has been recently confirmed by the direct measurement of Ca^{2+} using Quin 2 by Rink *et al.* (1983). On the other hand, at lower concentrations of A23187 20K protein is selectively phosphorylated without significant phosphorylation of 40K protein. The selective and independent induction of protein kinase C activation and Ca^{2+} mobilization is schematically shown in Fig. 3.

The experiment given in Fig. 4 shows that, when platelets are stimulated by synthetic diacylglycerol, 40K protein is phosphorylated almost equally in the presence and absence of Ca^{2+} ionophore. However, serotonin is

not sufficiently released from dense bodies by the addition of diacylglycerol alone, and the full response may be
observed only when diacylglycerol and Ca^{2+} ionophore are
added simultaneously. An essentially similar result has
been obtained with TPA instead of diacylglycerol
(Yamanishi *et al.*, 1983). Ca^{2+} ionophore alone at the
concentration used does not cause the formation of endogenous diacylglycerol, nor does it evoke the release of
serotonin. In these experiments the concentration of
A23187 (0.2 - 0.4 μM) is critical since at concentrations
greater than 0.5 μM, this Ca^{2+} ionophore alone causes the
phosphorylation of 40K protein as well as the release of
serotonin. This is presumably due to a large increase in
the Ca^{2+} concentration which may activate phospholipases
non-specifically. Likewise, it is important to use no more
than 30 μg/ml synthetic diacylglycerol or 10 ng/ml TPA,
because at higher concentrations (for instance, 50 μg/ml
or 50 ng/ml, respectively) these compounds alone induce
a significant release of serotonin without a large increase
in the Ca^{2+} concentration, probably by acting as membrane fusogens or perturbers and/or as weak Ca^{2+}
ionophores by generating superoxide.

In a similar experiment with synthetic diacylglycerol
or TPA and Ca^{2+} ionophore, it is possible to show that
either protein kinase C activation or Ca^{2+} mobilization is
not a sufficient requirement and both are synergistically
essential for N-acetylglucosaminidase release from platelet
lysosomes (Kajikawa *et al.*, 1983). The involvement of
two synergistic pathways, protein phosphorylation and
Ca^{2+} mobilization, may explain the agonist selectivity that
is observed for the release reactions from different
platelet granules. Although there appears no difference
in the sensitivity of serotonin and lysosomal enzyme
release to Ca^{2+} concentration (Knight *et al.*, 1982), the
balance of the two pathways may exert differential control
over different processes within a single activated platelet
(Kajikawa *et al.*, 1983).

The synergistic effects of synthetic diacylglycerol
and Ca^{2+} ionophore are not confined to the platelet, but
may be observed for many other cell types. For example,
rat mast cells and neutrophils actively release histamine
and lysosomal enzymes, respectively, only in the simultaneous presence of these two compounds, and both protein

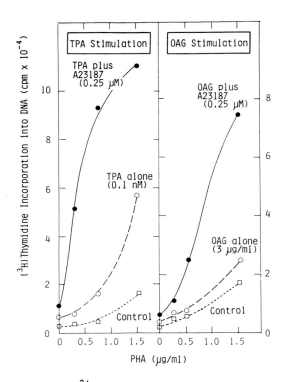

<u>Figure 5.</u> Effects of Ca^{2+} ionophore and tumor-promoting phorbol ester or synthetic diacylglycerol on phytohemagglutinin-induced DNA synthesis of macrophage-depleted peripheral lymphocytes. Macrophage-depleted peripheral lymphocytes were incubated for 72 h at 37°C with various concentrations of phytohemagglutinin in the presence or absence of A23187, TPA and 1-oleoyl-2-acetyl-glycerol as indicated. [^{3}H]Methyl thymidine (2 μCi) was added 16 h before harvesting. The cells were harvested and acid-precipitable radioactivity was determined. Detailed experimental procedures will be described elsewhere. OAG, 1-oleoyl-2-acetyl-glycerol; and PHA, phytohemagglutinin.

phosphorylation and Ca^{2+} mobilization are needed to elicit the full physiological responses.

SYNERGISM WITH CALCIUM FOR GROWTH RESPONSE

Stimulation of the receptors that induce inositol phospholipid breakdown often causes cellular proliferation and differentiation. Plausible evidence available to date indicates that protein kinase C is a receptor protein of tumor-promoting phorbol esters, and that pleiotropic action of the tumor promoters such as TPA may be mediated through the action of protein kinase C (Castagna *et al.*, 1982, Yamanishi *et al.*, 1983; Kikkawa *et al.*, 1983). It is generally accepted that several mitogens and growth factors such as epidermal growth factor may induce rapid turnover of inositol phospholipids in their target tissues.

In the next set of experiments shown in Fig. 5, macrophage-depleted peripheral lymphocytes are incubated with A23187 and either synthetic diacylglycerol or TPA in the presence of various concentrations of phytohemag-glutinin. Then, the radioactive thymidine that is incor-porated into DNA during this incubation period is deter-mined by using a conventional procedure. It is found that, under limited conditions, either synthetic diacyl-glycerol or TPA alone is not sufficient, and the supple-ment of A23187 may be necessary for the complete signal transduction to cause rapid incorporation of thymidine into DNA. The Ca^{2+} ionophore alone shows no effect under the given conditions. The results suggest that protein kinase C activation and Ca^{2+} mobilization may substitute for the action of Interleukin I and play synergistic roles in phytohemagglutinin-induced prolif-eration of the peripheral lymphocytes. It is also worth noting that under these conditions at least three soluble proteins having molecular weights of about 82,000, 56,000 and 45,000 and one particulate protein having a molecular weight of 68,000 are phosphorylated in the stimulated lymphocytes, and that all these proteins serve as pre-ferred substrates for a purified preparation of protein kinase C. The substrate proteins for protein kinase C may not be confined to the membrane compartment.

ROLE OF CYCLIC NUCLEOTIDES

In many tissues such as platelets, mast cells, neutrophils and lymphocytes, the receptors that induce inositol phospholipid breakdown promote the activation of cellular function and proliferation as mentioned above, whereas the receptors that produce cyclic AMP usually antagonize such activation, although the mechanism of this

393

antagonistic action is unclear (Fig. 3). In some tissues such as hepatocytes, adipocytes and some endocrine cells, the two classes of receptors appear not to interact with each other but to function cooperatively or independently. It is possible to show, again with platelets as a test system, that the thrombin-induced inositol phospholipid breakdown, diacylglycerol formation, 40K protein phosphorylation and cellular response are all blocked concomitantly by the prior incubation with either prostaglandin E1 or dibutyryl cyclic AMP (Kawahara *et al.*, 1980; Takai *et al.*, 1982; Sano *et al.*, 1983). Similar inhibitory actions of cyclic AMP are observed for the chemoattractant-induced inositol phospholipid labeling and lysosomal enzyme release of rat neutrophils, and also for the plant lectin-induced inositol phospholipid labeling and thymidine incorporation into DNA of human peripheral lymphocytes (Kaibuchi *et al.*, 1982b). These inhibitory actions of cyclic AMP most likely extend to the mobilization of Ca^{2+}, presumably through activation of cyclic AMP-dependent protein kinase (Takai *et al.*, 1982). In contrast, in hepatocytes, inositol phospholipid breakdown which is provoked by α-adrenergic stimulators is not blocked by β-adrenergic stimulators, glucagon or by dibutyryl cyclic AMP (Kaibuchi *et al.*, 1982b). These stimulators are all known to cause glycogenolysis in hepatic tissues.

Although the precise signal pathway for elevating cyclic GMP remains obscure, it is possible that arachidonic acid peroxide and prostaglandin endoperoxide serve as activators for guanylate cyclase (Graff *et al.*, 1978). In fact, the signal-induced inositol phospholipid turnover is usually associated with arachidonic acid release and cyclic GMP production. Thus, the phospholipid degradation, Ca^{2+}, arachidonic acid and cyclic GMP appear to be integrated together in a single receptor cascade system. However, cyclic GMP-dependent protein kinase shows very similar, if not identical, catalytic properties to those of cyclic AMP-dependent protein kinase, and it is possible that cyclic GMP acts as a negative, rather than a positive, messenger, providing an immediate feedback control that prevents over-response. In support of this proposal, again in human platelets stimulated by thrombin, all subsequent cellular processes mentioned above are inhibited by the prior incubation with sodium nitro-

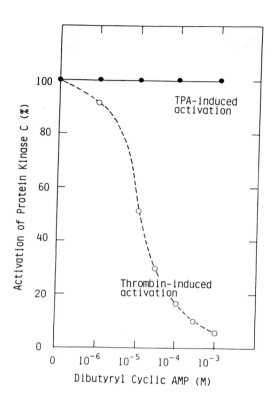

Figure 6. Effects of dibutyryl cyclic AMP on activation of protein kinase C in platelets stimulated by tumor-promoting phorbol ester and natural extracellular signal. Human platelets were labeled with $^{32}P_i$, and then stimulated by either TPA or thrombin in the presence of various concentrations of dibutyryl cyclic AMP. The activation of protein kinase C in platelets was estimated by measuring the phosphorylation of its specific substrate, 40K protein. Detailed experimental procedures were described elsewhere (Yamanishi et al., 1983).

prusside which induces cyclic GMP, or with 8-bromo cyclic GMP (Takai et al., 1981). In short, cyclic AMP and cyclic GMP do not antagonize each other but similarly inhibit the receptor-linked degradation of phospholipid in membranes and thereby cellular functions and proliferation by counteracting the activation of protein kinase C.

Kaibuchi et al.

The experiment shown in Fig. 6 is designed to illustrate that the activation of protein kinase C induced by natural extracellular messengers such as thrombin is profoundly inhibited by cyclic AMP, whereas that induced by tumor-promoting phorbol esters such as TPA is entirely resistant to the feedback control by this cyclic nucleotide. Unlike diacylglycerol, the tumor promoters may stay in membranes for prolonged periods of time, since the diterpenes are metabolized only slowly. Perhaps the tumor promoters relieve the cells of contact inhibition, which is well known to induce cyclic AMP formation.

CONCLUSION

Although the experimental support outlined above has resulted from the studies with a limited number of specific tissues, it is plausible that protein kinase C distributed widely in tissues and organs has a potential to play a crucial role in signal transduction for a variety of hormones, neurotransmitters, secretagogues, chemo-attractants, growth factors and many other biologically active substances which elicit inositol phospholipid turn-over in their respective target tissues. When cells are stimulated, this protein kinase is transiently activated by diacylglycerol which is produced in the membrane during the signal-induced turnover of inositol phospholipids. This receptor-mediated activation of the enzyme is biologi-cally independent of Ca^{2+} because its Ca^{2+} sensitivity is modulated. Available evidence suggests that the protein kinase activation and Ca^{2+} mobilization, which is associ-ated with the phospholipid turnover in a manner as yet uncertain, may act synergistically to elicit full physio-logical cellular responses. The precise target proteins of protein kinase C in each tissue may be clarified by further investigations.

ACKNOWLEDGEMENTS

We are grateful to Mrs. S. Nishiyama and Mrs. K. Kikkawa for their skillful secretarial assistance. This investigation has been supported in part by research grants from the Research Fund of the Ministry of Educa-tion, Science and Culture, the Intractable Diseases Divi-sion, Public Health Bureau, the Ministry of Health and Welfare, a Grant-in-Aid of New Drug Development from the Ministry of Health and Welfare, the Science and Technology Agency, the Yamanouchi Foundation for Research on Metabolic Disorders, and the Mitsuhisa Cancer Research Foundation, Japan.

REFERENCES

Berridge, M.J. (1981) *Mol. Cell. Endocrinol.* *24*, 115-140.

Castagna, M., Takai, Y., Kaibuchi, K., Sano, K., Kikkawa, U. and Nishizuka, Y. (1982) *J. Biol. Chem.* *257*, 7847-7851.

Graff, G., Stephenson, J.H., Glass, D.B., Haddox, M.K. and Goldberg, N.D. (1978) *J. Biol. Chem.* *253*, 7662-7676.

Ieyasu, H., Takai, Y., Kaibuchi, K., Sawamura, M. and Nishizuka, Y. (1982) *Biochem. Biophys. Res. Commun.* *108*, 1701-1708.

Inoue, M., Kishimoto, A., Takai, Y. and Nishizuka, Y. (1977) *J. Biol. Chem.* *252*, 7610-7616.

Kaibuchi, K., Takai, Y. and Nishizuka, Y. (1981) *J. Biol. Chem.* *256*, 7146-7149.

Kaibuchi, K., Sano, K., Hoshijima, M., Takai, Y. and Nishizuka, Y. (1982a) *Cell Calcium* *3*, 323-335.

Kaibuchi, K., Takai, Y., Ogawa, Y., Kimura, S., Nishizuka, Y., Nakamura, T., Tomomura, A. and Ichihara, A. (1982b) *Biochem. Biophys. Res. Commun.* *104*, 105-112.

Kaibuchi, K., Takai, Y., Sawamura, M., Hoshijima, M., Fujikura, T. and Nishizuka, Y. (1983) *J. Biol. Chem.* *258*, 6701-6704.

Kajikawa, N., Kaibuchi, K., Matsubara, T., Kikkawa, U., Takai, Y., Nishizuka, Y., Itoh, K. and Tomioka, C. (1983) *Biochem. Biophys. Res. Commun.* *116*, 743-750.

Kawahara, Y., Takai, Y., Minakuchi, R., Sano, K. and Nishizuka, Y. (1980) *Biochem. Biophys. Res. Commun.* *97*, 309-317.

Kikkawa, U., Takai, Y., Tanaka, Y., Miyake, R. and Nishizuka, Y. (1983) *J. Biol. Chem.* *258*, 11442-11445.

Kishimoto, A., Takai, Y., Mori, T., Kikkawa, U. and Nishizuka, Y. (1980) *J. Biol. Chem.* *255*, 2273-2276.

Knight, D.E., Hallam, T.J. and Scrutton, M.C. (1982) *Nature* *296*, 256-257.

Lyons, R.M., Stanford, N. and Majerus, P.W. (1975) *J. Clin. Invest.* *56*, 924–936.

Michell, R.H., Kirk, C.J., Jones, L.M., Downes, D.P. and Creba, J.A. (1981) *Phil. Trans. R. Soc. B* *296*, 123–137.

Rink, T.J., Sanchez, A. and Hallam, T.J. (1983) *Nature* *305*, 317–319.

Sano, K., Takai, Y., Yamanishi, J. and Nishizuka, Y. (1983) *J. Biol. Chem.* *258*, 2010–2013.

Swilem, A.M.F., Hawthorne, J.N. and Azila, N. (1983) *Biochem. Pharmacol.* *32*, 3873–3875.

Takai, Y., Kishimoto, A., Iwasa, Y., Kawahara, Y., Mori, T. and Nishizuka, Y. (1979a) *J. Biol. Chem.* *254*, 3692–3695.

Takai, Y., Kishimoto, A., Kikkawa, U., Mori, T. and Nishizuka, Y. (1979b) *Biochem. Biophys. Res. Commun.* *91*, 1218–1224.

Takai, Y., Kaibuchi, K., Matsubara, T. and Nishizuka, Y. (1981) *Biochem. Biophys. Res. Commun.* *101*, 61–67.

Takai, Y., Kaibuchi, K., Sano, K. and Nishizuka, Y. (1982) *J. Biochem.* *91*, 403–406.

Yagi, K., Yazawa, M., Kakiuchi, S., Ohshima, M. and Uenishi, K. (1978) *J. Biol. Chem.* *253*, 1338–1340.

Yamanishi, J., Takai, Y., Kaibuchi, K., Sano, K., Castagna, M. and Nishizuka, Y. (1983) *Biochem. Biophys. Res. Commun.* *112*, 778–786.

THE ROLE OF PHOSPHOPROTEIN B-50 IN

PHOSPHOINOSITIDE METABOLISM IN BRAIN

SYNAPTIC PLASMA MEMBRANES

W.H. Gispen, C.J. Van Dongen, P.N.E. De Graan,
A.B. Oestreicher, and H. Zwiers

Division of Molecular Neurobiology, Rudolf Magnus
Institute of Pharmacology and Institute of Molecular
Biology, State University of Utrecht,
Utrecht, The Netherlands

SUMMARY

In rat brain synaptic plasma membranes there is a Ca^{2+}-dependent, cyclic nucleotide-independent protein kinase that is inhibited by $ACTH_{1-24}$. Evidence suggests that this protein kinase is very similar to protein kinase C. One of its substrate proteins is the nervous tissue-specific protein B-50, which is predominantly localized in membranes of the presynaptic region of neurones. A variety of data suggest that this phosphoprotein may play a regulatory role in the conversion of phosphatidyl-*myo*-inositol 4-phosphate (PIP) into phosphatidyl-*myo*-inositol 4,5-bisphosphate (PIP_2). These data were obtained using specific anti-B-50 immunoglobulins, pre-phosphorylation experiments, peptide and neurotransmitter modulation in tissue slices and subcellular fractions. Preliminary results of experiments on the interaction of purified PIP kinase and B-50 phosphoprotein further support this notion.

INTRODUCTION

A generally accepted concept is that the cyclic phosphorylation and dephosphorylation of a protein alters its stereoconformation and hence its function (Weller,

1979). If the substrate protein is a structural component of the membrane, changes in the degree of phosphorylation may result in selective changes in ion permeability of that membrane. If the substrate protein is an enzyme, the degree of phosphorylation may determine its catalytic activity.

Protein-bound phosphate has a rapid turnover in many species and tissues: the highest rate is observed in brain. Furthermore, the endogenous protein phosphorylating capacity is most enriched in the subcellular fraction of brain tissue that contains the synaptic plasma and vesicle membranes. In view of this enrichment in synaptic regions and the fact that protein phosphorylation is involved in changes in synaptic efficacy and in neurotransmission in certain neuronal networks, it has been postulated that certain phosphoproteins may play a key role in the chemical communication between cells by altering the properties of the membranes involved in this process.

The present communication reviews the evidence available to date on the role of the brain-specific phosphoprotein B-50 in brain synaptic plasma membranes.

PHOSPHOPROTEIN B-50

When a synaptic plasma membrane (SPM) fraction is incubated in the presence of $[\gamma-^{32}P]ATP$, phosphorylation of a great number of proteins is observed (Rodnight, 1982). As these membranes contain a variety of acceptor proteins, protein kinases, phosphoprotein phosphatase(s) and [Na-K]ATPase, the phosphorylation profile is very much determined by the experimental conditions used. Depending on the brain region studied and amount of membranes used in the incubation, $ACTH_{1-24}$ in concentrations of $10^{-7} - 10^{-4}$ M inhibits the phosphorylation of some relatively low molecular weight protein bands (M_r 17-48 kDa; Mahler et al., 1982; Zwiers et al., 1982). For at least one protein (B-50) the evidence indicated that the reduced phosphorylation in the presence of ACTH has not been due to enhanced protein phosphatase activity but to inhibited protein kinase activity (Zwiers et al., 1978). Therefore, we concentrated on the characteristics of this protein/protein kinase complex in rat brain SPM.

Treatment of SPM with 0.5% Triton X-100 in 75 mM KCl solubilized approximately 15% of the B-50 protein kinase activity and preserved the sensitivity of the system to ACTH-like peptides (Zwiers et al., 1979). Further purification consisted of column chromatography over DEAE cellulose, differential precipitation by ammonium sulphate saturation (yielding a highly enriched ASP_{55-80} protein fraction) and two-dimensional electrophoresis on polyacrylamide slab gels (Zwiers et al., 1980). The B-50 protein has an M_r of 48 kDa and an IEP of 4.5. Based on several properties the protein B-50 may be identical to protein band γ-5 (Gower and Rodnight, 1982), F_1 (Routtenberg, 1982), protein 47K (Hershkowitz et al., 1982) or protein P54p (Ca) (Mahler et al., 1982). Recently, using iodinated, purified B-50 as a tracer in a radioimmunoassay with anti-B-50 immunoglobulins, we estimated the amount of B-50 in SPM from rat brain to be approximately 4-6 μg/mg total SPM protein.

Dialysis of the fraction precipitating with ammonium sulfate between 55 and 80% saturation (ASP_{55-80}) was one of the steps involved in the final characterization procedure. We noted that after dialysis the endogenous phosphorylation of the B-50 protein was markedly enhanced. This suggested the presence of a small molecular weight entity that interfered with the B-50 protein kinase activity. Therefore, the dialysis was analyzed in detail and a phosphorylation-inhibiting peptide was isolated and characterized (Zwiers et al., 1980). The peptide has an apparent M_r of 1600 - 1650, consists of about 15 amino acids and is enriched in basic amino acid residues. There is good evidence that specific proteolysis of the B-50 protein is responsible for the generation of the phosphorylation-inhibiting peptide and a protein B-60 with an M_r of 46 kDa (Zwiers et al., 1980, 1982). The proteolytic activity in the ASP_{55-80} protein fraction is enhanced by the addition of Ca^{2+} and/or calmodulin. However, only in the absence of Ca^{2+} did we observe a specific cleavage of B-50 into the peptide and B-60. Whether this specific cleavage of B-50 is of importance to its role in SPM functioning remains to be determined.

Using two-dimensional analysis and anti-B-50-antisera, B-50 protein could only be detected in the particulate fractions of rat brain homogenate and not in sub-

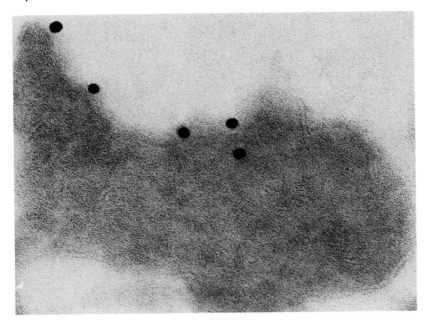

Figure 1. Cryo-section of an isolated synaptosome, immunoreacted for B-50 protein, using affinity-purified anti-B-50 IgG, diluted 1/50 and protein A-coated gold particles (diameter 16 nm) (Leunissen et al., in preparation). Magnification: 250,000 x.

cellular fractions of other tissues studied (adrenal cortex, skeletal muscle, liver, kidney and lung). Immunostaining of B-50-like proteins in brain homogenates of various vertebrate species revealed the presence of B-50 in human, rat, mouse, hamster, rabbit, cow and chick brain. B-50 immunoreactivity was absent in homogenates from Xenopus, goldfish and trout (Kristjansson et al., 1982; Oestreicher et al., in preparation).

The immunostaining profile at the light microscopic level in various brain areas revealed dense staining in regions rich in synaptic contacts, whereas white matter and cell perikarya were virtually without immunostaining (Oestreicher et al., 1981). Recently we studied the ultrastructural localization of B-50 in ultra-thin cryo-sections of fixed rat hippocampus tissue and isolated synaptosomes, by means of highly specific anti-B-50 immunoglobulins combined with protein A-gold staining procedures. We demonstrated the presence of B-50 immunoreactivity at presynaptic sites of nerve terminals *in situ* and in close association with the inside of the plasma membrane of the synaptosome (Fig. 1). Thus the localization studies revealed that the protein B-50 is specific for nervous tissue and can be found in pre-synaptic regions throughout the brain. The latter con-clusion is further supported by the studies of Sorensen *et al*. (1981) using subcellular fractionation techniques and endogenous B-50 phosphorylation.

B-50 KINASE AND KINASE C

The endogenous phosphorylation of the B-50 protein in SPM appeared to be insensitive to cAMP and cGMP (Zwiers *et al*., 1976). B-50 is phosphorylated by a Ca^{2+}-dependent protein kinase (Gispen *et al*., 1979), whereas most likely calmodulin does not act as transducer of the Ca^{2+} ions to activate the kinase (Rodnight, 1982; Sorensen and Mahler, 1983). B-50 kinase could be co-purified with its substrate B-50 (see above) and was found to have a M_r of 70 kDa and an IEP of 5.5 (Zwiers *et al*., 1980).

Another cyclic nucleotide-insensitive, calcium-requiring protein kinase, kinase C, has been isolated from rat brain (Inoue *et al*., 1977; Kikkawa *et al*., 1982). Although this kinase was originally isolated from the soluble fraction of rat brain homogenate (Inoue *et al*., 1977), it has been found in the particulate fraction as well (Kuo *et al*., 1980). The activity of kinase C is stimulated by either partial proteolysis by trypsin or a Ca^{2+}-dependent protease isolated from rat brain (Inoue *et al*., 1977) or by any of several phospholipids, such as phosphatidylserine (PS) (Takai *et al*., 1979a). It would appear that in the presence of Ca^{2+} the soluble kinase binds to membranes resulting in its activation (Takai *et al*., 1979b).

During our studies on the B-50 protein kinase, we noted several apparent similarities between this kinase and kinase C. Since a distinguishing characteristic of B-50 kinase is its ability to phosphorylate B-50 protein, we have compared several kinases for their ability to phosphorylate B-50. Of the kinases tested, only kinase

403

TABLE I

Comparison of B-50 Kinase and Kinase C

	B-50 kinase	Kinase C
Tissue localization		
Brain	+	+
Other tissues	not checked	+
Subcellular localization		
Membrane	+	+
Soluble	not checked	+
Molecular weight by SDS-PAGE	70 kDa	68 kDa*/82 kDa*
Isoelectric point	5.5	5.5
Substrates		
Endogenous	B-50	multiple
Exogenous	histones	histones, B-50
Metal requirements		
Ca^{2+}	0.1 - 1 mM	varies with assay conditions
Mg^{2+}	10 mM	varies with substrate
Activation by		
cAMP	−	−
cGMP	−	−
Phospholipids	+	+
Protease	+	+
Inhibition by		
ACTH(1-24)	+	+
Chlorpromazine	+	+

*Schatzman et al. (1983).
**Kikkawa et al. (1982).

C was able to do so. Furthermore, when added to SPM, kinase C was able to phosphorylate B-50, but not several other proteins. In Table I, the various properties of B-50 kinase and kinase C are compared. In view of the activation by PS and protease, the Ca^{2+} requirement etc., we concluded that B-50 kinase is a phospholipid-sensitive protein kinase that is very similar to kinase C (Aloyo *et al.*, 1983). It may well be that there exist many lipid-sensitive protein kinases, each having its own specific endogenous substrate, *e.g.*, B-50 protein kinase or fibrogen kinase. On the other hand, the same enzyme may exist in many tissues as reported by Kuo *et al.* (1980) and the restricted localization of the substrate proteins may determine which proteins are phosphorylated when the kinase C is activated.

PHOSPHOPROTEIN B-50 AND FORMATION OF PIP$_2$ IN RAT BRAIN

In a series of experiments we observed that ACTH-like peptides influenced the metabolism of phosphoinositides in subcellular fractions of rat brain. As we had shown that $ACTH_{1-24}$ influenced the phosphorylation of certain brain membrane proteins as well, it was investigated whether the peptide-induced changes in protein and lipid phosphorylation were related. The first studies were carried out in the ASP_{55-80} protein fraction, containing the B-50 protein kinase/B-50 substrate protein complex. We were unable to detect phosphoinositides in this protein fraction. When exogenous PIP and $[\gamma-^{32}P]ATP$ were added the only two labeled compounds appeared to be the B-50 phosphoprotein and PIP_2. Addition of PI did not result in formation of PIP. Apparently this protein fraction in addition to the protein kinase activity also contains PIP kinase activity.

By preincubation of ASP_{55-80} with the labeled ATP for different periods, the B-50 protein was phosphorylated to a different degree. If subsequently PIP was added as exogenous lipid substrate, it was observed that the more phosphorylated B-50 was present, the less PIP was converted to PIP_2. Such an inverse relationship between the labeling of B-50 and PIP_2 was further evidenced by the use of $ACTH_{1-24}$. Addition of ACTH to the ASP_{55-80} fraction inhibited the phosphorylation of

Gispen et al.

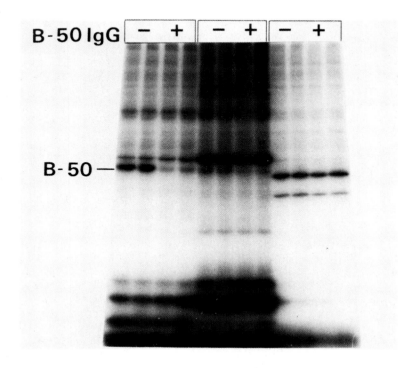

Figure 2. Inhibition of B-50 phosphorylation by anti-B-50 IgG in various subcellular fractions of rat brain. A: SPM; B: vesicles and myelin; C: mitochondria.

B-50 and concomitantly enhanced the formation of PIP_2. As the ASP_{55-80} did not contain detectable phosphodiesterase activity (Van Dongen, unpublished observations), at least under these circumstances the effect of $ACTH_{1-24}$ cannot be attributed to an inhibition of the hydrolysis of formed PIP_2. Further studies using a lysed synaptosomal fraction revealed a similar effect of $ACTH_{1-24}$. In that preparation, however, a concomitant decrease in the labeling of PA was found. In SPM no ACTH-induced inhibition of PA labeling could be established, underscoring previous data suggesting that the PA and PIP_2 effects are not necessarily directly related. Recently, Abdel-Latif and coworkers (1983) using

$ACTH_{1-24}$ and subcellular fractions from iris muscle, also observed an inverse relationship between the phosphorylation of certain protein bands and the formation of PIP_2. Also these authors reported no effect of $ACTH_{1-24}$ on protein phosphatase or phosphodiesterase activities in their preparation. In our laboratory we have used yet another approach to strengthen the proposed relationship between the degree of phosphorylation of the B-50 protein and the labeling of PIP_2 in rat brain SPM. By using the anti-B-50 immunoglobulins, we were able to specifically block the phosphorylation of B-50 in these membranes (Fig. 2). Either the B-50/anti-B-50 interaction takes place at the phosphorylatable site in B-50 or the introduction of the antibody is such a steric hindrance that the B-50 kinase inefficiently phosphorylates the B-50. The results of these experiments showed again that a lesser degree of phosphorylation of B-50 is paralleled by an increased labeling of PIP_2.

As reviewed extensively by a variety of authors elsewhere, transmitter-receptor activation may result in or be paralleled by changes in the phosphorylation of certain phosphoproteins and the hydrolysis of PIP_2. We have started studies to investigate whether the B-50 kinase/B-50/PIP-kinase/PIP_2 sequence could be influenced by modulators other than ACTH. In the first series of experiments we tested the effects of dopamine on endogenous phosphorylation of hippocampal proteins and phosphoinositides. Treatment of hippocampal slices with 5 x 10^{-4} M dopamine resulted in a 45% increase in *post hoc* phosphorylation of B-50 in SPM and a 40% decrease in the *post hoc* labeling of PIP_2. These effects of dopamine on PIP_2 labeling could be blocked by co-incubation with haloperidol (Jork *et al.*, 1984). These results not only support the proposed inverse relationship between the phosphorylation of B-50 and that of PIP_2, they also are first evidence that certain transmitter-receptor activation may involve changes in the multifunctional kinase complex described above.

The available evidence for the inverse relationship between the phosphorylation of B-50 and that of PIP-kinase activity is summarized in Table II. In view of these data we proposed that the protein B-50 could be a

TABLE II

Indirect Evidence for an Inverse Relationship
Between the Phosphorylation of B-50 and PIP-Kinase
Activity in Rat Brain Membranes

Treatment	Phosphorylation of B-50	DPI-Kinase Activity
Phosphorylation of B-50	⬆	⬇
ACTH$_{1-24}$	⬇	⬆
Anti-B-50 IgG	⬇	⬆
Dopamine*	⬆	⬇

*Added to hippocampal slices followed by post hoc phosphorylation.

regulatory factor in PIP kinase activity in certain synapses in the brain. Attempts have been undertaken to test this hypothesis in a more direct manner by studying the activity of a highly purified brain-PIP kinase preparation in the presence of exogenous B-50 with different degrees of phosphorylation.

B-50 was extracted from rat brain membranes by alkaline extraction and purified by ammonium sulphate precipitation and flat-bed isoelectric focusing (Oestreicher et al., 1983). The purified protein shows microheterogeneity upon isoelectric focusing in a narrow pH gradient (pH 3.5 - 5.0). As visualized by two-dimensional gel electrophoresis, B-50 resolved into 4 clearly separated forms which differ slightly in isoelectric point. It appeared that the forms are in part mutually convertible by exhaustive phosphorylation (using protein kinase C) and dephosphorylation (using E. coli alkaline phosphatase) (Fig. 3). We concluded therefore that the isoforms represent differently phosphorylated forms of the B-50 protein (Zwiers, in preparation).

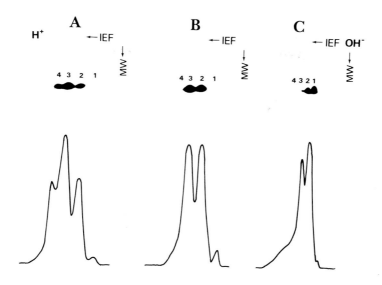

<u>Figure 3</u>. Upper part: protein-staining pattern of B-50 (3 µg) after two-dimensional separation. Only the B-50-containing part of the two-dimensional gels is shown. The direction of isoelectric focusing was from right to left and the separation of SDS gels from top to bottom. Panel A: incubation of B-50 at 30°C with 5 µg protein kinase C for 60 min; panel B: incubation with buffer; panel C: incubation with 4 µg alkaline phosphatase (<u>E</u>. <u>coli</u>) for 60 min. Lower part: densitometric tracings (at 650 nm) of protein-staining patterns shown in the upper part of the figure.

We have used this procedure on a preparative scale to obtain batches of B-50 enriched in the phosphorylated form and the dephosphorylated form. These batches of B-50 were tested for their ability to affect the activity of a highly purified preparation of PIP kinase, obtained from the soluble fraction of rat brain tissue. The purification procedure involved ammonium sulphate precipitation (yielding an ASP_{20-40} protein fraction), followed by DEAE-column chromatography. The peak fraction of PIP kinase activity was enriched 18-fold and consisted of one major protein of M_r 45 kDa, most likely representing the PIP kinase (Van Dongen et al., in preparation). Incubation of this preparation with 1 µg of either the phospho-form-enriched B-50 or the dephospho-form-enriched B-50

409

Gispen et al.

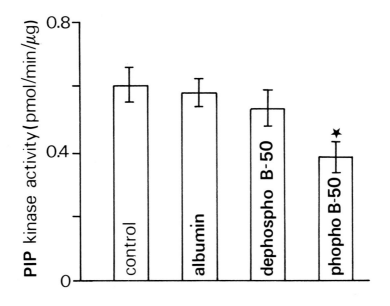

<u>Figure</u> 4. Effect of degree of phosphorylation of B-50 on PIP-kinase activity. A highly purified brain PIP-kinase fraction was obtained from brain cytosol after ammonium sulphate precipitation and DEAE-column chromatography. The PIP kinase present in the assay is estimated to be 1 μg. Additions: 1 μg of either albumin, dephospho- or phospho-B-50 protein. PIP-kinase activity was measured in the presence of 10 μg albumin and expressed as pmol phosphate transferred from ATP to PIP. n= 6; *p < 0.05 compared to control by Student <u>t</u> test.

resulted in different activities of PIP kinase. The dephosphorylated batch of B-50 had no effect, whereas the same amount of the phosphorylated batch of B-50 significantly inhibited the activity of PIP kinase (Fig. 4). To diminish nonspecific protein/protein interactions these experiments were performed in the presence of 10 μg bovine serum albumin. Therefore it would appear that there are conditions under which the degree of phosphorylation of B-50 may indeed govern the activity of brain PIP kinase.

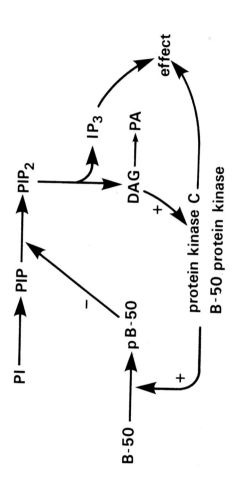

Figure 5. Model of the regulatory role of B-50 in the receptor-mediated polyphosphoinositide response.

CONCLUDING REMARKS

In the present paper we reviewed evidence indicating that the phosphorylation of the nervous tissue-specific protein B-50 is a regulatory event in the control of brain membrane PIP kinase activity. At present, we assume that there is a multifunctional enzyme complex in synaptic membranes which consists of at least a calcium/lipid-sensitive protein kinase (kinase C), B-50 protein and PIP kinase. In Fig. 5, a model is presented outlining the regulatory role that B-50 could play in the receptor-mediated polyphosphoinositide response.

Hydrolysis of PIP_2 results in the production of inositol trisphosphate (IP_3) and diacylglycerol (DG). Currently, the IP_3 is considered as a putative second messenger that modulates intracellular calcium mobilization (see the chapter by M. Berridge, this volume). The DG is known to stimulate the protein kinase C/B-50 kinase (see above and chapter by Y. Nishizuka, this volume). Subsequently, enhancement of B-50 phos-phorylation by the activated protein kinase C will inhibit the formation of PIP_2 by inhibiting the PIP kinase. Hence, the degree of phosphorylation of B-50 may serve as a negative feedback control in the production and further hydrolysis of PIP_2.

ACTH-like peptides released from peptidergic terminals can locally influence this feedback control by affecting the activity of protein kinase C/B-50 kinase. Such a mechanism could account for the presynaptic modulation of various types of neurotransmission by neuropeptides.

REFERENCES

Abdel-Latif, A.A. (1983) in: "Handbook of Neurochemistry" (A. Lajtha and A.A. Abdel-Latif, eds.) Vol. 3, pp. 91-131, Plenum Publ. Co., New York.

Aloyo, V.J., Zwiers, H. and Gispen, W.H. (1983) J. Neurochem. 41, 649-653.

Gispen, W.H, Zwiers, H., Wiegant, V.M., Schotman, P. and Wilson, J.E. (1979) Adv. Exptl. Med. Biol. 116, 199-224.

Gower, H. and Rodnight, R. (1982) *Biochim. Biophys. Acta 716*, 45-52.

Hershkowitz, M., Heron, D., Samuel, D. and Shinitzky, M. (1982) *Progr. Brain Res. 56*, 419-434.

Inoue, M., Kishimoto, A., Takai, Y. and Nishizuka, Y. (1977) *J. Biol. Chem. 252*, 7610-7616.

Jork, R., De Graan, P.N.E., Van Dongen, C.J., Zwiers, H., Matthies, H. and Gispen, W.H. (1984) *Brain Res., 291*, 73-81.

Kikkawa, U., Takai, Y., Minakuchi, R., Inohara, S. and Nishizuka, Y. (1982) *J. Biol. Chem. 257*, 13341-13348.

Kristjansson, G.I., Zwiers, H., Oestreicher, A.B. and Gispen, W.H. (1982) *J. Neurochem. 39*, 371-378.

Kuo, J.F., Andersson, R.G.G., Wise, B.C., Mackerlova, L., Salomonsson, I., Brackett, M.L., Katoh, N., Shoji, M. and Wrenn, R.W. (1980) *Proc. Natl. Acad. Sci. USA 77*, 7039-7043.

Mahler, H.R., Kleine, L.P., Ratner, N. and Sorensen, R.C. (1982) *Progr. Brain Res. 56*, 27-48.

Oestreicher, A.B., Zwiers, H., Schotman, P. and Gispen, W.H. (1981) *Brain Res. Bull. 6*, 145-153.

Oestreicher, A.B., Van Dongen, C.J., Zwiers, H. and Gispen, W.H. (1983) *J. Neurochem. 41*, 331-340.

Rodnight, R. (1982) *Progr. Brain Res. 56*, 1-25.

Routtenberg, A. (1982) *Progr. Brain Res. 56*, 349-374.

Schatzman, R.C., Raynon, R.L., Fritz, R.B. and Kuo, J.F. (1983) *Biochem. J. 209*, 435-443.

Sorensen, R.G., Kleine, L.P. and Mahler, H.R. (1981) *Brain Res. Bull. 7*, 57-61.

Sorensen, R.G. and Mahler, H.R. (1983) *J. Neurochem. 40*, 1349-1365.

Takai, Y., Kishimoto, A., Iwasa, Y., Kawahara, Y., Mori, T., Nishizuka, Y., Tamura, A. and Fukii, T. (1979a) *J. Biochem. 86*, 575-578.

Takai, Y., Kishimoto, A., Iwasa, Y., Kawahara, Y., Mori, T. and Nishizuka, Y. (1979b) *J. Biol. Chem. 254*, 3692-3695.

Weller, M. (1979) *"Protein Phosphorylation"*, PION Ltd., London.

Zwiers, H., Veldhuis, D., Schotman, P. and Gispen, W.H. (1976) *Neurochem. Res. 1*, 669-677.

Zwiers, H., Wiegant, V.M., Schotman, P. and Gispen, W.H. (1978) *Neurochem. Res. 3*, 455-463.

Gispen et al.

Zwiers, H., Tonnaer, J., Wiegant, V.M., Schotman, P. and Gispen, W.H. (1979) *J. Neurochem.* *33*, 247-256.
Zwiers, H., Verhoef, J., Schotman, P. and Gispen, W.H. (1980) *FEBS Lett.* *112*, 168-172.
Zwiers, H., Jolles, J., Aloyo, V.J., Oestreicher, A.B. and Gispen, W.H. (1982) *Progr. Brain Res.* *56*, *405-417.*

PURIFICATION OF PROTEIN KINASE C AND PHORBOL ESTER RECEPTOR USING POLYACRYLAMIDE-IMMOBILIZED PHOSPHATIDYLSERINE

Charles R. Filburn and Tsutomu Uchida

Laboratory of Molecular Aging, National Institute on Aging, NIH, Gerontology Research Center, Baltimore City Hospitals, Baltimore, MD

SUMMARY

A Ca^{2+} + phospholipid-dependent protein kinase and a phorbol ester receptor were purified from a cytosolic extract of rabbit renal cortex by ion exchange chromatography on DEAE-cellulose and affinity chromatography on polyacrylamide-immobilized phosphatidylserine. Protein kinase activity was monitored by assaying Ca^{2+} + phospholipid-dependent phosphorylation of histone H1, phorbol ester binding by measuring specific binding of [3H]-phorbol dibutyrate in the presence of Ca^{2+} and phosphatidylserine. Both activities eluted together from the DEAE-cellulose column, bound to the affinity column in the presence of Ca^{2+}, and eluted symmetrically from the affinity column upon application of EGTA. Recovery from the affinity column was the same for both activities (usually 20-30%) and resulted in a high degree of purification.

INTRODUCTION

Recent studies indicate that protein kinase C activity and phorbol ester binding activity are present in the same protein (Kikkawa et al., 1983; Niedel et al., 1983). Since phorbol esters stimulate this protein kinase (Castagna et al., 1982; Niedel et al., 1983), and produce some of the same effects in cells as hormones which perturb phosphoinositide metabolism and increase cellular levels of diacylglycerol and Ca^{2+} (Berridge, 1983; Dicker and Rozengurt, 1980; Thomas *et al.*, 1983; Putney *et al.*, 1983), both of which regulate this protein kinase (Kaibuchi *et al.*, 1981), further study of the regulation and physiological functions of this enzyme in various tissues is warranted. Present purification methods involve several steps and result in relatively low recovery of the protein kinase (Kikkawa *et al.*, 1982; Wise *et al.*, 1982). We have prepared an affinity column from poly-acrylamide-immobilized phosphatidylserine and have observed highly specific, Ca^{2+}-dependent binding of protein to the gel, thus permitting rapid, high recovery of a highly purified protein kinase C – phorbol ester receptor.

MATERIALS AND METHODS

Materials. Phosphatidylserine was from Avanti and Supelco. Phenylmethylsulfonylfluoride (PMSF) was from Calbiochem. Histone H1 (III-S), diolein, leupeptin and phorbol myristate acetate (PMA) were from Sigma. [γ-^{32}P]ATP and [^{3}H]phorbol dibutyrate (PDB) were from New England Nuclear.

Methods. A cytosolic extract of rabbit renal cortex was obtained by centrifuging at 35,000 x *g* for 30 min a 10% homogenate prepared with 0.25 M sucrose, 20 mM Tris pH 7.5, 5 mM dithiothreitol, 2 mM EDTA, 2 mM EGTA, and 0.1 mM PMSF. Cyclic AMP was added to 1 μM and the extract applied to a 1.5 x 25 cm DEAE-cellulose column equilibrated with homogenizing buffer lacking sucrose but containing 1 μM cyclic AMP. A linear 0-0.4 M gradient of KCl was applied in 500 ml and 5.0 ml fractions were collected and assayed. Fractions con-taining the peak of protein kinase and phorbol ester binding activities were pooled and divided into aliquots, some of which were either stored frozen at -70°C or applied to the affinity column.

Leupeptin was added to a 10-15 ml aliquot of the pooled fractions to 100 μM and the mixture pumped into a closed 0.5 ml mixing chamber along with a solution con-taining 10 mM MES, pH 6.5, 14 mM CaCl$_2$, 5 mM dithio-threitol, 0.1 mM PMSF and 300 mM KCl. The resulting mixture was pumped directly onto the affinity column. This sample was followed by 15 ml column buffer, then 45 ml column buffer containing 0.1 mM $CaCl_2$, and finally with buffer containing 2 mM EGTA in place of $CaCl_2$. Fractions of 3.0 ml were collected; plastic tubes were used for fractions eluted with EGTA.

An affinity column gel containing phosphatidylserine was prepared using a mixture of cholesterol (25 mg) and phosphatidylserine (5 mg) dissolved in chloroform and evaporated from a glass scintillation vial with nitrogen. After addition of 0.5 ml ethanol the vial was capped and boiled briefly to disperse the lipid. A 5.0 ml aliquot of 15% acrylamide-5% BIS was immediately added to the dispersed lipid and mixed vigorously, followed by 50 μl 0.14% ammonium persulfate, 2.5 μl TEMED, and additional 50 μl of ammonium persulfate. After further mixing, the solution was transferred to a glass test tube, covered with parafilm and aluminum foil and left overnight at room temperature. The tube was broken to remove the gel, which was then rinsed with water, minced and homogenized in 20-30 ml water with three passes of a Dounce homogenizer. After 5 min of settling, the cloudy supernatant was aspirated and the settled particles washed by repeated cycles of suspension in water, settling, and aspiration. Washed gel particles were finally resuspended in 5 mM MES, pH 6.5, 5 mM dithiothreitol, 1 mM $CaCl_2$, 0.1 mM PMSF, 200 mM KCl, packed in a siliconized glass column, and equilibrated with 5-10 column volumes of buffer.

Protein kinase was assayed by measuring incorporation of $[\gamma-^{32}P]ATP$ into histone H1 in a reaction mixture containing 25 mM Tris, pH 7.5, 200 μg/ml histone H1, 5 mM $MgCl_2$, 20 μM ATP, 1-2 x 10^6 cpm $[\gamma-^{32}P]ATP$, 200 μM EGTA and EDTA (cytosol and DEAE-cellulose eluate) or 200 μM EGTA (affinity column eluate), about 50 μg/ml phosphatidylserine, 0.5 μg/ml diolein, 0.5 mM free Ca^{2+} and 10 μl of enzyme in a total of 100 μl. Reactions were from 1 min at 30°C and were stopped by adding 1 ml 10% TCA, 5 mM NaH_2PO_4, 2 mM ATP, followed by 100 μl 0.63% bovine serum albumin. Precipitated protein was centrifuged, redissolved with 1 N NaOH, reprecipitated and collected on glass fiber filters for counting.

Phorbol ester binding was measured in a mixture containing 25 mM Tris, pH 7.5, 10 mM magnesium acetate, 1.4 mM CaCl$_2$, 0.4 mM EGTA and EDTA (DEAE-cellulose eluate) or 0.4 mM EGTA (affinity column eluate), 0-50 mM KCl, 4 mg/ml bovine serum albumin, 100 µg/ml phosphatidylserine (Avanti), 20 nM [^{3}H]PDB, about 3 µM PMA in 200 µl total volume. A total volume of 400 µl was used for assessing dependence of binding on PDB concentration. Incubations were for 2 h on ice. Bound PDB was separated from free PDB by filtering the mixture through Whatman GF/C glass filters. Tubes and filters were washed five times with 1 ml 20 mM Tris, pH 7.5, 10 mM magnesium acetate, 1 mM CaCl$_2$.

Protein was measured by a dye-binding assay (Reed and Northcote, 1981) with bovine serum albumin as standard.

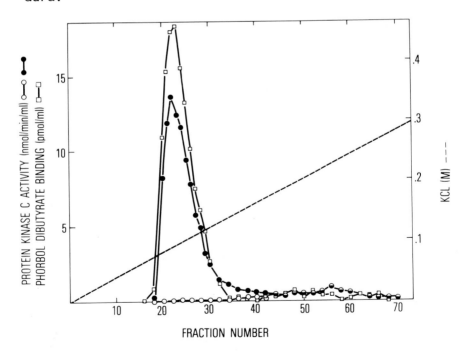

Figure 1. Elution profile of protein kinase C and [^{3}H]PDB binding from DEAE-cellulose.

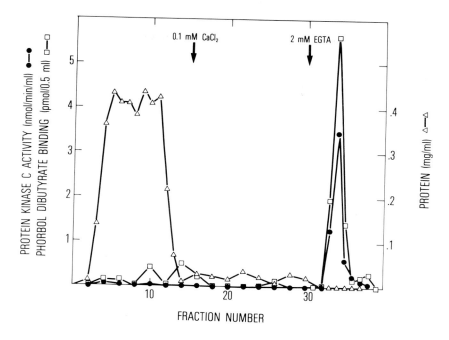

<u>Figure 2</u>. Affinity chromatography of pooled DEAE-cellulose fractions on polyacrylamide-immobilized phosphatidylserine.

RESULTS AND DISCUSSION

Protein kinase C activity and PDB-binding activity present in the rabbit renal cortex co-eluted from DEAE-cellulose at approximately 0.1 M KCl (Fig. 1). When a portion of the pool fractions of peak activities was applied to phosphatidylserine-acrylamide affinity column in the presence of $CaCl_2$, very little of either activity eluted from the column, while all of the readily detectable protein passed through the column (Fig. 2). Upon addition of EGTA, however, protein kinase C and PDB binding activity co-eluted in a single, sharp peak that contained a very low level of protein that was difficult to measure. Recovery of activity from the affinity column was comparable for both activities, usually 20-30% of the applied activity, with a very large increase in specific activities (Table I). Graphic analysis of the dependence of PDB

419

TABLE I

Purification of Renal Cortical Protein Kinase C –
Phorbol Ester Receptor.

	Protein Kinase C	[³H]PDB Binding
	$(nmol \times min^{-1} \times mg^{-1}$ protein)	(pmol/mg protein)
DEAE-cellulose pooled fractions	3.05	8.64
Affinity column peak fraction	1442 (473)	4680 (542)

binding on PDB concentration (Fig. 3) showed a K_D of 0.5 nM. Protein kinase C activity was very unstable, with complete loss of activity in 24 h if stored in glass tubes or frozen. Use of plastic tubes and glycerol retarded the loss, while inclusion of bovine serum albumin at 1 mg/ml fully stabilized the enzyme for weeks if stored on ice.

In summary, use of polyacrylamide-immobilized phosphatidylserine as an affinity medium provides a simple, rapid method for purification of both protein kinase C and phorbol ester binding activities. The applicability of the technique described may vary from tissue to tissue depending on the relative difficulty of preventing proteolytic degradation of the protein(s) by Ca^{2+}-dependent protease(s). Further study of the purified enzyme-receptor is needed for assessment of possible tissue differences and regulation by Ca^{2+}, lipids and phorbol esters.

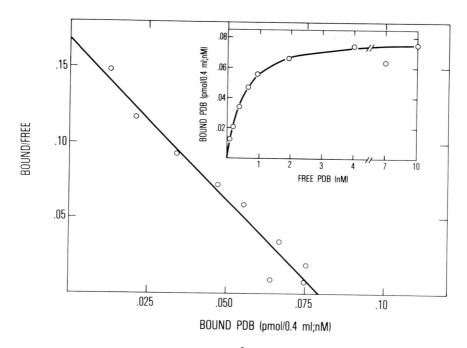

Figure 3. Scatchard analysis of [³H]PDB binding to phorbol ester receptor purified by affinity chromatography.

REFERENCES

Berridge, M.J. (1983) *Biochem. J.* *212*, 849–858.

Castagna, M., Takai, Y., Kaibuchi, K., Sano, K., Kikkawa, U. and Nishizuka, Y. (1982) *J. Biol. Chem.* *257*, 7847–7851.

Dicker, P. and Rozengurt, E. (1980) *Nature* *287*, 607–612.

Kaibuchi, K., Takai, Y. and Nishizuka, Y. (1981) *J. Biol. Chem.* *256*, 7146–7149.

Kikkawa, U., Takai, Y., Minakuchi, R., Inohara, S. and Nishizuka, Y. (1982) *J. Biol. Chem.* *257*, 13341–13348.

Kikkawa, U., Takai, Y., Tanaka, Y., Miyake, R. and Nishizuka, Y. (1983) *J. Biol. Chem.* *258*, 11442–11445.

Niedel, J.E., Kuhn, L.J. and Vanderbark, G.R. (1983) *Proc. Natl. Acad. Sci. USA* *80*, 36–40.

Putney, J.W., Jr., Burgess, G.M., Halenda, S.P., McKinney, J.S. and Rubin, R.P. (1983) *Biochem. J.* *212*, 483–488.

Reed, S.M. and Northcote, D.H. (1981) *Anal. Biochem.* *116*, 53–64.

Thomas, A.P., Marks, J.S., Coll, K.E. and Williamson, J.R. (1983) *J. Biol. Chem.* *258*, 5716–5725.

Wise, B.C., Raynor, R.L. and Kuo, J.F. (1982) *J. Biol. Chem.* *257*, 8481–8488.

SECOND MESSENGER ROLE OF INOSITOL TRISPHOSPHATE FOR MOBILIZATION OF INTRACELLULAR CALCIUM IN LIVER

John R. Williamson, Andrew P. Thomas
and Suresh K. Joseph

Department of Biochemistry and Biophysics,
University of Pennsylvania School of Medicine,
Philadelphia, Pa 19104

SUMMARY

Previous work has established that stimulation of hepatocytes by α_1-adrenergic agents and vasoactive peptides results in a mobilization of intracellular Ca^{2+}, which is accompanied by the rapid hydrolysis of phosphatidylinositol 4,5-bisphosphate (PIP_2) to diacylglycerol and *myo*-inositol 1,4,5-trisphosphate (IP_3). Present data using the fluorescent Ca^{2+} indicator Quin 2 showed that vasopressin increased the cytosolic free Ca^{2+} from 150 nM to a maximum value of about 400 nM within 7 s while IP_3 increased 3-fold over 5 min. The maximum increase of cytosolic free Ca^{2+} corresponded to an increase in IP_3 concentration of about 0.6 µM. Very similar dose response relationships with vasopressin were observed for changes of PIP_2 breakdown, IP_3 formation and cytosolic free Ca^{2+} when these parameters were expressed as rates. Addition of IP_3 to saponized cells supplemented with MgATP caused a rapid Ca^{2+} release, complete within 5 s, followed by a reuptake, which was associated with IP_3 dephosphorylation. Half-maximal and maximal Ca^{2+} release were obtained at IP_3 concentrations of 0.1 µM and 1 µM, respectively. The maximal amount of Ca^{2+} mobilized was 500 pmol/mg cell dry weight, and was found to originate solely from non-mitochondrial vesicular stores. The kinetics, dose response, specificity and amount of

423

Ca^{2+} release induced by IP_3 strongly support its postulated role as a second messenger involved in the hormonal mobilization of Ca^{2+} from intracellular stores.

INTRODUCTION

In liver the glycogenolytic action of α_1-adrenergic agonists and vasoactive peptide hormones is mediated by an increase of the cytosolic free Ca^{2+} concentration, which causes an allosteric activation of phosphorylase *b* kinase (Exton, 1981; Williamson et al., 1981). The subsequent increase of phosphorylase *a* activity is observed with hepatocytes incubated in Ca^{2+}-free medium, indicating that the initial increase of cytosolic Ca^{2+} is due to the mobilization of Ca^{2+} from intracellular storage sites (Blackmore et al., 1982; Joseph and Williamson, 1983). The presence of external Ca^{2+} sustains the hormone effect, but does not affect the initial kinetics of Ca^{2+} mobilization. The plasma membrane, mitochondria and endoplasmic reticulum have all been implicated as potential sources of the hormone-releasable calcium (Blackmore et al., 1979; Joseph and Williamson, 1983; Kimura et al., 1982), but until recently the mechanism of the effect has remained obscure. With the observation that breakdown of PIP_2 rather than phosphatidylinositol was one of the earliest biochemical events following binding of a variety of Ca^{2+}-mobilizing agonists to receptors on the plasma membranes of liver (Berridge et al., 1983; Creba et al., 1983; Litosch et al., 1983; Rhodes et al., 1983; Thomas et al., 1983), the water soluble product IP_3 became the prime candidate for an intracellular second messenger involved in alterations of calcium homeostasis. The direct effect of IP_3 as a Ca^{2+}-mobilizing agent was first demonstrated by Streb et al. (1983) using 'leaky' pancreatic acinar cells. Our studies have extended observations of IP_3 effects to saponin-permeabilized hepatocytes (Joseph et al., 1984), and provide strong evidence for the role of IP_3 as a second messenger for the action of Ca^{2+}-mobilizing hormones in liver.

RESULTS AND DISCUSSION

Figure 1 shows the results of an experiment in which hepatocytes were incubated at 37°C in a high K^+ medium, containing 110 mM KCl, 10 mM NaCl, 5 mM $KHCO_3$, 1 mM

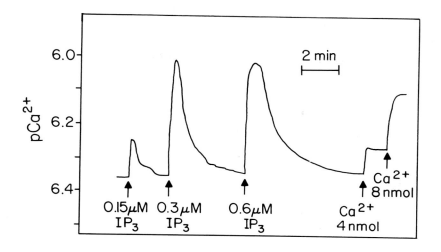

Figure 1. Ca^{2+} release induced by IP$_3$ in saponin-permeabilized hepatocytes. Hepatocytes (17 mg dry wt/ml) were incubated as described in the
text and Joseph et al. (1984). Saponin (15 µg/mg cell dry wt) was added
to permeabilize the cells and the changes in medium free Ca^{2+} concentrations were measured using a Ca^{2+} electrode.

KH$_2$PO$_4$, 20 mM HEPES, 20 mM creatine phosphate, 0.3
mM MgCl$_2$, 3 mM MgATP^{2-} and 10 units/ml of creatine
kinase, pH 7.2. The Ca^{2+} concentration in the medium
was monitored by a Ca^{2+} electrode, calibrated by the
procedure of Bers (1982). After 5 min of incubation
following addition of 15 µg/mg cell dry weight of saponin,
the Ca^{2+} concentration was decreased to 0.45 µM (pCa^{2+}
of 6.35). Addition of 0.15 µM IP$_3$ caused a rapid release
of Ca^{2+} followed by a slow reaccumulation. No Ca^{2+}
release was observed with inositol 4,5-bisphosphate or
inositol 2-phosphate and no electrode response was
observed after IP$_3$ addition in the absence of hepatocytes
(not shown). Sequential additions of IP$_3$ produced cycles
of Ca^{2+} release and reuptake in which the amount of Ca^{2+}
released (quantitated by an internal calibration) and the
time required for Ca^{2+} reuptake depended on the amount
of IP$_3$ added.

The Ca^{2+} electrode provided a very sensitive, noise-
free monitor of changes in free Ca^{2+} of the medium, but
suffered from having a response time in the order of

425

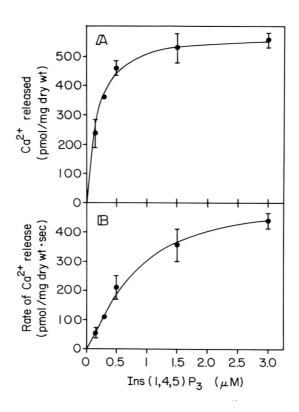

Figure 2. Effect of IP_3 ($Ins(1,4,5)P_3$) on the amount (A) and rate (B) of Ca^{2+} release from saponin-permeabilized hepatocytes. Hepatocytes (5 mg dry wt/ml) were incubated in high K^+ medium, pH 7.2, containing 75 μM Quin 2, 3 μg/ml of oligomycin, 0.5 μg/ml of rotenone, 5 mM succinate, 0.3 mM Mg^{2+}, 3 mM MgATP, 10 mM creatine phosphate and 10 units/ml of creatine kinase followed by addition of 75 μg/ml of saponin. Ca^{2+} changes were monitored by Quin 2 fluorescence.

1-2 s. Further studies were conducted using the more rapidly responding fluorescent Ca^{2+} indicator Quin 2 both as a Ca^{2+} buffer and a Ca^{2+} indicator (Tsien, 1980; Tsien et al., 1982). Figure 2A shows the concentration dependency of Ca^{2+} release from saponized hepatocytes by IP_3 in which a maximal Ca^{2+} release of about 500 nmol/mg cell dry weight was produced with 1 μM IP_3 and half-maximal release with 0.1 µM IP_3. Figure 2B shows that the rate of Ca^{2+} release by IP_3 was also concentration dependent, reaching a maximum value of about 400 pmol/mg cell dry weight/s and with a half-maximal rate at 0.6 µM IP_3.

The cytosolic free Ca^{2+} concentration of hepatocytes is in the range of 150-200 nM (Charest et al., 1980; Murphy et al., 1980; Thomas et al., 1984), but various studies with Ca^{2+}-permeable cells have shown that with these concentrations, an ATP-dependent uptake of Ca^{2+} occurred only into a non-mitochondrial, ATP-dependent vesicular pool, from which calcium could be released by Ca^{2+}-ionophores but not by uncoupling agents (Burgess et al., 1984; Joseph et al., 1984; Streb and Schulz, 1983). The reason for the lack of mitochondrial Ca^{2+} cycling at physiological Ca^{2+} concentrations is currently under investigation; this cycling appears to be due to loss of polyamines from the permeabilized cells. Spermine (0.3 mM) affects the properties of the mitochondrial Ca^{2+} transport systems and lowers the steady state extra-mitochondrial free Ca^{2+} in the presence of Mg^{2+} (Nicchitta and Williamson, 1984). In the present studies this problem was circumvented by using an external Ca^{2+} concentration with saponin-permeabilized cells in the range of 700 µM where mitochondrial Ca^{2+} uptake could be observed. For these studies the medium Ca^{2+} was both buffered and monitored by Quin 2. Figure 3A shows the results of an experiment in which saponin-permeabilized cells were first incubated in the presence of 5 mM succinate, 3 µg/ml of oligomycin, 0.5 µg/ml of rotenone plus the creatine phosphate ATP-regenerating system in order to achieve a steady state with respect to the medium Ca^{2+} concentration. This produced saturation of the endoplasmic reticulum calcium pool, while the mitochondria contained 2-3 nmol of Ca^{2+}/mg cell dry weight. Subsequent addition of 3 nmol of Ca^{2+} showed that it was removed from the medium, and addition of 0.5 µM IP_3 caused a rapid Ca^{2+} release and subsequent reuptake similar to that of Figure 1. Addition of the nonfluorescent uncoupling agent 1799 produced a release of Ca^{2+}, thereby verifying the presence of a mitochondrial Ca^{2+} pool in the permeabilized cells.

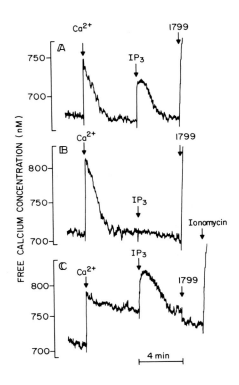

Fig. 3. Characterization of the IP_3-sensitive Ca^{2+} pool in saponin-permeabilized hepatocytes. Hepatocytes were incubated as in Fig. 2 except for an elevation of the medium Ca^{2+}. Additions to 1.6 ml of medium were: Ca^{2+}, 3nmol; IP_3, 0.5 μM; 1799, 5 μM; ionomycin, 5 μg/ml (final concentrations). In trace A, succinate, rotenone and MgATP were present to allow Ca^{2+} uptake into both mitochondrial and non-mitochondrial Ca^{2+} pools. In trace B, oligomycin was present and MgATP omitted. In trace C, succinate was omitted and 1 μM ruthenium red was added.

428

In Figure 3B, MgATP was omitted from the incubation medium, and since oligomycin was present, the mitochondria were unable to generate ATP. Under these conditions the mitochondria were able to take up Ca^{2+}, as shown by the removal of a 3 nmol addition of Ca^{2+}, but IP_3 was no longer able to cause a release of Ca^{2+}. Addition of the nonfluorescent Ca^{2+} ionophore ionomycin caused no Ca^{2+} release (not shown), while the large Ca^{2+} release induced by 1799 confirmed the presence of calcium in the mitochondria.

Finally, Figure 3C shows the results obtained after omission of succinate and addition of 1 µM ruthenium red to prevent mitochondrial Ca^{2+} uptake. When an additional amount of 3 nmol of Ca^{2+} was added, it was not sequestered, but addition of 0.5 µM IP_3 caused a similar cycle of Ca^{2+} release and reuptake as in Figure 3A. Subsequent additions of 1799 and ionomycin confirmed that under these conditions Ca^{2+} was sequestered only in a non-mitochondrial pool. Thus, these data clearly show that IP_3 does not release Ca^{2+} from the mitochondrial pool in permeabilized hepatocytes and that the observed IP_3-induced Ca^{2+} release originates entirely from a non-mitochondrial, vesicular compartment. The observations that the rate of Ca^{2+} release is less sensitive to the IP_3 concentration than the amount of Ca^{2+} released (Fig. 2), and that the maximal amount of Ca^{2+} released is less than the total Ca^{2+} content of the endoplasmic reticulum (Joseph and Williamson, 1983), suggest that IP_3 caused a recruitment in the opening of Ca^{2+} efflux channels in a specialized fraction of the endoplasmic reticulum.

In order to prove convincingly that IP_3 serves as a physiological Ca^{2+}-mobilizing second messenger, it must be shown that it can be formed fast enough and in sufficient amounts in the intact cell to cause the observed hormone-induced Ca^{2+} release. Figure 4A shows that the IP_3 content of intact hepatocytes increased significantly above control levels within 5 s after addition of 20 nM vasopressin. However, maximum changes, approaching a 3-fold increase, were not obtained until after 2 min of stimulation, when phosphorylase *a* was already maximal (Thomas *et al.*, 1984). Figure 4B shows a representative fluorometric trace for changes of the cytosolic free Ca^{2+} obtained from hepatocytes loaded with about 1 mM Quin 2

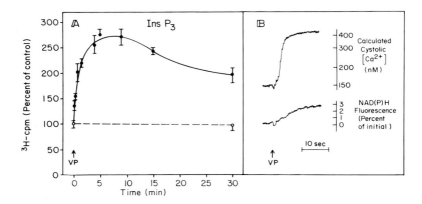

<u>Figure 4</u>. Effect of vasopressin on the change of IP_3 content and the cytosolic free Ca^{2+} in isolated hepatocytes. Hepatocytes (5 mg dry wt/ml) were incubated in Krebs bicarbonate medium containing 1.3 mM Ca^{2+} (A) or 0.65 mM Ca^{2+} (B). [^{3}H]Inositol was incorporated into inositol lipids and the [^{3}H]IP_3 subsequently assayed as described by Thomas <u>et al</u>. (1984). Cytosolic free Ca^{2+} was measured in cells loaded with Quin 2 as described by Thomas <u>et al</u>. (1984). The concentration of vasopressin used was 20 nM in A and 30 nM in B.

and stimulated with 30 nM vasopressin. Simultaneous measurement of the pyridine nucleotide fluorescence showed that the NAD(P)H fluorescence increase was slower and smaller than that of the Quin 2-Ca^{2+} fluorescence increase at comparable sensitivities of the fluorometer (c.f. Charest et al., 1983). After calibration of the Quin 2 fluorescence (Tsien et al., 1982), the cytosolic free Ca^{2+} was determined to increase from 150 nM to a maximum of about 400 nM after vasopressin. The change of the cytosolic free Ca^{2+} was very rapid, with a half maximal increase being observed after 3 s and a maximal increase after 7 s. Further studies showed that although the peak increase of the cytosolic free Ca^{2+} occurred at relatively low vasopressin concentrations, there was a concentration-dependent increase in the rate at which this peak Ca^{2+} concentration was attained as well as a decrease in the lag period before the cytosolic free Ca^{2+} started to increase after vasopressin addition (Thomas et al., 1984).

It is clear from a comparison of Figures 4A and 4B that the peak increase of cytosolic free Ca^{2+} occurred after relatively small changes of IP_3 and long before peak IP_3 levels were attained in the cell. From a comparison of a number of experiments with vasopressin concentrations varied over the range from 0.3 to 30 nM, it was found that the increase of IP_3 which coincided temporally with the peak increase of cytosolic free Ca^{2+} was approximately constant at 10-15% above the control IP_3 level with all concentrations of vasopressin above 1 nM (Thomas et al., 1984). Furthermore, the increase of IP_3 above basal values achieved at the time of the peak increase of the cytosolic free Ca^{2+} during vasopressin stimulation is calculated to be 1.2 ± 0.3 pmol/mg cell dry weight, or 0.6 μM. This value may be compared with a maximum increase of IP_3 with 20 nM vasopressin of about 10 μM (Thomas et al., 1984), but approximates the IP_3 concentration required to elicit a maximal Ca^{2+} release from permeabilized hepatocytes (c.f. Fig. 2). These data suggest that as observed with cAMP, most of the basal IP_3 in non-stimulated hepatocytes may be in a bound form, and that with maximal hormone stimulation it is produced and accumulated in far greater amounts than is required for expression of its maximum physiological response. In fact, the increase of IP_3 in hepatocytes after phenylephrine stimulation is barely detectable, despite an increase of the cytosolic free Ca^{2+} similar to that observed with vasopressin. The maximum amount of phenylephrine-induced Ca^{2+} release from hepatocytes incubated in Ca^{2+}-free medium was found to be about 200 pmol/mg cell dry weight (Joseph and Williamson, 1983), which would require an intracellular IP_3 concentration of only 0.1 μM.

As a further confirmation of the role of IP_3 as the intracellular Ca^{2+} mobilizing second messenger, vasopressin dose-response relationships between the rate of decrease of PIP_2, the rate of increase of IP_3 in intact hepatocytes and the rate of increase of cytosolic free Ca^{2+} are compared in Figure 5. The close correspondence between these parameters strongly suggests a cause and effect relationship.

Williamson et al.

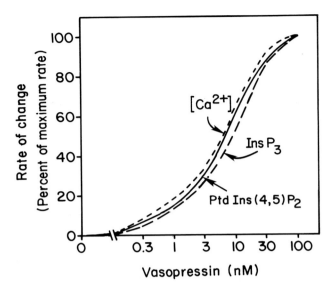

Figure 5. Comparison of vasopressin dose-response curves for the rates of change of PIP$_2$ (PtdIns4,5-P$_2$) breakdown, the rate of appearance of IP$_3$ (InsP$_3$) and the rate of increase of the cytosolic free Ca^{2+} ([Ca^{2+}]). For optimal comparison, each dose response curve has been normalized by plotting the changes as a percentage of the value obtained for the highest concentration of vasopressin used (100 nM). Absolute values for the changes in each parameter are presented by Thomas et al. (1984).

ACKNOWLEDGEMENTS

This work was supported by National Institutes of Health Grants AM-15120 and AA-05662.

REFERENCES

Berridge, M.J., Dawson, R.M.C., Downes, C.P., Heslop, J.P. and Irvine, R.F. (1983) Biochem. J. 212, 473-482.
Bers, D. (1982) Am. J. Physiol. 242, C404-C408.
Blackmore, P.F., Dehaye, J.P. and Exton, J.H. (1979) J. Biol. Chem. 254, 6945-6959.

Blackmore, P.F., Hughes, B.P., Shuman, E.A. and Exton, J.H. (1982) *J. Biol. Chem.* *257*, 190–197.

Burgess, G.M., McKinney, J.S., Fabiato, A., Leslie, B.A. and Putney, J.W., Jr. (1983) *J. Biol. Chem.* *258*, 15336–15345.

Charest, R., Blackmore, P.F., Berthon, B. and Exton, J.H. (1983) *J. Biol. Chem.* *258*, 8769–8773.

Creba, J.A., Downes, C.P., Hawkins, P.T., Brewster, G., Michell, R.H. and Kirk, C.J. (1983) *Biochem. J.* *212*, 733–747.

Exton, J.H. (1981) *Mol. Cell. Endocrinol.* *23*, 233–264.

Joseph, S.K. and Williamson, J.R. (1983) *J. Biol. Chem.* *258*, 10425–10432.

Joseph, S.K., Thomas, A.P., Williams, R.J., Irvine, R.F. and Williamson, J.R. (1984) *J. Biol. Chem.* *259*, 3077–3081.

Kimura, S., Kugai, N., Tada, R., Kojima, I., Abe, K. and Ogara, E. (1982) *Horm. Metab. Res.* *14*, 133–138.

Litosch, I., Lin, S.-H. and Fain, J.N. (1983) *J. Biol. Chem.* *258*, 13727–13732.

Murphy, E., Coll, K.E., Rich, T.L. and Williamson, J.R. (1980) *J. Biol. Chem.* *255*, 6600–6608.

Nicchitta, C.V. and Williamson, J.R. (1984) *Fed. Proc.* *43*, 1575.

Rhodes, D., Prpic, V., Exton, J.H. and Blackmore, P.F. (1983) *J. Biol. Chem.* *258*, 2770–2773.

Streb, H. and Schulz, I. (1983) *Am. J. Physiol.* *245*, 6347–6357.

Streb, H., Irvine, R.F., Berridge, M.J. and Schulz, T. (1983) *Nature* *306*, 67–69.

Thomas, A.P., Marks, J.S., Coll, K.E. and Williamson, J.R. (1983) *J. Biol. Chem.* *258*, 5716–5725.

Thomas, A.P., Alexander, J. and Williamson, J.R. (1984) *J. Biol. Chem.* *259*, 5574–5584.

Tsien, R.Y. (1980) *Biochemistry* *19*, 2396–2404.

Tsien, R.Y., Pozzan, T. and Rink, T.J. (1982) *J. Cell. Biol.* *94*, 325–334.

Williamson, J.R., Cooper, R.H. and Hoek, J.B. (1981) *Biochim. Biophys. Acta* *639*, 243–295.

DISCUSSION

Summarized by John N. Fain

Section of Physiological Chemistry, Brown University,
Box G, Providence, RI   02912

Dr. Michell suggested that the original view that phosphoinositide breakdown resulted in formation of diacylglycerol which migrated to the endoplasmic reticulum where it was converted to phosphatidic acid and phosphatidylinositol in equilibrium with the total intracellular pool may well be incorrect. Instead he postulated that the site of stimulated phosphatidic acid and phosphatidylinositol formation might be some compartment not in equilibrium with total phosphatidic acid and phosphatidylinositol. There is very little available information in most mammalian cells on the site of (or mechanisms responsible for) the marked increase in ^{32}P uptake into phosphatidic acid and phosphatidylinositol observed after the addition of hormones that increase phosphoinositide breakdown and elevate the cytosolic Ca^{2+}.

Dr. Putney and others suggested several years ago that phosphatidic acid might be the second messenger for Ca^{2+} movement. Dr. Williamson asked whether in view of the recent evidence for the role of IP_3 in the release of Ca^{2+} from the endoplasmic reticulum there is any role left for phosphatidic acid. Dr. Putney said he believes that phosphatidic acid might still play a role since, in many cells, stimuli result in a marked decrease in phosphoinositides and an increase in phosphatidic acid. Dr. Fain pointed out that in blowfly salivary glands there is no increase in phosphatidic acid formation after addition of 5-HT, yet the glands increase the entry of extracellular Ca^{2+}. Dr. Putney inquired as to the fate of diacylglycerol formed during phosphoinositide breakdown in blowfly salivary glands. Dr. Fain indicated that it transiently accumulates during hormone stimulation and then appears to be degraded to fatty acids and glycerol.

Dr. Shukla (Univ. Texas, San Antonio) asked whether there was any evidence for a direct phosphorylation of inositol 1-phosphate to give inositol bis- and tris-phosphates, since more IP_3 was formed than could be accounted for by the net decrease in PIP_2. Dr. Putney indicated that he thinks PIP and PI are converted to PIP_2 just prior to breakdown and that all breakdown of phosphoinositides might be secondary to conversion to PIP_2. Therefore, there should be no correspondence between net changes in PIP_2 and the formation of IP_3.

Dr. Dibner (Dupont Glenolden Lab.) asked whether there is any evidence that IP_3 might have a feedback effect on phosphoinositide breakdown. Dr. Putney responded that permeabilized cells respond to IP_3 addition with a release of Ca^{2+} but it is not possible to obtain effects of hormones on phosphoinositide breakdown in these permeabilized cells.

Dr. Michell pointed out that the time courses for IP_2 and IP_3 formation shown by Dr. Berridge were remarkably parallel and asked whether there was any evidence as to whether PIP and PIP_2 are both degraded concurrently by the same enzyme. Dr. Berridge suggested that it is totally irrelevant whether PIP and PIP_2 are being degraded concurrently. There is no evidence according to Dr. Berridge that IP_2 or IP release Ca^{2+} from intracellular organelles, so there is no known function at the moment for IP_2. However, the breakdown of PI, PIP and PIP_2 will all result in formation of diacylglycerol. Dr. Hawthorne suggested that it does not make a lot of sense to have a hormone trigger the formation of both an active and inactive second messenger (IP_3 and IP_2).

Dr. Berridge believes that the kinases and phosphatases involved in the interconversions of PI to PIP and PIP_2 provide multiple sites for regulation. The energy cost for conversion of PI to PIP_2 is supposedly counterbalanced by the advantage to the cell of being able, at any moment, to regulate PIP_2 formation.

The finding that IP_3 is sequentially degraded was cited by Dr. Berridge as further proof that IP_3 is a regulatory compound of great significance. Dr. Berridge pointed out that large effects of hormones on IP_3 accumulation are only seen with heroic concentrations of hormones. All the receptors are activated under these circumstances and there is an impressive increase in IP_3. However, with physiological concentrations of hormones the increases in IP_3 are not particularly impressive. These increases in IP_3 may be more than enough for the cell but not enough for the investigator to measure readily. Furthermore, the diacylglycerol and IP_3 may work together in a synergistic fashion since both messengers are formed from phosphoinositide breakdown and this may be especially significant at low hormone concentrations.

Dr. Fain noted that all studies of IP_3 accumulation in blowfly salivary glands were performed in the absence of LiCl which inhibits breakdown of IP_3 to inositol. Most other systems require the addition of LiCl to cells to allow detection of IP_3. Dr. Berridge replied that he did not use LiCl in his studies because he wanted to study the system in its natural state.

Dr. Fain pointed out that arachidonic acid does not appear to have any effect on blowfly salivary glands. Dr. Berridge agreed with this since Dr. Bridges at Cambridge has been able to grow houseflies on a synthetic arachidonic acid free diet (they cannot synthesize this fatty acid) and they appear to function normally. However, houseflies cannot reproduce despite having a normal phosphoinositide breakdown in response to 5-HT.

Dr. Perlman (Univ. Illinois) asked about the affinity of IP_3 for Ca^{2+}. There does not appear to be much information available on this point. There was a theory several years ago that PIP_2, which has a high affinity for Ca^{2+}, during its breakdown released Ca^{2+} which was bound to PIP_2 on the inner surface of the plasma membrane. However, there must be an amplification factor of at least 10 according to Dr. Berridge to account for the rise in cytosolic Ca^{2+}. It has been calculated that each mole of IP_3 released is responsible for an elevation of 10 moles of Ca^{2+}. The biggest objection to the theory that PIP_2 breakdown releases Ca^{2+} is the finding that PIP_2 has an equally high affinity for Mg^{2+} and the intracellular free Mg^{2+} is generally so much higher than free Ca^{2+} that Mg^{2+} should be preferentially bound to PIP_2.

Over the past 10 years, cyclic inositol phosphate and then phosphatidic acid were postulated to act as Ca^{2+} ionophores. These theories were supplanted by the idea that breakdown of PIP_2 releases sufficient Ca^{2+} to elevate cytosolic Ca^{2+}. Currently the remarkable effects of IP_3 on the release of Ca^{2+} from permeabilized cells have focused all attention on the role of this compound as the second messenger for regulation of release of bound Ca^{2+} stores to the cytosol. Dr. Berridge proposed that IP_3 might interact with receptors and this results in release of Ca^{2+} rather than IP_3 acting as a Ca^{2+} carrier.

The question still remains of whether hormones increase phosphoinositide breakdown by activating an enzyme or by exposing PIP_2 for degradation by cytosolic enzymes. It is as yet unclear whether breakdown of PIP and PIP_2 occur concurrently in fly salivary gland or whether all PI breakdown is through PIP and PIP_2.

Following Dr. Rubin's presentation, Dr. Fain raised the question of whether arachidonic acid release from phosphoinositides occurred through a phospholipase A_2 mechanism under physiological conditions. Most recent studies suggest that breakdown of phosphoinositides occurs via a phosphodiesterase to yield diacylglycerol followed by breakdown of the diacylglycerol by diacylglycerol lipase and monoacylglycerol lipase to release arachidonic acid. Dr. Rubin suggested that there are multiple modalities for regulation of phospholipid breakdown. Dr. Irvine pointed out that there is no phosphoinositide-specific phospholipase A_2. The preferred substrate for the intestinal phospholipase A_2 is phosphatidylethanolamine. Dr. Irvine further suggested that a potent stimulator of phospholipase A_2 activity on phosphatidylcholine is diacylglycerol. Therefore, diacylglycerol could be a link between the two enzymatic events.

Dr. Johnston (Univ. Texas, Dallas) asked Dr. Nishizuka whether he had any information on the function of the 40K protein which is phosphorylated in platelets by protein kinase C. The answer was that we have no idea of the function of the 40K protein.

Dr. Silvia Corvera Behar (UNAM, Mexico) asked whether the phorbol ester effects were readily reversed after the addition of calcium chelators. Dr. Nishizuka replied that the binding of phorbol ester is not readily reversed since it is very tight. Dr. Nishizuka also pointed out that protein kinase C can be activated after cleavage by a calcium-activated protease which requires only 1 μM Ca^{2+} for activity. Whether this protease has any physiological function remains to be proven.

Dr. Michell raised the question of why the cytosolic protein kinase C is not activated by diacylglycerol produced in the endoplasmic reticulum as a normal intermediate in many biosynthetic reactions and as an intermediate during triacylglycerol degradation.

After Dr. Gispen's talk Dr. Cuatrecasas (Wellcome Research Laboratories) asked whether the 50K protein of Dr. Gispen is the 50K postsynaptic protein of brain or the catalytic subunit of the Ca^{2+}-calmodulin protein kinase. The answer was that it is neither protein. Dr. Gispen further stated that the protein he discussed is localized exclusively to the pre-synaptic plasma membrane.

PART V

PHOSPHOINOSITIDES AND ARACHIDONIC ACID MOBILIZATION

MECHANISMS FOR EICOSANOID PRECURSOR UPTAKE

AND RELEASE BY A TISSUE CULTURE CELL LINE

P.W. Majerus, E.J. Neufeld and M. Laposata

Division of Hematology-Oncology, Departments of Medicine and Biological Chemistry, Washington University School of Medicine, St. Louis, MO 63110

SUMMARY

Mouse fibrosarcoma cells ($HSDM_1C_1$) that synthesize prostaglandin E_2 (PGE_2) in response to bradykinin and calcium ionophore A23187 were adapted to grow in lipid-free medium to produce arachidonate deficiency. Such cells produce PGE_2 in response to agonists only after repletion with arachidonate. Repleted cells release arachidonate from phosphatidylinositol (80%) and phosphatidylcholine (20%). Release from A23187-treated cells is qualitatively different, indicating that alternative mechanisms are involved in activation of cells by this non-physiological agonist. A variant cell line deficient in arachidonoyl-CoA synthetase was isolated after mutagenesis and radiation suicide selection. This cell line is not only deficient in arachidonate uptake but also in release of arachidonate in response to agonists. These results indicate that arachidonoyl-CoA synthetase is required for eicosanoid homeostasis. This enzyme metabolizes only those fatty acids that are potential eicosanoid precursors and accounts for the rapid and high affinity uptake of these fatty acids into phosphatidylinositol and phosphatidylcholine. Fatty acids are most susceptible to release by agonists immediately after uptake into the cell. Subsequently, they redistribute into metabolically inert phospholipid pools.

INTRODUCTION

Arachidonate is converted to a variety of oxygenated mediators including prostaglandins, thromboxanes, prostacyclins, leukotrienes, and other hydroxylated forms of arachidonic acid. The capacity to produce these substances depends on cellular mechanisms for uptake and release of arachidonate from phospholipids. The initial enzymes that oxygenate arachidonate are constitutively active within cells and therefore the production of products is controlled by the availability of arachidonate. Arachidonic acid is taken up rapidly from medium and efficiently esterified into phospholipids, so that most cells have free fatty acid present only in trace quantities. Stimulation of a cell by an appropriate agonist for which it has receptors leads to the rapid liberation of arachidonate from phospholipids.

It has been difficult to define the mechanisms for arachidonate release from stimulated cells because only a small fraction of phospholipid arachidonate is released when cells are stimulated (Majerus *et al.*, 1983). In the case of platelets, approximately 10% of the cell's total is released. However, in most other cells only 1% to 3% of cellular arachidonate is liberated in response to physiological agonists. Thus it is difficult to determine which cellular phospholipids liberate arachidonate upon stimulation by measuring arachidonate before and after stimulation; one must subtract two large numbers from each other to get a small difference. Upon stimulation of platelets with thrombin, phosphatidylinositol loses 30% to 60% of its arachidonate and phosphatidylcholine appears to lose from 0% to 10%. These two phospholipids account for essentially all of the arachidonate liberated from platelets (Neufeld and Majerus, 1983). Phosphatidylethanolamine and phosphatidylserine do not contribute significantly to the process. In the case of cultured cells, it has been much more difficult to elucidate the mechanisms and pathways of arachidonate release.

PHOSPHATIDYLINOSITOL IS THE MAIN SOURCE OF ARACHIDONATE RELEASED FROM AN ESSENTIAL FATTY ACID DEFICIENT CELL LINE AFTER BRADYKININ STIMULATION

In an attempt to produce a better experimental system, we have developed a tissue culture cell line with essential fatty acid deficiency (Laposata, 1982). A variety of cultured cell lines grow in lipid-free medium, but in most cases they do not metabolize arachidonate nor do they synthesize 5,8,11-eicosatrienoic acid, the fatty acid diagnostic of essential fatty acid deficiency. Therefore, we set out to induce essential fatty acid deficiency in a cell line known to metabolize arachidonate. We chose $HSDM_1C_1$ murine fibrosarcoma cells because they produce large amounts of PGE_2 in response to bradykinin (Levine et al., 1972). We adapted these cells to grow in lipid-free medium, and in this way it was possible to deplete the cells of arachidonate completely. Such cells appear to grow normally but no longer produce PGE_2 in response to bradykinin. When cells are repleted by addition of either linoleic acid (the precursor of arachidonate) or arachidonic acid itself, the capacity to produce PGE_2 is restored. In these experiments, we observed that during the initial time course of uptake of arachidonate into depleted cells, essentially all of the arachidonic acid was incorporated into two phospholipids: phosphatidylinositol and phosphatidylcholine. At times of incubation up to 20 minutes, less than 10% of the label was found in other phospholipids, such as phosphatidylethanolamine and phosphatidylserine, or in neutral lipids. In this way, it was possible to specifically label two cellular phospholipids. During this period of limited arachidonate distribution, the capacity to produce PGE_2 in response to bradykinin was restored to approximately 50% of the maximal level. Thus, charging cellular phospholipid pools of only phosphatidylcholine and phosphatidylinositol was sufficient to substantially restore the capacity to produce PGE_2.

Additionally, we observed that the fraction of radio-labeled arachidonate that is liberated upon stimulation of these cells decreased with increasing time of repletion. When control $HSDM_1C_1$ cells (not lipid depleted) are treated with a maximal concentration of bradykinin,

445

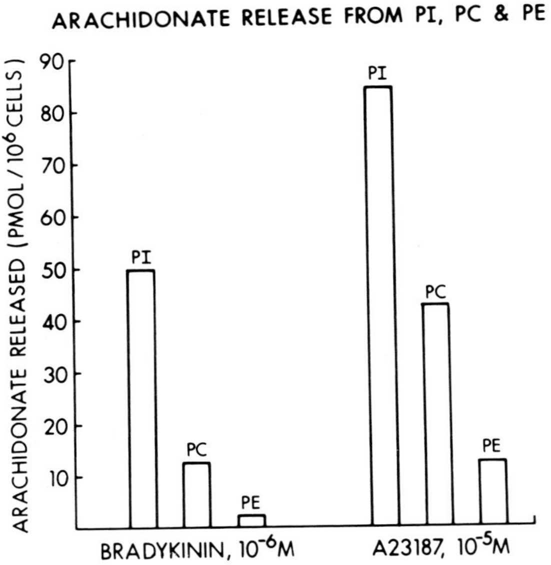

ARACHIDONATE RELEASE FROM PI, PC & PE

Figure 1. Release of arachidonate from phospholipids of HSDM$_1$C$_1$ cells in response to bradykinin and ionophore. Essential fatty acid deficient cells in 100 mm petri dishes were repleted for 15 min with 1 µM [^{14}C]-arachidonate. After several rinses to remove excess isotope, buffer, bradykinin (10^{-6} M), or A23187 (10^{-5} M) was added for 15 min at 37°C. Medium was removed, the cells were precipitated on the plate with cold TCA and the precipitate collected for lipid extraction. Tracer amounts of [^{3}H]-labeled PI, PC, and PE were added to the extracts for recovery determination and the phospholipids separated by HPLC (Prescott and Majerus, 1981). Arachidonate released from PI, PC, and PE was determined by subtraction of phospholipid cpm in agonist-treated plates from phospholipid cpm in buffer-treated plates. Data represent the mean of two experiments.

approximately 3 to 8% of cellular arachidonate is liberated. In an experiment with arachidonate depleted cells where repletion with [^{14}C]arachidonate was carried out for 15 min, we observed that bradykinin-induced release was 13.7 ± 2.1% (N=8). After 30 min of repletion, there was a significantly smaller fraction of arachidonate liberated, 11.5 ± 1.5% (N=8) (P<0.02). Since the cells contain no unlabeled arachidonate, incubation with [^{14}C]- or [^{3}H]-labeled arachidonic acid resulted in incorporation and subsequent liberation at exactly the same specific activity as that added. Therefore, measurement of radioactivity is directly equated to mass. It appears that as arachidonate is taken up and esterified it is placed in a cellular compartment where it is susceptible to liberation by agonists. With longer times of incubation, redistribution of fatty acids to metabolically inert pools appears to occur.

When cells were incubated with [^{14}C]arachidonate for 15 min, then washed and stimulated with bradykinin, the pattern of release from phospholipids shown in Figure 1 was obtained. As noted, approximately 80% of the arachidonate liberated comes from phosphatidylinositol. In this experiment, 35% of the total arachidonate incorporated into phosphatidylinositol was liberated as compared to 10% of that in phosphatidylcholine. An insignificant amount of arachidonate was released from phosphatidylethanolamine. Thus, it appears that under the influence of the agonist bradykinin, arachidonate release is largely from phosphatidylinositol. This cell line has been shown previously to contain the diacylglycerol lipase pathway for arachidonate release from phosphatidylinositol which involves three enzymes: phospholipase C, diacylglycerol lipase, and monoacylglycerol lipase (Bell, 1980). The liberation of arachidonate from phosphatidylcholine is presumed to proceed by a phospholipase A$_2$ (McKean et al., 1981), although no enzyme with specificity for eicosanoid precursor fatty acids has been found in cell extracts. That substantial release of arachidonate occurs from phosphatidylethanolamine after stimulation with A23187 indicated that this non-physiological agonist cannot simply be considered a mimic of physiological mechanisms. It appears that this agent produced more drastic changes in cellular phospholipid metabolism although even in this case phosphatidylinositol remains the major source of arachidonate.

PRODUCTION OF MUTANT CELL LINE LACKING ARACHIDONOYL-CoA SYNTHETASE

We have previously shown 'in studies with platelets that the initial uptake of eicosanoid precursor fatty acids

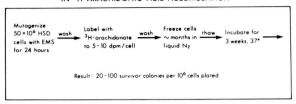

SUICIDE SELECTION FOR CELLS DEFECTIVE
IN ³H-ARACHIDONIC ACID ACCUMULATION

<u>Figure 2</u>. Method for selection of mutant cells defective in arachidonoyl-CoA synthetase. EMS: ethyl methanesulfonate.

depends on a recently described acyl-CoA synthetase that is relatively specific for arachidonic acid as opposed to other long-chain fatty acids (Wilson, *et al.*, 1982; Neufeld, *et al.*, 1983). In an effort to determine whether the arachidonoyl-CoA synthetase is involved in arachidonate homeostasis and arachidonate release, we produced mutants in this enzyme by the technique of radiation suicide (Neufeld *et al.*, 1984a). The strategy for this experiment is outlined in Figure 2. Mutagenesis was carried out using ethyl methanesulfonate followed by labeling of the cells with highly radioactive, tritiated arachidonic acid to an activity of 5 to 10 dpm/cell, allowing selection for defective arachidonate uptake by radiation suicide. This technique has been used previously in mammalian cells to isolate mutants in glycolysis utilizing tritiated 2-deoxyglucose (Pouyssegur *et al.*, 1980). After cells were incubated with 75 nM tritiated arachidonate for 2 h, they were washed and subsequently frozen in dimethyl sulfoxide. The cells were stored in liquid nitrogen for 2 to 4 months and then thawed. The cells that had been labeled with tritiated arachidonate appeared normal for 2 days in culture, but then began to show signs of radiation damage, such as multi-nucleated giant cells and cell death. After 2 to 3 weeks, most of the labeled cells had died. Only 100 colonies survived per million cells plated. In order to screen survivors of the selection for defective arachidonate uptake, we used the polyester filter replicate plating technique described by Raetz *et al.* (1982). Replicas from dishes containing up to 300 colonies were assayed autoradiographically for incorporation of tritiated arachidonate into cellular phospholipids. In this assay, polyester replica filters are incubated with radioactive arachidonate and then washed and stained with Coomassie Blue. After autoradiography, the films are compared to the stained colonies. Unlabeled colonies then are collected and cloned and grown up for assay for arachidonate uptake.

Utilizing this technique, we have isolated 7 mutant cell lines that we have designated EPU (eicosanoid precursor uptake). These cells occur at a frequency of about 1% of survivors of the suicide selection. No EPU mutant colonies have been detected in cell populations that underwent suicide selection but no mutagenesis, nor were any detected in samples treated with mutagen and frozen but never exposed to tritiated arachidonate selection. One mutant clone, designated EPU-1, has been studied extensively. EPU-1 cells take up arachidonate from lipid-free medium at approximately one-third the rate of normal cells. The defect is specific to arachidonate in that oleate uptake is the same in mutant and parental cell lines. The lesion in arachidonate accumulation was also apparent in control, i.e., lipid-containing medium. Under these circumstances, arachidonate uptake was approximately half as fast as the parental rate, whereas the uptake of several other fatty acids was normal or even enhanced compared to the parent.

Microsomes prepared from EPU-1 cells were assayed for arachidonoyl-CoA synthetase and the non-specific long-chain acyl-CoA synthetase. EPU-1 had less than 10% of the control level of arachidonoyl-CoA synthetase but had normal amounts of the non-specific enzyme. Thus, the defect in arachidonate uptake could be ascribed to deficiency of arachidonoyl-CoA synthetase. A mixing experiment demonstrated that the missing activity was due to lack of a functional enzyme and not to either the presence of an inhibitor in the mutant or to the presence of an acyl-CoA hydrolase. It is difficult to be certain whether the mutant completely lacks the arachidonate-specific enzyme or whether it contains trace amounts. Our assay for arachidonoyl-CoA synthetase measures radioactive arachidonoyl-CoA formation in the presence of unlabeled, excess oleate to inhibit the non-specific enzyme. Arachidonate is also a substrate for the latter enzyme, so that at high concentrations of arach-

449

idonate, a small amount of arachidonoyl-CoA is formed by competition with unlabeled oleate for the non-specific synthetase.

ARACHIDONOYL-CoA SYNTHETASE IS NECESSARY FOR A NORMAL EICOSANOID PRECURSOR HOMEOSTASIS IN HSDM$_1$C$_1$ CELLS

HSDM$_1$C$_1$ cells are able to convert linoleate to arachidonate (Laposata, *et al.*, 1982), and EPU mutant cells also have this ability. When cells were grown with [^{14}C]linoleate for 24 h, both parental and mutant cells converted 4.5% of the added linoleate to arachidonate. It was therefore conceivable that the mutant cells would have completely normal endogenous arachidonate levels despite their inability to take up arachidonate from the extracellular environment. To examine this possibility, we extracted and analyzed the lipids of EPU cells grown in control medium. The fatty acid compositions of mutant and control cells were quite different and indicated that the EPU-1 cells were unable to maintain arachidonate homeostasis. The EPU mutant was strikingly deficient in arachidonate in phosphatidylinositol, phosphatidylcholine, and phosphatidylethanolamine. In three experiments, arachidonate ranged from 7% to 33% of the normal levels in phosphatidylcholine, 50% to 70% in phosphatidylinositol, and 20% to 33% in phosphatidylethanolamine. To compensate for this deficiency, the mutant cells contained high levels of oleate and linoleate in these lipids. Phosphatidylinositol in the mutant also contained 5,8,11-eicosatrienoic acid. This fatty acid is produced by desaturation and elongation of oleate and is characteristic of cells deficient in essential fatty acids. It is not found in normal cells maintained in medium containing linoleate. Thus, the mutant cells, despite an adequate supply of precursors, cannot maintain normal arachidonate levels, demonstrating that arachidonoyl-CoA synthetase is required for normal polyunsaturated fatty acid metabolism.

In several experiments, we demonstrated that the level of arachidonate in phosphatidylcholine was consistently lower than would be predicted on the basis of the overall cellular deficiency in arachidonate. We therefore examined the rate of arachidonate and linoleate turnover in the major lipid pools of mutant and control cells. By pulse chase experiments, we showed that arachidonate turnover in phosphatidylcholine is greatly enhanced in the mutant compared to the control cells, while linoleoyl phosphatidylcholine turnover in the two cell types is identical. The half life for linoleate turnover was approximately 24 h in both cell lines; the half life for arachidonate was 8 h in the parent and 1 h in the mutant cells. In lipid species other than phosphatidylcholine, the turnover of arachidonate was similar in control and mutant cells.

STUDIES OF AGONIST-INDUCED ARACHIDONATE RELEASE

We used control and mutant cells that were completely depleted of essential fatty acids for studies of arachidonate release. We then repleted them with either radiolabeled linoleate or arachidonate. Agonist-induced arachidonate release was markedly deranged in the mutant cells. The results of one such experiment are shown in Figure 3. In this case, cells were grown in $[^{14}C]$-linoleate to allow desaturation and conversion to arachidonate. Upon subsequent challenge with bradykinin, the mutant cells produced less than 10% of the normal amount of PGE_2 in response to bradykinin, despite the fact that cyclooxygenase is normal in these cells as evidenced by equivalent conversion of exogenously added arachidonate to PGE_2 by EPU and parental cells (Fig. 3).

This defect was also demonstrated by stimulation of cells labeled with arachidonate. The time course and magnitude of arachidonate released from briefly repleted cells was markedly different comparing the mutant and parental cell lines. A 10-fold reduction in arachidonate release was observed 1 min after bradykinin stimulation, and after 20 min, the reduction in release by the mutant was approximately half that of the parental line. When the briefly repleted cells were stimulated with the calcium ionophore A23187, we found surprisingly that EPU-1 released a larger proportion of total cellular arachidonate than the parental cell line. However, production of eicosanoids were again deranged since a smaller-than-normal fraction of the liberated arachidonate was actually converted to PGE_2. In the case of cells labeled for 1 h with arachidonate, the parent converted 50% to PGE_2 as compared to 7% in the mutant. When cells were labeled to

451

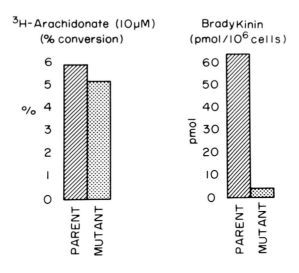

<u>Figure 3</u>. PGE$_2$ synthesis is diminished in EPU-1 cells. Essential fatty acid-deficient mutant and control cells were labeled to steady state (48 h) with [^{14}C]linoleate (5 μM, 5 x 10^4 cpm/nmol) in 24-well plates. The cells were washed with buffer, then incubated for 5 min at 37°C with buffer, 10^{-6} M bradykinin or 10 μM [^{14}C]arachidonate. The medium was acidified with formic acid and extracted with ethyl ether. The extract was analyzed for PGE$_2$ production by C18 reverse phase HPLC (Neufeld <u>et al</u>, 1984a). Each value is the mean of triplicate wells. In this experiment, EPU-1 cells contained half as much arachidonate as control cells.

a steady state, the fraction of arachidonate liberated decreased; the mutant liberated approximately twice as much arachidonate as the parental line; and the parent converted 2/3 of released arachidonate into PGE$_2$, whereas the mutant converted only 1/3.

These data suggest that the defect in the EPU mutant cell prevents it from filling its agonist sensitive arachidonate pools properly. In bradykinin stimulated cells, release is very low, whereas in ionophore-treated cells release is normal or increased but not properly directed to the cyclooxygenase. The agonist-sensitive pool could be a specific phospholipid species, such as phosphatidylinositol, or it could be a unique cellular compartment. The distribution of arachidonate in various lipid species suggests that phosphatidylinositol, by the magnitude of its labeling and release, must be involved in this deranged process in some manner. We are presently investigating the possibility that physical compartments containing arachidonic acid are different in parent and mutant. These experiments utilize electron microscopy and autoradiography to measure the time course of arachidonate uptake into specific membranes in control and mutant cells. The experiments presented above suggest that the arachidonoyl-CoA synthetase is involved in the specific arachidonate uptake, esterification, and liberation process associated with eicosanoid production. The inability of EPU cells to maintain normal arachidonate levels, despite their ability to desaturate linoleate, suggests that arachidonoyl-CoA synthetase is important to the process by which cells retain polyunsaturated fatty acid in phospholipids. This enzyme is probably uniquely important in cells such as platelets and endothelial cells which lack desaturase enzymes and are, therefore, required to take up arachidonate from plasma to allow for eicosanoid production.

It is possible that EPU mutant cells are pleiotropic and have other enzymatic defects. However, at least three systems known to be part of the prostaglandin synthesis machinery are normal. These are cyclooxygenase, the PI-specific phospholipase C, and diacylglycerol lipase. In addition, the bradykinin receptor coupling system is normal by inference since some release is accomplished by this mechanism over prolonged time periods. Furthermore, the release of arachidonate in response to the ionophore A23187 is abnormal in mutant cells indicating that the defect is not limited to bradykinin. The fact that faulty arachidonoyl CoA synthetase results in defective prostaglandin synthesis suggests that inhibition of this enzyme with pharmacologic agents might be useful therapeutically in diseases mediated by eicosanoids. If the diminished prostaglandin synthesis of the EPU mutants is due to the lack of an agonist-sensitive pool of arachidonate, it will be important to determine the nature of this pool.

453

FATTY ACID STRUCTURE REQUIREMENTS FOR THE
ACTIVITY OF ARACHIDONOYL-CoA SYNTHETASE

Since arachidonoyl-CoA synthetase appears to be an important component in eicosanoid metabolism, we have carefully examined the fatty acid substrate specificity of this enzyme using enzyme from platelet membranes and from bovine brain. A large variety of positional isomers and chain length analogues of arachidonic acid were synthesized, labeled with carbon-14 in the 1-position (Neufeld et al., 1984b). These fatty acids were then used in experiments in which their ability to act as substrates or their ability to competitively inhibit the metabolism of arachidonate was studied. In this way, we have compared the fatty acid specificity of arachidonoyl-CoA synthetase to cyclooxygenase and the 5-lipoxygenase which is required for leukotriene production. The results of these comparisons summarized in Table I. The chain length specificity for arachidonoyl CoA synthetase, comparing a series of 8,11,14-trienoic acids was C19 > C18 = C20 >> C21 > C22. Inhibition of activity by positional isomers of arachidonate indicated that arachidonate itself was the most potent inhibitor. Nearly as inhibitory was 6,9,12,15-eicosatetraenoic acid. 7,10,13,16-Eicosatetraenoic acid was much more inhibitory than 4,7,10,13-eicosatetraenoic acid. Although 6,9,12,15-eicosatetraenoic acid was a potent inhibitor, it was a poor substrate with less than 1/3 the V_{max} of arachidonate. It appears that the enzyme counts double bonds from the carboxyl terminus. Thus, as counted from the methyl terminus, we found that several n-6,9,12 fatty acids were ineffective as inhibitors, whereas all methylene interrupted tri- and tetraenoic fatty acids which contain $\Delta 8$ and $\Delta 11$ double bonds are potent inhibitors. The $\Delta 11$ double bond is best associated with activity: the novel fatty acid 5,11,14-eicosatrienoic acid was a moderately potent inhibitor, whereas 5,8,14-eicosatrienoic acid was ineffective. Most interestingly, 13-methyl, 8,11,14-eicosatrienoic acid did not inhibit the enzyme. This fatty acid does not inhibit cyclooxygenase since the 13-methyl hydrogen is the hydrogen abstracted by this first enzyme in prostaglandin and thromboxane synthesis. Therefore arachidonoyl-CoA synthetase has a specificity that, though not identical to cyclooxygenase, is very similar. It is apparent that arachidonoyl-CoA synthetase is capable of

454

TABLE I

Comparison of Substrate Requirements for Cyclooxygenase, 5-Lipoxygenase and Arachidonoyl-CoA Synthase

Substrate Requirements	Eicosanoid Precursor Metabolizing Enzymes		
	Cyclooxygenase	5-Lipoxygenase	Arachidonoyl-CoA Synthase
20 carbon FA:			
Dienes	No	No	No
Δ5,8,11	No	Yes	Yes
Δ8,11,14	Yes	No	Yes
Δ11 important	Yes	Yes	Yes
13 methyl Δ8,11,14	No	---	No
Counts double bonds from	$-CH_3$	HOOC-	HOOC-

(From Beerthuis et al., 1971; Beerthuis et al., 1978; Jakschik et al., 1980; Lands, 1979; Struijk et al., 1966).

specifically esterifying all the substrates that might be utilized by either the lipoxygenase or cyclooxygenase enzymes. Substrates lacking Δ11 double bonds or dienoic 20-carbon fatty acids are not metabolized by this enzyme or eicosanoid producing enzymes (Table I). Thus, it appears that a major function of the arachidonoyl-CoA synthetase is to specifically charge cellular pools that are important for the production of eicosanoid mediators. Thus far, however, there is no evidence that this enzyme uniquely esterifies arachidonate that is destined to be incorporated into phosphatidylinositol as opposed phosphatidylcholine. Elucidation of the mechanisms by which eicosanoid precursor fatty acids are specifically incorporated into phosphatidylinositol may provide the missing link in understanding eicosanoid production.

ACKNOWLEDGEMENTS

The authors thank Teresa Bross for technical assistance and David B. Wilson and Stephen M. Prescott for helpful suggestions.

This research supported by Grants HLBI 14147 (Specialized Center for Research in Thrombosis) and HL 16634 from the National Institutes of Health, and in part by National Institutes of Health Research Service Award, GM 07200, Medical Scientist, from the National Institute of General Medical Sciences.

REFERENCES

Beerthuis, R.K., Nugteren, D.H., Pabon, H.J.J. and van Dorp, D.A. (1968) *Rec. Trav. Chim. Pays. Bas.* *87*, 461-480.
Beerthuis, R.K., Nugteren, D.H., Pabon, H.J.J., Steenhoek, A. and van Dorp, D.A. (1971) *Rec. Trav. Chim. Pays. Bas.* *90*, 943-960.
Bell, R.L., Baenziger, N.L. and Majerus, P.W. (1980) *Prostaglandins* *20*, 269-274.
Jakschik, B.A., Sams, A.R., Sprecher, H. and Needleman, P. (1980) *Prostaglandins* *20*, 401-410.
Lands, W.E.M. (1979) in: *"Positional and Geometric Fatty Acid Isomers"* (E.A. Emken and H.J. Dutton, eds.) American Oil Chemists, Champaign, IL.

Laposata, M., Prescott, S.M., Bross, T.E. and Majerus, P.W. (1982) *Proc. Natl. Acad. Sci. USA* *79*, 7654–7658.

Levine, L., Hinkle, P.M., Voelkel, E.F. and Tashjian, A.J., Jr. (1972) *Biochem. Biophys. Res. Commun.* *47*, 888–896.

Majerus, P.W., Prescott, S.M., Hofmann, S.L., Neufeld, E.J. and Wilson, D.B. (1983) *Adv. Pros. Thromb. Leuk. Res.* *11*, 45–52.

McKean, M.L., Smith, J.B. and Silver, M.J. (1981) *J. Biol. Chem.* *256*, 1522–1524.

Neufeld, E.J. and Majerus, P.W. (1983) *J. Biol. Chem.* *258*, 2461–2467.

Neufeld, E.J., Wilson, D.B., Sprecher, H. and Majerus, P.W. (1983) *J. Clin. Invest.* *72*, 214–220.

Neufeld, E.J., Bross, T.E. and Majerus, P.W. (1984a) *J. Biol. Chem.* *259*, 1986–1992.

Neufeld, E.J., Sprecher, H., Evans, R.W. and Majerus, P.W. (1984b) *J. Lipid Res.* In press.

Pouyssegur, J., Franchi, A., Salomon, J.C. and Silvestre, P. (1980) *Proc. Natl. Acad. Sci. USA* *77*, 2698–2701.

Prescott, S.M. and Majerus, P.W. (1981) *J. Biol. Chem.* *256*, 579–582.

Raetz, C.R.H., Wermuth, M.M., McIntyre, T.M., Esko, J.D. and Wing, D.C. (1982) *Proc. Natl. Acad. Sci. USA* *79*, 3223–3227.

Struijk, C.B., Beerthuis, R.K., Pabon, H.J.J. and van Dorp, D.A. (1966) *Rec. Trav. Chim. Pays. Bas.* *84*, 1233–1250.

Wilson, D.B., Prescott, S.M. and Majerus, P.W. (1982) *J. Biol. Chem.* *257*, 3510–3515.

THE MOBILIZATION OF ARACHIDONATE AND

METABOLISM OF PHOSPHOINOSITIDES IN

STIMULATED HUMAN PLATELETS

Susan E. Rittenhouse

Brigham and Women's Hospital
Harvard Medical School, Boston MA   02115

SUMMARY

A variety of agonists, when added to human platelets, induce phosphoinositide turnover mediated by phospholipase C. Among such agonists are thrombin, collagen, the Ca^{2+} ionophores A23187 and ionomycin, prostaglandin H_2/thromboxane A_2 (PGH_2/TXA_2) and a stable analogue, U46619. Secretion of stored substances, including ADP, accompanies phosphoinositide turnover. Stimulation of platelets by thrombin, collagen, and Ca^{2+} ionophores also results in the liberation of arachidonic acid, which is converted by platelet cyclooxygenase to PGH_2/TXA_2. However, PGH_2/TXA_2 and U46619 are themselves poor inducers of arachidonic acid release.

Thrombin stimulates a phospholipase C-mediated transient accumulation of diacylglycerol which is not inhibited by either blocking PGH_2/TXA_2 formation or removing ADP. Collagen-induced formation of diacylglycerol is partially dependent upon TXA_2/PGH_2 and ADP, whereas activation of phospholipase C by ionophores is completely dependent upon these species. In contrast, making PGH_2/TXA_2 and ADP unavailable does not impair the elevation of cytoplasmic Ca^{2+} (monitored by Quin 2) and activation of phospholipase A_2 caused by ionophores. Thus, Ca^{2+} ionophores stimulate phospholipase A_2, but activate phospholipase C only indirectly, via PGH_2/TXA_2 and ADP effects. We conclude that phospholipase C is not activated by a Ca^{2+} flux in the platelet, and suggest that stimulation of phospholipase C is contingent upon a receptor-coupled event.

INTRODUCTION

Much attention has been focused recently on the routes by which the stimulation of phosphoinositide-specific phospholipase C (PLC) in platelets may lead to the mobilization of arachidonic acid from phospholipid stores. Arachidonic acid is an interesting fatty acid because it is the major substrate for cyclooxygenase and lipoxygenase enzymes, the products of which exert a wide array of biological effects. It has been proposed that arachidonic acid is liberated in stimulated platelets primarily as a function of coupled diacylglycerol (DG) and monoacylglycerol lipases which act once DG is formed via PLC (Bell *et al.*, 1979; Prescott and Majerus, 1983). A contradictory proposal (Lapetina *et al.*, 1981; Billah and Lapetina, 1982a) is that phosphatidylinositol serves as a source of arachidonic acid only through phospholipase A_2 (PLA_2). We have suggested that PLC and lipases account for the initial release of arachidonic acid, but that mobilized Ca^{2+} resulting from these changes activates PLA_2, which liberates more arachidonic acid-rich pools of several phospholipids (Rittenhouse-Simmons, *et al.*, 1981; Rittenhouse, 1984a). Others consider the activation of PLC and PLA_2 to be separate, parallel events (Broekman, *et al.*, 1980; McKean, *et al.*, 1981).

It is not our intention to contribute to this controversy in the present discussion. Rather, we will address the question of how the metabolites of arachidonic acid, secretion products, and Ca^{2+} affect the turnover of phosphoinositides which is catalyzed by PLC. In the platelet, arachidonic acid is converted by cyclooxygenase to PGH_2, which in turn is transformed by thromboxane synthetase to TXA_2, a labile compound which has been shown to cause platelet aggregation and secretion (Hamberg *et al.*, 1975). Normal platelets also contain a population of storage granules, called "dense granules", which are reservoirs for a non-metabolic pool of ADP as well as Ca^{2+} and 5-hydroxytryptamine (5HT), amongst other things. Secretion of these substances is quite rapid in response to platelet stimuli, closely following

460

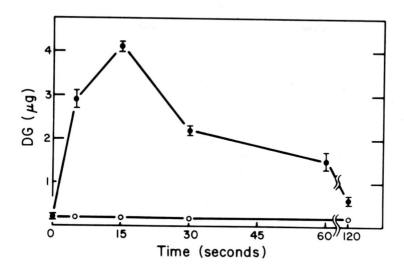

Figure 1. Generation of DG in response to thrombin. Washed human platelets (10^{10}/ml) were incubated with thrombin (0.17 U/10^8 cells) at 37°C. Incubations were stopped with $CHCl_3$/CH_3OH. DG was quantitated by densitometry of charred spots after thin-layer chromatography. Reproduced from The Journal of Clinical Investigation 63, 582 (1979) by copyright permission of The American Society for Clinical Investigation.

increases in DG. ADP, and to a much lesser degree, 5HT, will induce platelet aggregation, whereas Ca^{2+}, added in the absence of an agonist, will not. Therefore, we considered it possible that these substances which are released by activated platelets might contribute significantly to the stimulated metabolism of phosphoinositides.

TURNOVER OF PHOSPHOINOSITIDES

Thrombin

Our initial observation that the incubation of human platelets with thrombin leads to a transient accumulation of DG is illustrated in Figure 1 (Rittenhouse-Simmons, 1979). It is seen that the response is quite a rapid one, and is barely detectable 120 s after the addition of

461

thrombin. The fatty acid composition of this DG and the kinetics for the disappearance of phosphatidylinositol (PI) are consistent with the action of PLC on PI. Clearly, DG is metabolized further via DG kinase and lipase, which can explain the finding that accumulated DG accounts for only 10-20% of the PI lost. An extended incubation leads to a partial regeneration of PI (Rittenhouse-Simmons, *et al.*, 1977; Broekman, *et al.*, 1981; Prescott and Majerus, 1981) consistent with the operation of the "phosphoino-sitide cycle", which is inhibited by the presence of Ca^{2+} in the incubation medium. Ca^{2+} does not, however, affect the initial formation of DG or breakdown of PI. We also observed that acetylsalicylic acid (ASA), an irrever-sible inhibitor of platelet cyclooxygenase, does not interfere with the changes in PI and DG induced by thrombin, nor do low (<2μg/ml) concentrations of indo-methacin, another cyclooxygenase inhibitor. Similarly, secretion is unaffected. Further, the addition of the 5HT antagonist, cinnanserin, or removal of ADP with either apyrase or creatine phosphate/creatine phosphokinase (CP/CPK) has little effect on thrombin-initiated changes. By these criteria, thrombin is a potent agonist.

Recently, it has been reported that PI is not the only phosphoinositide to decrease when platelets are exposed to thrombin. The level of PIP_2 is depressed within seconds of the addition of thrombin, preceding changes in PI, and *myo*-inositol trisphosphate becomes detectable (Billah and Lapetina, 1982b; Agranoff, *et al.*, 1983). Subsequently, PIP_2 is regenerated. We have also described PLC activity, resolved from phosphatase activity, in the soluble fraction of platelet sonicates (Rittenhouse, 1983). The enzyme is active with respect to all three phosphoinositides, requires Ca^{2+}, displays substrate affinities in the order PI > PIP > PIP_2, and maximum rates in the order PIP > PIP_2 > PI. This order of substrate preference appears to differ from that observed for physiologically stimulated cells, and the reasons for this discrepancy are currently under inves-tigation. There is sufficient activity with respect to PIP_2 in our preparations to account for the amount of PIP_2 reported to be lost in stimulated platelets. At present, it is uncertain whether PI is acted upon directly by PLC or only via transformation to PIP and PIP_2.

Finally, we had observed that loss of PI and formation of DG were strongly inhibited by increased concentrations of cyclic AMP in intact platelets (Rittenhouse-Simmons, 1979), although we have found no effect of this compound on the PLC activity of broken platelets. Remarkably, cyclic AMP has not been found to block the initial drop in PIP_2 following the addition of thrombin, but to inhibit the regeneration of PIP_2 (Billah and Lapetina, 1982c). Therefore, one can conclude that unsupplemented PIP_2 contributes only a minor portion of the DG which accumulates in response to thrombin.

Collagen

Like thrombin, which is an hydrolysis product of circulating prothrombin, collagen is an important physiologic stimulus for platelets. Collagen is present in the subendothelium and becomes exposed to platelets at lesions in blood vessels. We have reported (Rittenhouse and Allen, 1982) that collagen stimulates PLC in human platelets, manifested by a decrease in PI, transient formation of DG, and release of *myo*-inositol. This response is not as rapid as that observed for thrombin, and is paralleled by a slower secretory response.

The changes induced by collagen are dependent in part on cyclooxygenase activity, as shown in Figure 2. The inhibitory effects of ASA or indomethacin can be overcome by the addition of PGH_2 (shown) or U46619, a synthetic, stable analogue of PGH_2. We have found more recently that removal of ADP by CP/CPK also inhibits collagen-induced responses, shown in Table II. Arachidonic acid is released (making possible PGH_2/TXA_2 formation) and some lysoPI is formed in response to collagen. These alterations are also inhibited by ASA, and more so by ASA+CP/CPK. Thus, a positive feedback effect is exerted by PGH_2/TXA_2, leading to more phospholipase activation, arachidonic acid release, and secretion. However, such amplification apparently must be expressed in the presence of collagen, since prostaglandins or analogues are relatively ineffective by themselves in producing this effect. As is the case when thrombin is a stimulus, cyclic AMP is inhibitory to the effects of collagen, blocking both secretion and phosphoinositide turnover.

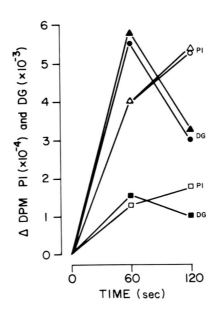

Figure 2. Effect of ASA or ASA+PGH$_2$ on collagen-induced PI hydrolysis and DG formation. Washed platelets (2 x 10^9/ml), labeled with [^{3}H]arachidonic acid, were incubated with collagen (100 μg/ml, ●); collagen + ASA, ▪); or collagen + ASA + PGH$_2$ (2 μM, ▲). The increases in counts in PI and DG were determined vs. stimulus-free controls. Reproduced from The Journal of Clinical Investigation 70, 1221 (1982) by copyright permission of The American Society for Clinical Investigation.

PGH$_2$ is a strong potentiator of the effects of collagen. Stimulation is synergistic, *i.e.*, small quantities of
PGH$_2$/TXA$_2$, which themselves have little effect on platelet levels of DG and free arachidonic acid, have a large
effect in conjunction with collagen. A major implication of
these findings is that the strongest effects of PGH$_2$/TXA$_2$
are at the site of injury in the blood vessel wall, where
collagen and ADP are available as synergistic cohorts.
However, the mechanism of such potentiation is as yet
unknown.

TABLE I

Release of Arachidonic Acid and Formation of Phosphatidic Acid in Human Platelets Expressed to Various Stimuli

Stimulus	Δ cpm [^{3}H]	
	AA	PA
Ionomycin	5614 $\pm$ 578	-137 $\pm$ 116
U46619	124 $\pm$ 60	1620 $\pm$ 261
Ionomycin+U46619	6044 $\pm$ 835	1958 $\pm$ 302

Gel-filtered platelets (1.5 x 10^9 ml) whose phospholipids contained [^{3}H]arachidonic acid, were incubated with ASA and CP/CPK, followed by a 60 s incubation with buffer, ionomycin (1 µM), U46619 (8 µM), or iono-mycin + U46619. Incubations were terminated with $CHCl_3/CH_3OH/HCl$, and the lipids resolved by thin-layer chromatography. Counts present as free arachidonic acid+metabolites (AA) and phosphatidic acid (PA) were determined and increases over stimulus-free controls were calculated.

U46619

Extending the studies just described, we have examined the activation of platelets by higher concentrations of U46619 than those used in the collagen potentiation experiments. It is not clear at this point whether U46619 interacts with TXA_2 "receptors" or whether PGH_2/U46619 and TXA_2 have separate active receptors. This issue will have to be addressed using an array of antagonists. In the present discussion, no argument is being raised to favor PGH_2 as the active species in platelets. We are merely taking advantage of the ability of U46619 and PGH_2/TXA_2 to behave similarly in overcoming the inhibitory effects of ASA and in causing platelet activation. We have observed that platelets exposed to U46619 undergo secretion, a transient loss in PIP_2, a loss in PI, and show increased levels of DG and phosphatidic acid (Rittenhouse, 1984b). However, very little arachidonic acid is released. ASA has no detectable effect on these changes; however agents which remove ADP are partially inhibitory. It is unlikely that U46619 is interfering with PLA_2 or DG lipase, since, when added in conjunction with stimuli which cause arachidonic acid release, U46619 does not inhibit such release (*e.g.*, Table I). In fact, as noted above, PGH_2/TXA_2 and U46619 promote collagen-induced release of arachidonic acid. Thus, it appears that U46619 is a poor activator rather than an inhibitor. Nonetheless, it is possible to achieve significant activation of PLC and secretion using this stimulus. Some of these results are shown in Figure 3 and Table II.

Ca^{2+} Ionophores

It has been known for several years that the Ca^{2+} ionophore A23187 causes the release of arachidonic acid and a loss of PI and phosphatidylcholine in human platelets (Rittenhouse-Simmons and Deykin, 1977; Pickett, *et al.*, 1977; Rittenhouse-Simmons and Deykin, 1978). In addition, it has been demonstrated that the concentration of Ca^{2+} in platelet cytoplasm rises in response to stimuli such as thrombin (Rink, *et al.*, 1982), and that PLC is dependent upon Ca^{2+}. We therefore explored the possibility that PLC could be activated by a flux in Ca^{2+}.

We found initially that, although exposure of human platelets to A23187 elicits some activation of PLC, A23187 is a very inefficient stimulus in comparison with thrombin (Rittenhouse-Simmons, 1981). For comparable amounts of PI hydrolyzed, A23187 stimulates the formation of one fourth to one sixth the amount of DG as does thrombin. Similar results are obtained when the yield of phosphatidic acid is measured. Yet, A23187 does not interfere with the kinetics of DG accumulation when added with thrombin, *i.e.*, it does not appear to accelerate the breakdown of DG. Therefore, we have examined the activation of platelets by Ca^{2+} more closely. As shown in Table II, A23187 promotes substantial formation of lysoPI (lacking arachidonic acid) in comparison with the other stimuli examined thus far. This indicates an activation of PLA_2. ASA has a strong inhibitory effect on the generation of PA and DG, while leaving the production of lysoPI and free arachidonic acid virtually unperturbed in platelets responding to A23187. The inhibitory effects of ASA

TABLE II

Effect of Inhibitors on Stimulated Increases in Lipid Metabolites

Stimulus	PA	lysoPI	AA	DG
	(+Stimulus)/(−Stimulus)			
Thrombin	39 ± 5	1.8 ± 0.2	15 ± 2	6.1 ± 0.9
+ASA	38 ± 4	1.9 ± 0.3	16 ± 3	6.3 ± 0.8
+ASA+CP/CPK	37 ± 4	1.7 ± 0.1	14 ± 1	5.9 ± 0.8
Collagen	11 ± 1	2.8 ± 0.3	7.5 ± 0.6	--
+ASA	3.1 ± 0.2	1.4 ± 0.1	2.3 ± 0.2	--
+ASA+CP/CPK	1.9 ± 0.2	1.2 ± 0.1	1.6 ± 0.1	--
U46619	20 ± 2	1.1 ± 0.2	1.8 ± 0.2	4.1 ± 0.3
+ASA	20 ± 3	1.2 ± 0.2	1.7 ± 0.3	3.9 ± 0.4
+ASA+CP/CPK	15 ± 1	1.0 ± 0.1	1.5 ± 0.2	2.8 ± 0.3
A23187	8.1 ± 0.9	5.8 ± 0.2	21 ± 1	3.1 ± 0.4
+ASA	2.3 ± 0.4	5.6 ± 0.3	20 ± ±	2.4 ± 0.3
+ASA+CP/CPK	1.1 ± 0.1	5.6 ± 0.3	18 ± 2	1.2 ± 0.2

Gel-filtered human platelets (1.5 x 10^9 ml), labeled with [^{3}H]arachidonic acid and ^{32}P, were incubated with buffer, thrombin (2 U/ml), A23187 (0.5 μM), or U46619 (8 μM) for 30 s, or collagen (100 μg/ml) for 90 s, in the presence or absence of inhibitors. Lipids were resolved and quantitated. Results are expressed as the values for stimulated cells divided by the values for stimulus-free controls ± SE. Thus, a value of 1.0 indicates no change from stimulus-free controls. AA = arachidonic acid + oxygenated metabolites; ASA = aspirin; CP/CPK = creatine phosphate/creatine phosphokinase.

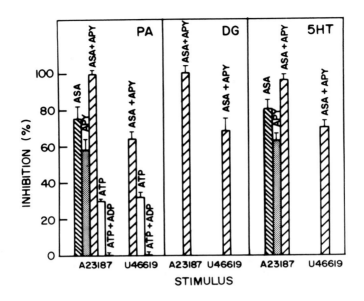

Figure 3. Inhibition of phospholipase C activation and secretion. [^{3}H]Arachidonic acid/^{32}P-labeled gel-filtered platelets were incubated for 15 s or 60 s with A23187 (0.5 μM) or U46619 (8 μM) in the presence of ASA, apyrase (APY), ASA + APY, ATP, or ATP + ADP. Inhibition of formation of [^{32}P]PA (phosphatidic acid, 60 s) or [^{3}H]DG (15 s) was determined. In parallel experiments, platelets labeled with [^{14}C]5HT (taken into the dense granules) were incubated for 60 s with the above stimuli and sedimented. Inhibition of release of [^{14}C]5HT to the supernatant was determined.

can be overcome by the addition of U46619. When platelets are incubated with both ASA and CP/CPK (or apyrase, not shown), the accumulations of DG and PA are eliminated. These results and the effects of inhibitors on secretion are illustrated in Figure 3. ATP, which antagonizes the binding of ADP, is also inhibitory, and its effects can be overcome by ADP. ASA and apyrase have a strong inhibitory effect on secretion caused by A23187, but this can be overcome with larger amounts of A23187. In contrast, PLC activation, as monitored by DG and phosphatidic acid, is still blocked. We have also observed a drop in PIP_2 followed by an increase for platelets exposed to A23187, and these changes are

TABLE III

Extent of Inhibition of Lipid Changes and Cytoplasmic Ca^{2+} Increases Induced by Ionomycin (Ionophore)

Inhibitor	Ca^{2+}	PA	lysoPI	AA
		% Inhibition		
ASA	0 ± 6	80 ± 5	0 ± 3	0 ± 8
ASA+CP/CPK	0 ± 5	100 ± 6	6 ± 3	2 ± 8

Human platelets were loaded with Quin 2 fluorescent probe as described by Rink et al. (1982) and exposed to ionomycin in the presence and absence of ASA or ASA + CP/CPK. Increased fluorescence indicating elevated cytoplasmic Ca^{2+} was compared and the degree of inhibition determined. The effects of inhibitors on ^{32}P or [^3H]arachidonic acid-labeled lipid changes were determined in Table I. AA = arachidonic acid + metabolites; PA = phosphatidic acid.

inhibited by ASA+CP/CPK or ASA+apyrase. The response of platelets to ionomycin, another Ca^{2+} ionophore, is similar to that for A23187 (see also Table I). We have monitored fluxes in cytoplasmic Ca^{2+} using the fluorescent probe, Quin 2, and have found that ASA or ASA+CP/CPK have no effect on increased cytoplasmic Ca^{2+} levels induced by ionomycin (Rittenhouse and Horne, 1984; Table III). Thus, a rise in cytoplasmic Ca^{2+} and activation of PLA_2 can be dissociated from activation of PLC.

CONCLUSION

The findings described in the foregoing discussion indicate that physiologic stimuli activate platelet PLC. In parallel, secretion is elicited. Although thrombin, collagen, PGH_2, TXA_2, (U46619), and ADP all cause turnover of phosphoinositides, thrombin is the only agonist which can be shown to be relatively independent of the effects of the other agonists. Thus, provided that enough thrombin can be generated at a site of injury in vivo, human platelets should be able to function. These findings are consistent with the fact that relatively few people have bleeding disorders after taking aspirin. However, as shown in Table II, cyclooxygenase products and ADP are very important contributors to the full effects of collagen, and vice versa (Rittenhouse and Allen, 1982).

As our data with ionophores indicate, an elevation of cytoplasmic Ca^{2+}, *per se*, does not cause net hydrolysis of phosphoinositides by PLC. These findings may be relevant to other tissues for which Ca^{2+} ionophores have been reported to stimulate PLC. Consequently, it would be well to investigate the possibility that secondary stimuli may be involved in such tissues. Despite the fact that platelet PLC is *dependent* upon Ca^{2+}, it is not *activated* by Ca^{2+}. Presumably sufficient Ca^{2+} is available at the membrane of resting platelets to permit PLC to function initially in conjunction with an appropriate stimulus, like thrombin. The question of what mechanism does account for the net increase in PLC activity in platelets exposed to such stimuli still remains. The answer may, for example, involve an altered presentation of substrate to PLC. Further, our results do not tell us whether the breakdown of phosphoinositides causes the mobilization of Ca^{2+}. Recently evidence derived from experiments with pancreatic acinar cells may be pertinent to this issue. The studies indicate that *myo*-inositol trisphosphate (resulting from the action of PLC on PIP_2) can cause the release of internal Ca^{2+} (Streb, *et al*, 1983). Similarly, increases in platelet *myo*-inositol trisphosphate may account for the rise in platelet cytoplasmic Ca^{2+}. However, it appears that a Ca^{2+} flux is not a completely satisfactory mechanism for connecting a stimulus to the secretory response in platelets. Some event associated with phosphoinositide breakdown is associated as well with the secretion of dense granule constituents. The event does not seem to be merely a Ca^{2+} flux, since it is possible to induce a consistent flux in cytoplasmic Ca^{2+} whether or not PLC is activated (using ionophores) and have secretion occur only in conjunction with PLC activity (Rittenhouse and Horne, 1984c). We suspect that this event is potentiating the effects of Ca^{2+} (*higher* levels of Ca^{2+}, produced by ionophore, can lead to secretion without PLC activity; Rittenhouse, 1984a,b) and may be

470

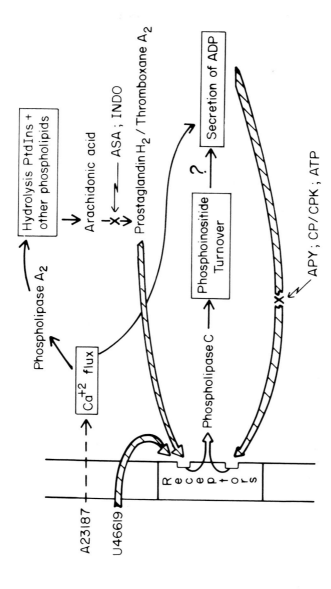

Figure 4. The route by which Ca^{2+} ionophore causes the activation of phospholipase C and secretion. Acetylsalicylic acid (ASA) and indomethacin (INDO) inhibit cyclooxygenase; apyrase (APY) and creatine phosphate/creatine phosphokinase (CP/CPK) convert ADP to non-stimulatory products; ATP competes with ADP for binding sites; U46619 is a stable structural analogue of PGH_2. Specific subcellular locales are as yet undefined. PtdIns = PI = phosphatidylinositol.

the activation of protein kinase C by DG, described by Takai, *et al.*, (1979, and this volume). A summary of these observations is presented in Figure 4.

Among the numerous questions raised by our attempts to answer questions are: (1) What are the major prerequisites for the mobilization of arachidonic acid (if activation of PLC by PGH_2/TXA_2 or U46619 is insufficient)? (2) How does cyclic AMP inhibit the turnover of phosphoinositides by PLC? (3) Does PLC act only upon PIP_2 in intact platelets? (4) Related to this, what allows PLC of intact platelets to use PIP_2 as a substrate, given the preference of the solubilized enzyme for PI and PIP? We hope that a resolution of some of these issues will be reached by experiments currently underway; undoubtedly, many more questions will take their place.

ACKNOWLEDGEMENTS

This work was completed during the tenure of an Established Investigator award from the American Heart Association, and was supported in part by grants HL 27897 and 29316 from the National Heart, Lung, and Blood Institute.

REFERENCES

Agranoff, B.W., Murthy, P. and Seguin, E.B. (1983) *J. Biol. Chem.* *258*, 2076-2978.

Bell, R.L., Kennerly, D.A., Stanford, N. and Majerus, P.W. (1979) *Proc. Natl. Acad. Sci. USA* *76*, 3238-3241.

Billah, M.M. and Lapetina, E.G. (1982a) *J. Biol. Chem.* *257*, 5196-5200.

Billah, M.M. and Lapetina, E.G. (1982b) *J. Biol. Chem.* *257*, 12705-12708.

Billah, M.M. and Lapetina, E.G. (1982c) *Biochem. Biophys. Res. Commun.* *109*, 217-222.

Broekman, M.J., Ward, J.N. and Marcus, A.J. (1980) *J. Clin. Invest.* *66*, 275-283.

Broekman, M.J., Ward, J.N. and Marcus, A.J. (1981) *J. Biol. Chem.* *256*, 8271-8274.

Hamberg, M., Svensson, J. and Samuelsson, B. (1975) *Proc. Natl. Acad. Sci. USA* *72*, 2994-2998.

Lapetina, E.G., Billah, M.M. and Cuatrecasas, P. (1981) *Nature* *292*, 367–369.

McKean, M.L., Smith, J.B. and Silver, M.J. (1981) *J. Biol. Chem.* *256*, 1522–1524.

Pickett, W.C., Jesse, R.L. and Cohen, P. (1977) *Biochim. Biophys. Acta* *486*, 209–213.

Prescott, S. and Majerus, P.W. (1981) *J. Biol. Chem.* *256*, 579–582.

Prescott, S. and Majerus, P.W. (1983) *J. Biol. Chem.* *258*, 764–769.

Rink, T.J., Smith, S.W. and Tsien, R.Y. (1982) *FEBS Letter* *148*, 21–26.

Rittenhouse-Simmons, S. (1979) *J. Clin. Invest.* *63*, 580–587.

Rittenhouse-Simmons, S. (1980) *J. Biol. Chem.* *255*, 2259–2262.

Rittenhouse-Simmons, S. (1981) *J. Biol. Chem.* *256*, 4153–4155.

Rittenhouse, S.E. (1983) *Proc. Natl. Acad. Sci. USA* *80*, 5417–5420.

Rittenhouse, S.E. (1984a) in: "*Advances in Inflammation Research, Vol. 8*" (G. Weissman, ed.) Raven Press, New York, In Press.

Rittenhouse, S.E. (1984b) *Biochem. J.* in press.

Rittenhouse, S.E. and Allen, C.L. (1982) *J. Clin. Invest.* *70*, 1216–1224.

Rittenhouse-Simmons, S. and Deykin, D. (1977) *J. Clin. Invest.* *60*, 495–498.

Rittenhouse-Simmons, S.and Deykin, D. (1978) *Biochim. Biophys. Acta* *543*, 409–422.

Rittenhouse-Simmons, S. and Deykin, D. (1981) in: "*Platelets in Biology and Pathology, Vol. 2*" (J. Gordon, ed.) pp. 349–372, Elsevier/North-Holland, Amsterdam.

Rittenhouse, S.E. and Horne, W.C. (1984) Manuscript in review.

Rittenhouse-Simmons, S., Russell, F.A. and Deykin, D. (1977) *Biochim. Biophys. Acta* *488*, 370–380.

Streb, H., Irvine, R.F., Berridge, M.J. and Schulz, I. (1983) *Nature* *306*, 67–68.

Takai, Y., Kishimoto, A., Kikkawa, U., Mori, T. and Nishizuka, Y. (1979) *Biochem. Biophys. Res. Commun.* *91*, 1218–1224.

THE RELEVANCE OF INOSITIDE DEGRADATION AND PROTEIN KINASE C IN PLATELET RESPONSES

Eduardo G. Lapetina

The Wellcome Research Laboratories
Research Triangle Park, NC 27709

SUMMARY

The interaction of a wide range of cell activators with specific membrane receptors induces the phosphodiesteratic cleavage (phospholipase C) of the phosphoinositides. The breakdown of the phosphoinositides produces formation of 1,2-diacylglycerol and myo-inositol phosphates. The 1,2-diacylglycerol is rapidly phosphorylated to phosphatidic acid by the action of 1,2-diacylglycerol kinase. Both 1,2-diacylglycerol and phosphatidic acid remain inside the cell and could have important roles as cellular modulators. 1,2-Diacylglycerol and phorbol esters activate a phospholipid- and Ca^{2+}-dependent protein kinase C which is implicated in transmembrane signalling, tumor promotion and cellular differentiation. On the other hand, phosphatidic acid is a Ca^{2+}-ionophore and a fusogen at low Ca^{2+} concentration. The fact that there is a close correlation between the amount of phosphatidic acid formed (reflecting stimulation of phospholipase C and formation of 1,2-diacylglycerol), protein kinase C activity and the degree of platelet activation indicates that these reactions have a key role in the transduction and amplification of the signals. Furthermore, phosphatidic acid might be related to the initiation of the release of arachidonic acid from membrane phospholipids such as phosphatidylcholine, phosphatidylethanolamine and phosphatidylinositol by the action of phospholipases of the A_2-type. The liberated arachidonic acid is converted to active products of the cyclooxygenase and lipoxygenase activities.

PROPERTIES OF PHOSPHOLIPASE C IN PLATELETS

In platelets, phospholipase C is present mainly in the soluble fraction (Mauco et al., 1979; Rittenhouse-Simmons, 1979; Billah et al., 1979; Siess and Lapetina, 1983). Platelet phospholipase C may be a single soluble phosphodiesterase attacking phosphatidylinositol, phosphatidylinositol 4-monophosphate and phosphatidylinositol 4,5-bisphosphate (Rittenhouse, 1983). Degradation of all three inositides has been observed soon after stimulation of platelets (Billah and Lapetina, 1982, 1983; Lapetina, 1983). One of the questions that immediately arises is how the cytosolic activity interacts with the different inositides that are bound to platelet membranes. It seems that the cytosolic activity degrades the membrane-bound substrates which are only made available to the enzyme during the activation of platelets. The several lines of evidence that support this conclusion are: (1) lack of membrane-bound activity in platelets; (2) no binding of the cytosolic activity to the membranes during stimulation; (3) the stimulation of intact platelets with thrombin does not change the activity of the cytosolic enzyme; (4) the inhibitory effect of prostacyclin on phospholipase C degradation of phosphatidylinositol in thrombin-stimulated platelets is not exerted directly on the cytosolic enzyme; prostacyclin might, therefore, affect directly the relation of the substrate and the platelet receptor for agonists such as thrombin; (5) the cytosolic enzyme acts on exogenous solubilized substrate or membrane inositides which are previously solubilized by deoxycholate; thus it could be inferred that in normal conditions of stimulation the substrate is "disclosed" for phospholipase C attack; (6) the cytosolic enzyme is not activated but inhibited by deoxycholate; (7) thrombin or fibrinogen do not stimulate phospholipase C activity of a cytosolic fraction on membrane-bound substrate; and (8) the mobilization of Ca^{2+} to raise Ca^{2+} concentration in the cytosol for maximal activity of the enzyme might also be a fundamental step for the activated degradation of the inositides in platelets.

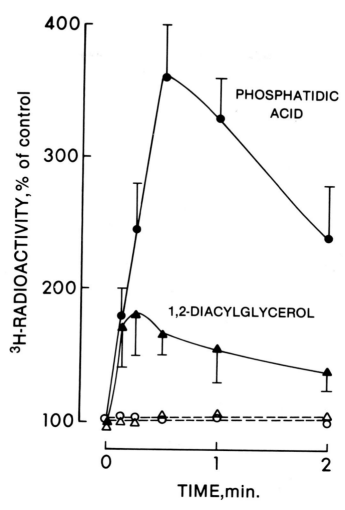

Figure 1. Exogenous arachidonic acid (5 μM) stimulates the formation of [^{3}H]1,2-diacylglycerol (▲) and [^{3}H]phosphatidic acid (●) in human platelets prelabeled with [^{3}H]arachidonic acid. Results (mean ± S.D.) are expressed as a percentage of the control values from four different experiments. The <u>open</u> <u>symbols</u> refer to [^{3}H]1,2-diacylglycerol (Δ) and [^{3}H]phosphatidic acid (O) in aspirin-treated platelets as previously described (Siess <u>et al</u>., 1983b). Total radioactivity for [^{3}H]1,2-diacylglycerol and [^{3}H]phosphatidic acid in unstimulated controls was 205 ± 15 and 610 ± 10 cpm, respectively.

The regulatory action for degradation of the inositides may then not involve the enzyme directly, but rather the "unmasking" of the substrate that might be closely associated to the specific receptors in the platelet membranes.

ACTIVATION OF PHOSPHOLIPASE C AND PROTEIN PHOSPHORYLATION IN THE STIMULATED PLATELET

An early response that follows activation of platelets with stimuli such as thrombin, collagen, ADP, arachidonic acid, platelet-activating factor or ionophore A23187 is the formation of phosphatidic acid. Phosphatidic acid is produced by the sequential actions of phospholipase C that degrades the inositide lipids and the phosphorylation of the resultant 1,2-diacylglycerol by 1,2-diacylglycerol kinase (for references see Lapetina and Siess, 1983). Figure 1 indicates that the addition of exogenous unlabeled arachidonic acid to washed human platelets prelabeled with [3H]arachidonic acid stimulates the rapid and transient formation of [3H]1,2-diacylglycerol and [3H]phosphatidic acid. Maximal formation of 1,2-diacylglycerol precedes that of phosphatidic acid and occurs concomitantly with degradation of phosphatidylinositol (Siess *et al.*, 1983b). The formation of phosphatidic acid parallels one of the initial platelet responses which is the change of platelet shape from discoid, smooth cells to spiny spheres. Platelet shape change precedes two other important platelet responses such as the secretion of the content of platelet granules and platelet aggregation. It is then possible to dissociate the early platelet shape change from secretion and aggregation. Platelet shape change could be monitored by a decrease of light transmission using an aggregometer and by microscopy (Siess *et al.*, 1983b). We have observed a strict correlation between the degree of stimulation that induces shape change and the formation of phosphatidic acid (Figure 2). No release reaction or aggregation is observed under those circumstances in which platelets are stimulated with micromolar concentrations of arachidonic acid and in the presence of EGTA.

The activation of phospholipase C, as reflected by the formation of phosphatidic acid during platelet shape change, induced by arachidonic acid (Siess *et al.*,

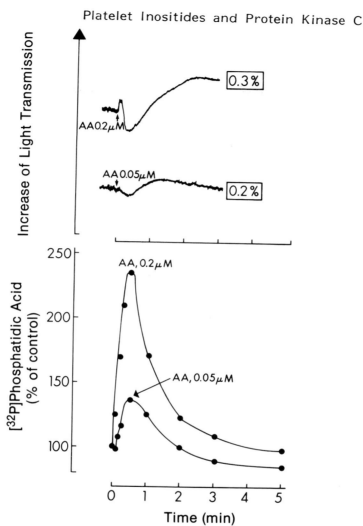

Figure 2. Arachidonic acid (AA) stimulates formation of [^{32}P]phosphatidic acid and shape change in human platelets. Platelet shape change is indicated by a decrease of light transmission and narrowing of the oscillation tracing which is produced by discoid platelets (upper panel). At various times as indicated 0.1 ml aliquots were directly transferred into tubes containing 3.75 volumes of chloroform/methanol (1:2) for lipid extraction and phosphatidic acid was estimated as previously described (Siess et al., 1983b) (lower panel). Release of [^{3}H]serotonin was measured after 3 min of stimulation and the values (% of total) are indicated inside the rectangles.

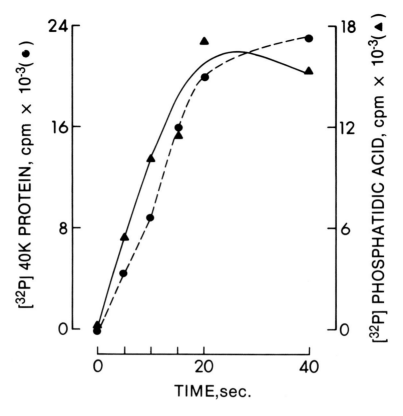

Figure 3. Time course of platelet-activating factor (PAF) -induced phosphorylation of a 40,000-dalton (40K) protein and formation of phosphatidic acid during platelet shape change. The radioactivity in phosphatidic acid (▲) and 40,000-dalton protein (●) was present in 1 ml of platelet suspension which contained 7.5×10^8 platelets. The ordinate scale shows increases from basal, etc. Basal values for phosphatidic acid and for the 40,000-dalton protein were 13,243 and 16,920 cpm, respectively. Other details as in Lapetina and Siegel, 1983.

1983b), platelet-activating factor (Lapetina and Siegel, 1983) or thrombin (Lapetina and Siess, 1983), occurs in parallel with the phosphorylation of a 40,000-dalton protein by a lipid-dependent protein kinase which is recognized as protein kinase C (Nishizuka, 1983).

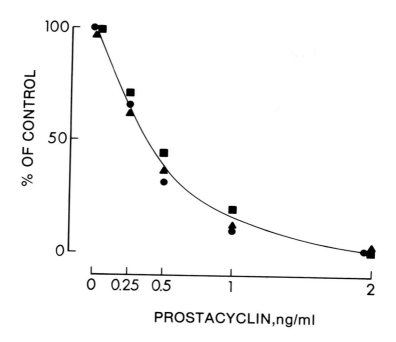

<u>Figure 4</u>. Effect of different concentrations of prostacyclin on the platelet-activating factor (PAF) -induced platelet shape change (■), phosphorylation of a 40,000-dalton (40K) protein (●), and formation of phosphatidic acid (▲). Washed human platelets prelabeled with [32]P (1.0 ml samples) were preincubated for 3 min in the aggregometer tubes at 37°C while stirring with different concentrations of prostacyclin, as indicated, and then stimulated with 0.1 μM PAF. Two platelet samples (0.2 ml each) were taken just before addition of PAF and 15 s after addition of PAF for separation of proteins and phosphatidate, respectively. Shape change was recorded, and the extent of inhibition of shape change induced by prostacyclin was measured in mm in relation to samples stimulated with PAF in the absence of prostacyclin. Results in all cases are expressed as % of changes induced by PAF in the absence of prostacyclin. The extent of stimulation of the phosphorylation of the 40,000-dalton protein and phosphatidic acid induced by 0.1 μM PAF during 15 s incubations was 194% and 193%, respectively. For other details, see Lapetina and Siegel, 1983.

Figure 3 presents an example of the stimulation of human platelets by platelet-activating factor which induces platelet shape change with simultaneous formation of phosphatidic acid and phosphorylation of a 40,000-dalton protein. Both phosphorylated products reach maximal formation at about 20 sec (Figure 3).

We have studied the effect of various concentrations of prostacyclin in an attempt to differentiate the various responses; *i.e.*, shape change, formation of phosphatidic acid and phosphorylation of the 40,000-dalton protein. One of such experiments is depicted in Figure 4 which indicates that the dependence of the various responses on prostacyclin concentration is identical. This suggests that there is a common site of action for prostacyclin in eliciting each of them.

MECHANISMS FOR ACTIVATION OF PHOSPHOLIPASE C IN PLATELETS

Platelet cyclooxygenase activity converts arachidonic acid to endoperoxides which are then rapidly converted to thromboxanes. It seems that stimulation of phospholipase C in platelets can be induced both by an endoperoxide-dependent mechanism and by an endoperoxide-independent mechanism. Arachidonic acid and collagen stimulate phospholipase C and phosphatidic acid formation via the initial production of endoperoxides. Indomethacin or aspirin, which block cyclooxygenase activity, also inhibit the formation of phosphatidic acid, protein phosphorylation and shape change of platelets stimulated with collagen or arachidonic acid (Siess, *et al.*, 1983a,b).

In contrast, thrombin or platelet-activating factor stimulate phosphatidic acid formation directly by an endoperoxide-independent mechanism (Siess, *et al.*, 1983a; Lapetina and Siegel, 1983). Phosphatidic acid formation may be due to the direct interaction of thrombin or platelet-activating factor with their respective receptors with the immediate activation of phospholipase C, a process which is not affected by inhibitors of arachidonate metabolism such as aspirin or indomethacin.

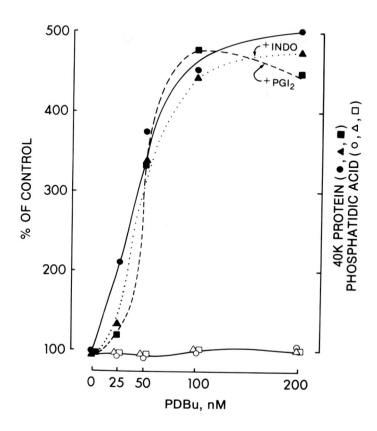

<u>Figure 5</u>. Effect of various concentrations of phorbol-12,13-dibutyrate (PDBu) on phosphorylation of a 40,000-dalton (40K) protein. Action of indomethacin and prostacyclin. Samples (1.0 ml) of suspensions of washed human platelets (7.5 x 10[8]/ml) prelabeled with ^{32}P were preincubated without additions (0,●) or with the addition of 10 μM indomethacin (ino; Δ,▲) or 4 ng/ml of prostacyclin (PGI$_2$; □,■) in the aggregometer tubes at 37°C for 3 min and subsequently stimulated with various concentrations of PDBu for 90 s. A sample (0.1 ml) was quenched for separation of proteins on 11% polyacrylamide gels and the remaining sample (0.9 ml) was extracted with chloroform/methanol for subsequent chromatographic separation of phosphatidic acid. Results are expressed as % of unstimulated controls. <u>Closed</u> <u>symbols</u> indicate [32P] 40K protein while <u>open</u> <u>symbols</u> represent [32P]phosphatidic acid.

PLATELET PROTEIN KINASE C

In platelets, protein kinase C phosphorylates a protein with an approximate molecular weight of 40,000-daltons (40K protein) (Nishizuka, 1983). It is recognized that 40K protein phosphorylation is dependent on phosphatidylserine and Ca^{2+} and is markedly enhanced by unsaturated 1,2-diacylglycerol. The fact that in most cells 1,2-diacylglycerol is produced by phospholipase C degradation of the inositol phospholipids in response to a wide variety of cell activators suggests the importance of phospholipase C activation as a primary step in the transmission of cellular signals. Protein kinase C is also stimulated by phorbol ester. Tumor-promoting phorbol esters bind to the platelet membrane and directly activate protein kinase C without inducing increased phosphatidylinositol turnover (Nishizuka, 1983). It has been suggested that the phorbol esters activate protein kinase C by substituting for 1,2-diacylglycerol. Figure 5 shows that phorbol esters activate phosphorylation of the 40K protein without the formation of phosphatidic acid. The 40K protein phosphorylation is not affected by indomethacin (inhibitor of cyclooxygenase) or prostacyclin (increases cyclic AMP levels). Phosphorylation of the 40K protein seems to be triggered by a direct action of phorbol esters on the protein kinase C.

The phosphorylation of the 40K protein by phorbol esters does not affect the subsequent action of thrombin on phospholipase C and A_2. Phosphorylation of the 40K protein by phorbol esters does not change the thrombin-induced formation of $[^{32}P]$phosphatidic acid (Figure 6) or the formation of $[^{14}C]$arachidonate metabolites derived from cyclooxygenase and lipoxygenase activities (not shown) in platelets that have been prelabeled with $[^{14}C]$arachidonate. These observations suggest that the phosphorylation of the 40K protein does not affect the activities of enzymes such as phospholipase C, 1,2-diacylglycerol kinase, phospholipase A_2, cyclooxygenase and lipoxygenase.

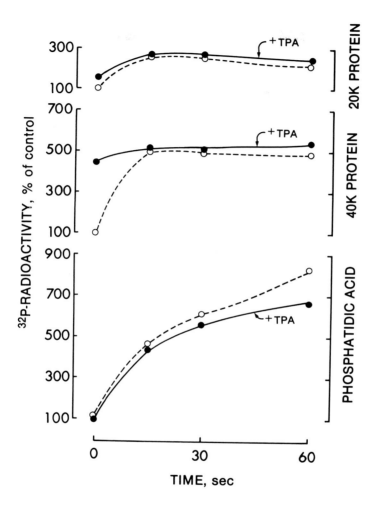

<u>Figure 6</u>. Consecutive stimulation of platelets with 12-0-tetradecanoyl-phorbol-13-acetate (TPA) and thrombin. Samples of platelets as described in Fig. 5 were treated for 2 min without or with TPA (560 nM) and then stimulated with thrombin (0.5 units/ml) for different times as indicated. O---O, refers to samples only stimulated with thrombin; ●---●, refers to platelets sequentially stimulated with TPA and thrombin. Other details as in Fig. 5.

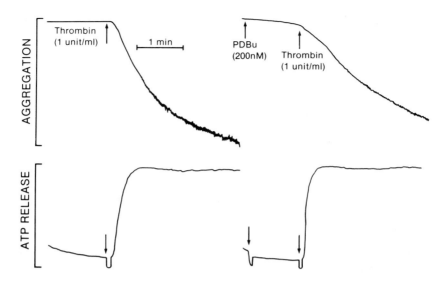

Figure 7. Effect of thrombin and phorbol-12,13-dibutyrate (PDBu) on platelet aggregation and release of ATP. Aggregation and release of ATP measured by the luciferase reaction were simultaneously monitored using a lumiaggregometer (Chromolog). Samples of platelets (1 ml, 7.5 x 10[8] cells) were stimulated with thrombin, 1 unit/ml, or, with PDBu, 200 nM, followed by thrombin, 1 unit/ml.

40K PROTEIN PHOSPHORYLATION IS NOT A SUFFICIENT STIMULUS FOR PLATELET ACTIVATION

Maximal phosphorylation of the 40K protein could be observed within the first minute of phorbol ester stimulation with practically no release of serotonin or ATP (Figure 7). Longer incubations do show some release of serotonin but this release is not correlated with the degree of 40K protein phosphorylation which is essentially complete after one minute. A previous report shows that thrombin or TPA induce equivalent phosphorylation of the 40K protein, but thrombin is a much more efficacious inducer of the release of serotonin than high concentrations of phorbol ester (Castagna et al., 1982). It seems then that kinase C phosphorylation of the 40K protein is not a sufficient signal to trigger platelet responses. Other reactions must occur during the time that elapses between the rapid phosphorylation of the 40K

486

protein and the onset of platelet responses. These reactions could be related to the activation of phospholipases C or A_2 since phorbol esters will not induce the formation of 1,2-diacylglycerol, phosphatidic acid, or arachidonate metabolites. It is important to notice that the subsequent action of thrombin or ionophore A23187 on platelets pretreated with phorbol esters induces formation of phosphatidic acid which parallels further phosphorylation of myosin light chain (20K protein).

In our experiments, release of [^{3}H]serotonin is induced by combining phorbol esters and submaximal concentrations of the Ca^{2+}-ionophore A23187 (Figure 8) and this suggests that both protein phosphorylation and Ca^{2+}-mobilization are essential for the secretory reaction. Ca^{2+} might be acting at a subsequent step to drive the release of [^{3}H]serotonin. Platelet activators such as thrombin, collagen, platelet-activating factor and arachidonic acid stimulate the phosphatidylinositol cycle which provides both formation of 1,2-diacylglycerol for activation of kinase C and mobilization of Ca^{2+}.

FUNCTIONAL ROLE OF KINASE C

It would be important to have an effective inhibitor of the phosphorylation of the 40K protein to evaluate the functional role of the kinase C phosphorylation in platelets. We have tried unsuccessfully to inhibit the phorbol ester-induced phosphorylation of the 40K protein in intact platelets with various concentrations (10 to 100 µM) of compounds such as indomethacin, aspirin, prostacyclin, BW755, retinal, caffeic acid, neomycin sulphate, trifluoperazine, chlorpromazine and polymyxin. However, the effect of platelet physiological stimuli such as arachidonic acid, platelet-activating factor or thrombin on the phosphorylation of the 40K protein could be inhibited by prostacyclin or agents that increase cyclic AMP (Nishizuka, 1983; Ieyasu et al., 1982; Lapetina and Siegel, 1983; Siess et al., 1983b). In all of these cases, the action of cyclic AMP elevation is not confined to protein kinase C-induced phosphorylation since all other responses such as phosphorylation of myosin light chain (20K protein), formation of 1,2-diacylglycerol, phosphatidic acid, cyclooxygenase and lipoxygenase metabolites, as well as shape change, release reaction and aggregation are also inhibited.

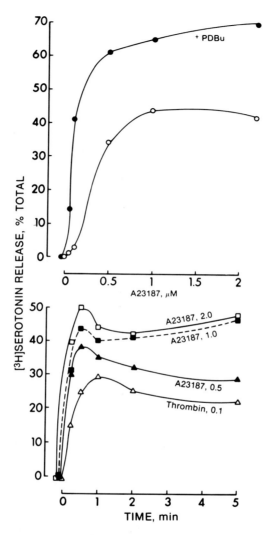

Figure 8. Potentiation by phorbol-12,13-dibutyrate (PDBu) and ionophore A23187 of the release of [^{3}H]serotonin of human platelets. Lower panel shows the time-course of the release of [^{3}H]serotonin by thrombin, 0.1 unit/ml (Δ) and ionophore A23187, 0.5 μM (▲), 1.0 μM (■) and 2.0 μM (□). Upper panel shows the release of [^{3}H]serotonin induced by different concentrations of ionophore A23187 (O) during 2 min incubation. It also shows the effect of preincubation for 30 s with PDBu, 100 nM, on the release of [^{3}H]serotonin induced by ionophore A23187 (O).

We have observed a strict temporal and quantitative correlation between 40K protein phosphorylation and formation of phosphatidic acid during the shape change of human platelets stimulated with platelet-activating factor (Lapetina and Siegel, 1983), arachidonic acid (Siess *et al.*, 1983b), or thrombin (Lapetina and Siess, 1983). This might reflect the stimulation of kinase C by 1,2-diacylglycerol formed by the activation of phospholipase C by these agents. That would indicate that activation of phospholipase C is the initial response, whilst phosphorylation of the 40K protein is an epiphenomenon of the formation of 1,2-diacylglycerol.

FORMATION OF ARACHIDONATE METABOLITES FOLLOWS ACTIVATION OF PHOSPHOLIPASE C

Several years ago we found that the phosphodiesteratic cleavage of the inosities precedes the activation of phospholipase A_2 that induces the liberation of arachidonic acid and formation of endoperoxides and thromboxanes (Lapetina and Cuatrecasas, 1979; Lapetina, 1982). This sequential activation of phospholipases C and A_2 has recently been corroborated with a different experimental approach. We have studied biochemical reactions that occur during shape change of platelets. In activated platelets, shape change precedes aggregation and release of granule content. We have observed that during the shape change induced by platelet-activating factor or low concentrations of thrombin there is formation of phosphatidic acid and protein phosphorylation without metabolism of arachidonic acid (Lapetina and Siegel, 1983; Lapetina and Siess, 1983). Trifluoperazine, which inhibits phospholipases of the A_2 type, does not affect formation of phosphatidic acid or protein phosphorylation (Lapetina and Siegel, 1983). Neither are these responses affected by indomethacin or aspirin which effectively inhibit cyclooxygenase and the consequent formation of active endoperoxides and thromboxanes (Lapetina and Siegel, 1983; Lapetina and Siess, 1983). It seems then that the liberation of arachidonic acid is not relevant to the induction of platelet shape change; however, metabolism of arachidonic acid seems to be important during the initiation of platelet aggregation. At the point where aggregation induced by thrombin or platelet-activating factor starts, arachidonic acid is metabolized through

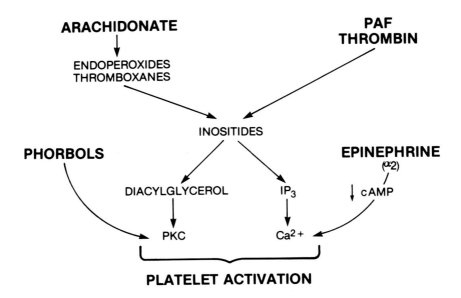

Mechanism for platelet activation. The scheme is discussed in the last section of this chapter. Epinephrine exerts its action through stimulation of α_2-receptors and directly produces aggregation of platelets without activation of the phosphodiesteratic cleavage of the inositol phospholipids. PKC, protein kinase C; IP_3, myo-inositol-1,4,5-trisphosphate; PAF, platelet-activating factor.

platelet cyclooxygenase to endoperoxides and thromboxanes A_2 (Lapetina and Siess, 1983). Endoperoxides and thromboxanes could act as positive feedback promoters within the same cell, or are released, and activate phospholipase C in other platelets, which induces further formation of 1,2-diacylglycerol and phosphatidic acid, amplifying in this way further liberation of arachidonic acid. It seems then that platelet cyclooxygenase has a critical role in inducing further stimulation of phospholipase C and triggering aggregation and release reaction in stimulated platelets.

OVERVIEW OF PLATELET ACTIVATION

Figure 9 tries to integrate the information discussed in this chapter in relation to platelet activation. Arachidonic acid induces the phosphodiesteratic cleavage of the inositol phospholipids through the formation of endoperoxides or thromboxanes while platelet-activating factor or thrombin activate directly. The phosphodiesteratic cleavage of the inositodes leads to formation of 1,2-diacylglycerol and *myo*-inositol phosphates. 1,2-Diacylglycerol activates kinase C phosphorylation of the 40K protein. 1,2-Diacylglycerol is further phosphorylated to phosphatidic acid and could then activate phospholipases of the A_2 type (Lapetina, 1982). One of the *myo*-inositol phosphates (*i.e.*, *myo*-inositol 1,4,5-trisphosphate) might trigger Ca^{2+} mobilization (Streb *et al.*, 1983). Kinase C activation and Ca^{2+} mobilization are both needed for platelet activation. Phorbol esters directly activate kinase C which is not enough to produce platelet activation. Platelet responses are observed when kinase C activation occurs in parallel with Ca^{2+} mobilization. Recent observations (Siess *et al.*, 1984) indicate that epinephrine produces platelet aggregation through stimulation of α_2-receptors and without stimulation of phospholipase C. Epinephrine requires exogenous Ca^{2+} to produce platelet aggregation. Epinephrine can directly induce exposure of fibronectin receptors to which fibrinogen binds in the presence of Ca^{2+} or Mg^{2+} leading to platelet aggregation (Siess *et al.*, 1984).

REFERENCES

Billah, M.M. and Lapetina, E.G. (1982) *J. Biol. Chem.* *257*, 12705-12708.

Billah, M.M. and Lapetina, E.G. (1983) *Proc. Natl. Acad. Sci. USA* *80*, 965-968.

Billah, M.M., Lapetina, E.G. and Cuatrecasas, P. (1980) *J. Biol. Chem.* *255*, 10227-10231.

Castagna, M., Takai, Y., Kaibuchi, K., Sano, K., Kikkawa, U. and Nishizuka, Y. (1982) *J. Biol. Chem.* *257*, 7847-7851.

Ieyasu, H., Takai, Y., Kaibuchi, K., Sawamura, M. and Nishizuka, Y. (1982) *Biochem. Biophys. Res. Commun.* *108*, 1701-1708.

Lapetina, E.G. (1982) *Trends Pharmacol. Sci.* *3*, 115-118.

Lapetina, E.G. (1983) *Life Sciences* *32*, 2069-2082.

Lapetina, E.G. (1984) *Biochem. Biophys. Res. Commun.* *120*, 37-44.

Lapetina, E.G. and Cuatrecasas, P. (1979) *Biochem. Biophys. Acta* *573*, 394-402.

Lapetina, E.G. and Siegel, F.L. (1983) *J. Biol. Chem.* *258*, 7241-7244.

Lapetina, E.G. and Siess, W. (1983) *Life Sciences* *33*, 1011-1018.

Mauco, G., Chap, H. and Douste-Blazy, L. (1979) *FEBS Letter* *100*, 367-370.

Nishizuka, Y. (1983) *Trends Biochem. Sci.* *8*, 13-16.

Rittenhouse-Simmons, S. (1979) *J. Clin. Invest.* *63*, 580-587.

Rittenhouse, S.E. (1983) *Proc. Natl. Acad. Sci. USA* *80*, 5417-5420.

Siess, W. and Lapetina, E.G. (1983) *Biochim. Biophys. Acta* *752*, 329-338.

Siess, W., Cuatrecasas, P. and Lapetina, E.G. (1983a) *J, Biol. Chem.* *258*, 4683-4686.

Siess, W., Siegel, F.L. and Lapetina, E.G. (1983b) *J. Biol. Chem.* *258*, 11236-11242.

Siess, W., Weber, P. and Lapetina, E.G. (1984) *J. Biol. Chem.* *259*, 8286-8292.

Streb, H., Irvine, R.F., Berridge, M.J. and Schulz, I. (1983) *Nature* *306*, 67-68.

REGULATION OF ARACHIDONIC ACID RELEASE FOR

PROSTAGLANDIN PRODUCTION DURING PARTURITION

John M. Johnston, Chiaki Ban, and Maurizio M. Anceschi

Departments of Biochemistry, Obstetrics-Gynecology,
Cecil H. and Ida Green Center for Reproductive
Biology Sciences, University of Texas Health Science
Center, Dallas, TX 75235

SUMMARY

The sources of arachidonic acid for the production of prostanoids during early labor have been identified as (diacyl)phosphatidylethanolamine and phosphatidylinositol. The enzymatic mechanisms that account for the selective mobilization of arachidonic acid in amnion tissue from these two glycerophospholipids are described. Based on the activation/inhibition of the enzymes involved in these processes, we suggest a central role for Ca^{2+} in the regulation of prostanoid biosynthesis in this tissue. Platelet-activating factor (PAF) has been identified in amniotic fluid from women in labor, in the urine and tracheal fluid of the newborn, and in amnion tissue. Increased PAF concentrations were found in the amnion tissue obtained at term from women in labor. PAF biosynthesis and hydrolysis has been demonstrated in amnion tissue. LysoPAF:acetylCoA acetyltransferase was activated by Ca^{2+} at micromolar concentrations. A biochemical model for the regulation of arachidonic acid release and prostanoid formation in relation to parturition is presented.

INTRODUCTION

Prostaglandins that are produced in largest amounts by uterine and intra-uterine tissues during human parturition are PGE_2 and $PGF_{2\alpha}$. The obligate precursor of

493

these prostaglandins is free arachidonic acid. However, in mammalian tissues little arachidonic acid is present in a free form and most is found in an esterified form, from which it must be released before it can be utilized for prostaglandin biosynthesis. It is known that the enzymatic release of free arachidonic acid is of major importance in the regulation of prostaglandin biosynthesis (Lands and Samuelsson, 1968; Vonkeman and van Dorp, 1968). Despite this importance of free arachidonic acid availability little was known of the origin of the arachidonic acid used to synthesize the prostanoids involved in labor. Several years ago we observed that the concentration of free fatty acids in amniotic fluid increased during labor and the increase in arachidonic acid concentration was disproportionately large compared to other fatty acids (MacDonald *et al.*, 1974). We also demonstrated that the microsomal fraction prepared from intrauterine tissues have a high capacity for prostaglandin formation from free arachidonic acid (Okazaki *et al.*, 1981a). We proposed at that time that the source of this arachidonic acid was the fetal membranes and there is now much evidence in support of an involvement of the fetal membranes in prostaglandin production during the initiation and maintenance of labor.

Our investigations of the biochemical mechanisms by which arachidonic acid is mobilized from human fetal membranes during parturition were undertaken in an attempt to accomplish 3 basic objectives. These were: (1) to identify the source(s) of arachidonic acid for prostaglandin production in human fetal membranes; (2) to define the enzymatic mechanism(s) by which free arachidonic acid is released from esterified forms in these tissues; and (3) to characterize the regulation of arachidonic acid mobilization during parturition.

RESULTS AND DISCUSSION

We have previously reported (Schwarz *et al.*, 1975) that fetal membranes are rich in esterified arachidonic acid and that during early labor the concentration of esterified arachidonic acid of these tissues decreases (Okita *et al.*, 1982a). In subsequent experiments, we established that the arachidonic acid is mobilized from glycerophospholipids. We found that ethanolamine-con-

494

taining lipids accounted for more than half of all arachidonic acid esterified in the glycerophospholipid fraction. Phosphatidylinositol was also rich in arachidonic acid. When these fatty acid analyses were carried out using fetal membranes obtained during early stages of labor (Okita et al., 1982a), two glycerophospholipids were found to lose selectively arachidonic acid. These were (diacyl)phosphatidylethanolamine and phosphatidylinositol. In amnion tissue, the arachidonic acid content of (diacyl)-phosphatidylethanolamine and phosphatidylinositol decreased by 42% and 35%, respectively. No change in the arachidonic acid content of the plasmalogen ethanolamine fraction could be detected.

We next investigated the enzymatic mechanisms by which free arachidonic acid is released from (diacyl)-phosphatidylethanolamine and phosphatidylinositol in fetal membranes. We initially investigated the release of arachidonic acid from glycerophospholipids as catalyzed by phospholipase A_2. We and others have found that human intrauterine tissues contain an active phospholipase A_2 (Gustavii, 1972; Schultz et al., 1975; Grieves and Liggins, 1976). The enzyme from chorioamnion which was recovered in the cytosolic and microsomal fractions had a pH optimum of 8 (Okazaki et al., 1978). Enzymatic activity was low in the absence of Ca^{2+} ions and was stimulated by their presence. The alkaline pH optimum and Ca^{2+} requirement would suggest that this enzyme is not of lysosomal origin. To examine the substrate specificity of this phospholipase A_2, the rates of fatty acid release from a variety of molecular species of either phosphatidylcholine or phosphatidylethanolamine were measured (Okazaki et al., 1978). The rate of reaction with 1-palmitoyl, 2-arachidonoyl glycerophosphoethanolamine was approximately 4 times that observed with the corresponding molecular species of phosphatidylcholine. Furthermore, phosphatidylethanolamine containing arachidonic acid at the sn-2 position was a better substrate than phosphatidylethanolamine containing oleic acid at the sn-2 position. The preference of chorioamnion phospholipase A_2 for phosphatidylethanolamine containing arachidonic acid was still evident when a range of different molecular species of phosphatidylcholine and phosphatidylethanolamine were examined as substrates either individually or as mixtures. When rat liver mitochondria were employed as the enzyme source, no preferential release of arachidonic acid compared to oleic acid from phosphatidylethanolamine was observed. Thus, the phospholipase A_2 in fetal membranes and perhaps other tissues that produce significant amounts of prostanoids have a unique phospholipase A_2 which preferentially hydrolyzed arachidonoyl containing glycerophospholipids.

The phospholipase A_2 activity in fetal membranes did not hydrolyze phosphatidylinositol. However, phosphatidylinositol-specific phospholipase C activity was found in these tissues (Di Renzo et al., 1981). The products of phospholipase C action were shown to be inositol phosphates and diacylglycerol. It was found that the phospholipase C of amnion tissue was active optimally in vitro at pH 7, was dependent on Ca^{2+} and was specific for phosphatidylinositol. No specificity of the phospholipase C for arachidonoyl containing phosphatidylinositol was demonstrable. Further support for the physiologic importance of phosphatidylinositol-specific phospholipase C activity was the observation that the diacylglycerol content of amnion tissue increased by 2-fold during early labor (Okita et al., 1982b). Furthermore, the fatty acid composition of the diacylglycerols was almost identical to the phosphatidylinositol fraction in this tissue.

Phosphatidylinositol-specific phospholipase C action by itself would not release arachidonic acid from phosphatidylinositol but rather would produce a diacylglycerol that is rich in arachidonic acid. Since we had previously shown that the phosphatidylinositol fraction was enriched with arachidonoyl esters (Okita et al., 1982a), this diacylglycerol must then be metabolized further to release arachidonic acid. We had previously noted the formation of monoacylglycerol in addition to diacylglycerol at longer incubation time periods (Di Renzo et al., 1981). This observation provided the first evidence that these tissues contained a diacylglycerol lipase.

Diacylglycerol lipase activity was characterized with reference to the biochemical mechanism by which arachidonic acid is released from diacylglycerol (Okazaki et al., 1981b). Diacylglycerol containing a [^{14}C]palmitoyl group in the sn-1 position and [^{3}H]oleoyl in the sn-2 position was synthesized. The rates of formation of the free fatty

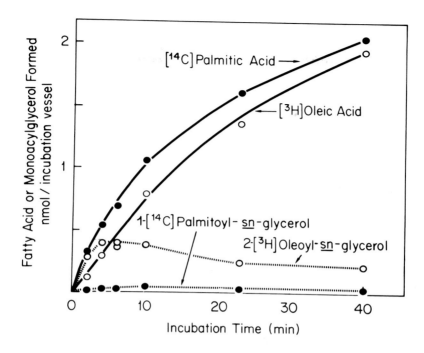

<u>Figure 1</u>. The rate of hydrolysis of 1-[^{14}C]palmitoyl-2-[^{3}H]oleoyl-sn-glycerol employing the microsomal fraction of decidua vera tissue. (Data from Okazaki <u>et al</u>., 1981b, with permission from The American Society of Biological Chemistry).

acid and monoacylglycerols were ascertained employing a microsomal fraction obtained from amnion or decidua vera tissues. The results of these experiments are illustrated in Figure 1. The release of [^{14}C]palmitic acid was greater than that of [^{3}H]oleic acid at all incubation times. In addition, the rates of formation of 2-[^{3}H]oleoyl-containing monoacylglycerol is considerably greater than that of 1-[^{14}C]palmitoyl monoacylglycerol which was almost undetectable at all time periods. These findings are consistent with a reaction sequence in which the fatty acid at the sn-1 position of diacylglycerol is released followed by the hydrolysis of the fatty acid at the sn-2 position. Furthermore, when various diacylglycerols which contained different fatty acids in the sn-2 position were used as substrates, a preferential hydrolysis of diacylglycerol containing arachidonic acid in the sn-2 position was observed (Okazaki *et al.*, 1981b). Monoacylglycerol lipase activity was also detected in amnion, chorion laeve, and decidua vera tissues. Activity of this enzyme appeared to be located predominantly in the cytosolic fraction and the specific activity of monoacylglycerol lipase was approximately 7 times greater than that of diacylglycerol lipase. This latter observation would explain why monoacylglycerol does not accumulate in great quantities during the hydrolysis of diacylglycerols. The monoacylglycerol lipase activity in fetal membranes and decidua vera hydrolyzed preferentially 2-arachidonoyl glycerol. Based on the subcellular distribution of the activities of diacylglycerol lipase and monoacylglycerol lipase, as well as other characteristics of these lipases, it was concluded that these enzymatic activities are attributable to two distinct enzymes. The di- and monoacylglycerol lipase activities were not Ca^{2+} sensitive. Similar results have recently been reported for the metabolism of diacylglycerols in platelets (Marjerus *et al.*, 1983).

The diacylglycerol produced in the amnion can also be phosphorylated by diacylglycerol kinase to produce phosphatidic acid. Diacylglycerol kinase activity was demonstrated in fetal membranes, and decidua vera tissues (Okazaki *et al.*, 1981b). Diacylglycerol kinase activity in the absence of added Ca^{2+} was found to have a higher affinity for diacylglycerol (Km = 0.6 mM) than did diacylglycerol lipase (Km = 2.1 mM) (Sagawa *et al.*, 1982). The Vmax of diacylglycerol lipase, however, was higher than that of diacylglycerol kinase (152 versus 61 nmol x h^{-1} x mg^{-1} protein). Ca^{2+} did not affect the diacylglycerol lipase activity but decreased the activity of diacylglycerol kinase. In Figure 2 are illustrated the results of experiments in which diacylglycerol lipase and diacylglycerol kinase activities were measured simultaneously, in the presence and absence of added Ca^{2+} at two concentrations of diacylglycerol (Sagawa *et al.*, 1982). At the lower concentration of diacylglycerol (0.1 mM) and in the absence of added Ca^{2+}, diacylglycerol was utilized primarily in the diacylglycerol kinase catalyzed reaction. This was also the case even in the presence of Ca^{2+} (5 mM) and diacylglycerol (0.1 mM). When the higher concentrations of Ca^{2+} (5 mM) and diacylglycerol (1mM)

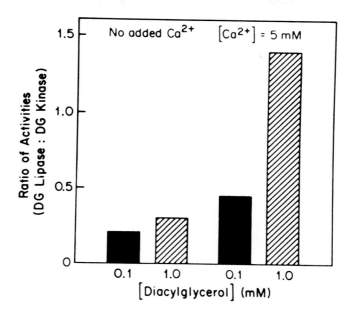

<u>Figure 2</u>. Effect of Ca^{2+} on diacylglycerol utilization in human amnion. 1-Palmitoyl-2-[^{3}H]oleoyl-<u>sn</u>-glycerol (0.1 or 1.0 mM) was incubated with Mg^{2+} (10 mM), ATP (7.5 mM), and the 750 x <u>g</u> supernatant fraction of amnion tissue. Radiolabeled fatty acid released diacylglycerol lipase and monoacylglycerol lipase from the <u>sn</u>-2 position of diacylglycerol and phosphatidic acid formed by diacylglycerol kinase action were separated by thin layer chromatography. The ratio of fatty acid to phosphatidic acid was considered as the ratio of activities of diacylglycerol lipase to diacylglycerol kinase (DG lipase:DG kinase). In the absence of Ca^{2+}, when the diacylglycerol concentration was either 0.1 or 1.0 mM, the ratio of DG lipase:DG kinase, was 0.20 and 0.30, respectively. When these incubations were conducted in the presence of Ca^{2+} (5 mM), the ratio of DG lipase:DG kinase increased to 0.42 and 1.41, respectively. (Data from Sagawa <u>et al</u>, 1982, with permission from The American Society of Biological Chemistry.

Johnston et al.

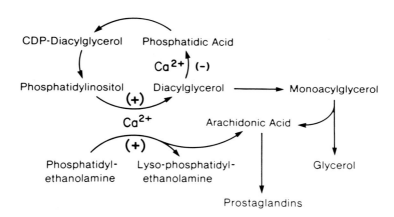

Figure 3. Pathways for the mobilization of arachidonic acid from (diacyl)phosphatidylethanolamine (involving phospholipase A_2) and phosphatidylinositol (involving phospholipase C, diacylglycerol lipase and monoacylglycerol lipase). The proposed points of regulation by Ca^{2+} are indicated by + (stimulation) and by - (inhibition). From Bleasdale and Johnston, 1984 with permission.

were utilized, however, diacylglycerol was metabolized largely by diacylglycerol lipase. These observations are consistent with the view that when diacylglycerol concentration and Ca^{2+} concentration are low, activity of diacylglycerol kinase (and recycling of diacylglycerol) is preferred over that of diacylglycerol lipase. When the diacylglycerol content of amnion increases during early labor (Okita et al., 1982b) and the cytosolic concentration of Ca^{2+} is proposed to increase then diacylglycerol lipase activity predominates.

The experiments described regarding the mobilization of arachidonic acid for prostaglandin production are depicted in the model proposed in Figure 3 (Bleasdale and Johnston, 1984). Arachidonic acid is mobilized from (diacyl)phosphatidylethanolamine by the action of phospholipase A_2 and from phosphatidylinositol by phospholipase C, diacylglycerol lipase and monoacylglycerol lipase. The enzymatic pathways depicted in Figure 3 can account for the selective mobilization of

500

arachidonic acid from phosphatidylethanolamine and phosphatidylinositol that occurs during early labor.

More recently we have investigated the regulation of the enzymatic mechanisms described for the release of arachidonic acid in amnion tissue during early labor. Developmental changes in the activities of these enzymes

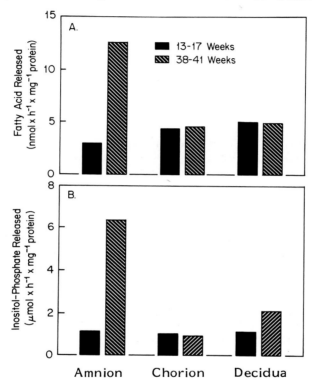

Figure 4. Activity of phospholipase A_2 and phospholipase C in fetal membranes and decidua vera obtained either early or late in gestation. The specific activities of (A) phospholipase A_2 and (B) phospholipase C were measured in amnion, chorion laeve and uterine decidua vera tissues obtained either at 13-17 weeks or 38-41 weeks of gestation. Phospholipase A_2 was assayed employing the 750 x g supernatant fraction of each tissue. Phospholipase C activity was assayed using the 105,000 x g supernatant fraction of each tissue. (Data from Okazaki et al., 1981c, with permission of The Society for the Study of Reproduction).

were first considered. The activities of phospholipases A_2 and C, diacylglycerol lipase, monoacylglycerol lipase and diacylglycerol kinase in fetal membranes and decidua vera obtained either early (13-17 weeks) or late (38-41 weeks) in gestation were assayed under optimal conditions (Okazaki et al., 1981c). The specific activities of phospholipase A_2 and phospholipase C increased markedly during gestation only in amnion tissue (Fig. 4). Although developmental changes in the specific activities of phospholipases A_2 and C were observed, the activities of these two enzymes in fetal membranes and decidua vera obtained at term before labor were not significantly different from those in the corresponding tissues obtained at term after labor.

Since we had previously demonstrated that both phospholipase A_2 and phospholipase C require Ca^{2+} for activity (Okazaki et al., 1978; Di Renzo et al., 1981) and Ca^{2+} inhibits the recycling of diacylglycerol back to phosphatidylinositol (Sagawa et al., 1982), we next considered the possibility that Ca^{2+} may serve a regulatory function in the mobilization of arachidonic acid from phosphatidylethanolamine and phosphatidylinositol. The proposed points at which Ca^{2+} may influence the mobilization of arachidonic acid are also illustrated in Figure 3. The amount of Ca^{2+} necessary to support maximal activities of phospholipases A_2 and C is much greater than that found in the cytosol of cells. For Ca^{2+} to serve as a regulator, therefore, the cytosolic concentration of Ca^{2+} in amnion cells must vary under stimulation and/or the sensitivity of the phospholipases for Ca^{2+} must be modified.

Since polyamines had been shown previously to affect the activity of phospholipase C from rat brain (Eichberg et al., 1981), the influence of polyamines on the lipases of amnion tissue in relation to Ca^{2+} activation was investigated (Sagawa et al., 1983). In the presence of Ca^{2+} (1 mM), the hydrolysis of phosphatidylinositol (2 mM) by phospholipase C was increased greatly by spermine or spermidine but not by putrescine or cadaverine (Sagawa et al., 1983). Polyamine concentration and metabolism are known to be altered in placenta and amniotic fluid as gestation advances (Porta et al., 1978; Russell et al., 1978) and the concentrations of polyamines in the serum of pregnant women are greater than those of non-pregnant women (Russell *et al.*, 1978). The activity of polyamine oxidase in the decidua vera and serum of pregnant women increases as gestation advances (Illei and Morgan, 1979a; 1979b). The potential exists, therefore, for polyamines to participate in the regulation by Ca^{2+} of arachidonic acid mobilization from phosphatidylinositol.

Other mechanisms by which Ca^{2+} could participate in the regulation of arachidonic acid mobilization involve Ca^{2+}-dependent phosphorylation. Nishizuka and colleagues (Takai *et al.*, 1981) have described the presence of protein kinase C in a number of tissues. Kinase C is phospholipid-requiring and Ca^{2+}-dependent. A possible role for protein kinase C in the regulation of arachidonic acid mobilization for thromboxane A_2 formation in platelets has been suggested (Takai *et al.*, 1981). It was found that the Ca^2 requirement of protein kinase C for activity is lowered in the presence of diacylglycerol (Takai *et al.*, 1979). Since a role for Ca^{2+} in the regulation of phospholipase A_2, phospholipase C, and diacylglycerol kinase in the amnion was proposed (Sagawa *et al.*, 1982), and since it was known that the diacylglycerol content of amnion tissue increases during early labor (Okita *et al.*, 1982b), we examined human fetal membranes and decidua vera tissue for protein kinase C activity. Protein kinase C activity was indeed detected. The enzyme has now been partially purified from these tissues (Okazaki *et al.*, 1984). In addition to requiring Ca^{2+}, the protein kinase C from amnion was found to require phospholipid for activity. Protein kinase C in amnion was activated by diolein, however, only at concentrations of Ca^{2+} of 10^{-5} M (Fig. 5). In these experiments, histone was used as an *in vitro* substrate. At least two amnion proteins with molecular weights of 41,000 and 47,000 are phosphorylated by protein kinase C (Okazaki *et al.*, 1984). Whether or not either one of these proteins directly modulate the activity of the phospholipases A_2 and C as has been described by Flower and Blackwell (1979) or Hirata (1981) awaits future investigations.

Recently we have identified platelet-activating factor (PAF) (1-0-alkyl, 2-acetyl-*sn*-glycero-3-phosphocholine) in human amniotic fluid (Billah and Johnston, 1983). PAF was first identified in basophils as an hypotensive agent

503

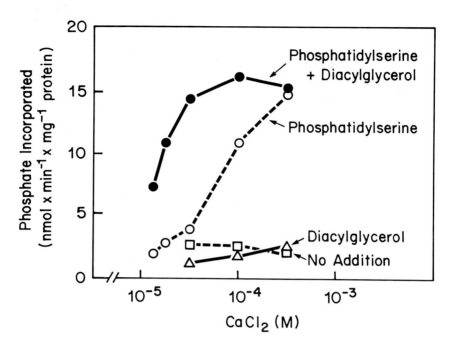

Figure 5. The specific activity of protein kinase C of amnion in the presence of Ca^{2+} at various concentrations and diolein (diacylglycerol) 0.5 μg/tube (Δ—Δ), phosphatidylserine, 10 μg/tube (O—O), diacylglycerol, 0.5 μg/tube plus phosphatidylserine, 10 μg/tube (●—●) or no addition (□---□). H_1 Histone (50 μg/tube) was utilized as the phosphate acceptor. The total volume of the assay was 0.1 ml. (Data from Okazaki *et al.*, 1984, with permission from Academic Press).

(Demopoulos *et al.*, 1979; and Blank *et al.*, 1979; Beneveniste *et. al.*, 1979). PAF is known to induce a rapid increase in the cytosolic concentration of Ca^{2+} in platelets (and some other blood cells) resulting in an activation of phospholipases, mobilization of arachidonic acid and increased prostanoid production (Roukin *et al.*, 1983). We have identified PAF in the amniotic fluid obtained from 14 of 24 women who were in active labor (Billah and Johnston, 1983). PAF was undetectable in all samples of amniotic fluid from women at term but not in labor.

PAF in amniotic fluid may originate from several sources. It was detected in the lamellar body fraction of amniotic fluid, in the urine of newborn infants and in amnion tissue (Billah and Johnston, 1983). Complementing these observations was the detection of activity of 1-0-alkyl-*sn*-glycero-3-phosphocholine:acetyl CoA acetyltransferase (acetyltransferase) in human amnion, chorion laeve, decidua vera, fetal kidney and fetal lung (Ban,

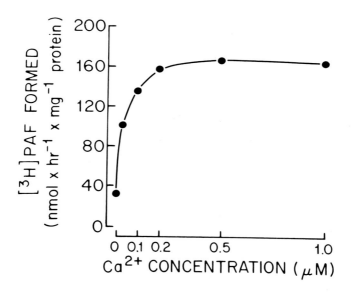

Figure 6. The effect of Ca^{2+} on the biosynthesis of platelet-activating factor (PAF) from lyso-PAF and [^{14}C]acetyl CoA as catalyzed by acetyltransferase obtained from amnion tissue.

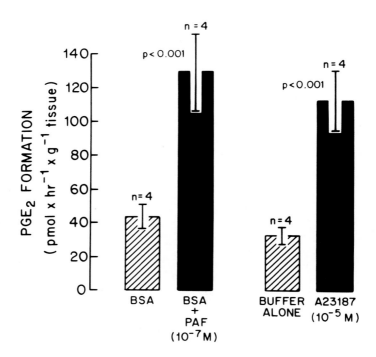

Figure 7. The effect of PAF and a Ca^{2+} ionophore (A23187) on the efflux of PGE_2 from discs of amnion tissue. Amnion tissue discs were maintained in pseudo-amniotic fluid that either contained bovine serum albumin (BSA) (when the effects of PAF were investigated) or lacked BSA (when the effects of A23187 were investigated).

Billah and Johnston, unpublished observations). It has previously been reported that in the adult rat, the kidneys, lungs and brain had the highest specific activity of acetyltransferase (Wykle *et al.*, 1980). Acetyltransferase activity in amnion tissue was stimulated markedly by Ca^{2+} at low concentrations (Fig. 6). A similar Ca^{2+} activation of the acetyltransferase present in rat peritoneal cells (Ninio *et al.*, 1982) and alveolar macrophages (Albert and Snyder, 1983) has been reported. The mechanism of Ca^2 activation of acetyltransferase is presently under investigation. The activity of PAF-acetylhydrolase was

506

also detected in amniotic fluid and amnion tissue (Ban, Billah and Johnston, unpublished observations). The presence of acetylhydrolase activity in amniotic fluid may explain why PAF was not present in some samples of amniotic fluid obtained from women in labor and why it is undetectable in the amniotic fluid obtained from women not in labor. In addition, there was a 2- to 3-fold increase in the amount of PAF (μmol PAF/mol total phospholipids) in the amnion tissue obtained from women in labor. The addition of A23187 to discs of amnion tissue resulted in a 20-fold increase in the PAF concentration in amnion tissue (Billah, Di Renzo, Ban, Anceschi, Bleasdale and Johnston, unpublished observations).

When discs of amnion tissue (approx. 35 mg) were incubated *in vitro* in pseudo-amniotic fluid, PAF (10^{-7} M) stimulated greatly the release of prostaglandin E_2 to the incubation medium (Fig. 7) (Billah, Di Renzo, Ban, Anceschi, Bleasdale and Johnston, submitted for publication). This effect of PAF on prostaglandin E_2 production was similar to that elicited by a calcium ionophore, A23187. Thus PAF satisfied many of the requirements for a Ca^{2+} dependent regulator of prostaglandin production by the amnion. A precise definition of the role played by PAF during parturition awaits extension of the investigations of the metabolism of PAF by intra-uterine tissues and the identification of the major source of the PAF that appears in amniotic fluid during labor.

CONCLUSION

In this chapter we have defined the enzymatic mechanisms for the release of arachidonic acid from (diacyl) phosphatidylethanolamine and phosphatidylinositol. We have also focused on a possible role for Ca^{2+} in the regulation of these enzymatic mechanisms. From these *in vitro* observations it is evident that the maximal effects of Ca^{2+} are only obtained with Ca^{2+} concentrations which are in excess of the intracellular Ca^{2+} concentration with the exception of the lysoPAF:acetyl CoA acetyltransferase which was activated by changes in Ca^{2+} on the order of 1.0 μM. Mechanisms were described which increase the sensitivity to Ca^{2+}. The possible role of the Ca^{2+}-calmodulin complex on the modulation of the reaction se-

507

quence described for arachidonic acid release is yet to be established. Whether or not the mechanisms described account for the activation of the arachidonic acid cascade in relation to the initiation of human parturition as well as other biological systems awaits future investigations.

ACKNOWLEDGEMENTS

We gratefully acknowledge the editorial assistance of Ms. Dolly Tutton. The reported results have been supported by UPHS Grant HD11149 and the Robert A. Welch Foundation, Houston, Texas.

REFERENCES

Albert, D.H. and Snyder, F. (1983) *J. Biol. Chem. 258*, 97-102.
Beneveniste, J., Tence, M., Varenne, P., Bidault, J., Boullet, C. and Polonsky, J. (1979) *C.R. Acad. Sci. 289*,1037-1040.
Billah, M.M. and Johnston, J.M. (1983) *Biochem. Biophys. Res. Commun. 113*, 51-58.
Blank, M.L., Snyder, F., Byers, L.W., Brooks, B. and Muirhead, E.E. (1979) *Biochem. Biophys. Res. Commun. 90*, 1194-1200.
Bleasdale, J.E. and Johnston, J.M. (1984) in: *Reviews in Perinatal Medicine*, Vol. 5 (E.M. Scarpelli, ed.) pp. 151-191, Alan R. Liss, Inc., New York.
Demopoulos, C.A., Pinckard, R.N. and Hanahan, D.J. (1979) *J. Biol. Chem. 254*, 9355-9358.
Di Renzo, G.C., Johnston, J.M., Okazaki, T., Okita, J.R., MacDonald, P.C. and Bleasdale, J.E. (1981) *J. Clin. Invest. 67*, 847-856.
Eichberg, J., Zetusky, W.J., Bell, M.E. and Cavanagh, E. (1981) *J. Neurochem. 36*, 1868-1871.
Flower, R.J. and Blackwell, G.J. (1979) *Nature 278*, 456-459.
Grieves, S.A. and Liggins, G.C. (1976) *Prostaglandins 12*, 229-241.
Gustavii, B. (1972) *Lancet 2*, 1149-1150.
Hirata, F. (1981) *J. Biol. Chem. 256*, 7730-7733.
Illei, G. and Morgan, D.M.L. (1979a) *J. Obstet. Gynecol. 86*, 873-877.

Illei, G. and Morgan, D.M.L. (1979b) *J. Obstet. Gynecol.* *86*, 878–881.

Lands, W.E.M. and Samuelsson, B. (1968) *Biochim. Biophys. Acta* *164*, 426–429.

MacDonald, P.C., Schultz, F.M., Duenhoelter, J.H., Gant, N.F., Jimenez, J.M., Pritchard, J.A. Porter, J.C. and Johnston, J.M. (1974) *Obstet. Gynecol.* *44*, 629–636.

Majerus, P.M., Prescott, S.M., Hofmann, S.L., Neufeld, E.J. and Wilson, D.B. (1983) in: *Advances in Prostaglandin, Thromboxane, and Leukotriene Research"*, Vol. 11 (B. Samuelsson, R. Paoletti and P. Ramwell, eds.) pp. 45–52, Raven Press, New York.

Ninio, E., Mencia-Huerta, J.M., Heymans, F. and Beneveniste, J. (1982) *Biochim. Biophys. Acta* *710*, 23–31.

Okazaki, T., Okita, J.R., MacDonald, P.C. and Johnston, J.M. (1978) *Am. J. Obstet. Gynecol.* *130*, 432–438.

Okazaki, T. Casey, M.L., Okita, J.R., MacDonald, P.C. and Johnston, J.M. (1981a) *Am. J. Obstet. Gynecol.* *139*, 373–381.

Okazaki, T., Sagawa, N., Okita, J.R., Bleasdale, J.E., MacDonald, P.C. and Johnston, J.M. (1981b) *J. Biol. Chem.* *256*, 7316–7321.

Okazaki, T., Sagawa, N., Bleasdale, J.E., Okita, J.R., MacDonald, P.C. and Johnston, J.M. (1981c) *Biol. Reprod.* *25*, 103–109.

Okazaki, T., Ban, C. and Johnston, J.M. (1984) *Arch. Biochem. Biophys.* *229*, 27–32.

Okita, J.R., MacDonald, P.C. and Johnston, J.M. (1982a) *J. Biol. Chem.* *257*, 14029–14034.

Okita, J.R., MacDonald, P.C. and Johnston, J.M. (1982b) *Am. J. Obstet. Gynecol.* *142*, 432–435.

Porta, R., Servillo, L., Abbruzzese, A. and Pietra, G.D. (1978) *Biochem. Med.* *19*, 143–147.

Roukin, R., Tense, M., Mencia-Huerta, J.M., Arnoux, B., Ninio, E. and Beneveniste, J. (1983) *Lymphokines 8*, 240–276.

Russell, D.H., Giles, H.R., Christian, C.D. and Campbell, J.L. (1978) *Am. J. Obstet. Gynecol.* *132*, 649–652.

Johnston et al.

Sagawa, N., Okazaki, T., MacDonald, P.C. and Johnston, J.M. (1982) *J. Biol. Chem.* *257*, 8158-8162.

Sagawa, N., Bleasdale, J.E. and Di Renzo, G.C. (1983) *Biochim. Biophys. Acta* *752*, 153-161.

Schultz, F.M., Schwarz, B.E., MacDonald, P.C. and Johnston, J.M. (1975) *Am. J. Obstet. Gynecol.* *123*, 650-653.

Schwarz, B.E., Schultz, F.M., MacDonald, P.C. and Johnston, J.M. (1975) *Obstet. Gynecol.* *46*, 564-568.

Takai, Y., Kishimoto, A., Kikkawa, U., Mori, T. and Nishizuka, Y. (1979) *Biochem. Biophys. Res. Commun.* *91*, 1218-1224.

Takai, Y., Kishimoto, A., Kawahara, Y., Minakuchi, R., Sano, K., Kikkawa, U., Mori, T., Yu, B., Kaibuchi, K. and Nishizuka, Y. (1981) *Adv. Cyclic Nucleotide Res.* *14*, 301-313.

Vonkeman, H. and Van Dorp, D.A. (1968) *Biochim. Biophys. Acta* *164*, 430-432.

Wykle, R.L., Malone, B. and Snyder, F. (1980) *J. Biol. Chem.* *255*, 10256-10260.

IS PHOSPHATIDYLINOSITOL INVOLVED IN THE RELEASE OF FREE FATTY ACIDS IN CEREBRAL ISCHEMIA?

G.Y. Sun, W. Tang, S.F-L. Huang and L. Foudin

Sinclair Comparative Medicine Research Farm and
Biochemistry Department, University of Missouri,
Columbia, MO   65201

SUMMARY

Since acyl groups of phosphatidylinositol (PI) are enriched in stearic (18:0) and arachidonic (20:4) acids, and these fatty acids are preferentially released during cerebral ischemia, a metabolic relationship between PI and the free fatty acid (FFA) release is suggested. However, several lines of evidence have pointed against a direct involvement of the PI in this event. In our studies, ischemia elicited an increase in the level of diacylglycerols (DG) enriched in 18:0 and 20:4, but the level of PI in rat brain synaptosomes was not changed. Furthermore, active hydrolysis of PI by phosphodiesterase could not be demonstrated in synaptosomes under physiologically feasible conditions, although incubation of the same fraction resulted in a two-fold increase in DG. When rats pre-labeled with $^{32}P_i$ were subjected to post-decapitative ischemic treatment, radioactivity of phosphatidyl-inositol 4,5-bisphosphate (PIP$_2$) and phosphatidate (PA) in synaptosomes was decreased, but that of PI increased. These results suggest that an increase in enzymic degradation of these lipids coupled with hydrolysis of DG by DG lipase may account for the increase in DG and FFA enriched in 18:0 and 20:4 during the early phase of cerebral ischemia.

511

FREE FATTY ACID RELEASE IN CEREBRAL ISCHEMIA

Bazan (1970, 1971) first demonstrated that the FFA pool in brain increases significantly during ischemia and electroconvulsive shock. Subsequently, a similar increase in brain FFA was found in other convulsive conditions induced by drugs, acute hypoxia and hypoglycemia (Bazan, 1976; Siesjo *et al.*, 1982; Rodriguez de Turco *et al.*, 1983; Agardth *et al.*, 1981; Strosznajder *et al.*, 1983). The ischemia-induced FFA release can be attenuated to some extent by anesthetic agents (Shiu and Nemoto, 1981). Examination of the FFA pool with time of ischemic treatment revealed a biphasic mode of increase for 18:0 and 20:4, and the rate of initial release (1 min) was three times more rapid than that in the remaining period (Tang and Sun, 1982). A similar biphasic mode of FFA release was also shown in the gerbil brain after ischemia induced by bilateral ligation of the common carotid arteries (Yoshida *et al.*, 1983). This observation suggests that more than one mechanism is responsible for the FFA release.

Many hypotheses have been suggested for the biochemical mechanism of the ischemia-induced FFA release in brain. There is general agreement that these FFA are derived mainly from the membrane phospholipids. The level of triacylglycerol in brain is too low to account for this release, and based on the acyl group profile of this lipid (Sun, 1970), it is not possible to explain the preferential release of 18:0 and 20:4 (Sun, 1970). On the other hand, the phosphoinositides (PI, PIP and PIP_2) in brain are enriched in 18:0 and 20:4 (Keough *et al.*, 1972), and a similar fatty acid profile is shown for DG (Sun, 1970). The similarity in acyl group profiles suggests that these lipids may contribute directly or indirectly to the FFA release during cerebral ischemia. In mature rats, an increase in brain DG level has been shown to accompany the FFA increase during ischemia and other convulsive states (Tang and Sun, unpublished results; Rodriguez de Turco *et al.*, 1983). The DG released during ischemia are also enriched in 18:0 and 20:4 (Fig. 1). Unlike the FFA release, the DG level reached a peak at 2 min and then showed no further increase (Fig. 2). The same small transient increase in DG was also found in the gerbil model in which ischemia

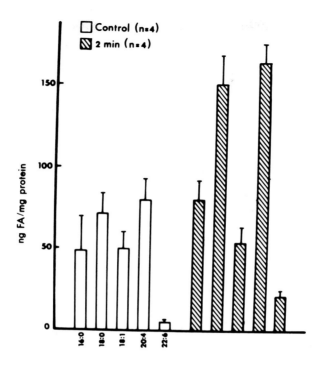

Figure 1. The fatty acids of DG in rat cerebral cortex, comparing controls and rats subjected to a 2 min post-decapitative ischemic treatment. Note that DG enriched in 18:0 and 20:4 were preferentially released during ischemia.

was induced by ligation of the carotid arteries (DeMedio *et al.*, 1980). This characteristic behavior of DG suggests that the lipid undergoes rapid metabolic turnover even during the ischemic treatment.

One of the major difficulties in identifying the lipid(s) responsible for the ischemia-induced lipid change is the small pool of FFA and DG as compared to the large pool of phospholipids in the membrane. Results from our study indicated that approximately 300 ng of FFA/mg protein were released from the rat brain during the first

513

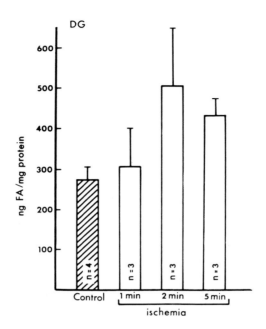

Figure 2. The level of DG in rat cerebral cortex with respect to time of ischemic treatment. The level is expressed as total fatty acid of DG (ng)/mg protein.

minute of post-decapitative ischemia, and approximately 700–900 ng of FFA/mg were released during the entire 5 min period (Tang and Sun, 1982). The increase in DG corresponded to approximately 200–300 ng of fatty acids/mg protein. On the other hand, fatty acids liberated from individual phospholipids of brain homogenate are normally in the µg/mg protein range; e.g., 47 µg for phosphatidylcholine (PC), 9.5 µg for PI, and 2.8 µg for PA. It is apparent that only a small pool of the phospholipids is involved in the FFA release, and this pool is probably metabolically very active. Therefore, special procedures are needed to identify the pool and to study the metabolic changes due to the ischemic treatment.

METABOLISM OF PHOSPHATIDYLINOSITOL IN BRAIN

Previous studies *in vivo* have shown that labeled arachidonic acid is preferentially taken up by PI and PC in brain within minutes after intracerebral injection (Sun and Yau, 1976). This uptake activity is due to the presence of an active acyl-CoA:lysophospholipid acyltransferase(s) which preferentially transfers the arachidonoyl group to lysoPC and lysoPI (Baker and Thompson, 1973; Corbin and Sun, 1978). Among the arachidonoyl-labeled phospholipids, PI in the synaptosomal fraction underwent the most rapid turnover (Sun and Su, 1979). The fast turnover pool normally constitutes only a small portion of the total PI in the tissue. During carbamylcholine stimulation, a condition which also elicits an increase in the FFA pool, there is a decrease in the incorporation of labeled arachidonate into PI in synaptosomes and a concomitant increase in labeled DG (Su and Sun, 1977). Although the exact biochemical involvement which resulted in the observed lipid changes *in vivo* could not be ascertained, either a direct or indirect metabolic relationship between PI and DG is illustrated in the brain stimulation study.

The biochemical phenomenon underlying the acetylcholine-mediated increase in PI turnover in exocrine glands and nervous tissue has been studied extensively in the past (Hokin and Hokin, 1954; Hokin-Neaverson, 1977). Initially, the hydrolysis of PI by a specific Ca^2-dependent phosphodiesterase was suggested as the initiation point for these cellular events, especially those involved in the coupled stimulus-secretion response (Michell, 1975; Michell *et al.*, 1977). PI metabolism was also suggested in synaptic functions involving neurotransmitter release (Hawthorne and Pickard, 1979). However, using 1-acyl-2-[^{14}C]arachidonoyl-GPI as substrate for incubation with synaptosomes, we were not able to detect PI phosphodiesterase activity in this fraction unless deoxycholate and Ca^{2+} were present in the incubation medium (Der and Sun, 1981). Limited PI hydrolysis under physiologically feasible conditions in the presence

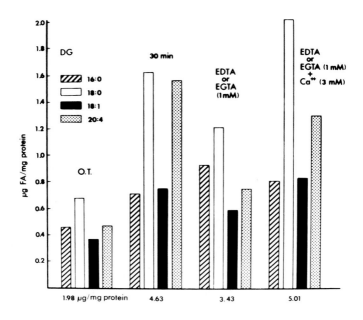

<u>Figure 3</u>. The fatty acids of DG in rat brain synaptosomes before incuba-
tion (O.T.) and after incubation for 30 min, incubation in the presence
of either 1 mM EGTA or 1 mM EDTA, and incubation in the presence of 1 mM
EGTA or EDTA and 3 mM Ca^{2+}. Numbers at bottom of graph represent total
fatty acids in μg/mg protein. Synaptosomes were isolated from brain
homogenates containing no Ca^{2+}-chelating agents.

of Ca^{2+} was demonstrated with synaptosomes in which PI
was prelabeled with [^{14}C]arachidonate, in spite of the
fact that a large amount of PI was hydrolyzed when
detergent was added (Manning and Sun, 1983). On the
other hand, incubation of synaptosomes elicited a doubling
in DG enriched in 18:0 and 20:4 (Fig. 3). In this
experiment, synaptosomes were isolated in the absence of
Ca^{2+} chelating agents, and therefore would contain endogenous Ca^{2+} for mediating the DG release. This is shown
by the fact that EDTA or EGTA (1 mM) partially inhibited
the DG release. The Ca^{2+}-dependency of DG release from
synaptosomes is further illustrated in results shown in
Figure 4 in which addition of EDTA (1-3 mM) completely

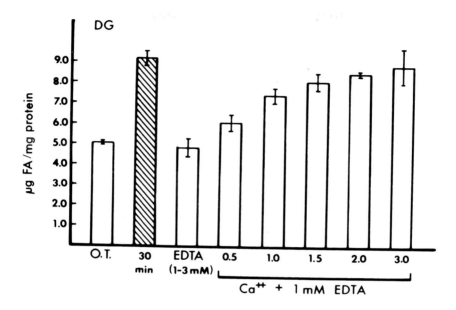

<u>Figure 4</u>. The effects of EDTA and Ca^{2+} on DG release in rat brain synaptosomes during incubation at 37°C for 30 min. The effect of Ca^{2+} was tested by adding various concentrations of Ca^{2+} to synaptosomes containing 1 mM EDTA.

abolished the release, and addition of Ca^{2+} in the presence of EDTA restored the release. Since experiments with labeled PI in synaptosomes have already indicated the absence of a physiologically active phosphodiesterase for hydrolysis of the PI, the DG formed is probably due to breakdown of PA and polyphosphoinositides. However, neither of these phospholipids were appreciably labeled during incubation of synaptosomes with labeled arachidonate.

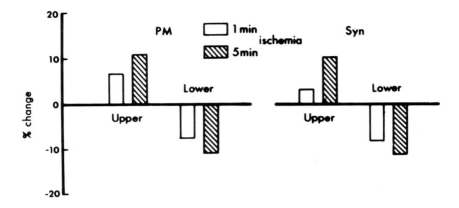

Figure 5. Percent change in radioactivity of upper (aqueous) and lower (organic) phases of a lipid extraction of rat brain plasma membrane (PM) and synaptosomal (Syn) fractions after post-decapitative ischemic treatment. Young rats were injected i.p. with the $^{32}P_i$ to label the phospholipids in brain. After ischemic treatment, the brain subcellular fractions were prepared, and lipids were extracted with acidified chloroform-methanol. Aliquots of the upper and lower phase were taken for measurement of radioactivity. Results are expressed as the percent change of radioactivity relative to controls. (Note: Since the brain homogenates were subjected to subcellular fractionation, the heads were not immersed in liquid nitrogen after decapitation, but brain tissues were removed as rapidly as possible.)

METABOLISM OF PHOSPHOINOSITIDES IN CEREBRAL ISCHEMIA

In order to investigate the metabolism of phospholipids in cerebral ischemia, young rats were injected intraperitoneally with $^{32}P_i$ 12 h prior to post-decapitative ischemic treatment in order to prelabel the phospholipids. The control and ischemic brain samples were homogenized in 0.32 M sucrose containing 50 mM Tris-HCl (pH 7.4) and 1 mM EDTA. The brain homogenates were subjected to differential and sucrose gradient centrifugation, to obtain the plasma membrane and synaptosomal fractions. Lipids from the ^{32}P-labeled membranes were extracted with acidic chloroform-methanol 2:1 (v/v) in order to include extraction of the polyphosphoinositides. Individual phospholipids were separated by two-dimensional thin-layer chromatography (TLC), and lipid spots were taken for measurement of radioactivity. Results indicated that the proportion of radioactivity recovered in the organic phase was decreased by 8 and 12% for the 1 and 5 min ischemic samples, respectively, whereas the proportion of radioactivity for the upper phase was increased (Fig. 5). Similar results were obtained for both plasma membrane and synaptosomal fractions. The result suggests that labeled phospholipids in brain membranes were hydrolyzed during ischemia, and some of the products could be recovered in the aqueous phase upon extraction with the acidic chloroform-methanol 2:1 (v/v).

Using this procedure for labeling the rat brain lipids, all phospholipids including PA and all of the phosphoinositides were labeled. In general, radioactivity was higher in PC and phosphatidylethanolamine (PE) than in the acidic phospholipids. The ischemic treatment induced obvious changes in distribution of radioactivity of the phospholipids in both synaptosomes and plasma membrane fractions. Most noticeably, there was a decrease in radioactivity of the polyphosphoinositides and PA, and substantial changes had already occurred at 1 min after the onset of ischemia. Since the proportion of radioactivity in PE relative to that of total lipids in each fraction was not appreciably altered, this phospholipid was used to compare changes in radioactivity of other phospholipids. The results showed large decreases in polyphosphoinositides/PE and PA/PE ratios with respect to the ischemic treatment. The PC/PE and phosphatidylserine (PS)/PE ratios were decreased slightly, but the PI/PE ratio was increased. In both membrane fractions, the increase in PI/PE ratio was more obvious at 5 min than at 1 min after the ischemic treatment (Fig. 6). Based on the results obtained from this experiment a scheme depicting the enzymic routes which can explain the dramatic decrease in polyphosphoinositides and PA is shown in Fig. 7. It is probable that under normal conditions, the ATP-mediated reactions predominate in maintaining a cyclic equilibrium among these lipids.

Not only was the radioactivity of PI increased during ischemia, but the level of PI in the plasma membrane fraction was also elevated (Fig. 8). This result was

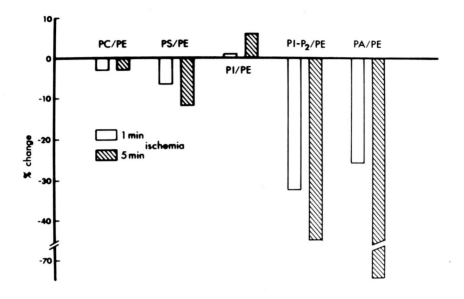

<u>Figure 6</u>. Percent change in radioactivity of individual phospholipids of rat brain plasma membrane (PM) fraction with respect to post-decapitative ischemic treatment. Phospholipids from each fraction were separated by two-dimensional TLC. Radioactivity of individual phospholipids was normalized with respect to that of PE. Experiment procedure was same as described in Figure 5.

obtained from an experiment in which rats were subjected to post-decapitative ischemia and the plasma membrane fraction was isolated from the brain homogenates. The membrane lipids, were extracted, phospholipids were separated by two-dimensional TLC, and individual phospholipid spots were assayed for phosphorus content. In a separate experiment, quantitative analysis of PA in brain homogenates also revealed a small decrease in level due to the ischemic treatment (Fig. 9). However, since several PA pools are known to be present in brain, the small change in level is probably due to the fact that not all of the PA pools are involved in the ischemic event.

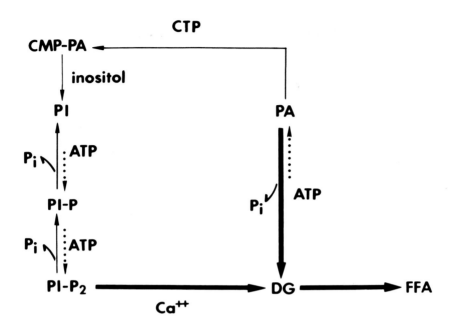

<u>Figure 7</u>. A scheme depicting the metabolic pathways explaining the degradation of PIP$_2$ and PA and their possible involvement in the increase in DG and FFA during early phase of cerebral ischemia.

A rapid post-mortem disappearance of polyphospho-inositides in brain tissue has been observed previously (Dawson and Eichberg, 1965; Eichberg and Hauser, 1967), although this observation was not linked to the FFA release phenomenon. Hauser *et al.* (1971) showed that the loss of polyphosphoinositides is, in general, an event associated with the gray matter, and little change was found in the brain stem which is comprised largely of myelin. This observation is quite similar to that noted for the FFA release in brain (Bazan, 1971). In an earlier study, Gonzalez-Sastre *et al.* (1971) correlated post-mortem treatment to the decrease in specific activity of [^{32}P]PIP$_2$, but not PIP. In our study, these two fractions were too close to be separated.

521

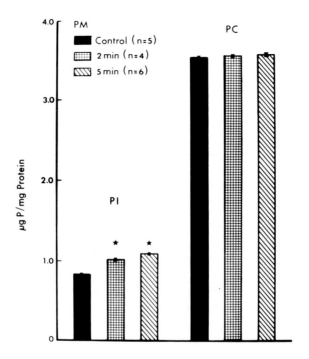

Figure 8. The level of PI and PC in plasma membrane (PM) fraction of rat brain with respect to post-decapitative ischemic treatment (2 and 5 min). Lipids from the PM fractions were separated by two-dimensional TLC, and individual phospholipids were assayed for their phosphorus content. Results are mean ± S.D. from 4-6 animals in each group. The increase in PI level in the 2 and 5 min ischemic groups is significant (p<0.05) based on one way analysis of variance.

The relative increase in radioactivity and mass of PI may be associated partly with the accelerated breakdown of polyphosphoinositides via the phosphomonoesterase route. On the other hand, since ATP availability in cerebral tissue is rapidly depleted during ischemia, inability of PI to become phosphorylated can also lead to its accumulation. Regardless of the explanation, PI is probably not contributing to the rapid increase in DG and

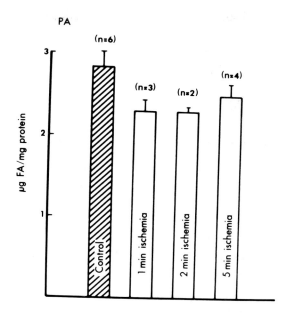

PA

Figure 9. The level of PA in rat cerebral cortex homogenate with respect to post-decapitative ischemic treatment. Lipids were extracted from brain homogenates and were separated by two-dimensional TLC. The PA spots were analyzed for fatty acid content by GLC. Results are mean ± S.D. of μg FA/mg protein with an internal standard.

FFA (18:0 and 20:4) during the initial phase of cerebral ischemia. However, this does not exclude the possibility that some PI may be hydrolyzed by phospholipase A_2, and that the lysophospholipids formed are further degraded by lysophospholipase. This latter possibility was invest-igated in a recent study *in vitro* with synaptosomes prelabeled with [14C]arachidonate in the phospholipids (mainly PI and PC). Results indicated that both arach-idonoyl-PC and PI are hydrolyzed by endogenous phospholipase A_2 in synaptosomes (Sun *et al.*, 1984). The hydrolysis of PI is Ca^{2+}-dependent. A major concern

523

about the involvement of this mechanism in the early events of ischemia is the slow rate of hydrolysis (formation of products is linear up to 1 h of incubation). Therefore, unless regulated otherwise, the phospholipase A_2 process is not able to account for the rapid increase in brain FFA during the initial period of the ischemic insult. Nevertheless, this does not preclude its involvement in the slow phase of the FFA release phenomenon.

METABOLISM OF POLYPHOSPHOINOSITIDES, PHOSPHATIDIC ACID AND DIACYLGLYCEROL IN SYNAPTOSOMES

If FFA release from cerebral ischemia is mediated by the pathways indicated in Fig. 7, it is necessary to demonstrate the presence of enzymes participating in the metabolic conversions. These studies are important for understanding the characteristic properties of the enzymes involved in the proposed mechanism. We have recently discussed some of the difficulties in investigating enzymes using membrane phospholipids as substrates (Sun et al., 1984). Because of these limitations, the active participation of polyphosphoinositides and PA in synaptosomes has not been explored in detail. Previous attempts from our laboratory have failed to effectively label the arachidonoyl group of PIP and PIP_2 for studying their metabolism. However, a recent study by Van Rooijen et al. (1983) has demonstrated that the membrane fractions of nerve-endings contain a membrane-bound, Ca^{2+}-stimulated phosphodiesterase which hydrolyzes the polyphosphoinositides. This result is in good agreement with our earlier observation that although arachidonoyl-labeled PI in synaptosomes did not yield labeled DG upon incubation in the presence of Ca^{2+}, an obvious increase in DG enriched with 18:0 and 20:4 can be observed in the course of the incubation.

The disappearance of labeled PA during ischemia can be attributed to its conversion to PI via CDP-DG or to an enhanced degradation to DG by PA phosphohydrolase. The factors regulating metabolism of PA in brain have not been investigated in detail. If hydrolysis of the poly-phosphoinositides and PA are contributing to the coupled increase in DG and FFA during cerebral ischemia, an active DG lipase must be present in the brain subcellular fractions to mediate the FFA release. Indeed, several DG lipases have been demonstrated in brain (Cabot and Gatt, 1976; 1977; 1978), and enzymic activity is found in microsomes, plasma membranes and cytosol fractions (Faroqui *et al.*, 1984). The moderate increase of DG enriched with 18:0 and 20:4 during ischemia suggests that this lipid is metabolically very active, and consequently may play the role of an intermediate in contributing to the FFA release. Further experiments are needed to investigate the properties of the DG lipase and its response in the ischemic condition.

ACKNOWLEDGEMENTS

The authors thank Dianne Torres for preparing the manuscript and figures. This research program is supported in part by research grant NS 16715 from Department of Health and Human Services.

REFERENCES

Argardth, C.-D., Chapman, A.G., Nilsson, B. and Siesjo, B.K. (1981) *J. Neurochem.* *36*, 490-500.

Baker, R.R. and Thompson, W. (1973) *J. Biol. Chem.* *248*, 7060-7065.

Bazan, N.G. (1970) *Biochim. Biophys. Acta 218*, 1-10.

Bazan, N.G. (1971) *Lipids 6*, 211-212.

Bazan, N.G. (1976) in: *"Function and Metabolism of Phospholipids in Central and Peripheral Nervous Systems"* (G. Porcellati, L. Amaducci and C. Galli, eds.) pp. 317-335, Plenum Press, New York.

Cabot, M.C. and Gatt, S. (1976) *Biochim. Biophys. Acta 431*, 105-115.

Cabot, M.C. and Gatt, S. (1977) *Biochemistry 16*, 2330-2334.

Cabot, M.C. and Gatt, S. (1978) in: *"Enzymes of Lipid Metabolism"* (S. Gatt, L. Freysz and P. Mandel, eds.) pp. 101-112, Plenum Press, New York.

Corbin, D.R. and Sun, G.Y. (1978) *J. Neurochem. 30*, 77-82.

Dawson, R.M.C. and Eichberg, J. (1965) *Biochem. J. 96*, 634-643.

DeMedio, G.E., Goracci, G., Horrocks, L.A., Lazarewicz, J.W. and Trovarelli, G. (1980) *Ital. J. Biochem. 29*, 412-432.

Sun et al.

Der, O.M. and Sun, G.Y. (1981) *J. Neurochem.* *36*, 355-362.

Eichberg, J. and Hauser, G. (1967) *Biochim. Biophys. Acta* *144*, 415-422.

Faroqui, A.A., Pendley, C.E., Taylor, W.A. and Horrocks, L.A. (1984) in: "*Physiological Role of Phospholipids in the Nervous System*" (J.N. Kanfer, L.A. Horrocks and G. Porcellati, eds.) Raven Press, New York (in press).

Gonzalez-Sastre, F., Eichberg, J. and Hauser, G. (1971) *Biochim. Biophys. Acta* *248*, 96-104.

Hauser, G., Eichberg, J. and Gonzalez-Sastre, F. (1971) *Biochim. Biophys. Acta* *248*, 87-95.

Hawthorne, J.N. and Pickard, M.R. (1979) *J. Neurochem.* *32*, 5-14.

Hokin, M.R. and Hokin, L.E. (1954) *J. Biol. Chem.* *209*, 549-558.

Hokin-Neaverson, M. (1977) in: "*Function and Biosynthesis of Lipids*" (N.G. Bazan, R.R. Brenner and N.M. Giusto, eds.) pp. 429-446, Plenum Press, New York.

Keough, K.M.W., MacDonald, G. and Thompson, W. (1972) *Biochim. Biophys. Acta* *270*, 337-347.

Manning, R. and Sun, G.Y. (1983) *J. Neurochem.* *41*, 1735-1743.

Michell, R.H. (1975) *Biochim. Biophys. Acta* *415*, 81-147.

Michell, R.H., Jones, L.M. and Jafferji, S.S. (1977) *Biochem. Soc. Trans.* *5*, 77-81.

Rodriguez de Turco, E.B., Morelli de Liberti, S. and Bazan, N.G. (1983) *J. Neurochem.* *40*, 252-259.

Shiu, G.K. and Nemoto, E.M. (1981) *J. Neurochem.* *37*, 1448-1456.

Siesjo, B.K., Ingvar, M. and Westerberg, E. (1982) *J. Neurochem.* *39*, 796-802.

Strosznajder, J., Tang, W. and Sun, G.Y. (1983) *Neurochem. Int.* (in press).

Su, K.L. and Sun, G.Y. (1977) *J. Neurochem.* *29*, 1059-1063.

Sun, G.Y. (1970) *J. Neurochem.* *17*, 445-446.

Sun, G.Y. and Su, K.L. (1979) *J. Neurochem.* *32*, 1053-1059.

Sun, G.Y. and Yau, T.M. (1976) *J. Neurochem.* *27*, 87-92.

Sun, G.Y., Tang, W., Huang, S.F-L. and Kelleher, J. (1984) in: *"Physiological Role of Phospholipids in the Nervous System"* (L.A. Horrocks, J.N. Kanfer and G. Porcellati, eds.) Raven Press, New York (in press).

Tang, W. and Sun, G.Y. (1982) *Neurochem. Int.* 4, 269-273.

Van Rooijen, L.A.A., Seguin, E.B. and Agranoff, B.W. (1983) *Biochem. Biophys. Res. Commun.* 112, 919-926.

Yoshida, S., Inoh, S., Asano, T., Sano, K., Shimasaki, H. and Veta, N. (1983) *J. Neurochem.* 40, 1278-1286.

ENDOGENOUS PHOSPHOLIPID METABOLISM IN STIMULATED HUMAN PLATELETS: A LINK BETWEEN INOSITIDE BREAKDOWN AND CALCIUM MOBILIZATION

M. Johan Broekman

Divisions of Hematology/Oncology, Departments of
Medicine, New York Veterans Administration Medical
Center and Cornell University Medical College,
New York, NY 10010

SUMMARY

The time course of changes in endogenous phospholipids, induced by ionophore A23187, collagen and thrombin, was compared to arachidonate (AA) oxygenation and changes in chlortetracycline (CTC) fluorescence, an indicator of membrane-bound calcium. The data indicate that AA oxygenation, and thus phospholipase A_2 activity, preceeds changes in phosphatidylinositol (PI) and phosphatidic acid (PA), and suggest that thrombin-induced calcium release from membrane sites may be responsible for activation of this enzyme. In contrast to thrombin, release of membrane-bound calcium upon ionomycin addition was minimal, suggesting that ionophores release calcium from another site. A difference in pattern of phospholipid hydrolyses with A23187, notably a slight lag period and a lack of PA generation, is consistent with this conclusion. In contrast, the initial breakdown of endogenous phosphatidylinositol 4,5-bisphosphate (PIP_2) upon platelet stimulation paralleled changes in CTC-fluorescence, suggesting a close link between PIP_2 breakdown and release of membrane calcium. Calculations indicate that cytoplasmic free Ca^{2+} in stimulated human platelets could be derived from calcium bound to stimulus-sensitive PIP_2.

INTRODUCTION

Stimulation of human platelets with physiological (thrombin, collagen) and nonphysiological agonists (ionophores) leads to release to the medium of contents of intracellular organelles, and culminates in aggregation. These phenomena are accompanied by intracellular events, mobilization of membrane-bound calcium (Gerrard, 1981), phospholipid metabolism (Bell, 1979; Billah, 1982; Broekman, 1980, 1981; Lapetina, 1981; Rittenhouse, 1979, 1981), and oxygenation of AA to eicosanoids (Bressler, 1979; Pickett, 1976). The sequence of events and the question whether they are causally related or occur in parallel in response to the initial stimulus is important. The present data suggest that stimulus-specific phospholipid metabolism, notably of the phosphoinositides, is correlated to release of calcium from membrane-associated sites, a response of platelets to stimulation with thrombin, but not other agents such as AA or ionophore.

EXPERIMENTAL PROCEDURES

Aliquots of 5×10^9 washed platelets/2 ml were incubated (Broekman, 1980, 1981). Stimuli were added in 100 μl 154 mM NaCl after 4 min preincubation. Lipids were extracted and chromatographed (Broekman, 1980, 1981), except for inositides, which were separated (Schacht, 1978) after extraction essentially as described by Cohen (1971). Lipid P (Broekman, 1980) and the "burst" of O_2 consumption due to AA oxygenation (Bressler, 1979) were measured. For CTC fluorescence measurements platelet-rich plasma was incubated (25 min, 37°C) with 50 μM CTC. After platelet washing, changes in CTC fluorescence upon addition of stimulus were recorded at 37°C in a Perkin-Elmer fluorescence spectrophotometer (*cf.* Serhan, 1983), without stirring, in order to measure the initial effects of agonists without superimposition of aggregation effects.

RESULTS AND DISCUSSION

Arachidonate Oxygenation Induced by Ionophore A23187

The "burst" of O_2 consumption by stimulated platelets is a consequence of O_2 utilization during oxygenation

530

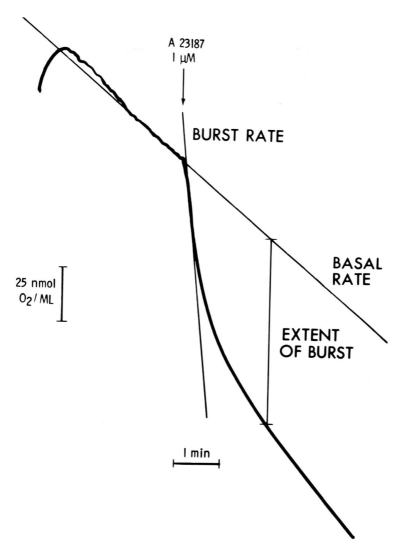

<u>Figure 1</u>. "Burst" of O_2 consumption upon platelet stimulation. Addition of 1 μM A23187 interrupts basal (mitochondrial) O_2 consumption. Following a very slight lag period, a sharp increase in the downward slope of the curve indicates arachidonate oxygenation (Pickett, 1976; Bressler, 1979).

of released or added AA (Bressler, 1979). Thus the shape of the O_2 consumption curve reflects the time course of this oxygenation and by implication measures the kinetics of release of AA. Ionophore A23187 causes an extensive O_2 burst (Fig. 1). Close examination reveals a short, but definite lag period preceding the burst, not occurring upon thrombin or AA stimulation (Bressler, 1979; Pickett, 1976), while with collagen a lag of 15-25 seconds is observed (Bressler, 1979).

Time Course of Changes in Endogenous
Platelet Phospholipids

Ionophore A23187-induced changes in endogenous amounts of PI, PA, phosphatidylethanolamine (PE), phosphatidylcholine (PC), lysoPE and lysoPC (Fig. 2) differed from thrombin-induced changes (Broekman, 1980, 1981): 1. A slight but definite lag period occurred consistently. 2. A23187 induced extensive decreases in PI, but, in contrast to thrombin and collagen, *not* the characteristic increases in PA (Broekman, 1980, 1981). 3. The increases in lysoPE and lysoPC, as well as the decrements in PE and PC generated by 2 μM A23187 were greater than observed with thrombin at doses of maximal stimulation (Broekman, 1980). In addition to changes in PI and PA (Broekman, 1980), collagen led to increases in lysoPE and lysoPC, and decreases in PC and PE (data not shown), indicating phospholipase A_2 activity. The lag in formation of lysoPC and lysoPE corresponded to the lag in the O_2 "burst".

AA (50 μM) causes a large O_2 burst (Bressler, 1979). However, changes in endogenous phospholipid were minimal, and limited to slight but consistent decreases in PI and equivalent increases in PA (data not shown). Since no endogenous generation of free AA was required to produce thromboxane A_2, I speculate that in this case the changes in PI and PA were a consequence of aggregation, rather than a direct effect of cell stimulation.

Kinetics and Stimulus-Dependent Changes
in CTC Fluorescence

The above results, and previous investigations (Broekman, 1980, 1981), indicate that thrombin induced

532

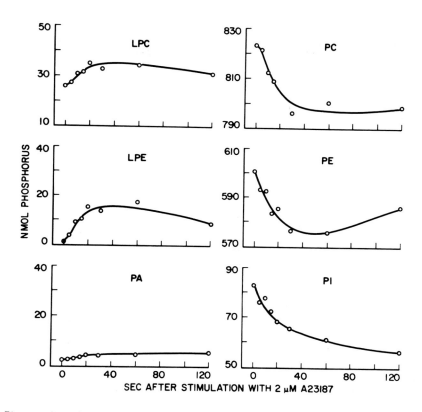

<u>Figure 2</u>. Ionophore A23187 affects platelet phospholipid content. LysoPC and lysoPE rise and PC and PE drop, with a slight lag in lysoPC, lysoPE and PC. There is no change in PA with A23187, in contrast to thrombin (Broekman, 1980; 1981).

the most immediate response of the stimuli studied. The effects of thrombin could be due to direct activation of phospholipase activities, or to activation by a second messenger such as Ca^{2+}. To study this question, the release of membrane-bound calcium was measured by recording fluorescence of CTC loaded platelets upon platelet stimulation. Fig. 3 demonstrates the changes in fluorescence upon addition of thrombin, AA, and ionomycin on CTC fluorescence. Thrombin caused the most immediate, most rapid and most extensive change in fluorescence of the three stimuli, indicating that thrombin

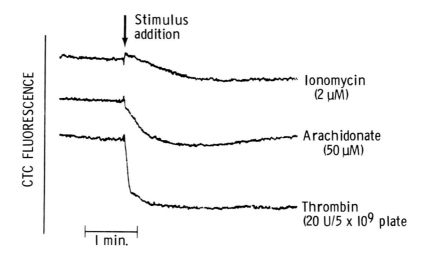

Figure 3. Time course of chlortetracyclin (CTC) fluorescence upon platelet stimulation. See text for details. Note the sharp and extensive drop in CTC fluorescence upon thrombin addition, the minute response to ionomycin, and the intermediate effects of arachidonate.

causes very rapid intracellular release of calcium from membrane sites, and suggesting that this free cytoplasmic Ca^{2+} then might initiate other platelet responses.

In preliminary studies (Broekman, 1983; 1984), the decrease in amounts of endogenous PIP_2 was found to be extremely rapid (5-10 seconds) and extensive (up to 40% of baseline levels of 77 pmol/10^8 platelets). If PIP_2 were to bind calcium in a 1:1 ratio, and the net cytoplasmic volume of platelets were 1/5 their total volume, i.e., 1 femtoliter, the amount of Ca^{2+} released into the cytoplasm upon platelet stimulation could raise total cytoplasmic Ca^{2+} to 0.4 mM, sufficient to raise free Ca^{2+} to 1-2 μM (cf. Rink, 1982). Obviously, this speculation requires more study, and alternative explanations (Streb, 1983) might also account for a link between PIP_2 metabolism and calcium fluxes. Nevertheless, the possibility that PIP_2 binds calcium at a stimulus sensitive site may be an attractive working hypothesis.

ACKNOWLEDGEMENTS

Supported by grants from the Veterans Administration, the American Heart Association, the New York Heart Association, and the National Institutes of Health (HL 29034, HL 18828 SCOR, RR 05396).

Dr. Broekman is an Established Investigator of the American Heart Association.

REFERENCES

Bell, R.L., Kennerly, D.A., Stanford, N. and Majerus, P.W. (1979) *Proc. Natl. Acad. Sci. USA* 76, 3238-3241.

Billah, M.M. and Lapetina, E.G. (1982) *J. Biol. Chem.* 257, 12705-12708.

Bressler, N.M., Broekman, M.J. and Marcus, A.J. (1979) *Blood* 53, 167-178.

Broekman, M.J. (1983) *Blood* 62, Suppl. 1, 251a (abstract 908).

Broekman, M.J. (1984) *Biochem. Biophys. Res. Commun.* 120, 226-231.

Broekman, M.J., Ward, J.W. and Marcus, A.J. (1980) *J. Clin. Invest.* 66, 275-283.

Broekman, M.J., Ward, J.W. and Marcus, A.J. (1981) *J. Biol. Chem.* 256, 8271-8274.

Cohen, P., Broekman, M.J. Verkley, A., Lisman, J.W.W. and Derksen, A. (1971) *J. Clin. Invest.* 50, 762-772.

Gerrard, J.M., Peterson, D.A. and White, J.G. (1981) in: "*Platelets in Biology and Pathology*" (J.L. Gordon, ed.) Vol. 2, pp 407-436, Elsevier/North Holland, Amsterdam.

Lapetina, E.G., Billah, M.M. and Cuatrecasas, P. (1981) *J. Biol. Chem.* 256, 5037-5040.

Pickett, W.C. and Cohen, P. (1976) *J. Biol. Chem.* 251, 2536-2538.

Rink, T.J., Smith, S.W. and Tsien, R.Y. (1982) *J. Physiol.* 324, 53P-54P.

Rittenhouse-Simmons, S. (1979) *J. Clin. Invest.* 63, 580-587.

Rittenhouse-Simmons, S. (1981) *J. Biol. Chem.* 256, 4153-4155.

Serhan, C.N., Broekman, M.J., Korchak, H.M., Smolen, J.E., Marcus, A.J. and Weissman, G. (1983) *Biochim. Biophys. Acta* *762*, 420-428.

Streb, H., Irvine, R.F., Berridge, M.J. and Schulz, I. (1983) *Nature* *306*, 67-69.

PHOSPHATIDYLINOSITOL TURNOVER IN MADIN-DARBY CANINE KIDNEY CELLS: COMPARISON OF STIMULATION BY A23187 AND 12-*O*-TETRADECANOYL-PHORBOL-13-ACETATE

Larry W. Daniel

Department of Biochemistry, Bowman Gray School of
Medicine of Wake Forest University,
Winston-Salem, NC 27103

SUMMARY

Treatment of Madin-Darby canine kidney cells (MDCK) with either A23187 or 12-*O*-tetradecanoyl-phorbol-13-acetate (TPA) causes a rapid deacylation of [^{3}H]arachidonic acid from membrane phospholipids and an increased production of [^{3}H]prostaglandins (Daniel *et al.*, 1984). The present study revealed that A23187 in contrast to TPA stimulated a rapid increase in the incorporation of $^{32}P_i$ or [^{3}H]myo-inositol into phosphatidylinositol (PI). However, TPA did increase the incorporation of $^{32}P_i$ into phosphatidylcholine. Neither A23187 nor TPA stimulated the incorporation of $^{32}P_i$ into phosphatidylinositol 4-phosphate (PIP) or phosphatidylinositol 4,5-bisphosphate (PIP$_2$). Present and previous data indicate that phosphatidylinositol turnover is not a required event for the stimulation of arachidonic acid release and prostaglandin synthesis in TPA stimulated MDCK cells.

INTRODUCTION

Deacylation of arachidonic acid by MDCK cells is stimulated by both TPA and A23187 (Daniel *et al.*, 1984). The major donor of arachidonic acid with either stimulus is phosphatidylethanolamine (PE) although phos-

537

phatidylcholine (PC) and PI are also deacylated (Daniel *et al.*, 1981). These results differ from those obtained with a variety of other cells including platelets (McKean *et al.*, 1981). Platelets have been demonstrated to derive a portion of the released arachidonic acid from the sequential hydrolysis of PI by a PI-specific phospholipase C and diacylglycerol lipase (Bell *et al.*, 1979). PI hydrolysis has been proposed to be a regulatory event which may participate in Ca^{2+} dependent cellular responses. PI hydrolysis thus has been suggested to be an initial step in the response of platelets to thrombin which subsequently stimulates phospholipases A_2 that deacylate other phospholipids (Billah *et al.*, 1980). The present study was designed to determine if PI turnover was a required event for the deacylation of arachidonic acid in the MDCK cell.

EXPERIMENTAL PROCEDURES

Radiolabeled myo-[2-^{3}H]inositol, 5 Ci/mmol was purchased from Amersham Corp. Carrier free $^{32}P_i$ was purchased from ICN Chemical Co., Irvine, CA. PIP and PIP_2 standards were from Sigma, Saint Louis, MO. Sources of other chemicals were as described (Daniel *et al.*, 1984).

Cells were cultured as described previously (Daniel *et al.*, 1984) and treated with TPA or A23187 in Hepes buffered saline (CaCl$_2$, 100 mg/l; KCl, 200 mg/l, MgSO$_4$, 59.2 mg/l; NaCl, 8 g/l; phenol red, 17 mg/l; and N-2-hydroxyethylpiperazine-N'-L-ethane sulfonic acid, 20 mM) or in cell culture medium.

Phospholipids were extracted as described by Billah and Lapetina (1982) and separated on Silica Gel 60 thin layer plates (E. Merck, Darmstadt, West Germany) which were prerun in a solvent system of methanol/water (2:1 v/v) containing 1% potassium oxalate. The phospholipids were identified by comparing their migration to known standards. Radioactivity was measured by autoradiography using Kodak SB-5 film or by scintillation counting (Daniel *et al.*, 1981).

538

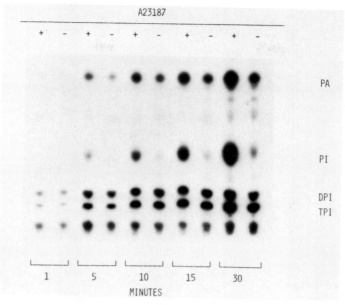

<u>Figure 1</u>. Effect of A23187 on $^{32}P_i$ incorporation into phospholipids. MDCK cells were incubated with or without A23187, 10 µM, in Hepes-buffered saline containing 100 µCi/ml $^{32}P_i$. At the indicated times, the cells were extracted and the lipid extract was analyzed by thin-layer chromatography as described in Experimental Procedures. DPI, phosphatidylinositol 4-phosphate; TPI, phosphatidylinositol 4,5-bisphosphate.

RESULTS

The incorporation of $^{32}P_i$ into the cellular phospholipids of MDCK cells treated with 10 nM TPA, or 10 µM A23187 was determined since these concentrations were previously shown to result in a marked increase in deacylation of [^3H]arachidonic acid (Daniel *et al.*, 1984). These experiments determined that A23187 but not TPA stimulated the incorporation of $^{32}P_i$ into PI and PA. Results of a representative experiment are shown in Fig. 1. Under these conditions the increased incorporation of $^{32}P_i$ into PA and PI was demonstrable within 5 min. There was a rapid synthesis of PIP and PIP$_2$; however, the synthesis of these phospholipids was not increased by A23187 treatment. When the cells were labeled with $^{32}P_i$ to a constant specific activity the PIP

Daniel

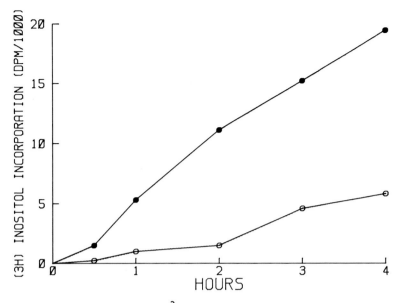

<u>Figure 2</u>. Stimulation of [³H]inositol incorporation into cellular phospholipids by A23187. MDCK cells stimulated with A23187 (10 μM) (●—●) or control cells (○—○) were incubated with [³H]inositol (0.5 μCi/ml) in Hepes buffered saline.

and PIP_2 comprised <3% of the total cellular phospholipid. Therefore, these phospholipids are quantitatively minor components of the MDCK cells; however, they appear to undergo rapid synthesis and degradation in both A23187 stimulated and unstimulated cells.

A23187 also caused an increase in the incorporation of [³H]inositol into PI (Fig. 2). The increased incorporation was detectable within 30 min and at 4 h after stimulation was 4-fold higher than the incorporation by control cells. Increased ³²P$_i$ and [³H]inositol incorporation could be demonstrated in experiments which used Hepes-buffered saline (Fig. 1 and Fig. 2) and in experiments which used complete medium containing serum (data not shown). TPA did not increase the incorporation of [³H]inositol (data not shown).

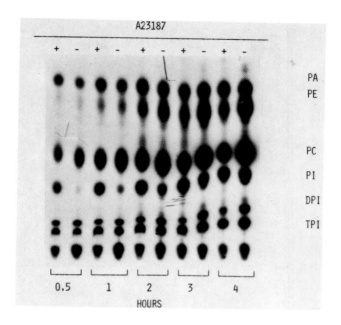

Figure 3. Effect of TPA pretreatment on the incorporation of $^{32}P_i$ into phospholipids. MDCK cells were treated with TPA (10 nM) for 1 h in serum free medium then incubated in medium without TPA for 12 h. The medium was then removed and the cells were incubated with or without A23187 (10 µM) in Hepes buffered saline containing 100 µCi/ml of $^{32}P_i$.

To determine if TPA treatment had effects on the A23187 induced stimulation of PI synthesis, cells were pretreated with TPA for 12 h and then labeled with $^{32}P_i$ for various times in the presence or absence of A23187. TPA did not have an effect on the A23187 stimulated incorporation of $^{32}P_i$ into PI (Fig. 3). However, the TPA pretreatment did increase the incorporation of $^{32}P_i$ into PC (see Figs. 1 and 3).

DISCUSSION

The studies presented herein demonstrate that treatment of MDCK cells with A23187 in contrast to TPA results in an increased incorporation of $^{32}P_i$ or [^3H]-inositol into PI. Therefore, although both A23187 and

541

Daniel

TPA stimulate the deacylation of arachidonic acid from cellular phospholipids, they have very different effects on the synthesis and turnover of phospholipids. These data also indicate that PI turnover is not a required signal for the deacylation of arachidonic acid and prostaglandin synthesis by the MDCK cells. However, TPA and A23187 may be initiating a cascade of events at different points. Recently, Nishizuka (1983) has proposed that the bio-logical effect of PI turnover is to provide diacylglycerol which stimulates protein kinase C. TPA has been shown to stimulate the kinase directly (Castagna *et al.*, 1982) without a requirement for diacylglycerol formation.

The present data is consistent with the hypothesis that TPA stimulation results in the activation of protein kinase C without the requirement for diacylglycerol production due to PI turnover. However, TPA in con-trast to A23187 stimulates PC synthesis in the MDCK cells. Paddon and Vance (1980) have also found that TPA stimulates the synthesis of PC in HeLa cells. These findings underscore the pleiotropic nature of cellular responses to TPA and indicate that there may be addi-tional mechanisms of action for the phorbol diesters.

ACKNOWLEDGEMENTS

This work was supported by Grant R80770 from the Environmental Protection Agency, CA12197 from the National Cancer Institute, AM11799 from the National Institutes of Health, a grant from the North Carolina United Way and by the Bowman Gray School of Medicine. The author wishes to thank Gwen Charles for editorial assistance in the preparation of this manuscript.

REFERENCES

Bell, R.L., Kennerly, D.A., Stanford, N. and Majerus, P.W. (1979) *Proc. Natl. Acad. Sci. U.S.A. 76*, 3238-3241.
Billah, M.M. and Lapetina, E.G. (1982) *J. Biol. Chem. 257*, 12705-12708.
Billah, M.M., Lapetina, E.G. and Cuatrecasas, P. (1980) *J. Biol. Chem. 255*, 10227-10231.

Castagna, M., Takai, Y., Kaibushi, K., Sano, K., Kikkawa, V. and Nishizuka, Y. (1982) *J. Biol. Chem.* *257*, 7847-7851.

Daniel, L.W., King, L. and Waite, M. (1981) *J. Biol. Chem.* *256*, 12830-12835.

Daniel, L.W., Beaudry, G.A., King, L. and Waite, M. (1984) *Biochim. Biophys. Acta, 792*, 33-38.

McKean, M.L., Smith, J.B. and Silver, M.J. (1981) *J. Biol. Chem.* *256*, 1522-1524.

Nishizuka, Y. (1983) *Trends Biochem. Sci. 8*, 13-16.

Paddon, H.B. and Vance, D.E. (1980) *Biochim. Biophys. Acta 620*, 336-340.

DISCUSSION

Summarized by Bernard W. Agranoff

Mental Health Research Institute, Neuroscience Laboratory
University of Michigan, 1103 E. Huron,
Ann Arbor, MI 48109

This session dealt with the release of arachidonate in various tissues and its possible relationship to the stimulated breakdown of inositol lipids. This possibility has long suggested itself because of the enrichment in arachidohate in the sn-2 position of inositol lipids. In a question by Dr. Berridge regarding a possible analogy between the "agonist-sensitive" pool for arachidonate release and that for inositides, Dr. Majerus emphasized that while very little inositide breakdown is needed to produce a physiological secretory response in systems described thus far in the conference, platelets, polymorphonuclear leukocytes, and the cultured cells in which arachidonate release is being studied are characterized by a profound breakdown of inositides. Thus, in the former case a few percent of inositol lipid might breakdown, while more than half of the inositides could breakdown in the latter case. While this does not preclude a common mechanism, the quantitative difference should not be ignored. Dr. Zelenka asked whether any arachidonate metabolites are necessary for growth of the mutant cells. Dr. Majerus responded that when grown in minimal medium, the cells had a slightly lower doubling time and cell density than the parent cells. They also differ in morphology. Dr. Graff (A.H. Robins Co.) inquired as to whether an aberrant cyclooxygenase or alternatively, a PGH isomerase, might be present. Dr. Majerus responded that added arachidonate is converted to PGE_2 by mutant cells in a manner similar to that of parent cells with a similar K_m for the cyclooxygenase. Possible substitution by PGH_2 has not been examined.

Dr. Rittenhouse described the action of a number of agonists of platelet activation, pointing out that while they all could lead to DG formation, they may not share a common mechanism. Dr. Berridge cited evidence that collagen can activate secretion without raising intracellular Ca^{2+}. The possibility that DG could be liberated without concomitant production of IP_3 was discussed as an explanation for these findings. Dr. Rittenhouse cited experimental evidence which discounted a number of possible sources of DG, such as triacylglycerol and PA. Dr. Berridge referred to a possible explanation suggested by Dr. Hokin-Neaverson, that PIP breakdown could have such an effect, since DG and IP_2 rather than IP_3 would be released. It was agreed that more evidence is needed to resolve this matter. In response to a question by Dr. Shukla (Univ. Texas, San Antonio), Dr. Rittenhouse stated that about 85% of PI broke down via phospholipase C, while about 10% appeared to go via the phospholipase A_2 pathway. Breakdown was seen not only in ^{32}P-labeled lipids, but in arachidonate-labeled lipids as well. A question was raised by Dr. Cockcroft as to whether all of the PI breakdown was via PIP_2, or whether PI breakdown could still account for the primary agonist-triggered event. Dr. Rittenhouse responded that, in her opinion, at present the two possibilities were of equal probability.

The question arose as to whether arachidonate metabolism associated with parturition could involve the $PI-PIP_2$ cycle. Dr. Michell mentioned work by Dr. Kirk that was indicative that [^3H]inositol-prelabeled rat uterus released IP_2 and IP_3 in response to oxytocin. Dr. Johnston noted that the amount of PI breakdown would appear inadequate to explain his results, but acknowledged that an arachidonate flux through the inositides could still occur. Concern and caution were expressed (by Drs. Rittenhouse, Williamson, Fain) in regard to proposed roles of Ca^{2+} in intracellular regulatory mechanisms, principally because of the relatively high levels of Ca^{2+} required for enzyme activation in comparison to that found intracellularly.

Dr. Sun and colleagues presented evidence that appearance of arachidonate and stearate in brain following ischemia could result from breakdown of polyphosphoinositides and subsequent hydrolysis of DG. Dr. Rubin pointed out that hydrolysis of PI to lysoPI with rapid resynthesis could be occurring under these circumstances. Dr. Eichberg stated that rapid breakdown of PIP_2 and PIP had previously been demonstrated (Eichberg and Hauser), yet in the present studies only PIP_2 appeared to breakdown. Dr. Sun noted that her studies were based on ^{32}P labeling rather than chemical amounts of the lipids, and this might, in part, account for the difference in results.

The hypothesis presented by Dr. Broekman that PIP_2 breakdown elicited by thrombin released membrane-bound Ca^{2+} which could then activate phospholipase A_2 in platelets was criticized on a number of grounds. It was pointed out that in the case of liver cells at least, the amount of Ca^{2+} liberated was far too great to be accounted for without invocation of an amplification step (Michell), and furthermore, that there was evidence for IP_3 release in platelets (Berridge, Agranoff), although Dr. Broekman has not yet observed its presence. It was also noted that the kinetics of PIP_2 breakdown could be accounted for by flux through PIP_2 (Majerus). Dr. Putney noted that in the resting state, PIP_2 is rapidly turning over, primarily via monoesterase degradation, so that the stimulated breakdown by phosphodiesterase does not reflect so much a change in the level of PIP_2 as it does in production of IP_3 and DG. It was generally agreed that many of the questions raised could be resolved experimentally.

Evidence that PI turnover may not be required for increased arachidonate release came from studies by Dr. Daniel on stimulation of canine kidney cells by A23187 and by phorbol ester. Both caused arachidonate release, while only the ionophore caused increased ^{32}P incorporation into PC. Other participants (Drs. Rubin and Davis) reported similar findings and Dr. Davis suggested that phorbol esters might stimulate the diesteratic phospholipase C cleavage of PC.

In summary, a precise relationship between stimulated inositide turnover and arachidonate release remains elusive. Evidence at present appears to suggest that Ca^{2+} mobilization required for phospholipase A_2 activation requires signal amplification. For the moment, the search for parsimony in interpretation of results must yield to the improbabilities that all agonists act in a similar fashion, or that different tissues respond to a given agonist in exactly the same way.

PART VI

PHOSPHOINOSITIDE METABOLISM IN THE NERVOUS SYSTEM

INOSITOL, SORBITOL AND DIABETIC NEUROPATHY

J.N. Hawthorne, E.M. Smith, K.R. Gillon,
and F.A. Millar

Department of Biochemistry, Medical School,
Queen's Medical Centre, Nottingham NG7 2UH, U.K.

SUMMARY

There is a reduced concentration of free inositol in sciatic nerve from rats made diabetic by injection of streptozotocin and in the same nerve obtained post-mortem from diabetic patients. Phosphatidylinositol concentration is reduced in sciatic nerves from acutely diabetic rats. Endoneurial preparations from less severely diabetic animals show decreased polyphosphoinositide. These changes may contribute to the neuropathy shown initially in these animals by reduced conduction velocity. Aldose reductase inhibition by the drug sorbinil lowers sorbitol concentrations in nerves from diabetic animals and at the same time restores inositol concentration to normal. Genetically diabetic mice show reduced sciatic nerve conduction velocity and lack of inositol, but have no additional sciatic nerve sorbitol. This suggests that lack of inositol is more important than excess sorbitol in the development of conduction defects. There is also evidence that phosphoinositides may have some connection with the sodium pump ATPase. This enzyme has been assayed in microsomal fractions from sciatic nerve. It appears less active in nerves from diabetic rats. The assay method used with homogenates of nerve by other workers has not been successful in our hands.

551

TABLE I

Free and Lipid Inositol of Sciatic Nerve in Diabetes

		Free Inositol	Lipid Inositol	References
Mouse	Control	2.71	2.14	Gillon and Hawthorne, 1983b
	Diabetic strain C57 BL/KS (db/db)	2.07	1.43	
Wistar rat	Control	3.17	--	Greene et al., 1975
	Streptozotocin 70 mg/kg, 2 wks	2.19	--	
	Control	3.10	1.89	Palmano et al., 1977
	Streptozotocin 35 mg/kg, 13 wks	2.44	1.86	
Human	Control biopsy	--	1.7*	Brown et al., 1979
	Diabetic patients biopsy	--	0.5*	
	Control post-mortem	3.33	1.11	Mayhew et al., 1983
	Diabetic patients post-mortem	1.91	0.71	

*Sural nerve; combined phosphatidylinositol and phosphatidylserine figures (mg/100 mg dry weight). All other figures are μmol/g fresh weight.

552

INOSITOL AND INOSITOL LIPID CONCENTRATIONS IN PERIPHERAL NERVE OF DIABETIC ANIMALS AND PATIENTS

It is now well established that sciatic nerves of rodents made diabetic with streptozotocin contain less free *myo*-inositol than controls, as first shown by Greene *et al.* in 1975. Selected figures from various laboratories are given in Table I. We have recently shown similar loss of inositol in post-mortem nerve samples from diabetic patients (Mayhew *et al.*, 1983). The phospholipid changes are less clear at present but as the Table shows, most workers find less lipid inositol in peripheral nerve of diabetic animals or human subjects than in healthy controls.

We have now made preliminary studies of the individual phosphoinositide concentrations in endoneurial preparations obtained from sciatic nerve of streptozotocin-diabetic rats. This nerve preparation (Greene *et al.*, 1979) consists essentially of axons and their surrounding Schwann cells. It is free from epineurium (connective tissue and adipocytes) and the perineurial membrane. As will be seen from Table II, the endoneurial preparations from diabetic rats contained somewhat lower concentrations of polyphosphoinositides, but phosphatidylinositol concentrations were normal. The endoneurial preparations had been used for metabolic studies and had been incubated for 60 min before analysis of the phosphoinositides. Previous studies showed that under these conditions the tissue maintained ATP, phosphocreatine, glucose and inositol concentrations (Gillon and Hawthorne, 1983a). It is unlikely that lack of ATP is responsible for the reduced polyphosphoinositide levels in endoneurial preparations from diabetic animals. Other possible explanations involving changes in enzyme activity or effects of the sorbitol which accumulates in diabetes are discussed below.

ENZYMES OF PHOSPHOINOSITIDE METABOLISM IN SCIATIC NERVE OF DIABETIC RATS

Some years ago we showed that CDP-diacylglycerol: inositol phosphatidyltransferase was much less active in sciatic nerve of streptozotocin-diabetic rats than in con-

TABLE II

Phosphoinositides of Rat Sciatic Endoneurial Preparations

	Control (7)	Diabetic (8)
PIP_2	33.2 ± 0.8	30.5 ± 0.9*
PIP	11.2 ± 0.9	6.1 ± 0.4**
PI	19.1 ± 0.7	18.2 ± 1.2

Abbreviations: PIP_2, phosphatidylinositol 4,5-bisphosphate; PIP, phosphatidylinositol 4-phosphate; PI, phosphatidylinositol. Figures represent μg P/mg total lipid P ± S.E.; number of preparations in parenthesis. Differences from control, significance using Student's t-test: *$p < 0.05$, **$p < 0.001$. The diabetic rats had been given a single intraperitoneal injection of streptozotocin (35 mg per kg body weight) 6 wks previously. They had all shown a serum glucose concentration higher than 16 mM 2 days after this injection. Each endoneurial preparation had been incubated for 60 min at 37°C in the Krebs-Ringer bicarbonate buffer with bovine serum albumin, 5 mM glucose and 50 μM inositol as used previously for studies of inositol transport (Gillon and Hawthorne, 1983b). Phospholipids were extracted by the method of Griffin and Hawthorne (1978) and separated for analysis by phosphate determination using one-dimensional TLC (Bell et al., 1982).

trols (Whiting et al, 1979). The rats had been given 35 mg/kg body weight streptozotocin 12 weeks previously. No change was seen in PI kinase activity, but there was a decrease in PIP kinase, significant only at the $p < 0.1$ level. Assays were made using homogenates of the whole sciatic nerve and this may account for the variable results.

The reduced activity of CDP-diacylglycerol:inositol phosphatidyltransferase and the lack of free inositol probably account for the decreased concentrations of lipid inositol in nerves from diabetic animals (Table I). Natarajan et al. (1981) also report considerable reductions of PI concentration in sciatic nerve from rats made diabetic 20 weeks previously, using a single injection of streptozotocin (50 mg/kg).

It is possible that the reduced activity of nerve enzymes in diabetes is a consequence of impaired axonal flow. Since tubulin is involved in this process, the recent finding (Williams et al., 1983) that nonenzymatic glycosylation of this protein inhibits its polymerization could be important.

SORBITOL-INOSITOL INTERACTIONS IN SCIATIC NERVE OF DIABETIC ANIMALS

It is well known that sorbitol and fructose accumulate in peripheral nerves of diabetic animals, the former being produced from glucose by aldose reductase. We have recently studied the effects of sorbinil, an inhibitor of this enzyme, on sorbitol and inositol concentrations in sciatic nerve of streptozotocin-diabetic rats. Sorbinil (6-fluorospiro(chroman-4,4'-imidazolidine)-2',5' dione) is produced by Pfizer, U.S.A. The rats had received one intraperitoneal injection of streptozotocin 35 mg/kg). After six weeks, motor nerve conduction velocity was significantly less than in control rats (Gillon and Hawthorne, 1983b). Over a period of a further 9 weeks it was restored to normal by a daily dose of 25 mg/kg sorbinil given by gastric intubation. A similar beneficial effect was produced by a daily dose of 650 mg/kg inositol given in the same way. At the end of the 9 week period of treatment, sugars and polyols of sciatic nerve were analyzed. Glucose concentration rose from 2.56 μmol/g wet weight to 8.21 μmol/g and was unaffected by sorbinil or inositol administration. In nerves of diabetic rats, the sorbitol concentration was 2.90 μmol/g (controls 0.24 μmol/g). Sorbinil treatment reduced this concentration to 0.15 μmol/g but inositol had no significant effect. Sciatic nerve inositol concentration fell from 3.92 μmol/g in controls to 2.98 μmol/g in diabetic rats. Interestingly, it was restored to normal by sorbinil administration. This suggests a link between sorbitol and the active transport of inositol into peripheral nerve, though studies with endoneurial preparations *in vitro* showed only increased accumulation of inositol when the incubation medium contained 30 mM sorbitol (Gillon and Hawthorne, 1983a). This experiment does not provide a very good model of the diabetic state however, since in diabetic animals the sorbitol accumulates within the nerve rather than in the fluids around it. Administration of inositol to diabetic rats also restored the nerve concentration of this substance to normal values (Gillon and Hawthorne, 1983b; Gillon *et al.*, 1983).

Slowing of nerve conduction is an early sign of diabetic neuropathy and is seen in the genetically diabetic mouse (strain C57 BL/KS (db/db)). In comparison with normal mice, the concentration of glucose is greatly increased in sciatic nerve, but sorbitol and fructose are normal (Gillon and Hawthorne, 1983b). Inositol concentration, however, is reduced to 2.07 μmol/g in the diabetic mice compared with 2.71 μmol/g in control animals. This suggests that lack of inositol is more important in relation to diabetic neuropathy than excess of sorbitol. In view of the comments above, it may also suggest that the lack of inositol in sciatic nerve is not a consequence of sorbitol accumulation, but that it is due to some other factor in diabetes. If this is true, the aldose reductase inhibitor sorbinil corrects lowered inositol concentrations in nerves of diabetic animals by some interaction with this unknown factor and not simply by lowering sorbitol concentration.

PHOSPHOINOSITIDE METABOLISM IN NERVES OF DIABETIC ANIMALS

Endoneurial preparations from sciatic nerve of diabetic rats show decreased concentrations of PIP and PIP$_2$ (Table II). Treatment with sorbinil for 9 weeks at the dose used above restored the concentrations to normal, though inositol treatment had no effect (Table III). We have also used endoneurial preparations from diabetic and normal rats to measure ^{32}P incorporation *in vitro* into the phosphoinositides (Table IV). Labeling of the polyphosphoinositides over a 60 min period was significantly greater in the tissue from the diabetic animals.

TABLE III

Rat Sciatic Endoneurial Phosphoinositides After Sorbinil or Inositol Treatment

	Control	Diabetic	Diabetic + Sorbinil	Diabetic + Inositol
PIP_2	32.9 ± 1.8	28.8 ± 0.9	36.7 ± 1.0*	26.4 ± 0.5
PIP	11.1 ± 1.0	9.3 ± 0.2	13.9 ± 1.4**	8.4 ± 0.4
PI	20.2 ± 1.0	21.2 ± 0.9	25.1 ± 2.8	19.3 ± 0.3

Figures represent µg P/mg total lipid P and are means from about 8 rats ± S.E. Significantly different from untreated diabetic value: *$p <$ 0.001, **$p <$ 0.01. Doses of sorbinil and inositol are given in the text. Each was administered by gastric intubation daily over a 9 week period. Details of analytical methods are given with Table II.

The reduction in PI labeling was not statistically significant. The Table gives specific radioactivities (pmol ^{32}P incorporated per µmol phosphoinositide P) and results are thus difficult to compare with previous studies. Natarajan et al. (1981) expressed results as percentages of total phospholipid radioactivity and found increased labeling of PIP but decreased labeling of PIP_2 when whole sciatic nerve from diabetic animals was incubated for 60 min. Bell et al. (1982) found increased labeling of PIP_2 in sciatic nerve of diabetic rats and no appreciable change in the concentrations of the lipid.

At the present time, it is not possible to interpret these results using the enzyme determinations mentioned above (Whiting et al., 1979). In diabetic rats, sciatic nerve PIP kinase was less active than in controls, while PI kinase was not affected by diabetes. The increased labeling with ^{32}P of the polyphosphoinositides may reflect availability of labeled ATP rather than kinase activity. Pathways competing for this ATP may be less active so that more is available for synthesis of PIP_2. The phos-

TABLE IV

Labeling of Rat Sciatic Endoneurial
Phosphoinositides with ^{32}P *in vitro*

PIP$_2$	203 ± 16	252 ± 13*
PIP	170 ± 16	385 ± 22**
PI	53 ± 15	26 ± 7

The incubation was for 60 min in 2 ml of the medium described in Table II, to which 20 µCi [^{32}P]orthophosphate was added. The figures represent pmol ^{32}P incorporated per µmol lipid P and are means ± S.E. from 7 preparations. Significances of difference from control by Student's t test are as follows: *p < 0.05, **p < 0.001. Analytical methods and induction of diabetes are also detailed in Table II. Phospholipid spots were detected on TLC plates by iodine vapor and scraped off for digestion at 180°C in 0.4 ml 70% perchloric acid until clear. After addition of 2.1 ml of water, 0.5 ml of the digest was taken for scintillation counting of ^{32}P as described previously (Gillon and Hawthorne, 1983a).

phomonoesterase attacking this lipid does not show reduced activity in sciatic nerve of diabetic rats (Whiting et al, 1979), but the phosphodiesterase (phospholipase C) has not been estimated. This enzyme is of considerable interest because it is linked to the activation of receptors and may play a part in regulation of intracellular calcium ion concentration (for a recent review, see Hawthorne, 1983). Changes in the metabolic activity of polyphosphoinositides in peripheral nerve of diabetic animals could well be related to defective conduction mechanisms. White *et al*. (1974) showed that the incorporation of ^{32}P into these lipids of vagus nerve was increased by electrical stimulation over a period of 30 min. In the light of recent work, shorter time periods are likely to produce more dramatic and specific effects but we have not yet had the opportunity to stimulate in this way.

The restoration of sciatic nerve polyphosphoinositide concentrations to normal by sorbinil given to diabetic rats

(Table III) also suggest that these lipids are important in nerve conduction, since at the same time sorbinil corrects the slowing of motor nerve conduction velocity (Gillon *et al.*, 1983). Preliminary results however, show no effect of sorbinil on the increased labeling of polyphospho-inositides seen *in vitro* when endoneurial preparations are incubated with ^{32}P (Table IV).

THE SODIUM PUMP ATPase IN SCIATIC NERVE OF DIABETIC RATS

We do not understand why certain tissues such as brain and peripheral nerve maintain a high concentration of free inositol. There is a connection between energy expenditure to maintain high K^+ and low Na^+ concentrations and inositol concentration in isolated nerve endoneurium (Simmons *et al.*, 1982). Charalampous (1971) found a similar connection in cultures of an inositol-requiring KB cell strain. When the culture medium was deficient in inositol, these cells showed defective Na^+ efflux and K^+ influx. The Na^+/K^+ ATPase of a particulate fraction from the cells was assayed by measurement of ouabain-sensitive ATP hydrolysis. Its activity in the deficient cells was only 40% of normal.

Das *et al.* (1976) reported a reduction of Na^+/K^+ ATPase activity in sciatic nerve of streptozotocin-diabetic rats. This has been confirmed by Greene and Lattimer (1983) who also showed that the enzyme activity could be restored by feeding a diet containing 1% inositol.

Table V shows a similar reduction of this enzyme in rats which had been given a single injection of strepto-zotocin (either 35 mg or 60 mg/kg body weight).

In our hands, assay of the Na^+/K^+ ATPase by the NADH-linked method of Schwartz *et al.* (1969) gave variable results when sciatic nerve homogenates were used. There was a high background hydrolysis of ATP by enzymes not sensitive to ouabain. Problems were avoided by using a microsomal fraction prepared from epineurium-free sciatic nerve. The tissue was homo-genized in 0.25 M sucrose buffered to pH 7.7. The homogenate was then centrifuged at 1,100 × *g* for 5 min to remove debris and for a further 20 min at 7,700 × *g* to

TABLE V

Activity of Na^+/K^+ ATPase in Sciatic Nerve
Microsomal Fraction from Control and Diabetic Rats

	After 4 weeks	After 8 weeks
Control	14.5 ± 3.7	6.7 ± 3.1
Streptozotocin (35 mg/kg)	1.4 ± 3.9	5.1 ± 0.4
Streptozotocin (60 mg/kg)	0	2.4 ± 2.2

Figures represent mmol ATP hydrolyzed per mg protein per h and are means
± S.D. of 3 assays from one rat.

sediment mitochondria. The supernatant was then centrifuged at 55,000 x g for 30 min to obtain the microsomal
fraction used for assay of Na^+/K^+ ATPase. Figures in
Table V represent the ouabain-sensitive hydrolysis of
ATP.

A defective sodium pump could contribute to the
slowing of nerve conduction seen in diabetes. Since the
impairment appears to result from lack of free inositol it
will be reinforced, because inositol uptake itself is an
active process dependent upon the Na^+ gradient (Greene
and Lattimer, 1982; Gillon and Hawthorne, 1983a). Thus
there will be a cycle of damaging events: reduced uptake
of inositol, reduced Na^+/K^+ ATPase, further reduction of
inositol uptake as a consequence, and so on.

CONCLUSIONS

A study of inositol and phosphoinositides in peripheral nerve of diabetic animals has thrown some light on
possible functions of these compounds. As yet, we do
not know whether the conduction changes, and possibly
the demyelination seen later, are direct results of reduced
free inositol concentration, or whether changes in inositol
lipids are more important. The sodium-pump ATPase is a

plasma membrane enzyme activated by acidic phospho-lipids, but whether PI is specifically required seems doubtful. Mandersloot *et al.* (1978) consider that the kidney enzyme is activated by this lipid but other workers have reached different conclusions. So far, the polyphosphoinositides have not been studied as activators. It may be relevant that there is a deficiency of these lipids in rat sciatic endoneurial preparations from diabetic animals (Table II). Work with iris muscle (Akhtar and Abdel-Latif, 1982) suggests a connection between sodium ions and the polyphosphoinositides. Breakdown of these lipids caused by acetylcholine was only seen in the pres-ence of both Na^+ and Ca^{2+}.

It seems likely, then, that a detailed study of phos-phoinositides and inositol in peripheral nerve of diabetic animals will not only contribute to the understanding of diabetic neuropathy, but also help to elucidate the func-tions of these compounds generally.

REFERENCES

Akhtar, R.A. and Abdel-Latiff, A.A. (1982) *J. Neurochem.* *39*, 1374-1380.

Bell, M.E., Peterson, R.G. and Eichberg, J. (1982) *J. Neurochem.* *39*, 192-200.

Brown, M.J., Iwamori, M. Kishimoto, Y., Rapoport, B., Moser, H.W. and Asbury, A.K. (1979) *Annals Neurol.* *5*, 245-252.

Charalampous, F.C. (1971) *J. Biol. Chem.* *246*, 455-460.

Das, P.K., Bray, G.M., Aguayo, A.J. and Rasminsky, M. (1976) *Exp. Neurol.* *53*, 285-288.

Gillon, K.R.W. and Hawthorne, J.N. (1983a) *Biochem. J.* *210*, 775-781.

Gillon, K.R.W. and Hawthorne, J.N. (1983b) *Life Sci.* *32*, 1943-1947.

Gillon, K.R.W., Hawthorne, J.N. and Tomlinson, D.R. (1983) *Diabetologia* *25*, 365-371.

Greene, D.A. and Lattimer, S.A. (1982) *J. Clin. Invest.* *70*, 1009-1018.

Greene, D.A. and Lattimer, S.A. (1983) *J. Clin. Invest.* *72*, 1058-1063.

Greene, D.A., De Jesus, Jr., P.V. and Winegrad, A.I. (1975) *J. Clin. Invest.* *55*, 1326-1336.

Greene, D.A., Winegrad, A.I., Carpentier, J.L., Brown, M.J., Fukuma, M. and Orci, L. (1979) *J. Neurochem.* *33*, 1007-1018.

Griffin, H.D. and Hawthorne, J.N. (1978) *Biochem. J.* *176*, 541-552.

Hawthorne, J.N. (1983) *Bioscience Reports* *3*, 887-904.

Mandersloot, J.G., Roelofsen, B. and De Gier, J. (1978) *Biochim. Biophys. Acta* *508*, 478-485.

Mayhew, J.A., Gillon, K.R.W. and Hawthorne, J.N. (1983) *Diabetologia* *24*, 13-15.

Natarajan, V., Dyck, P.J. and Schmid, H.H.O. (1981) *J. Neurochem.* *36*, 413-419.

Palmano, K.P., Whiting, P.H. and Hawthorne, J.N. (1977) *Biochem. J.* *167*, 229-235.

Schwartz, A., Allen, J.C. and Harigaya, S. (1969) *J. Pharm. Exp. Ther.* *168*, 31-41.

Simmons, D.A., Winegrad, A.I. and Martin, D.B. (1982) *Science* *217*, 848-851.

White, G.L., Schellhase, H.U. and Hawthorne, J.N. (1974) *J. Neurochem.* *22*, 149-158.

Whiting, P.H., Palmano, K.P. and Hawthorne, J.N. (1979) *Biochem. J.* *179*, 549-553.

Williams, S.K., Howarth, N.L., Devenny, J.J. and Bitensky, M.W. (1982) *Proc. Natl. Acad. Sci. USA* *79*, 6546-6550.

ALTERED NERVE *MYO*-INOSITOL METABOLISM IN EXPERIMENTAL DIABETES AND ITS RELATIONSHIP TO NERVE FUNCTION

Douglas A. Greene and Sarah A. Lattimer

Diabetes Research Laboratories, Department of Medicine, School of Medicine, University of Pittsburgh, Pittsburgh, PA 15261

SUMMARY

The following picture of altered *myo*-inositol metabolism in diabetic peripheral nerve has recently emerged. Hyperglycemia, through competitive inhibition of sodium-dependent *myo*-inositol uptake and/or increased polyol (sorbitol) pathway activity, reduces nerve *myo*-inositol content, secondarily altering nerve phosphoinositide metabolism and impairing the function of the membrane-bound sodium-potassium ATPase. The resulting reduction in the transmembrane sodium gradient impairs nerve conduction and further reduces sodium gradient-dependent *myo*-inositol uptake, creating a self-reinforcing metabolic defect in diabetic peripheral nerve. Other sodium-gradient dependent processes such as amino acid uptake and intracellular water and electrolyte homeostasis may also be secondarily altered. These abnormalities may have potentially widespread pathophysiological implications, possibly leading to the later structural defects in diabetic peripheral nerve, which are thought to underlie neurological deficits in diabetic neuropathy.

INTRODUCTION

Metabolic abnormalities in peripheral nerve resulting from chronic insulin deficiency and/or hyperglycemia are thought to heavily influence the development of diabetic

563

neuropathy; until recently, the identity of the responsible metabolic factors remained shrouded in confusion and controversy (Clements, 1979). Ever since the mid-nineteen seventies, one proposed metabolic link between the diabetic state and altered nerve function has been *myo*-inositol (Greene *et al.*, 1975; Mayhew *et al.*, 1983). Stated in the broadest terms, the so-called *myo*-inositol hypothesis for diabetic neuropathy proposes that unidentified factors in the diabetic milieu reduce the normal 90-fold nerve-to-plasma *myo*-inositol concentration gradient, and that altered nerve *myo*-inositol levels secondarily impair nerve impulse conduction and other nerve functions, as well thereby contributing to the pathogenesis of diabetic neuropathy (Winegrad and Greene, 1976). This hypothesis is based on several observations in animals and humans. Tissue *myo*-inositol is reduced by 20-30% in the nerves of diabetic patients (Mayhew *et al.*, 1983). Acute experimental diabetes in the rat reduces nerve conduction velocity and nerve *myo*-inositol content by approximately 25% without changing plasma *myo*-inositol level (Greene *et al.*, 1975; Palmano *et al.*, 1977; Greene *et al.*, 1982). Insulin replacement that normalizes nerve conduction in the diabetic rat prevents the fall in nerve *myo*-inositol (Greene *et al.*, 1975). Pharmacological elevation of plasma *myo*-inositol prevents or reverses both the reduced nerve *myo*-inositol content and motor nerve conduction velocity in the diabetic rat without affecting hyperglycemia or the raised nerve glucose, fructose or sorbitol concentrations (Greene *et al.*, 1975; Mayer and Tomlinson, 1983).

Until very recently, both the mechanism(s) by which diabetes lowers nerve *myo*-inositol, and the mechanism(s) by which altered nerve *myo*-inositol metabolism impairs nerve conduction remained completely unknown. Within the last two years, the activity of the nerve Mg^{2+}-dependent, Na^+- and K^+-stimulated, adenosinetriphosphatase ("sodium-potassium ATPase," EC 3.6.1.3) and the plasmalemmal sodium gradient have emerged as central factors linking nerve *myo*-inositol to both the diabetic state and to nerve impulse conduction (Greene, 1983, Greene and Lattimer, 1982, Greene and Lattimer, 1983).

An alteration in sodium-potassium ATPase activity in diabetic nerve was postulated on the basis of three separate observations: (1) composite resting nerve energy utilization, attributed in large part to the sodium-potassium ATPase activity (Ritchie, 1967), is reduced in experimental diabetes (Greene and Winegrad, 1981); (2) membrane phosphoinositides have been implicated in membrane-bound sodium-potassium ATPase function in some mammalian tissues (Mandersloot et al., 1978); and (3) impairment of nerve conduction in the spontaneously diabetic BB-rat has been related to a decreased sodium

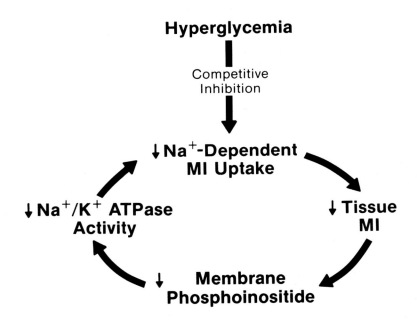

Figure 1. Proposed scheme interrelating glucose-mediated competitive inhibition of sodium-dependent myo-inositol (MI) uptake, phosphoinositide metabolism, and sodium-potassium ATPase activity in diabetic nerve. Either polyol (sorbitol) pathway activity or aldose reductase inhibitors modify the activity of this cycle by a mechanism which remains to be determined. The putative role of phosphoinositides in sodium-potassium ATPase activity has not been defined directly. From Greene and Lattimer, (1984a) with permission.

"in-current" with depolarization, possibly reflecting decreased sodium pump activity (Brismar and Sima, 1981). Therefore, we proposed that peripheral nerve *myo*-inositol depletion in experimental diabetes might secondarily derange sodium-potassium ATPase activity, presumably via a phosphoinositide mechanism, thereby reducing both nerve conduction and resting energy utilization (Greene, 1983). Furthermore, since carrier-mediated *myo*-inositol uptake in nerve is itself sodium-gradient dependent (Greene and Lattimer, 1982), a sodium-potassium ATPase defect might further impair sodium-dependent *myo*-inositol uptake, thereby initiating a self-reinforcing cycle of metabolic abnormality in diabetic peripheral nerve (Figure 1) (Greene, 1983).

Detailed understanding of these relationships in diabetic nerve requires appropriate consideration of each of the components of this "*myo*-inositol-sodium-potassium ATPase cycle".

DETERMINANTS OF NERVE TISSUE *MYO*-INOSITOL CONCENTRATIONS

In order to clarify the mechanisms responsible for the high concentration of *myo*-inositol in peripheral nerve, and its reduction by experimental diabetes, it was necessary to study the transport of *myo*-inositol into and out of peripheral nerve. The lack of suitable *in vitro* tissue preparations derived from peripheral nerve had traditionally precluded such studies; therefore, we developed and characterized an *in vitro* "endoneurial" preparation, comprised of an intact and defined fascicle segment from rabbit sciatic nerve from which all epineurial tissue and perineurial diffusion barriers are removed by limited preliminary collagenase digestion and microdissection. When the endoneurial preparation is incubated in a Krebs bicarbonate buffer pH 7.4 equilibrated with 95% O_2/5% CO_2 at 37°C with 5 mM glucose as sole exogenous substrate, the metabolic and ultrastructural integrity of nerve fibers is preserved (Greene *et al.*, 1979), thereby permitting valid *in vitro* transport studies.

The uptake of 2-[^{3}H]*myo*-inositol by the endoneurial tissue preparation was linear with respect to time for at least 15 minutes, and water-soluble radioactivity remained

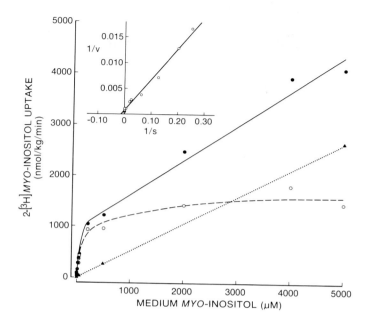

<u>Figure 2</u>. 2-[³H]myo-inositol uptake in the endoneurial preparation as a function of medium <u>myo</u>-inositol concentration in the presence (solid line) and absence (dotted line) of sodium. Sodium-dependent uptake (dashed line), the difference between uptake in the presence and absence of sodium, is shown as a double reciprocal plot in the inset. From Greene and Lattimer (1982) with permission.

in the free *myo*-inositol fraction. Uptake studies at various concentrations of *myo*-inositol demonstrated that myo-inositol uptake occurred by at least two distinct transport systems. A sodium- and energy-dependent saturable transport system was responsible for at least 95% of the measured uptake at medium *myo*-inositol concentrations approximating that in plasma. Sodium-dependent transport was completely saturable, and conformed to Michaelis-Menten kinetics, with a high apparent affinity for *myo*-inositol (K_t = 63 µM), but a relatively low maximum transport velocity (V_{max} = 1400 nmol/kg/min) (Figure 2).

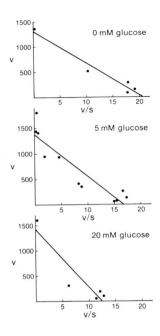

Figure 3. Eadie-Hofstee plot of 2-[^{3}H]<u>myo</u>-inositol uptake in the endo-neurial preparation at various medium glucose concentrations. The V_{max} of transport is given by the intercept of the vertical axis and the K_t by the slope of the plotted line. From Greene and Lattimer (1982) with permission.

Sodium-dependent transport was not influenced acutely by physiological concentrations of insulin *in vitro* but was inhibited by hyperglycemic concentrations of glucose added to the incubation medium. Kinetic analysis of inhibition revealed that glucose altered the apparent K_t but not the V_{max} of transport (Figure 3). Analysis by a Dixon plot suggested that the interaction between glucose and sodium-dependent 2-[^{3}H]*myo*-inositol uptake was primarily one of competitive inhibition, with an apparent K_i of approximately 30 mM. Thus, the carrier had an affinity for *myo*-inositol some 500 times greater than for glucose, but millimolar concentrations of glucose competitively inhibited sodium-dependent uptake of micromolar concentrations of *myo*-inositol (*e.g.* at 5 μM *myo*-inositol, raising medium glucose from 5 to 20 mM reduced sodium-dependent *myo*-inositol uptake by approximately 30%). The effect was relatively specific since similar concentrations of mannitol or fructose had no effect on sodium-dependent *myo*-inositol uptake. Thus, this specific high-affinity sodium-dependent carrier-mediated *myo*-inositol transport system probably contributes to the establishment and/or maintenance of the normal nerve-tissue-to-plasma *myo*-inositol concentration gradient. Hyperglycemic concentrations of glucose competitively inhibit *myo*-inositol uptake via this transport system (Greene and Lattimer, 1982), providing a potential mechanism by which diabetes could reduce nerve *myo*-inositol content. The low-transport-capacity, high-concentration-gradient characteristics of this transport system suggest that factors which influence passive efflux of *myo*-inositol might also greatly influence the *myo*-inositol concentration gradient (see below). Thus, competitive inhibition by glucose of sodium-dependent *myo*-inositol uptake provides a potential explanation for the low peripheral nerve tissue *myo*-inositol concentration in experimental diabetes (Greene and Lattimer, 1982).

IMPAIRED NERVE CONDUCTION IN ANIMAL DIABETES

Experimental animal models for diabetic peripheral nerve dysfunction originated with Eliasson's 1964 observation that acute alloxan or pancreatic diabetes in the rat reduced peripheral motor nerve conduction velocity (Eliasson, 1964). Although simulating the reversible conduction impairment of newly-diagnosed Type I diabetes (Gregersen, 1968), the original model was not pursued for nearly 10 years, because Eliasson could not establish either a histological or biochemical basis for the conduction defect, nor demonstrate an unequivocal cause-and-effect relationship with insulin deficiency (Eliasson, 1964, Eliasson, 1965, Eliasson, 1969). Since 1974, however, numerous studies have confirmed that conduction impairment in both the streptozocin and spontaneously-diabetic BB rat is a consequence of insulin deficiency (Greene *et al.*, 1975; Jakobsen, 1979; Sima, 1980), which occurs acutely in the absence of significant segmental demyelination or axonal degeneration (Sharma and Thomas, 1974; Brown *et al.*, 1980; Jakobsen, 1979).

Recent single-node voltage clamp studies in the acutely diabetic BB-Wistar rat have attributed impaired maximal nerve conduction velocity to a reduced sodium influx with depolarization (Brismar and Sima, 1981). An associated reduction in the resting membrane potential suggested that the diminished sodium "in"-current was attributable in part to a reduced resting transmembrane sodium gradient. Since the membrane sodium gradient reflects sodium-potassium ATPase activity, Brismar and Sima (1981) proposed that limited availability of metabolic energy to the sodium-potassium ATPase might be responsible for impaired nerve conduction in experimental diabetes. However, biochemical studies suggest that reduced energy utilization rather than reduced energy availability metabolically characterizes diabetic peripheral nerve (see below).

ENERGY METABOLISM IN DIABETIC PERIPHERAL NERVE: RELATIONSHIP TO SODIUM-POTASSIUM ATPase FUNCTION AND *MYO*-INOSITOL METABOLISM

The same lack of appropriate *in vitro* tissue preparations has traditionally precluded steady-state balance studies of normal and diabetic nerve metabolism [previously published *in vitro* metabolic studies with incubated whole nerve segments (Field and Adams, 1964, Field and Adams, 1965) were interpreted erroneously due to epineurial adipocyte contamination and unsuspected perineurial barriers to substrate diffusion]. The endoneurial preparation incubated with 5 mM glucose as sole substrate maintains nearly steady-state energy metabolism, with 95% of tissue energy production accounted for by oxidative glucose metabolism, and with greater than 60% of measured glucose uptake being metabolized oxidatively (Greene and Winegrad, 1979). Albumin-bound unesterified fatty acids, which probably do not penetrate unfenestrated endoneurial capillaries *in vivo*, cannot be used as substrate by the endoneurial preparation *in vitro* (Greene and Winegrad, 1981). Limited substrate selection often endows tissues with autoregulated metabolism wherein substrate utilization is regulated by intrinsic tissue energy demands rather than by extrinsic hormonal control or circulating substrate concentration. Hence, nerve glucose utilization is not unexpectedly quite independent of acute regulation by both insulin levels and glucose concentrations in the physiological range [although the latter become limiting when external glucose levels are reduced (Greene and Winegrad, 1979; Greene and Winegrad, 1981)].

Although neither insulin nor physiological glucose concentration directly or acutely influence overall glucose metabolism in normal peripheral nerve (Greene and Winegrad, 1979), acute experimental diabetes significantly alters peripheral nerve energy and substrate metabolism (Greene and Winegrad, 1981). Acute alloxan diabetes altered endoneurial metabolism, as assessed during incubation with 5 mM glucose, in several ways. Tissue free *myo*-inositol concentrations were significantly reduced. Respiration was reduced by 30–40% compared to non-diabetic endoneurial preparations despite the presence of augmented endogenous substrate stores in the form of increased tissue glucose and fructose; both tissue ATP concentration and P-creatine-to-creatine ratio were unchanged, indicating that metabolic substrate availability was not limiting for respiration (Greene and Winegrad, 1981). Tissue lactate was not increased despite a small but significant increase in lactate production which did not compensate for the diminished oxidative energy production so that total energy production was significantly lower in diabetic nerve preparations *in vitro*. The normal ATP concentration and P-creatine-to-creatine ratio suggested that energy production and energy utilization remained coupled, thereby suggesting a primary defect in diabetic nerve energy utilization (Greene and Winegrad, 1981). Insulin (100 μU/ml = 0.70 × 10^{-9} M) in the incubation medium had no effect on tissue respiration, P-creatine, ATP, P-creatine:creatine ratio, or total creatine pool.

The potential metabolic inhibitory effects of the very high tissue fructose concentrations in diabetic endoneuria were assessed by raising tissue fructose in nondiabetic endoneurial preparations by incubation with 20 mM fructose. Despite tissue fructose levels approximating the highest observed in diabetic preparations, the metabolic perturbations of the diabetic were not reproduced. Rather, 20 mM fructose replaced glucose as an energy substrate as tissue glucose fell profoundly. In contrast to normal endoneurial preparations, diabetic endoneurial preparations maintained stable energy levels during 2 h

571

incubations in the absence of exogenous substrate. This was attributed to the continued utilization of persisting fructose stores once tissue glucose fell to very low levels, and to a lower rate of energy utilization in the diabetic. Incubation of normal endoneurial preparations with 20 mM glucose raised tissue glucose concentration to that seen in diabetic endoneurial preparations, but did not reproduce the characteristic metabolic derangements of the diabetic except for a small increase in lactate production. Thus, neither elevated tissue fructose nor glucose concentration alone acutely reproduced the effects of experimental diabetes on endoneurial metabolism, suggesting the presence of an intrinsic defect in metabolic energy utilization in diabetic peripheral nerve (Greene and Winegrad, 1981).

Recently, the reduced energy utilization of diabetic peripheral nerve has been ascribed to an intrinsic alteration in nerve sodium-potassium ATPase function. Approximately half of resting steady-state energy utilization in peripheral nerve occurs via the sodium-potassium ATPase, which hydrolyzes ATP thereby deriving biochemical energy to generate the electrochemical gradients of sodium and potassium across the plasma membrane (Ritchie, 1967). Ouabain, a specific inhibitor of the sodium-potassium ATPase, can be employed to probe that component of resting tissue energy metabolism ascribed to sodium-potassium ATPase activity. Such studies in the diabetic endoneurial preparation incubated in 5 mM glucose have demonstrated that the reduction in steady-state resting energy utilization is confined to, and quantitatively equals, the ouabain-inhibitable respiratory fraction (Greene and Lattimer, 1984a), which in nerve is an expression of sodium-potassium ATPase activity (Ritchie, 1967). However, the expression of this respiratory defect in diabetic endoneurial preparations was influenced by alterations in medium osmolality: the addition of 5 mM sorbitol, mannitol, glucose or choline chloride to the incubation medium prevented expression of the defect (these compounds had no effect on nondiabetic endoneurial respiration) (Greene and Lattimer, 1984a).

Taken together, these data suggest that diabetes significantly decreases resting energy utilization by peripheral nerve *in vitro*. The reduction in nerve energy utilization is confined to and quantitatively equals the ouabain-inhibitable component of steady-state respiration, a measure in nerve of sodium-potassium ATPase activity. However, the defect in this enzymatic activity in diabetic nerve is either reversed or masked by alterations in the solute composition and/or osmolality of the incubation bath, by processes which remain unclear (Greene and Lattimer, 1984a).

INFLUENCE OF SODIUM-POTASSIUM ATPase ACTIVITY ON SODIUM-DEPENDENT *MYO*-INOSITOL UPTAKE

The peripheral nerve *myo*-inositol transport system is driven by the sodium gradient generated by the sodium-potassium ATPase (Greene and Lattimer, 1982). Therefore, it might be anticipated that the reduced sodium-potassium ATPase activity of diabetic peripheral nerve would secondarily impair sodium-dependent *myo*-inositol uptake. Hence, the physiological importance of this reduction in sodium-potassium ATPase activity and its presumed attendant alteration in the transmembrane sodium gradient were assessed by studying sodium-dependent 2-[^{3}H]*myo*-inositol uptake in normal and diabetic endoneurial preparations under various incubation conditions. Sodium-dependent *myo*-inositol uptake was equally reduced in nondiabetic and diabetic endoneurial preparations in 20 mM glucose, a glucose concentration which both competitively inhibits *myo*-inositol transport but also normalizes sodium-potassium ATPase activity in diabetic nerve. In the presence of 5 mM glucose + 15 mM mannitol (an osmotically active agent which normalizes diabetic nerve sodium-potassium ATPase but does not compete with *myo*-inositol), 2-[^{3}H]*myo*-inositol uptake was normal in both diabetic and nondiabetic nerve. In the presence of 5 mM glucose alone (where sodium-potassium ATPase is reduced in diabetic endoneuria but glucose is only minimally inhibitory to *myo*-inositol uptake), sodium-dependent *myo*-inositol uptake was reduced in diabetic but not normal nerve; under these conditions, diabetic nerve tissue glucose levels are normalized during the course of the incubation. These experiments demonstrate that sodium-dependent *myo*-inositol uptake remains reduced in diabetic peripheral nerve even in the face of normal tissue glucose concentrations as long as the defect

573

in sodium-potassium ATPase activity is expressed. Thus, sodium-gradient 2-[³H]*myo*-inositol uptake correlates with sodium-potassium ATPase activity under various experimental conditions (Greene and Lattimer, 1984a), suggesting that the sodium gradient is limiting for *myo*-inositol uptake and that the reduced diabetic nerve sodium-potassium ATPase activity significantly alters the transmembrane sodium gradient in peripheral nerve. These relationships thus set the stage for a self-reinforcing cyclic metabolic derangement involving *myo*-inositol and the sodium-potassium ATPase in diabetic peripheral nerve (Figure 1).

NATURE OF THE *MYO*-INOSITOL-ASSOCIATED DEFECT IN THE SODIUM-POTASSIUM ATPase

The sodium-potassium ATPase, which is a phospholipid-dependent enzyme, provides a potential mechanism to relate the defects in *myo*-inositol-phospholipid metabolism in diabetic peripheral nerve (Palmano *et al.*, 1977; Natarajan *et al.*, 1981; Clements and Stockard, 1980; Hothersall and McLean, 1979) to abnormal impulse conduction, *myo*-inositol uptake and energy utilization. In mammalian cells, the membrane-bound sodium-potassium ATPase is regulated by a variety of intrinsic intracellular modulators (including cytoplasmic sodium, potassium, magnesium, calcium, ATP, pH), by the characteristics of the plasma membrane into which it is embedded, and by extracellular sodium and potassium concentrations (Trachtenberg *et al.*, 1981). Although extrinsic hormonal modulation of the sodium-potassium ATPase has been described in some tissues (mediated by phosphorylation-dephosphorylation of the enzyme itself [Trachtenberg *et al.*, 1981] or an associated membrane protein [Lingham and Sen, 1982], or by regulation of ATPase protein subunit turnover [Trachtenberg *et al.*, 1981]), the sodium-potassium ATPase defect expressed by endoneuria from diabetic rabbits is not influenced acutely by insulin *in vitro* (Greene and Winegrad, 1981). Furthermore, this analagous defect in the diabetic rat is expressed in broken cell nerve homogenates, where the concentration of water-soluble modulators is optimized (Das *et al.*, 1977; Greene and Lattimer, 1983), and therefore is intrinsic to the sodium-potassium ATPase-membrane complex. The defect is entirely prevented or reversed by *myo*-inositol

supplementation *in vivo* , which prevents the characteristic falls in *myo*-inositol level and motor conduction
velocity in diabetic rat sciatic nerve (Greene and
Lattimer, 1983). Thus, the impairment of diabetic rat
sodium-potassium ATPase activity is a consequence of
altered diabetic nerve *myo*-inositol metabolism, which
secondarily alters the structure and/or function of the
sodium-potassium ATPase-membrane complex. Since
myo-inositol is a substrate for the synthesis of phosphatidylinositol (and the higher polyphosphoinositides),
which is an endogenous regulator of renal microsomal
sodium-potassium ATPase (Roelofsen, 1981; Mandersloot *et
al.*, 1978), it is tempting to speculate that the effect of
myo-inositol on nerve sodium-potassium ATPase is
mediated by this phospholipid. Support for this con-
tention comes from the recent observation of Simmons *et
al.* (1982) that free *myo*-inositol levels may limit phosphatidylinositol turnover in peripheral nerve.

Hence, insulin deficiency induces, and dietary
myo-inositol supplementation prevents, nerve conduction
slowing in experimental diabetes by a sodium-potassium
ATPase mechanism, most likely mediated by unidentified
changes in membrane phosphoinositide metabolism and/or
composition. Furthermore, since nerve action potential
generation and sodium-dependent *myo*-inositol uptake are
altered by the sodium-potassium ATPase reduction, it is
likely that other sodium gradient dependent processes are
also impaired in diabetic nerve. With the exception of
Ca^{2+}, H^+ and K^+ ions, all known samples of net extrusion
or accumulation of any substance against its concentration
gradient across an animal cell plasma membrane involves
the coupling of that movement to the electrochemical
gradient of sodium created by the sodium-potassium
ATPase (Kyte, 1981). Therefore, a defect in diabetic
peripheral nerve sodium-potassium ATPase activity could
lead not only to an acute reduction in nerve action potential generation, but also to multiple biochemical and
physiological abnormalities effecting any and all substrates
and metabolites actively transported across the cell membrane. In addition, recent observations suggest that
altered *myo*-inositol metabolism not only contributes to
impaired nerve action potential transmission, but also to a
defect in slow axonal transport in diabetic nerve (Mayer
and Tomlinson, 1983). Resultant long term metabolic and

electrolyte imbalances might well culminate in structural alterations in peripheral nerve which contribute to clinical diabetic neuropathy.

RELATIONSHIP BETWEEN THE *MYO*-INOSITOL-RELATED SODIUM-POTASSIUM ATPase DEFECT AND POLYOL (SORBITOL) PATHWAY ACTIVITY IN DIABETIC PERIPHERAL NERVE

Although the major pathways of nerve glucose metabolism are unaffected by high ambient glucose concentrations *in vitro* (Greene and Winegrad, 1981), several potentially important peripheral nerve intracellular biochemical alterations are directly attributable to elevated extracellular glucose levels. In tissues like peripheral nerve, where facilitated glucose transport is neither insulin-regulated nor rate-limiting for phosphorylation (Greene and Winegrad, 1979), *in vivo* tissue glucose concentrations reflect plasma levels (Stewart *et al.*, 1967; Winegrad *et al.*, 1972). Polyol pathway activity in turn is regulated by the intracellular glucose concentration, since the K_m values of aldose reductase and sorbitol dehydrogenase for their respective substrates, glucose and sorbitol, exceed tissue levels by at least an order of magnitude (Winegrad *et al.*, 1972; Kinoshita *et al.*, 1962a). Peripheral nerve glucose, sorbitol and fructose levels are raised in human (Mayhew *et al.*, 1983) and experimental diabetes, but fall quickly when plasma glucose is lowered with insulin (Stewart *et al.*, 1967), suggesting that sorbitol and fructose turn over rapidly. However, in the diabetic lens, sorbitol, which is poorly diffusible across cell membranes, accumulates in millimolar concentrations, which closely correlate with tissue overhydration, implying a possible important polyol osmotic contribution in the pathogenesis of the "sugar cataract" (Gabbay, 1975; Kinoshita *et al.*, 1962a; Kinoshita *et al.*, 1962b). Extrapolation of this osmotic hypothesis to diabetic nerve ignores the micromolar (vs. millimolar) sorbitol concentration range in nerve, which is unlikely to be osmotically significant, unless sorbitol were highly localized to a minute tissue fraction. Furthermore, increased tissue water is neither a constant nor necessary feature of diabetic nerve conduction impairment (Greene, 1983). Moreover, when tissue water accumulation occurs in diabetic peripheral nerve, it is thought to be extra-

rather than intracellular in location (Jakobsen, 1978) (as may also be the case in lens [Kuwabara *et al.*, 1969; Unakar *et al.*, 1981]), while sorbitol accumulation is generally assumed to be intracellular (Gabbay, 1975).

Although polyol osmotic effects are unlikely to contribute prominently to diabetic peripheral nerve dysfunction, other manifestations of increased polyol pathway activity may be of pathogenetic relevance. Recent studies in man and rats demonstrate that specific aldose reductase inhibitors may improve nerve function in diabetes (Yue *et al.*, 1982; Judzewitsch *et al.*, 1983; Mayer and Tomlinson, 1983; Fagius and Jameson, 1981), although the effects in humans are small.

Increased polyol (sorbitol) pathway activity and reduced tissue *myo*-inositol content are commonly invoked as alternate pathogenetic mechanisms for diabetic neuropathy; however, their possible interrelationship in diabetic nerve has only been explored recently. Several studies have demonstrated that aldose reductase inhibitors prevent the reduction in peripheral nerve *myo*-inositol content which usually accompanies acute experimental diabetes, while affecting neither plasma *myo*-inositol nor nerve glucose concentrations (Gillon and Hawthorne, 1983; Mayer and Tomlinson, 1983; Finegold *et al.*, 1983). These observations suggest that increased polyol pathway activity contributes to the fall in nerve *myo*-inositol content in acute experimental diabetes, or, alternatively, that aldose reductase inhibitors might have effects on nerve *myo*-inositol metabolism unrelated to aldose reductase activity (including possible direct effects on *myo*-inositol uptake, efflux, or the generation of the sodium gradient in diabetic nerve). The kinetic characteristics of sodium-dependent *myo*-inositol transport in peripheral nerve, particularly the low V_{max} relative to the high transmembrane *myo*-inositol concentration gradient, require that back diffusion of *myo*-inositol must be minimal, implying that factors which alter *myo*-inositol efflux might significantly affect tissue *myo*-inositol levels. Perhaps aldose reductase inhibition and/or polyol pathway activity may influence *myo*-inositol efflux from nerve as they are presumed to affect efflux from the ocular lens (Broekhuyse, 1968). This supposition has not yet been established experimentally in nerve.

577

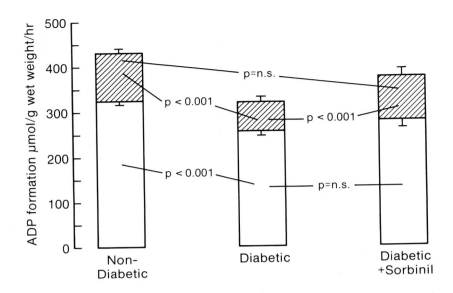

<u>Figure 4</u>. ATPase activity in crude homogenates of sciatic nerve from nondiabetic, untreated diabetic and aldose reductase inhibitor (sorbinil)-treated diabetic rats, measured enzymatically as ADP formation. Composite ATPase activity is shown by the height of the bars, and the sodium-potassium-stimulated component (sodium-potassium ATPase activity) by the cross-hatched area. From Greene and Lattimer (1984a) with permission.

These observations raised the possibility that the two most commonly cited alternate pathogenetic biochemical mechanisms for diabetic neuropathy may be closely interrelated, both exerting their effects via a *myo*-inositol-related sodium-potassium ATPase defect. Therefore, we directly examined the effect of aldose reductase inhibitor administration on sodium-potassium ATPase activity measured enzymatically in crude nerve homogenates from streptozocin-diabetic rats (Figure 4) (Greene and Lattimer, 1984b). As might be predicted, the 40% reduction in nerve sodium-potassium ATPase activity which occurred in the untreated diabetics was completely prevented by treatment with an aldose reductase inhibitor. These observations suggest that the effect of sorbinil administration which prevents the fall in *myo*-inositol also prevents the fall in sodium-potassium ATPase activity in diabetic nerve. Thus, normal sodium-potassium ATPase activity is preserved in diabetic nerve when the fall in nerve myo-inositol is prevented either by dietary myo-inositol supplementation or sorbinil administration.

In summary, recent studies in various animal models of diabetes and in vitro have revealed a constellation of metabolic abnormalities in diabetic peripheral nerve which provide a biochemically and physiologically plausible link between insulin deficiency and hyperglycemia, on the one hand, and peripheral nerve dysfunction (e.g., slowed nerve conduction) on the other. A newly-described myo-inositol-related sodium-potassium ATPase defect plays a central role in this pathogenic scheme. Hyperglycemia alters nerve myo-inositol metabolism by glucose-mediated competitive inhibition of sodium-dependent myo-inositol uptake and/or polyol pathway hyperactivity (by a mechanism which remains to be clarified). Reduced tissue myo-inositol content is thought to alter nerve membrane phosphoinositide metabolism (Simmons et al., 1982), which modifies the characteristics of the membrane-bound sodium-potassium ATPase by a molecular mechanism which remains to be elucidated. Impaired sodium-potassium ATPase activity reduces nerve energy utilization and the transmembrane sodium gradient, suppressing sodium gradient-dependent processes such as voltage-dependent membrane depolarization, sodium-dependent myo-inositol uptake, and possibly other sodium-dependent processes as well (e.g., amino acid uptake, calcium-sodium antiport, and intracellular pH regulation via sodium-hydronium ion antiport), which may be significant in the pathogenesis of diabetic neuropathy.

ACKNOWLEDGEMENTS

The authors wish to express their gratitude to Lorraine Weber, Jim Bensen, Jean Kim, Carol Korbanic, and Lisa Tamres for expert technical assistance. The studies reported here were supported in part by U.S.P.H.S. research grant RO1-AM2982, Juvenile Diabetes Foundation research grant 82R-231, and the Harry Soffer Memorial Research Fund of the University of Pittsburgh.

579

REFERENCES

Brismar, T. and Sima, A.A.F. (1981) *Acta Physiol. Scand. 113*, 499-506.

Broekhuyse, R.M. (1968) *Biochim. Biophys. Acta 163*, 269-272.

Brown, M.J., Sumner, A.J., Greene, D.A., Diamond, S.M. and Asbury, A.K. (1980) *Ann. Neurol. 8*, 169-178.

Clements, R.S. (1979) *Diabetes 28*, 604-611.

Clements, R.S. and Stockard, G.R. (1980) *Diabetes 29*, 227-235.

Das, P.K., Bray, G.M., Aguayo, A.J. and Rasminsky, M. (1976) *Exp. Neurol. 53*, 285-288.

Eliasson, S.G. (1964) *J. Clin. Invest. 43*, 2353-2358.

Eliasson, S.G. (1965) *Lipids 1*, 237-240.

Eliasson, S.G. (1969) *J. Neurol. Neurosurg. Psychiat. 32*, 525-529.

Fagius, J. and Jameson, S. (1981) *J. Neurol., Neurosurg. Psychiat. 44*, 991-1001.

Field, R.A. and Adams, L.C. (1964) *Medicine 43*, 275-279.

Field, R.A. and Adams, L.C. (1965) *Biochim. Biophys. Acta 106*, 474-479.

Finegold, D., Lattimer, S.A., Nolle, S., Bernstein, M. and Greene, D.A. (1983) *Diabetes 32*, 988-992.

Gabbay, K.H. (1975) *Ann. Rev. Med. 36*, 521-536.

Gillon, K.R.W. and Hawthorne, J.N. (1983) *Life Sciences 32*, 1943-1947.

Greene, D.A. (1983) *Metabolism 32* (Suppl. 1), 118-123.

Greene, D.A. and Lattimer, S.A. (1982) *J. Clin. Invest. 70*, 1009-1018.

Greene, D.A. and Lattimer, S.A. (1983) *J. Clin. Invest. 72*, 1058-1063.

Greene, D.A. and Lattimer, S.A. (1984a) *Amer. J. Physiol. 246*, E311-E318.

Greene, D.A. and Lattimer, S.A. (1984b) *Diabetes 33*, 712-716.

Greene, D.A. and Winegrad, A.I. (1979) *Diabetes 28*, 878-887.

Greene, D.A. and Winegrad, A.I. (1981) *Diabetes 30*, 967-974.

Greene, D.A., DeJesus, P.V., Jr., and Winegrad, A.I. (1975) *J. Clin. Invest. 55*, 1326-1336.

Greene, D.A., Winegrad, A.I., Carpentier, J-L., Brown, M.J., Fukuma, M. and Orci, L. (1979) *J. Neurochem. 33*, 1007-1018.

Greene, D.A., Lewis, R.A., Lattimer, S.A. and Brown, M.J. (1982) *Diabetes 31*, 573-578.

Gregersen, G. (1968) *Diabetologia 4*, 273-277.

Hothersall, J.S. and McLean, P. (1979) *Biochem. Biophys. Res. Comm. 88*, 477-484.

Jakobsen, J. (1978) *Diabetologia 14*, 113-119.

Jakobsen, J. (1979) *J. Neurol., Neurosurg., Psychiat. 42*, 509-518.

Judzewitsch, R.G., Jaspan, J.B., Polonsky, K.S., Weinberg, C.R., Halter, J.B., Halar, E., Pfeifer, M.A., Vukaninovic, C., Bernstein, L., Schneider, M., Liang, K-Y., Gabbay, K.H., Rubenstein, A.H. and Porte, D., Jr. (1983) *N. Engl. J. Med. 308*, 119-125.

Kinoshita, J.H., Merola, L.O. and Dikmak, E. (1962a) *Biochem. Biophys. Acta 62*, 176-178.

Kinoshita, J.H., Merola, L.O. and Dikmak, E. (1962b) *Eye Res. 1*, 405-410.

Kuwabara, T., Kinoshita, J.H. and Cogan, D.G. (1969) *Invest. Ophthalmol. 8*, 133-149.

Kyte, J. (1981) *Nature 16*, 201-204.

Lingham, R.B. and Sen, A.K. (1982) *Biochim Biophys Acta 688*, 475-485.

Mandersloot, J.G., Roelofsen, B. and De Gier, J. (1978) *Biochim. Biophys Acta 508*, 478-485.

Mayer, J.H. and Tomlinson, D.R. (1983) *J. Physiol. 340*, P25-P26.

Mayhew, J.A., Gillon, K.R.W. and Hawthorne, J.N. (1983) *Diabetologia 24*, 10-13.

Natarajan, V., Dyck, P.J. and Schmid, H.H.O. (1981) *J. Neurochem. 36*, 413-419.

Palmano, K.P., Whiting, P.H. and Hawthorne, J.N. (1977) *Biochemical J. 167*, 229-235.

Ritchie, J.M. (1967) *J. Physiol. 188*, 309-329.

Roelofsen, B. (1981) *Life Sci. 29*, 2235-2247.

Sharma, A.K. and Thomas P.K. (1974) *J. Neurol. Sci. 23*, 1-15.

Sima, A.A.F. (1980) *Acta Neuropathol. 51*, 223-227.

Simmons, D.A., Winegrad, A.I. and Martin, D.B. (1982) *Science 217*, 848-851.

Stewart, M.A., Sherman, W.R., Kurien, M.M., Moonsammy, G.I. and Wisgerhof, M. (1967) *J. Neurochem. 14*, 1057-1066.

Trachtenberg, M.C., Packey, D.J. and Sweeney, T. (1981) *Current Topics in Cellular Regulation* *19*, 159–217.

Unakar, N.J., Tsui, J.Y. and Harding, C.V. (1981) *Ophthalmic Res.* *13*, 20–35.

Winegrad, A.I. and Greene, D.A. (1976) *N. Engl. J. Med.* *295*, 1416–1421.

Winegrad, A.I., Clements, R.S., Jr. and Morrison, A.D. (1972) in: *"Handbook of Physiology Sec 7 Vol I"* (R.O. Greep, E.B. Astwood, D.F. Steiner, N. Freinkel and S.R. Geiger, eds.) pp. 457–471, Waverly Press, Baltimore.

Yue, D.K., Hanwell, M.A., Satchell, F.M. and Turtle, J.R. (1982) *Diabetes* *31*, 789–794.

CHANGES IN PERIPHERAL NERVE POLYPHOSPHO-

INOSITIDE METABOLISM IN EXPERIMENTAL DIABETES:

NATURE AND SIGNIFICANCE

J. Eichberg*, L. Berti-Mattera*, M.E. Bell*
and R.G. Peterson+

*Department of Biochemical and Biophysical Sciences,
University of Houston, Houston, TX 77004 and
+Department of Anatomy, Indiana University School of
Medicine, Indianapolis, IN 46223

SUMMARY

The incorporation of ^{32}P-orthophosphate into poly-phosphoinositides, especially phosphatidylinositol 4,5-bisphosphate (PIP_2), is substantially enhanced in isolated sciatic nerves from rats made diabetic by streptozotocin injection. Increased labeling of PIP_2 was apparent after two weeks, but not one week following induction of disease. The metabolic changes preceded demonstrable decreases in peripheral nerve conduction velocity. The extent of increased isotope incorporation into PIP_2 was correlated with progressively greater doses of strepto-zotocin that resulted in more severe hyperglycemia. The addition of insulin to nerve preparations incubated *in vitro* failed to reverse the metabolic alterations, but implantation of insulin pumps into animals soon after the onset of diabetes prevented the appearance of changes. The increased uptake of isotope was not detected in that portion of PIP_2 recovered in a purified myelin fraction of sciatic nerve. These findings indicate that altered peripheral nerve polyphosphoinositide metabolism is a relatively early consequence of diabetes and is linked to prolonged hyperglycemia. It is proposed that the respon-sive PIP_2 pool is located in the axolemma and is perhaps involved in the mechanism of nerve conduction.

INTRODUCTION

The rapid metabolic turnover of the polar head group of the polyphosphoinositides, though long known, has been without defined functional significance for many years. Recent observations in several laboratories have established that the phosphodiesteratic hydrolysis of PIP_2, and often PIP is greatly stimulated by receptor activation in a variety of cell types (cf. Sections 3 and 4 in this volume) and may be intimately involved in the regulation of intracellular cytosolic Ca^{2+} levels. These findings have provided fresh impetus for the investigations of the meaning of the dynamic metabolism of these compounds in nervous tissue.

Some years ago, the hypothesis was put forward that Ca^{2+} movements which occur during nerve impulse propagation along an axon are associated with polyphosphoinositide metabolism (Hawthorne and Kai, 1970; Abdel-Latif, 1983). The evidence in support of this idea arises from the finding of relatively high levels of polyphosphoinositides, especially PIP_2, in myelin or contiguous structures (Eichberg and Dawson, 1965), the presence of these compounds in other neuronal and glial membranes (Eichberg and Hauser, 1973), and their metabolic lability and affinity for divalent cations (Eichberg and Dawson, 1965; Hendrickson and Reinertson, 1971; Buckley and Hawthorne, 1972). Further, increases in the metabolism of the phosphate groups of PIP and PIP_2 have been reported to occur in poorly myelinated invertebrate or vertebrate nerve preparations subjected to electrical stimulation, (Birnberger et al., 1971; White et al., 1974; Tret'jak et al., 1977) and in isolated Noctuid moth auditory organs exposed to sound (Killian and Schacht, 1980). These observations, while suggestive of a link between nerve impulse conduction and polyphosphoinositide metabolism, constitute at most circumstantial evidence for such a connection.

A role for polyphosphoinositides in axonal function might be profitably investigated in a system in which one could ascertain whether altered metabolism of these substances is linked to perturbation of nerve conduction. For this reason, several years ago we initiated studies of phosphoinositide metabolism in experimental diabetic

584

neuropathy, a disease in which the velocity of nerve conduction is reduced. Our results to date indicate that abnormalities in polyphosphoinositide metabolism do occur in this disorder and that their appearance is clearly associated with its onset and may be prevented by insulin therapy.

STATUS OF *MYO*-INOSITOL AND PHOSPHOINOSITIDES IN DIABETIC NEUROPATHY

Diabetic neuropathy is now most commonly considered a disease with a metabolic etiology (Clements, 1979). Following its induction by drug injection in experimental animals, peripheral nerve changes noted within a few weeks include abnormal electrophysiological properties, especially decreased conduction velocity (Eliasson, 1964; Greene *et al.*, 1975), a reduced content of *myo*-inositol and increased concentrations of sorbitol and fructose (Stewart *et al.*, 1967). The lower concentration of cyclitol appears to be brought about by impaired active transport of the substance into nerve under hyperglycemic conditions via a carrier system for which glucose is a competitive inhibitor (Gillon and Hawthorne, 1983; Greene and Lattimer, 1982). It is of particular interest that supplementation of the diet with *myo*-inositol tends to restore peripheral nerve conduction velocity in diabetic rats to normal levels (Greene *et al.*, 1975; Greene *et al.*, 1983).

The amount of lipid-bound inositol was found to be decreased only in acute and not in chronic streptozotocin-induced diabetes (Palmano *et al.*, 1977). Indications of metabolic alterations have included reports that $[^3H]$-inositol incorporation into sciatic nerve lipids is depressed both *in vitro* and *in vivo* and that alterations occur in the activities of some enzymes in inositol lipid metabolism, most clearly, a decline in the activity of CDP-diacylglycerol:inositol phosphatidyltransferase in liver and sciatic nerve of experimentally diabetic rats (Kumara-Siri and Gould, 1980; Whiting *et al.*, 1977; Whiting *et al.*, 1979).

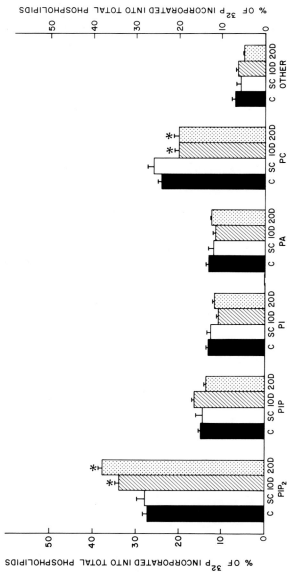

Figure 1. Distribution of ^{32}P incorporated into phospholipids of intact rat sciatic nerve. Isolated sciatic nerve segments were incubated in Krebs-Ringer-bicarbonate buffer containing glucose and inorganic ^{32}P for 2 h and incorporation of isotope into individual phospholipids was determined. Results are expressed as percent of radioactivity recovered into phospholipids ± S.E. C, Control rats (number of independent incubations = n = 27; number of rats from which nerve segments obtained = r = 16); SC, streptozotocin-control rats (10 weeks) (n = 7, r = 3); 10D, 10D, 10-week diabetic rats (n = 16, r = 8); 20D, 20-week diabetic rats (n = 14, r = 5). *p < 0.001 different from controls based on number of rats. Reproduced by permission from Bell et al. (1982).

INCORPORATION OF ^{32}P INTO PHOSPHOLIPIDS IN SCIATIC NERVE FROM NORMAL AND STREPTOZOTOCIN-DIABETIC RATS

In our initial studies, we injected female rats intraperitoneally with streptozotocin (60 mg/kg). Control animals received the sodium citrate buffer vehicle. At the time of death, frankly diabetic animals had serum glucose levels of 400–600 mg/100 ml. A small number of rats to which the drug had been administered failed to become diabetic and were included in the experimental protocol as a "streptozotocin control" group to evaluate possible extraneous effects of the agent. Sciatic nerve segments from rats with chronic diabetes were incubated *in vitro* with inorganic ^{32}P-orthophosphate and the uptake of isotope into individual phospholipids compared to weight-matched controls. A highly significant alteration in the pattern of ^{32}P$_i$ incorporation into phospholipids from that for both normal and streptozotocin control animals was observed in nerves from rats which had been diabetic for 10 and 20 weeks (Fig. 1) (Bell *et al.*, 1982) such that when expressed as percent label incorporated into total phospholipid, the proportion in PIP$_2$ rose appreciably and that in PC fell. When results were expressed as the quantity of isotope which entered phospholipids, the only consistent change was an elevation of isotope incorporation into PIP$_2$ (Table I) (Bell *et al.*, 1982). Similar results have been obtained using male rats. It is noteworthy that Natarajan *et al.* (1981) also reported changes in incorporation of ^{32}P into polyphosphoinositides in nerves from chronically diabetic rats; however, these investigators, for reasons which are not yet clear, found that the proportion of isotope in PIP rose whereas that in PIP$_2$ either fell or remained constant. We were unable to reproduce their results when we carried out incubations in the medium which they described (Bell *et al.*, 1982), so the explanation must lie elsewhere.

CORRELATION OF ONSET OF CONDUCTION VELOCITY DECREASE WITH PHOSPHOINOSITIDE METABOLISM CHANGES

In order to determine how soon the alterations in PIP$_2$ metabolism would be manifested, we investigated the

587

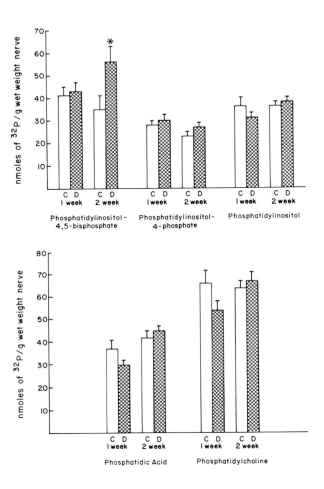

<u>Figure 2.</u> Incorporation of ^{32}P into sciatic nerve of rats 1 and 2 weeks after intraperitoneal injection of streptozotocin. Nerve segments were incubated as described in the legend to Figure 1. Results are expressed as average ± S.E. Numbers of independent incubations ranged from 20-28 (1 week) and from 8-12 (2 weeks). Number of rats used for each group was from 4-6. *p < 0.001 different from controls based on numbers of incubations.

incorporation of ^{32}P into nerve phospholipids at relatively short times after administration of streptozotocin. An increased uptake of isotope into PIP$_2$ was clearly evident after two weeks, but not after one week following onset of the disease (Fig. 2). More recently, the appearance by two weeks of abnormal PIP$_2$ metabolism as judged by ^{32}P incorporation has been confirmed using nerves from animals which were injected intravenously and in which overt diabetic symptoms were more pronounced (Berti-Mattera, Peterson and Eichberg, unpublished experiments).

To correlate these biochemical changes with electrophysiological alterations, we measured conduction velocities at successive time intervals after induction of disease (Bell *et al.*, 1982; Eichberg *et al.*, 1982). Significant reductions were observed at 10 and 20 weeks, but not at 2 weeks. However, decreases in peripheral nerve conduction velocities have been detected by other investigators as early as two weeks after diabetes was produced. Thus an alteration in PIP$_2$ metabolism is among the earliest abnormalities manifested in peripheral nerve and its time of appearance coincides with that of measurably reduced conduction velocity.

ALTERATIONS IN POLYPHOSPHOINOSITIDE METABOLISM ARE AFFECTED BY THE DEGREE OF HYPERGLYCEMIA AND ARE PREVENTED BY INSULIN THERAPY

To establish more firmly that the observed metabolic changes were a consequence of hyperglycemia and/or insulin deficiency, we performed experiments designed to regulate the level of serum glucose to which the nerve would be chronically exposed. In one approach, we decreased the dose of streptozotocin administered to as low as 30 mg/kg, a level at which only about 50% of the animals developed diabetes and those which did were characterized by serum glucose values approximately 60% of those usually obtained at twice the amount of injected drug. The extent of increased isotope incorporation into PIP$_2$ rose in a manner parallel with successively higher doses of streptozotocin and the degree of hyperglycemia achieved (Table I).

589

TABLE I

Effect of Streptozotocin Dose on Incorporation of ^{32}P into Phospholipids of Sciatic Nerve

	Control (6)	Diabetic (60 mg/kg) (6)	Control (2)	Diabetic (45 mg/kg) (6)	Control (3)	Diabetic (30 mg/kg) (3)
			(nmol ^{32}P incorporated /g wet weight nerve)			
PIP_2	18	30*	31	41*	35	43*
PIP	12	15	19	21	22	25
PI	9	9	20	18	29	24
PA	8	10	22	17	23	26
PC	20	19	33	31	41	43

Female rats were injected intraperitoneally with streptozotocin as indicated and nerve incubations performed as described in the legend to Figure 1. Numbers in parentheses indicate the number of rats. Differences from respective controls: *$p < 0.01$.

We have also examined the ability of insulin treatment to affect the alterations in sciatic nerve PIP_2 metabolism brought about by diabetes. Incubation of intact nerves with purified beef insulin (0.1 and 1.0 mU/ml medium) failed to have any effect on the increased labeling of the polyphosphoinositides (Table II). In another series of experiments, animals were injected intravenously with streptozotocin and within 72 hours thereafter were implanted subcutaneously with long-acting insulin pumps. Additional groups of rats were either not treated or received pumps containing NaCl following drug administration. Nerves were removed and incubations performed eight weeks later. Animals with uncompensated diabetes had very high serum glucose values and showed an exceptionally large increase in the proportion of radioactivity incorporated into PIP_2, and a smaller but significant rise in PIP labeling as well. Animals which had undergone prolonged insulin treatment following streptozotocin injection were hypoglycemic (ca. 60 mg glucose/ml serum) at the time of sacrifice. In these rats, the pattern of ^{32}P incorporation into phospholipids was virtually indistinguishable from that obtained for nerves from normals (Table II). When sciatic nerves from the normal, diabetic and insulin-treated animals were stripped of epineurium prior to incubation, the same results were obtained, although the proportion of isotope incorporated into PIP_2 was considerably higher for all groups. Thus, sustained availability of insulin which prevents hyperglycemia also prevents the alterations otherwise seen in polyphosphoinositide metabolism.

INVESTIGATIONS OF THE SUBCELLULAR LOCALIZATION OF THE METABOLIC CHANGES

If polyphosphoinositides are integral to the mechanism of nerve impulse propagation, it would seem plausible that the pool of these compounds involved in this process would be located in the axolemma, distinct from the bulk of PIP and PIP_2 in the myelin sheath, and constitute the fraction of these compounds which exhibit metabolic abnormalities in diabetes. In an approach to determine whether a portion of peripheral nerve PIP_2 which undergoes increased labeling is confined to a discreet subcellular site, normal and diabetic sciatic nerves were incubated with $^{32}P_i$ and myelin was then isolated according to the procedure of Wiggins *et al.* (1975) performed at alkaline pH to prevent the enzymatic hydrolysis of polyphosphoinositides (Eichberg and Hauser, 1973). In comparison with the radiolabeling pattern for whole nerve, a much higher proportion of phospholipid radioactivity in myelin was recovered in PIP_2 and substantially less was present in PC (Fig. 3). Whereas PIP_2 in whole nerve homogenate from diabetic animals showed increased radiolabeling as expected, there was no discernible difference in the proportion of isotope incorporated into this lipid in the purified myelin fraction. When this experiment was repeated using epineurium-free nerve, the same result was obtained. Examination of the denser heterogeneous non-myelin fraction revealed that a slightly greater fraction of radiolabel was present in PIP_2 in diabetic nerve as compared to control. In the course of lengthy subcellular fractionations, there is always the danger that, despite precautions, a highly labile portion of polyphospho-inositides may be lost. This possibility not withstanding, these findings suggest at the very least that the pool of PIP_2 which undergoes metabolic alterations in experimental diabetes is not associated with myelin.

DISCUSSION

Enhanced ^{32}P labeling of somatic peripheral nerve PIP_2 appears to be a characteristic feature of both acute and chronic experimental diabetic neuropathy. Maximally increased incorporation of isotope is apparent as soon as two weeks after the onset of diabetes and is therefore temporally correlated with the appearance of decreased peripheral nerve conduction velocity. Four pieces of evidence strongly link the metabolic changes to the effects of hyperglycemia or insulin deficiency. First, successively higher doses of streptozotocin produce progressively higher serum glucose levels and greater ^{32}P uptake into PIP_2. Second, injection of streptozotocin into animals which did not become diabetic caused no altera-tions in the labeling pattern of phospholipids, tending to rule out effects of the drug unrelated to diabetes. Third, rats which received drug via the intravenous route and developed more pronounced hyperglycemia than those injected intraperitoneally, also displayed more prominent PIP_2 radiolabeling changes and some increase in

TABLE II

Effect of Insulin on Incorporation of ^{32}P into Phospholipids of Sciatic Nerve from Streptozotocin-Diabetic Rats

% of ^{32}P incorporated into phospholipids

	Insulin *in vitro*			Insulin *in vivo*		
	Control	Diabetic	Diabetic +0.1 mU/ml Insulin	Control (5)	Diabetic (6)	Diabetic + Insulin Pump (6)
PIP$_2$	27	31*	35*	20	34*	20
PIP	11	11	12	12	16*	11
PI	11	11	11	15	12*	15
PA	15	14	13	15	14	14
PC	26	24	21*	26	18*	31*
PE	3	3	2	5	4	5

For experiments *in vitro*, nerve segments were incubated as described in the legend to Figure 1. The pattern of incorporation for control animals was unchanged by the presence of insulin. Insulin was administered *in vivo* using long acting insulin pumps as described in the text. Figures in parentheses are number of rats.

*p < 0.01 different from controls based on number of rats.

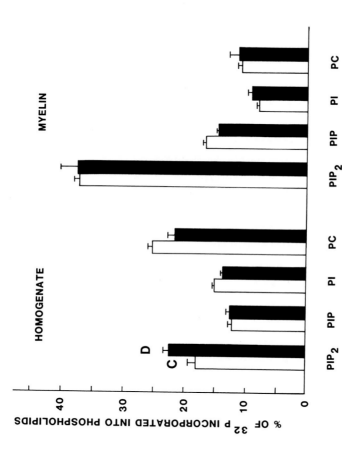

Figure 3. Distribution of ^{32}P incorporated into homogenate and purified myelin from normal and diabetic rats. Nerve segments were incubated as described in the legend to Figure 1 and the labeled nerve then fractionated to isolate myelin according to the procedure of Wiggins et al. (1975). Results are expressed as percent of radio-activity recovered in total phospholipid ± S.E. Nerves from three rats were utilized for each group. *p < 0.01 different from control.

PIP radioactivity as well (compare Fig. 1 and Table II). Fourth, insulin continuously supplied for many weeks from implanted pumps maintained streptozotocin-treated animals in a somewhat hypoglycemic state and completely prevented the metabolic alterations in polyphosphoinositide metabolism. Insulin therapy has previously been shown to restore or alleviate the conduction velocity deficit (Greene et al., 1982).

In contrast to the preventive effect of insulin administered in vivo, the presence of insulin in incubations of nerve in vitro failed to counteract the altered incorporation pattern. Such a result may not be surprising in that insulin supplied in this manner may not penetrate the multiple connective tissue barriers which exist in intact sciatic nerve. Alternatively, since carbohydrate metabolism of nerve, like that of brain, is essentially unresponsive to insulin, a compensatory effect of the hormone on energy production, even if it is accessible to the axon, is unlikely although pathways not involving glucose could conceivably be affected.

There are several indirect indications that an unusually rapidly metabolized pool of polyphosphoinositides might be associated with axonal and other non-myelin membranes. Whereas most nervous tissue polyphosphoinositides undergo rapid hydrolysis post mortem, the portion in myelin-deficient regions undergoes the most rapid breakdown, largely disappearing within a few minutes after death (Gonzales-Sastre et al., 1971). Further, the time course of equilibration of polyphosphoinositide P with brain acid soluble P is much more rapid in myelin-poor than in myelin-rich areas (Hauser et al., 1971). It is attractive to consider that such a pool of PIP_2 would be preferentially affected in diabetic neuropathy. The results of our preliminary subcellular fractionation studies are consistent with the conclusion that the fraction of PIP_2 which exhibits metabolic alterations in diabetic nerve is not present in myelin, where the preponderant amount of this substance is located. The heterogeneity of the non-myelin fraction, which is obtained together with purified myelin in the procedure used, precludes any statement concerning the site of the presumably small, responsive PIP_2 pool. Whether it resides in the axolemmal membrane could perhaps be determined by subcellular isolation of a fraction enriched in this structure (Yoshino *et al.*, 1983), provided *post mortem* degradation of polyphosphoinositides could be adequately controlled.

The mechanism underlying the changes in nerve PIP_2 metabolism associated with diabetes remain to be elucidated. The possibility that the diabetic state influences the specific activity of a radioactive ATP pool which serves as phosphate donor in the biosynthesis of PIP and PIP_2 cannot yet be excluded, but seems unlikely for two reasons. First, PIP_2 labeling is much more readily increased than that of PIP, whereas both compounds should be comparably affected if their common immediate precursor were involved. Second, the amount of ATP is not detectably different in incubated endoneurial preparations from normal and diabetic sciatic nerve (Greene and Winegrad, 1981).

Assay of enzymes responsible for the formation and degradation of PIP_2 has thus far shown a possible diminution of PIP kinase activity and no change in either PI kinase or PIP_2 phosphomonoesterase activity in diabetic peripheral nerve (Whiting *et al.*, 1979). The activity of PIP_2 phosphodiesterase, which is stimulated by a variety of receptor-agonist interactions remains to be examined and, if altered, may provide an indication as to whether increased biosynthesis or decreased breakdown is involved. In this connection, it is not obvious how a depressed level of CDP-diacylglycerol:inositol phosphatidyltransferase in diabetic nerve could contribute significantly to increased labeling of PIP_2.

The relationship of changes in polyphosphoinositide metabolism to other documented biochemical deficits in peripheral nerve function in diabetes is obscure. Recently, it was found that diabetic nerve displays substantially reduced Na^+-K^+-dependent ATPase activity, and that this deficiency may be ameliorated by dietary inositol supplementation (Das *et al.*, 1976; Greene and Lattimer, 1983). Diabetic nerve also accumulates Na^+ in the axoplasm (Brismar and Sima, 1981), an effect which may be attributed to the substantially reduced Na^+-K^+-dependent ATPase activity, and which could create electrophysiological disturbances, thereby compromising nerve conduction velocity. This has prompted the hypothesis (Greene, 1983; Greene and Lattimer, this volume), that since inositol transport into the nerve is Na^+-dependent, decreased sodium pump activity, as well as competition with glucose for entry, could lead to depressed tissue inositol levels and consequently to reduced formation of PI. Since there is evidence that endogenous PI can activate kidney Na^+-K^+-dependent ATPase (Mandersloot et $al.$, 1978), it is conceivable that one or more phosphoinositides are modulators of this enzyme in nerve. Therefore, we speculate that the metabolic imbalance in polyphosphoinositide metabolism which arises in diabetes could adversely affect Na^+-K^+-dependent ATPase activity. Investigation of this possibility will require isolation of this membrane-bound enzyme from nerve and evaluation of its function in different lipid environments.

ACKNOWLEDGEMENTS

We wish to thank Ms. Jaime Glasser for expert technical assistance. This work was supported by NIH grants AM-30577 and RR-07147 and by a grant from the American Diabetes Association.

REFERENCES

Abdel-Latif, A.A. (1983) in: "Handbook of Neurochemistry", 2nd edition, Vol. 3 (A. Lajtha, ed.) pp. 91-131, Plenum Press, New York.

Bell, M.E., Peterson, R.G. and Eichberg, J. (1982) J. Neurochem. 39, 192-200.

Birnberger, A.C., Birnberger, K.L., Eliasson, S.G., and Simpson, P.C. (1971) J. Neurochem. 18, 1291-1298.

Brismar, T. and Sima, A.A.F. (1981) Acta Physiol. Scand. 113, 499-506.

Buckley, J.T. and Hawthorne, J.N. (1972) J Biol. Chem. 247, 7218-7223.

Clements, R.S., Jr. (1979) Diabetes 28, 604-611.

Clements, R.S., Jr. and Stockard, C.R. (1980) Diabetes 29, 227-235.

Das, P.K., Bray, B.M. and Aguayo, A.J. (1976) Exp. Neurol. 53, 285-288.

Eichberg, J. and Dawson, R.M.C. (1965) *Biochem. J. 96*, 644-650.

Eichberg, J. and G. Hauser (1973) *Biochim. Biophys. Acta 326*, 210-223.

Eichberg, J., Bell, M.E. and Peterson, R.G. (1982) in: *"Phospholipids in the Nervous System. Volume I: Metabolism"* (L.A. Horrocks, ed.) pp. 271-282, Raven Press, New York.

Eliasson, S.G. (1964) *J. Clin. Invest. 43*, 2353-2358.

Gillon, K.R.W. and Hawthorne, J.N. (1983) *Biochem. J. 210*, 775-781.

Gonzales-Sastre, F., Eichberg, J. and Hauser, G. (1971) *Biochim. Biophys. Acta 248*, 96-104.

Greene, D.A. (1983) *Metabolism 32* Suppl 1, 118-123.

Greene, D.A. and Lattimer, S.A. (1982) *J. Clin. Invest. 70*, 1009-1018.

Greene, D.A. and Lattimer, S.A. (1983) *J Clin. Invest. 72*, 1058-1063.

Greene, D.A. and Winegrad, A.L. (1981) *Diabetes 30*, 967-974.

Greene, D.A., De Jesus, P.V. Jr. and Winegrad, A. (1975) *J. Clin. Invest. 55*, 1324-1336.

Greene, D.A., Lewis, R.A., Brown, M.J. and Lattimer, S.A. (1982) *Diabetes 31*, 573-578.

Hauser, G., Gonzales-Sastre, F. and Eichberg, J. (1971) *Biochim. Biophys. Acta 248*, 87-95.

Hawthorne, J.N. and Kai, M. (1970) in: *"Handbook of Neurochemistry"* Vol. 3 (A. Lajtha, ed.) pp. 491-508, Plenum Press, New York.

Hendrickson, H.S. and Reinertson, J.L. (1971) *Biochem. Biophys. Res. Commun. 44*, 1258-1264.

Hothersall, J.S. and McLean, P. (1979) *Biochem. Biophys. Res. Commun. 88*, 477-484.

Killian, P. and Schacht, J. (1980) *J. Neurochem. 34*, 709-712.

Kumara-Siri, M.H. and Gould, R.M. (1980) *Brain Res. 180*, 315-330.

Mandersloot, J.G., Roelofson, B. and De Gier, J. (1978) *Biochim. Biophys. Acta 508*, 478-485.

Natarajan, V., Dyck, P.J. and Schmid, H.H.O. (1981) *J. Neurochem. 36*, 413-419.

Palmano, K.P., Whiting, P.H. and Hawthorne, J.N. (1977) *Biochem. J. 167*, 229-235.

Stewart, M.A., Sherman, W.R., Kunen, M.M., Moonsammy, G.I. and Wisgerhof, M. (1967) *J. Neurochem.* *14*, 1057-1066.

Tret'jak, A.G., Limatenko, I.M., Kossova, G.V., Gulak, P.V. and Kozlov, Yu. P. (1977) *J. Neurochem.* *28*, 199-205.

White, G.L., Schellhase, H.U. and Hawthorne, J.N. (1974) *J. Neurochem.* *22*, 149-158.

Whiting, P.H., Bowley, M., Sturton, R.G., Pritchard, P.H., Brindley, D.N. and Hawthorne, J.N. (1977) *Biochem. J.* *168*, 147-153.

Whiting, P.H., Palmano, K.P., and Hawthorne, J.N. (1979) *Biochem. J.* *179*, 549-553.

Wiggins, R.C., Benjamins, J.A. and Morell, P. (1975) *Brain Res.* *89*, 99-106.

Yoshino, J.E., Griffin, J.W. and DeVries, G.H. (1983) *J. Neurochem.* *41*, 1126-1130.

RELATIONSHIP OF INOSITOL, PHOSPHATIDYLINOSITOL AND PHOSPHATIDIC ACID IN CNS NERVE ENDINGS

J.R. Yandrasitz, G. Berry and S. Segal

Division of Metabolism, Children's Hospital
Philadelphia, PA 19104

SUMMARY

Nerve endings were prepared from rat cerebra by differential centrifugation and a 5-step Ficoll gradient, and challenged with several agents during incubation with $^{33}PO_4$ and [^3H]inositol. The specific radioactivities of phospholipids and free inositol were determined in an attempt to discern the metabolic steps which are responsible for increased radiolabeling of phosphatidylinositol (PI) and phosphatidic acid (PA). While ACh greatly stimulated $^{33}PO_4$ incorporation into PI, comparison of PI specific radioactivities with that of precursor PA did not indicate a greatly enhanced rate of PI synthesis. In low calcium buffers the ACh stimulation of PA radiolabeling was more attenuated than that of PI. The neurotransmitter stimulated [^3H]inositol incorporation into PI much less than $^{33}PO_4$. Inositol transport was unaffected by ACh, but its specific radioactivity was decreased, partly by dilution with inositol from PI. Incorporation of inositol into PI by an exchange mechanism was not important in the action of ACh. Increasing $[K^+]_o$ greatly enhanced $^{33}PO_4$ incorporation into PA and PI, but probably by a Ca^{2+} dependent mobilization of diacylglycerol from a lipid other than these. The uptake and specific radioactivity of inositol were decreased by elevated K^+ in parallel with decreased ^3H radiolabeling of PI; this effect was not dependent on Ca^{2+} or Na^+. Elevated $[K^+]_o$ and ACh acted synergistically, indicating that the two agents act at the same subcellular sites but at different metabolic steps.

601

INTRODUCTION

The central nervous system (CNS) was one of the first tissues examined for the occurrence of the "phospholipid effect". Brain slices were found to respond with increased phospholipid turnover upon exposure to neurotransmitters such as acetylcholine (ACh) or norepinephrine (Hokin and Hokin, 1955; Abdel-Latif *et al.*, 1981) or to depolarizing stimuli such as elevated extracellular potassium (Yoshida and Quastel, 1962) or electric fields (Pumphrey, 1969). When subcellular fractions were examined, the major effect was found in nerve endings isolated as synaptosomes. This was true even when a muscarinic stimulus was applied *in vivo* (Lunt and Pickard, 1975) or to brain slices (Abdel-Latif *et al.*, 1981). Isolated nerve endings were also found to respond to *in vitro* stimulation by neurotransmitters (Schacht and Agranoff, 1972; Sneddon and Keen, 1970) or electrical depolarization (Hawthorne and Bleasdale, 1975).

In some respects the response of synaptosomes to acetylcholine is similar to the "phospholipid effect" found in other tissues. It has a dose dependence characteristic of the low affinity muscarinic receptor (Miller, 1977) which in the CNS, has a role in the presynaptic modulation of neurotransmitter release (Nordstrom and Bartfai, 1980). Turnover of phosphatidylinositol (PI) and phosphatidic acid (PA) is confined to the polar headgroups of phosphate and inositol, while ATP specific radioactivity (Schacht and Agranoff, 1974) and uptake of [³H]inositol (Warfield *et al.*, 1978) are unaffected by ACh. However, other studies have suggested that the "phospholipid effect" in synaptosomes involves more than an increased degradation of phosphoinositides. Stimulation of phosphate incorporation into PA is often much greater than into PI, and increased incorporation of [³H]inositol into PI is much less than incorporation of ³²P (Schacht and Agranoff, 1974). In prelabeled synaptosomes, PI appeared remarkably stable upon exposure to ACh (Lapetina and Michell, 1974; Schacht and Agranoff, 1973). In a galactose toxicity animal model for galactosemia, ACh stimulation of [³H]inositol incorporation into PI was diminished, while ³³PO₄ incorporation was essentially unaffected (Warfield and Segal, 1978; Berry *et al.*, 1981). These studies have suggested that PI and PA metabolism in synaptosomes might not be as tightly coupled as in extraneural tissues, and that the "phospholipid effect" in these organelles might involve more than one pathway of lipid metabolism. Our approach to these questions has been to examine several parameters of phospholipid and inositol metabolism by measuring radioactivity and quantities of important species in the same preparations, and to combine several stimuli which affect PI metabolism in order to delineate the pathways involved in stimulated phospholipid metabolism.

EXPERIMENTAL

Synaptosomes were prepared from rat brains by a combination of differential centrifugation and a 5-step Ficoll gradient (Warfield and Segal, 1974), and the upper two of three synaptosomal layers were collected. These 'light' synaptosomes are enriched in ACh metabolizing enzymes (DeRobertis et al., 1962) and show the greatest phospholipid response to ACh (Miller, 1977).

Uptake of [^{3}H]inositol was measured as a distribution ratio of the releasable label from synaptosomal pellets to the label in the medium (Warfield et al., 1978). Inositol as the pertrimethylsilylated derivative was measured by gas liquid chromatography (GLC). For specific radioactivity determinations, the peak was collected at a stream splitter. The same system was used for analysis of PI as glycerophosphoryl inositol obtained after a mild deacylation of the lipid extract as described by Cicero and Sherman (1973). This provides excellent resolution of PI from other lipids, and with the inclusion of monogalactosyl diacylglycerol as an internal standard, it provides very good quantitation of PI in small samples. Our other method for separation of phospholipids is high performance liquid chromatography (HPLC) (Yandrasitz et al., 1981). This provides a clean separation of PA, and allows easy and quantitative collection of lipids for subsequent analysis by liquid scintillation counting and phosphate assay. Lipids of interest elute as sharp well confined peaks as seen by the elution of [^{14}C]lecithin, and are free of contamination by inorganic phosphate (Yandrasitz et al., 1983).

RESULTS AND DISCUSSION

Acetylcholine produced a parallel increase in radio-labeling of PI with $^{33}PO_4$ and [^3H]inositol and a small decrease in PI levels (Yandrasitz and Segal, 1979). However, the radiolabeling effects for phosphate and inositol occurred on different scales. While ACh increased $^{33}PO_4$ incorporation into PI by more than 3-fold, [^3H]inositol incorporation increased by only 40% in the same samples. Consideration of the pathways of phospholipid metabolism indicates several ways in which this discrepancy might arise. Phosphate radiolabel in PI is a function of both the rate of synthesis from PA and of the changes in PA specific radioactivity. The latter depends on a mobilization of diacylglycerol, but this may come from lipids other than phosphoinositides, including PA itself; stimulation of phosphatidate phosphatase has been suggested as the mechanism for the "phospholipid effect" in synaptosomes (Schacht and Agranoff, 1974). Inositol enters synaptosomes very slowly, raising the possibility that radiolabel in free inositol might be limiting with increased phospholipid turnover. In addition, while $^{33}PO_4$ is incorporated into PI only by the synthetic pathway via CDP-diacylglycerol:inositol transferase, [^3H]inositol may be incorporated by this pathway or by Mn^{2+} dependent exchange catalyzed either by the transferase (Bleasdale and Wallis, 1981) or by a nucleotide independent enzyme (Takenawa et al., 1977). Our studies emphasized three possible mechanisms for the discrepancy in radiolabel incorporation into PI: That stimulated PA radiolabeling occurred independently of PI degradation, which we examined by determining the relationship of PA and PI in various metabolic states; that a baseline exchange incorporation of [^3H]inositol into PI might reduce the apparent magnitude of its incorporation by resynthesis; and that the handling of free [^3H]inositol limited its incorporation into PI.

Synaptosomal Free and Phospholipid Inositol

The CNS appears to meet its need for inositol by synthesis rather than by uptake from the peripheral circulation (Barkai, 1981). Synaptosomes, when incubated with a wide range of inositol concentrations, showed no evidence of saturable uptake characteristic of a

TABLE I

Effect of ACh on Synaptosomal Free Inositol

	Uptake D.R.	Specific Radio- activity dpm/nmol	Amount nmol/mg protein
Control	0.51 ± 0.04	2181 ± 22	9.66 ± 0.57
		Normalized to Control	
No ACh	1.00 ± 0.07	1.00 ± 0.01	1.00 ± 0.01
0.2 mM	1.03 ± 0.08	0.81 ± 0.05	1.08 ± 0.03

Incubation for 40 min with 0.02 mM myo-[^{3}H]inositol at a specific radioactivity of 366,000 dpm/nmol. Uptake was measured as a Distribution Ratio (DR) of label in synaptosomal cytosol to label in the incubation medium and corrected for zero time. Specific radioactivity was measured as dpm/GLC peak area and dpm/nmol estimated from peak areas of inositol standards. Levels were measured by GLC with respect to an internal standard of ribitol. Values are mean ± S.E. for 6 determinations (4 for uptake).

carrier mediated process, and entry of inositol was not inhibited by ouabain or metabolic inhibitors (Warfield *et al.*, 1978). Uptake of inositol achieved a distribution ratio of 0.5 after 40 min of incubation (Table I); a value of 1 indicates equilibration and higher numbers indicate active transport. Acetylcholine had no effect on inositol uptake in this and earlier studies (Warfield *et al.*, 1978). However, the specific radioactivity of free inositol declined about 20% with ACh. Part of this drop could be attributed to dilution resulting from a 10% increase in levels, but the origin of the remainder is unknown. This may reflect a shift of radioactivity to water-soluble inositol phosphates which were not measured in these experiments. Measurement of levels and specific radioactivities indicated the relationship between inositol and PI. The specific radioactivity of inositol was about 30 times greater than that of PI at the same time, indicating that

Yandrasitz et al.

TABLE II

Effect of $MnCl_2$ on Uptake of Myo-[^{3}H]Inositol

| | Uptake Relative to Parallel Controls | |
	0.001 mM Mn^{2+}	0.5 mM Mn^{2+}
Total	1.00 ± 0.01 (4)	1.54 ± 0.08 (8)
Soluble	1.00 ± 0.02 (8)	1.02 ± 0.04 (8)
Membranes	1.00 ± 0.02 (4)	5.09 ± 0.28 (4)
Membranes/Total	0.14 ± 0.01	0.48 ± 0.01

Distribution ratio of 0.02 mM inositol = 0.5 at 40 min. Measured as described in Table I and normalized to parallel control incubations which contained 0.001 mM Mn^{2+}. Values are mean ± S.E. for N in parentheses.

at least the bulk pool of inositol was not limiting for entry of radioactivity into PI. However, the levels of free inositol and PI were nearly the same. Thus a decline in PI levels due to increased degradation results in a comparable increase in free inositol and a dilution of its specific radioactivity.

To assess the role of CDP-diacylglycerol:inositol transferase in PI metabolism, we incubated synaptosomes with the specific cofactor Mn^{2+}. This resulted in a large increase in [^{3}H]inositol incorporation into PI with only a small effect on $^{33}PO_4$ radiolabel (Yandrasitz and Segal, 1979), characteristic of an exchange reaction. An effect of Mn^{2+} on free inositol was assessed by measurement of this pool (Table II). At 0.5 mM $MnCl_2$ there was no increase in the soluble radiolabel, while membrane activity was markedly increased. Surprisingly, there was also no decrease in free [^{3}H]inositol, which one would expect if internal pools were drawn upon to supply the large increase in lipid radioactivity. This result was confirmed by measurement of specific radioactivity, and suggested that

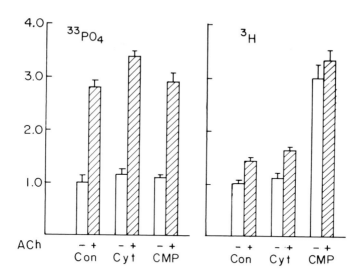

<u>Figure 1</u>. Effect of cytidine nucleotides on basal and acetylcholine-stimulated phosphatidylinositol metabolism. Synaptosomes were incubated with no nucleotide, (con), 1 mM cytidine, (Cyt), or 0.05 mM CMP, with or without 0.2 mM acetylcholine. Incorporation of $^{33}PO_4$ or [^{3}H]myo-inositol into PI were measured by GLC of PI derived glycerophosphoryl inositol, and values were normalized to control samples. The ordinate indicates specific radioactivity relative to parallel control (no nucleotide) incubations. Bars indicate the Mean ± S.E. for 6 determinations.

extrasynaptosomal inositol was used directly for the Mn^{2+} stimulated incorporation. An external or plasma membrane locus was also suggested by experiments with nucleotides which showed that exchange activity was due to CDP-di-acylglycerol:inositol transferase rather than a nucleotide independent exchange enzyme. Cytidine, which is taken up via nucleoside transport (Bender *et al.*, 1981), had no effect on the Mn^{2+} dependent labeling of PI, while CMP, which should be excluded from synaptosomes, had a profound effect on exchange (Berry *et al.*, 1983). At 0.5 mM $MnCl_2$, incorporation of [^{3}H]inositol into PI in the absence of added nucleotide was only about 3% of that which was obtained at 10 µM CMP. The K_m for CMP was about 2 µM, which is low enough to suggest that

Yandrasitz et al.

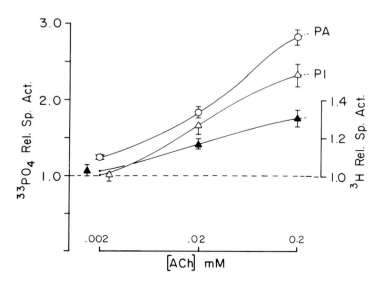

<u>Figure 2</u>.   Effect of acetylcholine on synaptosomal phospholipid turn-
over.   Synaptosomes were incubated for 40 min with no agent or with the
indicated concentrations of ACh in the presence of 0.2 mM eserine.
After separation of phospholipids by HPLC the specific radioactivity was
determined for $^{33}PO_4$ in PA (0) or in PI (Δ), and for [^{3}H]$\underline{myo}$-inositol in
PI ($\blacktriangle$).   These results were normalized to simultaneous control samples.
Bars indicate the S.E. for 6 determinations at 0.002, 12 at 0.02, and 20
at 0.2 mM acetylcholine.

exchange activity in the absence of added nucleotide
might be supported by endogenous levels of CMP.

We then combined these agents with ACh to investigate what role the inositol transferase and exchange
activity might have in basal and stimulated PI turnover.
The ACh-produced increment in PI radiolabeling by
[^{3}H]inositol was not enhanced by 1 mM $MnCl_2$, and was
hardly discernible above the Mn^{2+} elevated baseline
(Yandrasitz and Segal, 1979).   Manganese had little effect
on PI turnover as indicated by $^{33}PO_4$ incorporation;
however, the ACh effect was slightly enhanced.   This
was also seen with 1 mM cytidine (Fig. 1) which should
act inside the synaptosome.   In this case a small increase
in the ACh effect on inositol incorporation was also seen.

TABLE III

Effect of ACh on Synaptosomal PA and PI

	Control	0.2 mM ACh
PA (nmol/mg prot)	2.47 ± 0.31	2.74 ± 0.24 (1.11)
[^{33}P]PA (Sp. Act.)	307 ± 45	721 ± 94 (2.35)
Total ^{33}P	758.3 dpm/mg prot	
PI (nmol/mg prot)	9.91 ± 0.36	8.83 ± 0.42 (0.89)
[^{3}H]PI (Sp. Act.)	68.8 ± 3.6	92.9 ± 2.8 (1.35)
[^{33}P]PI (Sp. Act.)	37.5 ± 1.6	85.1 ± 19.6 (2.27)
Total ^{33}P	371.6 dpm/mg prot	
PA to PI conversion (nmol)	1.21	1.04

Synaptosomes were incubated for 40 min in the presence or absence of 0.2 mM ACh + eserine sulfate. Phospholipids were separated by HPLC and quantitated by phosphate assay. Specific radioactivity was calculated as dpm/nmol P. Values are mean ± S.E. for 12 determinations. Values in parentheses indicate the effect of ACh as a ratio of values in parallel control incubations. Phosphatidic acid (PA) to phosphatidylinositol (PI) is a rough estimate of this conversion obtained by dividing the total radiolabel in PI by the specific radioactivity of PA at the end of the incubation.

In contrast, CMP had no effect on ^{33}PO$_4$ incorporation into PI, and appeared to compete with the ACh-stimulated metabolism. We conclude from these studies that the inositol transferase reaction is not rate limiting for PI labeling during basal turnover, but may become so at high levels of ACh, and that inositol exchange with PI is not involved in the ACh-stimulated turnover.

Relationship between PA and PI

Phosphatidic acid and PI accounted for nearly all the ^{33}PO$_4$ incorporated into synaptosomal phospholipids, and the effect of ACh was confined to these two species (polyphosphoinositides were not examined). In our hands, ACh-stimulation of PA labeling was only slightly greater than that of PI, and this was only apparent at higher ACh concentrations (Fig. 2). While PA specific radioactivity was 10-fold that of PI, the latter contained 1/3 to 1/2 as much label, indicating that PI is a major fate for the $^{33}PO_4$ incorporated into PA (Table III). At 0.2 mM ACh, PI levels declined by about 1 nmol/mg protein which is about the same amount as the increase in free inositol. Phosphatidic acid levels increased slightly. Although this is not a significant increase, it does argue against a mechanism involving increased phosphatidate phosphatase activity. Inositol label in PI increased by about 35%. This increase would be 65% if a correction were made for the decrease in inositol specific activity discussed above. This is still less than the 2.3-fold increase in PI $^{33}PO_4$ label, but the phosphate label indicates both synthesis and the change in PA specific radioactivity. If a rough estimate is made of the conversion of PA to PI by dividing the total label in PI by the specific radioactivity of PA, it appears that with ACh there is in fact a 15% decline in PI synthesis despite and increase in $^{33}PO_4$ specific activity. That is, the increase in PA specific activity is more than enough to account for the increased label in PI, and the latter does not in itself indicate increased PI degradation and resynthesis.

The increased labeling of PA might be due to mobilization of diacylglycerol from lipids other than phosphoinositides, perhaps stimulated by calcium entry as described in lymphocytes (Allan and Michell, 1977). We therefore examined the effect of low calcium on basal and ACh-stimulated turnover. When calcium was buffered to about 10^{-6} M with EGTA (Raaflaub, 1960), the effect of ACh on phospholipid labeling was markedly diminished, (shaded) while basal turnover was only slightly affected (Fig. 3). Apparently, a large part of the $^{33}PO_4$ effect of ACh involves Ca^{2+} entry, and this is of significance for the postulated role of the synaptosomal "phospholipid effect" in inhibition of transmitter release. However, in low calcium media, the ACh effect for PA labeling was much more attenuated than for PI, particularly at 0.02 mM ACh. The data at low Ca^{2+} are more consonant with a primary effect of ACh on PI turnover, and suggest that in normal buffers there is a secondary effect on PA due to calcium entry.

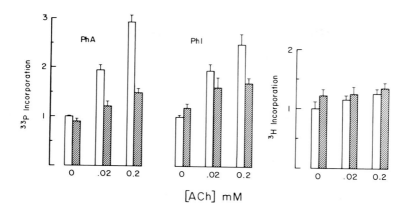

Figure 3. Calcium dependence of the acetylcholine effect on phospholipid labeling. Synaptosomes were incubated for 40 min with the indicated concentrations of acetylcholine = 0.2 mM eserine in normal incubation medium (open bars) or in calcium free buffers (shaded). Incorporation of $^{33}PO_4$ into PA and PI, and [^{3}H]myo-inositol into PI were measured and normalized to 1.3 mM calcium controls. Bars indicate the S.E. of 6 determinations at 0.02 mM and 16 at 0.2 mM ACh.

Potassium Depolarization

To further examine the relationship of PA and PI metabolism and the importance of calcium, we turned to another stimulus. Depolarization of synaptosomes by increasing medium potassium has been associated with both a "phospholipid effect" (Hawthorne and Bleasdale, 1975) and neurotransmitter release subsequent to calcium entry (Blaustein, 1975). Our protocol was very similar to that of Hawthorne and Bleasdale (1975) except that K^+ replaced Na^+, and we obtained essentially the same results for phosphate labeling, including a decline in PI labeling at higher [KCl] (Fig. 4). Since we also measured [^{3}H]inositol incorporation, we obtained the anomalous finding that inositol incorporation into PI declined in parallel with an increase in $^{33}PO_4$; this will be discussed below.

The effect of increased $[K^+]_o$ was confined to PA and PI and appeared to be a specific effect of potassium

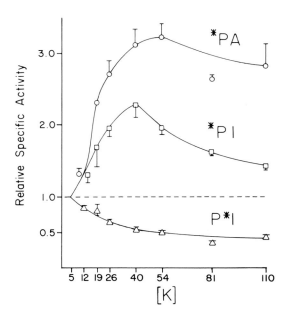

Figure 4. Effect of elevated potassium on phospholipid labeling. Synaptosomes were incubated for 40 min in the indicated concentrations of KCl. Incorporation of $^{33}PO_4$ into PA (0) and PI ($\square$), and [^{3}H]myo-inositol incorporation into PI ($\triangle$) were measured and normalized to values for control (5 mM K) samples. Bars indicate the range of two determinations at 19 and 81 mM K, or the S.E. for 4 determinations at 12 mM, 6 at 26 mM, and 10 determinations at the other points.

depolarization on phospholipid turnover. The same results were found in the presence of atropine to inhibit an action of released endogenous ACh, and in buffers where sodium was held constant at 105 mM with KCl replacing 40 mM Tris. The effect of increasing [K$^+$] may be contrasted with that of ACh. Stimulation of PA labeling was always much greater than that of PI. In fact PI radioactivity increased only as long as that of PA and then declined. This might be interpreted as a progressive inhibition of PI turnover which only becomes apparent after PA specific activity levels off.

TABLE IV

Effect of KCl on Synaptosomal Free Inositol

	Uptake D.R.	Specific Activity dpm/nmol	Levels nmol/mg Protein
Control	0.53 ± 0.06	2182 ± 114	9.66 ± 0.57
[KCl]		Normalized to Control	
5 mM	1.00 ± 0.10	1.00 ± 0.03	1.00 ± 0.02
19 mM	0.66 ± 0.14	0.67 ± 0.05	0.91 ± 0.08
26 mM	0.69 ± 0.04	0.71 ± 0.09	---
40 mM	0.61 ± 0.09	0.63 ± 0.05	0.91 ± 0.04
54 mM	0.57 ± 0.07	0.51 ± 0.09	
110 mM	0.40 ± 0.03	0.50 ± 0.01	

Synaptosomes were incubated with 0.02 mM [^{3}H]myo-inositol for 40 min in normal medium or in high potassium media. Measurements of inositol were made as in Table I. Results are mean ± range of 2 determinations or mean ± S.E. of at least 4 determinations at 5 and 40 mM KCl.

To understand the effect of potassium on tritium labeling of PI, we examined its effect on free inositol. Although other data indicate that inositol enters synaptosomes only by diffusion, increasing $[K^+]_o$ reduced this uptake (Table IV). In this case, the uptake and specific radioactivity of free and lipid inositol all declined in parallel and by the same degree – about 40% at 40 mM KCl. Thus comparison of free and lipid inositol pools does not suggest increased PI turnover with elevated potassium. The mechanism for this effect is currently unknown. Preliminary data suggests that increasing $[K^+]_o$ may affect only a pool of synaptosomal inositol, but one which is important for PI synthesis.

TABLE V

Effect of Potassium on Synaptosomal PA and PI

	Control	40 mM KCl
PA (nmol/mg prot)	2.33 ± 0.22	3.01 ± 0.45 (1.29)
[^{33}P]PA (Sp. Act.)	439 ± 77	1606 ± 514 (3.66)
Total ^{33}P	1023 dpm/mg prot	
PI (nmol/mg prot)	10.2 ± 0.4	9.2 ± 0.5 (0.90)
[^{33}P]PI (Sp. Act.)	32.3 ± 3.8	88.5 ± 30 (2.74)
Total ^{33}P	329.5 dpm/mg prot	
PA to PI Conversion (nmol)	0.75	0.51

Synaptosomes were incubated for 40 min in normal medium or one in which 40 mM KCl replaced that concentration of NaCl. Values are mean ± S.E. for 8 or 10 determinations. Other conditions as in Table III.

When the effects of 40 mM KCl on PA and PI were examined in more detail (Table V), we found that levels of neither lipid changed markedly, suggesting that the diacylglycerol which fuels the increase in PA labeling comes from another source. The specific radioactivity of both PA and PI was increased, but in contrast to the case with ACh, this increase was much greater for PA. If we again apply our crude correction for PA specific radioactivity to obtain a rate of PI synthesis, we find that this synthesis declines by more than 30% at 40 mM KCl, supporting other indications that elevated [K^+] inhibits PI turnover, and that the increase in label in PI is due solely to an effect on PA labeling.

Increasing extracellular potassium has several effects on synaptosomal phospholipids and inositol, all with the

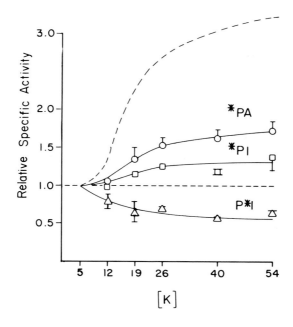

Figure 5. Effect of elevated potassium on phospholipid labeling in calcium free media. Synaptosomes were incubated in calcium free media containing the indicated concentrations of KCl for 40 min. Incorporation of labels into phospholipids was measured and normalized to 5 mM K controls. The dashed curve indicates the potassium effect on PA in a medium containing 1.3 mM $CaCl_2$ (see Figure 4). Symbols as in Figure 4. Bars indicate the range of 2 determinations at 12 and 19 mM or the S.E. of 6 determinations.

same concentration dependence. We examined whether these might all be due to a single message of depolarization induced calcium entry. When $[K^+]_o$ was increased while calcium was held low with EGTA, the $^{33}PO_4$ labeling effects were nearly abolished (Fig. 5); the dashed line indicates the effect on PA in a normal calcium buffer. In contrast to the case with ACh, the effects of increasing $[K^+]_o$ on PI and PA were attenuated in parallel by low Ca^{2+}_o, again indicating that the PI label merely follows that of PA. However, the decline in PI label above 40 mM KCl was not seen in low Ca^{2+} media, suggesting that inhibition of PI synthesis was due to calcium entry. The decline in [^{3}H]inositol labeling of PI with increasing $[K^+]_o$

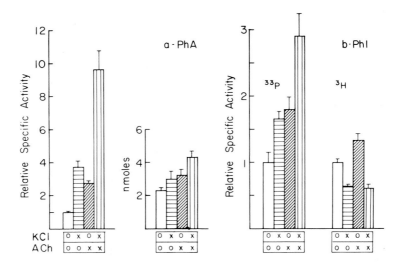

Figure 6. Effect of a combination of high $[K^+]_o$ and ACh on synaptosomal phospholipid metabolism. Synaptosomes were incubated with or without 0.2 mM ACh in normal (5 mM) or high (40 mM) KCl buffers. Incorporation of label into phospholipids was normalized to 5 mM K controls. Part A shows the effect on phosphatidic acid labeling and levels; Part B indicates the effect on both labels for phosphatidylinositol. Bars indicate the S.E. for 6 determinations.

was not affected by reduced calcium, and this is apparently a direct effect of potassium depolarization.

We thus have two stimuli which produce a synaptosomal "phospholipid effect", but which appear to act in different fashions. Acetylcholine and elevated $[K^+]_o$ do share however, a calcium dependent component which affects mainly PA labeling. It was of interest to examine whether they shared some pathways, and whether they acted at the same subcellular locus; it has been suggested that ACh might act postsynaptically (Fisher *et al.*, 1980, 1981), while depolarization acts on presynaptic structures (Pickard and Hawthorne, 1978). We therefore examined the result of combining both agents at their maximally effective concentrations in the same synaptosomal preparations. The results (Fig. 6) summarize our other studies. Both stimuli had comparable effects on PI $^{33}PO_4$ labeling but elevated $[K^+]_o$ was more effective than ACh in increasing PA label. Only ACh increased [^{3}H]inositol incorporation into PI. However, when the two stimuli were combined, their effects were neither competitive nor additive as might be expected if they acted at the same biochemical sites or on different subcellular loci. Rather ACh and elevated $[K^+]_o$ acted synergistically to increase phospholipid labeling. These data indicate that the two stimuli affect different biochemical steps in a common pathway within a common subcellular locus. The increase in PA levels with the combined stimuli was now significant, suggesting that neither agent caused an increase in phosphatidate phosphatase activity. Elevated potassium blocked the effect of ACh on incorporation of [^{3}H]inositol into PI, perhaps by depleting an essential inositol pool.

In summary, our approach to combining several stimuli and measuring several parameters in the same synaptosomal preparations has given us some understanding of the complexity of the pathways of phosphatidylinositol metabolism. We hope to continue, and to delineate the actual steps which are affected by neurotransmitter and depolarization in order to appreciate the role of the "phospholipid effect" in CNS function.

ACKNOWLEDGEMENTS

This work would not have been possible without the assistance of Shing Mei Hwang and Victoria Cipriano and the support of NIH grant HD 08536. Its presentation would be less comprehensible without discussions with Dr. Marvin Medow.

REFERENCES

Abdel-Latif, A.A., Yau, S.J. and Smith, J.P. (1981) *J. Neurochem.* *22*, 383-393.
Allan, D. and Michell, R.H. (1977) *Biochem. J.* *164*, 389-397.
Barkai, A.I. (1981) *J. Neurochem.* *34*, 1485-1491.
Bender, A.S., Wu, P.H. and Phillis, J.W. (1981) *J. Neurochem.* *37*, 1282-1290.
Berry, G., Yandrasitz, J.R. and Segal, S. (1981) *J. Neurochem.* *37*, 888-891.

Berry, G. Yandrasitz, J.R. and Segal, S. (1983) *Biochem. Biophys. Res. Commun. 112*, 817-821.

Blaustein, M.P. (1975) *J. Physiol. 247*, 617-655.

Bleasdale, J.E. and Wallis, P.A. (1981) *Biochim. Biophys. Acta 664*, 428-440.

Cicero, T.J. and Sherman, W.R. (1973) *Anal. Biochem. 54*, 32-39.

DeRobertis, E., Pellegrino de Iraldi A., Rodriguez deLores Arnaiz, G. and Salganicoff, L. (1962) *J. Neurochem. 9*, 23-35.

Fisher, S.K., Boast, C.A. and Agranoff, B.W. (1980) *Brain Res. 189*, 284-288.

Fisher, S.K., Frey, K.H. and Agranoff, B.W. (1981) *J. Neurosci. 1*, 1407-1413.

Hawthorne, J.N. and Bleasdale, J.E. (1975) *Molec. Cell. Biochem. 8*, 83-87.

Hokin, L.E. and Hokin, M.R. (1955) *Biochim. Biophys. Acta 16*, 229-237.

Lapetina, E.E. and Michell, R.H. (1974) *J. Neurochem. 23*, 283-287.

Lunt, G.G. and Pickard, M.R. (1975) *J. Neurochem. 24*, 1203-1208.

Miller, J.C. (1977) *Biochem. J. 168*, 549-555.

Nordstrom, O. and Bartfai, T. (1980) *Acta Physiol. 108*, 347-353.

Pickard, M.R. and Hawthorne, J.N. (1978) *J. Neurochem. 30*, 144-155.

Pumphrey, A.M. (1969) *Biochem. J. 112*, 61-70.

Raaflaub, J. (1960) *Methods Biochem. Anal. 3*, 301-325.

Schacht, J. and Agranoff, B.W. (1972) *J. Biol. Chem. 247*, 771-777.

Schacht, J. and Agranoff, B.W. (1973) *Biochem. Biophys. Res. Commun. 50*, 934-941.

Schacht, J. and Agranoff, B.W. (1974) *J. Biol. Chem. 249*, 1551-1557.

Sneddon, J.M. and Keen, P. (1970) *Biochem. Pharmacol. 19*, 1297-1306.

Takenawa, T., Saito, M., Nagai, Y. and Egawa, K. (1977) *Arch. Biochem. Biophys. 182*, 244-250.

Warfield, A., Hwang, S.-M. and Segal, S. (1978) *J. Neurochem. 31*, 957-960.

Warfield, A. and Segal, S. (1974) *J. Neurochem. 23*, 1145-1151.

Warfield, A.S. and Segal, S. (1978) *Proc. Natl. Acad. Sci. USA 75*, 4568-4572.

Yandrasitz, J.R. and Segal, S. (1979) *FEBS Letters* *108*, 279–282.

Yandrasitz, J.R., Berry, G. and Segal, S. (1981) *J. Chromatogr.* *225*, 319–328.

Yandrasitz, J.R., Berry, G. and Segal, S. (1983) *Anal. Biochem.* *135*, 239–243.

Yoshida, H. and Quastel, J.H. (1962) *Biochim. Biophys. Acta* *57*, 67–72.

POLYPHOSPHOINOSITIDE TURNOVER

IN THE NERVOUS SYSTEM

Bernard W. Agranoff and Lucio A.A. Van Rooijen

Neuroscience Laboratory, Mental Health Research Institute
University of Michigan, Ann Arbor, MI 48109

SUMMARY

Nerve endings prepared from guinea pig brain support a cholinergic stimulation of phospholipid labeling. In experiments on ^{32}P labeling of phosphatidate and phosphatidylinositol, it has been shown that the stimulation is muscarinic and postsynaptic. It appears that a low affinity form of the receptor mediates the stimulated labeling. Recent evidence in several tissues indicates that an early event in stimulated lipid labeling is the breakdown of polyphosphoinositides. While it is technically difficult to demonstrate ligand-stimulated breakdown of polyphosphoinositides in intact nerve tissue, an endogenous phosphodiesterase has been shown that cleaves prelabeled membrane polyphosphoinositides to diacylglycerol and inositol phosphates. The phosphodiesteratic cleavage products of phosphatidylinositol 4,5-bisphosphate, inositol trisphosphate, and diacylglycerol, are discussed in relation to their possible intracellular messenger function.

INTRODUCTION

Recent discoveries that the ligand-stimulated breakdown of polyphosphoinositides in a number of tissues may release candidates for second messengers has provoked much interest in stimulated lipid turnover, previously observed in the radiolabeling of phosphatidate (PA) and phosphatidylinositol (PI). It is safe to say that there have been more publications in this area in the past three years than in the preceding thirty. The stimulated turnover of phospholipids in the nervous system has special biomedical relevance in that it provides an insight into the biochemistry of CNS muscarinic receptors, which have been implicated in a number of neurological and psychiatric disorders (Pepeu and Ladinsky, 1981), including senile dementia of the Alzheimer type (Coyle et al., 1983). In addition, the therapeutic action of lithium in affective psychiatric disorders may have a biochemical basis in the demonstrated selective action of this cation on a phosphatase that degrades inositol mono-phosphate (IP) produced in phosphoinositide breakdown (Hallcher and Sherman, 1980). The stimulated turnover of nervous system phospholipids thus has promise as a pharmacological assay system for muscarinic effectors and antagonists which may have therapeutic value (Honchar et al., 1983).

It is this increasing recognition of the physiological significance of inositol lipid turnover that prompts a brief review of the history of our understanding of the inositol lipids. One of us (BWA) was drawn to this field by the discovery of Kennedy and coworkers that cytidine nucleo-tides mediated phosphatidylcholine (PC) and phosphatidyl-ethanolamine (PE) synthesis (Kennedy, 1957) and by the observation of Hokin and Hokin (1954) that cholinergic ligands stimulate labeling of two trace lipids, PA and PI. While the biosynthetic schemes for PE and PC might have predicted the existence of CDP-inositol, in analogy with CDP-choline and CDP-ethanolamine, there was no experi-mental evidence for the existence of the hypothetical inositol nucleotide, even though cytidine was found to stimulate PI labeling. Our eventual proposal that a novel liponucleotide, cytidine diphosphodiacylglycerol (CDP-DG), was the intermediate in PI synthesis (Agranoff et al., 1958) took into consideration the fact that ^{32}P in PA appeared in PI, a result predicted by the CDP-DG path-way and incompatible with the hypothetical CDP-inositol pathway. In reviewing the biosynthetic scheme for phospholipids as presently understood (Fig. 1), a number of points can be made relevant to the topic at hand. We digress briefly to point out the existence of an alternative pathway for the biosynthesis of PA which does not involve glycerophosphate (GP) as an intermediate.

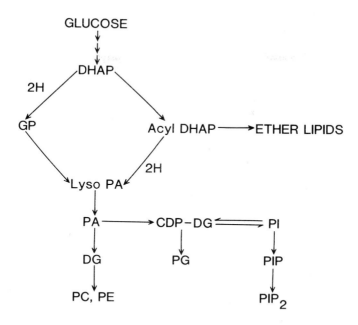

<u>Figure 1</u>   Biosynthetic steps in <u>de</u> <u>novo</u> synthesis of phospholipids. Dihydroxyacetone phosphate (DHAP) produced by glycolysis can be reduced by a NADH-dependent reductase to form glycerophosphate (GP), then acylated twice to form PA.   Alternatively, it can be acylated to acyl DHAP, then reduced by a NADPH-dependent reductase to form lyso PA, and again acylated to PA.   Acyl DHAP is the precursor of alkyl and alkenyl ether lipids.   PA is dephosphorylated prior to formation of PC or PE, but is converted to CDP-DG with its phosphate intact in the case of phosphatidylglycerol (PG) and PI synthesis.   Radiolabeling of lipids from [^{32}P]ATP in brief incubations is primarily by PI and PIP kinases, as well as DG kinase (see Fig. 3).

Instead of being reduced to GP, dihydroxyacetone phosphate (DHAP) is acylated to long-chain acyl DHAP (Hajra and Agranoff, 1968a).   This product is then reduced to lyso PA and again acylated to form PA (Hajra and Agranoff, 1968b).   Acyl DHAP is the biosynthetic precursor of all ether lipids (Hajra, 1983), and thus of platelet-activating factor (PAF; Hanahan *et al.*, 1980), the importance of which has received special recognition at this meeting.   Its possible function in the physiological

623

consequences of stimulated phospholipid labeling is addressed below. Of current interest, a deficiency in peroxisomal enzymes, including those of the acyl DHAP pathway, has recently been shown in a fatal genetic disorder characterized by mental retardation, as well as liver and kidney involvement (Heymans *et al.*, 1983).

It should also be noted that in the biosynthetic scheme illustrated in Figure 1, the synthesis of CDP-DG appears to be rate-limiting. It is therefore a step at which regulation may be physiologically mediated and is a likely site for pharmacological manipulation. Also, the CDP-DG:inositol phosphatidyltransferase step is reversible (Petzold and Agranoff, 1967), so that cellular CMP levels may play an important role, as has been indicated in lung tissue (Bleasdale, this volume). It is also worthy of note that galactinol has been identified as a reasonably potent blocker of the transferase (Benjamins and Agranoff, 1969), and would appear to be a superior inhibitor to the halogenated cyclitols.

Finally, we acknowledge contributions of chemists on whose structural contributions our biochemical and pharmacological knowledge is based. Eric Baer and H.O.L. Fischer developed the methods for the stereospecific synthesis of phospholipids. Theodore Posternak's treatise (1965) documents their contributions, as well as of the many carbohydrate chemists whose syntheses distinguished *myo*-inositol and its derivatives from its 8 isomers. The laboratories of Ballou (Brockerhoff and Ballou, 1961) and Dawson (Dittmer and Dawson, 1961) contributed to the elucidation of the structures of the polyphosphoinositides and their degradation products. Special recognition must be given to the late Jordi Folch-Pi, known best to later generations for his lipid extraction and chromatographic techniques, who discovered the polyphosphoinositides (Folch, 1942). His pioneering work in neurochemistry touched the lives of many of us here today. Regrettably, he did not live to see the recent rekindling of interest in the biochemistry of inositol lipids.

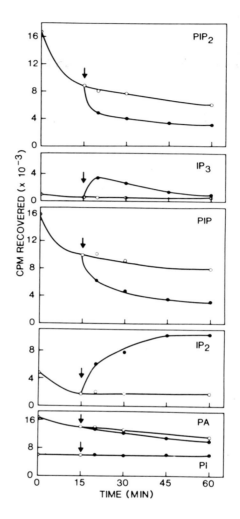

<u>Figure 2</u>. Phosphodiesteratic degradation of endogenous polyphospho-inositides in nerve ending membranes. ^{32}P-Prelabeled membranes were prepared from nerve endings (Van Rooijen <u>et al</u>., 1983). Upon incubation at 37°C, a transient monoesteratic loss of ^{32}P from PIP$_2$ and PIP is observed. After 15 min 1.5 mM Ca^{2+} (↓) is added and an additional rapid loss of ^{32}P occurs, with a concomitant increase in ^{32}P recovered in IP$_3$ and IP$_2$. Loss of [^{32}P]PI is not observed under these conditions, either in the absence or presence of Ca^{2+}.

Rooijen *et al.*, 1983). In the Ca^{2+}-activated membrane preparation, we sometimes see more radiolabeled IP_2 than can be accounted for by the phosphodiesteratic breakdown of PIP. While this result could be interpreted as support for a postulated phospholipase D-type phosphodiesteratic cleavage of PIP_2 to PA and IP_2 (see Cockroft, this volume), there are other, more likely sources of the IP_2. Since IP_3 breaks down, some IP_2 can be attributed to phosphatase activity, even in the absence of added Mg^{2+}. Also, monoesteratic degradation of PIP_2 to PIP could be followed by diesteratic breakdown, with production of IP_2. The definitive experiment to test the hypothesized phospholipase D activity would be to examine the stereo-specificity of the IP_2 produced. Phospholipase D action on PIP_2 would lead to D-*myo*-inositol 4,5-bisphosphate, while phospholipase C breakdown of PIP yields D-*myo*-inositol 1,4-bisphosphate. While it is as yet unproven, it is likely that phosphatase action on D-*myo*-inositol 1,4,5-trisphosphate yields the 1,4-bisphosphate, since only one monophosphate isomer (D-*myo*-inositol 1-phosphate) accumulates following inositol phospholipid breakdown.

Studies are in progress in our laboratory on the phosphodiesterase activity of a soluble high speed supernatant of beef brain against polyphosphoinositides. At pH 7.4, in the absence of detergents, and using denatured prelabeled membranes as substrate, we find no discrimination between Ca^{2+}-activated PIP and PIP_2 breakdown after over 29-fold purification of the diesterase. Under the incubation conditions used, there is no activity against PA or PI, in the presence of either Ca^{2+} or Mg^{2+}. It would appear then that the observed selective breakdown of PIP_2 seen *in vivo* is a reflection of differential substrate availability to an enzyme that degrades both PIP and PIP_2. While the activity has been demonstrated, it remains to be demonstrated that muscarinic agonists stimulate PIP_2 breakdown in nerve endings by means of this enzyme.

The finding that Li^+ blocks the breakdown of IP to P_i and inositol was used to advantage by Berridge *et al.* (1982) to produce a sensitive pharmacological assay of ligands that stimulate inositide breakdown. The addition of Li^+ greatly increases the accumulation of IP in

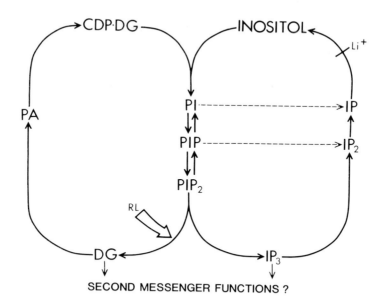

<u>Figure 3</u> Cyclic turnover of inositol lipids. The originally proposed cycle of PI → DG → PA → CDP-DG → PI is now extended to include the polyphosphoinositides. While phosphodiesteratic cleavage of PIP is also indicated, there is better evidence for stimulated PIP_2 breakdown following receptor-ligand (RL) activation. The cycle on the right demonstrates the sequential breakdown of IP_3 to inositol. The breakdown of IP is blocked by Li^+. Both inositol trisphosphate and diacylglycerol have been postulated as intracellular messengers.

carbachol-stimulated brain slices and the insect salivary gland (Berridge *et al.*, 1982). The IP is presumably a degradation product of IP_3 released upon PIP_2 break-down. It now becomes of interest to investigate the regulation of a possible inositol cycle (see Fig. 3). While inositol can be transported into the cell from the extra-cellular fluid or can be synthesized from glucose 6-phosphate, it appears that the breakdown of IP plays a major role in regulation of intracellular inositol levels.

THE ROLES OF DIACYLGLYCEROL
AND INOSITOL TRISPHOSPHATE

As noted above, when carbachol is added to nerve endings in the presence of $^{32}P_i$, the PA and PI formed are primarily of the ST/AR species. When $[^{32}P]ATP$ is added to non-radiolabeled nerve ending membranes under conditions of phosphodiesteratic cleavage of the polyphosphoinositides, the PA formed is mostly tetraenoic, indicating that ST/AR-DG had been released from PIP and PIP_2 (Van Rooijen, Hajra and Agranoff, in preparation). Thus, if arachidonate is released in the cycle: $PIP_2 \rightarrow DG \rightarrow PA \rightarrow\rightarrow\rightarrow PIP_2$, there must be rapid resynthesis from arachidonoyl CoA. The hypothesis that DG released from inositide breakdown has a second messenger function via stimulation of protein kinase C (Takai et al., 1982) is particularly attractive. If true, it would lift ligand-stimulated phospholipid radiolabeling from the realm of epiphenomena to that of causality. If, in fact, the DG produced by stimulated inositide turnover catalyzes protein kinase C, one might expect especially high activation by ST/AR-DG. This has not yet been demonstrated, although one might predict from published structure-function studies that ST/AR-DG will prove as active as several synthetic derivatives, including those having a short-chain substituent in the 2-position. It remains possible that the physiologically significant DG activator of the kinase is not ST/AR-DG. It might, for example, be 1-octadecyl,2-acetyl-sn-glycerol, a degradation product of PAF. While PAF is presently known to be an extracellular effector, it originates intracellularly and may also re-enter cells. It remains possible, then, that the PAF metabolite or some other DG-related molecule is the physiological activator of protein kinase C, and that the DG released from inositol lipids is not involved in the phosphorylation process.

IP_3 is also a contender for the role of a second messenger that is released (see Berridge, this volume). Whether IP_3 or related molecules such as its cyclic derivative is the active species (Agranoff and Seguin, 1974; Lapetina et al., 1975) is presently also conjectural.

Finally, we would like to speculate about the nature of what we now consider to be the first step in stimulated lipid radiolabeling -- the phosphodiesteratic breakdown of PIP_2 to DG and IP_3. Since the ligand-receptor interaction occurs on the extracellular surface of the plasma membrane and the lipid radiolabeling effects are inferred to be on the intracellular surface, it appears that we are dealing with a transmembrane mechanism. Its basis could be catalytic, *i.e.*, alteration of the active site of the phosphodiesterase, or substrate availability. If both substrate and enzyme are membrane-bound and have limited mobilities, it follows that when the membrane is relatively static, there is little available substrate. When the membrane is perturbed by a conformational change induced by the ligand-receptor interaction, previously unavailable lipid and substrate domains are approximated, and the rate of breakdown (and resynthesis) increases. The model does not speak to the physiological conse-quences of the breakdown; it simply states that local membrane perturbations will expose fresh substrate to enzyme and lead to increased lipid turnover. If IP_3, DG, or both are indeed second messengers, this may be how they are released. Future experiments need therefore to be directed at both the biochemistry of the receptor-ligand interaction, and the possible messenger functions of the products of inositide cleavage.

REFERENCES

Agranoff, B.W. and Seguin, E.B. (1974) *Prep. Biochem.* 4, 359-366.

Agranoff, B.W., Bradley, R.M. and Brady, R.O. (1958) *J. Biol. Chem. 233*, 1077-1083.

Agranoff, B.W., Murthy, P.P.N. and Seguin, E.B. (1983) *J. Biol. Chem. 258,* 2076-2078.

Akhtar, R.A. and Abdel-Latif, A.A. (1980) *Biochem. J. 192*, 783-791.

Benjamins, J. and Agranoff, B.W. (1969) *J. Neurochem. 16,* 513-527.

Berridge, M.J. (1983) *Biochem. J. 212,* 849-858.

Berridge, M.J., Downes, C.P. and Hanley, M.R. (1982) *Biochem. J. 206,* 587-595.

Billah, M.M. and Lapetina, E.G. (1982) *J. Biol. Chem. 257,* 12705-12708.

Birdsall, N.J.M. and Hulme, E.C. (1983) *Trends in Pharmacol. Sci. 4,* 459-463.

Brockerhoff, H. and Ballou, C.E. (1961) *J. Biol. Chem. 236,* 1907-1911.

Coyle, J.T., Price, D.L. and DeLong, M.R. (1983) *Science 219,* 1184-1189.

Dittmer, J.C. and Dawson, R.M.C. (1961) *Biochem. J. 81,* 535-539.

Fisher, S.K. and Agranoff, B.W. (1981) *J. Neurochem. 37,* 968-977.

Fisher, S.K., Boast, C.A. and Agranoff, B.W. (1980) *Brain Res. 189,* 284-288.

Fisher, S.K., Frey, K.A. and Agranoff, B.W. (1981) *J. Neurosci. 1,* 1407-1413.

Fisher, S.K., Klinger, P.D. and Agranoff, B.W. (1983) *J. Biol. Chem. 258,* 7358-7363.

Folch, J. (1942) *J. Biol. Chem. 146,* 35-44.

Hajra, A.K. (1983) in: *Ether Lipids: Biomedical Aspects,* (H. Mangold, and F. Paultauf, eds.), pp. 85-106, Academic Press, New York.

Hajra, A.K. and Agranoff, B.W. (1968a) *J. Biol. Chem. 243,* 1617-1622.

Hallcher, L.M. and Sherman, W.R. (1980) *J. Biol. Chem. 255,* 10896-10901.

Hanahan, D.J., Demopoulos, C.A., Liehr, J. and Pinckard, R.N. (1980) *J. Biol. Chem. 255,* 5514-5516.

Heymans, H.S.A., Schutgens, R.B.H., Tan, R., Van Den Bosch, H. and Borst, P. (1983) *Nature 306,* 69-70.

Hokin, M.R. and Hokin, L.E. (1954) *J. Biol. Chem. 209,* 549-558.

Honchar, M.P., Olney, J.W. and Sherman, W.R. (1983) *Science 220,* 323-325.

Kennedy, E.P. (1957) *Ann. Rev. Biochem. 26,* 119-148.

Lapetina, E.G., Seguin, E.B. and Agranoff, B.W. (1975) *Biochim. Biophys. Acta 398,* 118-124.

Michell, R.H. (1975) *Biochim. Biophys. Acta 415,* 81-147.

Pepeu, G. and Ladinsky, H., Eds. (1981) *Adv. Behav. Biol. Vol. 25: Cholinergic Mechanisms,* Plenum Press, New York.

Petzold, G.L. and Agranoff, B.W. (1967) *J. Biol. Chem. 242,* 1187-1191.

Posternak, T. (1965) *The Cyclitols,* Holden-Day, Inc., Publishers, San Francisco, California.

Rhodes, D., Prpic, V., Exton, J.H. and Blackmore, P.F. (1983) *J. Biol. Chem. 247,* 771-777.

Schacht, J. and Agranoff, B.W. (1974) *J. Biol. Chem. 249,* 1551-1557.

Schacht, J., Neale, E.A. and Agranoff, B.W. (1974) *J. Neurochem. 23,* 211-218.

Takai, Y., Minakuchi, R., Kikkawa, U., Sano, K., Kaibuchi, K., Yu, B., Matsubara, T. and Nishizuka, Y. (1982) *Prog. Brain Res. 56* (W.H. Gispen and A. Routtenberg, eds.), pp. 287-301.

Van Rooijen, L.A.A., Seguin, E.B. and Agranoff, B.W. (1983) *Biochim. Biophys. Res. Commun. 112,* 919-926.

Weiss, S.J., McKinney, J.S. and Putney, J.W. (1982) *Biochem. J. 206,* 555-560.

METABOLISM OF INOSITOL-CONTAINING PHOSPHOLIPIDS
IN THE SUPERIOR CERVICAL GANGLION OF THE RAT

Joel Horwitz, Sofia Tsymbalov and Robert L. Perlman
University of Illinois College of Medicine
Chicago, IL 60680

SUMMARY

Muscarine stimulates the incorporation of $^{32}P_i$ and [^{3}H]inositol into phosphatidylinositol in the superior cervical ganglion of the rat. Muscarine also increases the accumulation of [^{3}H]inositol-1-phosphate in ganglia in which the inositol-containing phospholipids were pre-labeled by incubation with [^{3}H]inositol. The production of [^{3}H]inositol-1-phosphate under these conditions pre-sumably reflects the activity of phospholipase C in the ganglion. This effect of muscarine is not dependent upon extracellular Ca^{2+}. The Ca^{2+} ionophore, ionomycin, and high K^+ cause only a small increase in the formation of [^{3}H]inositol-1-phosphate. Homogenates of the ganglion contain a phospholipase C activity that hydrolyzes the phosphatidylinositol phosphates. This phospholipase C requires Ca^{2+} and is stimulated by deoxycholate. This enzyme may be responsible for the muscarine-stimulated breakdown of inositol-containing phospholipids in the intact ganglion. Although the phospholipase C is depen-dent upon Ca^{2+} in vitro, muscarine presumably increases the activity of the enzyme by some mechanism other than by increasing the influx of Ca^{2+}.

The superior cervical ganglion (SCG) is the largest paravertebral sympathetic ganglion. This ganglion receives preganglionic fibers that originate in the spinal cord and contains the cell bodies of postganglionic neurons whose axons provide sympathetic innervation to the head and neck. The preganglionic fibers release

acetylcholine as a neurotransmitter. The postganglionic
neurons appear to contain both nicotinic and muscarinic
cholinergic receptors. The stimulation of nicotinic receptors produces a fast depolarization that is due to a
generalized increase in the permeability of the neuronal
membrane to cations. The activation of muscarinic
receptors leads to a slower depolarization that is thought
to result from the inhibition of a specific K^+ current, the
M current (Brown et al., 1981). The stimulation of
cholinergic receptors in the ganglion produces metabolic
as well as electrical effects. One well-documented effect
of cholinergic stimulation is to increase the rate of
catecholamine synthesis in the SCG (Ip et al., 1982). It
appears that there are distinct nicotinic and muscarinic
mechanisms for the activation of tyrosine hydroxylase in
the ganglion. Another well-documented effect of cholinergic stimulation is to increase the turnover of inositol-
containing phospholipids in the SCG (Hokin et al., 1960).
In particular, muscarinic agonists increase the incorporation of $^{32}P_i$ into phosphatidylinositol (PI) in the ganglion
(Lapetina et al., 1976).

In our initial experiments, we studied the effect of
the muscarine on the incorporation of $^{32}P_i$ into phospholipids in the ganglion. In these experiments, we incubated ganglia in medium containing $^{32}P_i$, in the presence
or absence of muscarine. We then extracted the phospholipids and separated them by thin-layer chromatography
(Jolles et al., 1981). Muscarine causes a large, rapid
and concentration-dependent increase in the incorporation
of $^{32}P_i$ into PI. This effect is visible within 2 min and is
produced by concentrations of muscarine as low as 0.1
µM. In a 10-min incubation period, 100 µM muscarine
produces a 6- to 15-fold increase in the labeling of PI
(Table I). Muscarine also increases the incorporation of
$^{32}P_i$ into phosphatidic acid, but, at short incubation
times, it does not affect the labeling of other phospholipids. The stimulation of PI labeling by muscarine is
essentially independent of extracellular Ca^{2+}. Bethanechol also increases the incorporation of $^{32}P_i$ into PI,
but is much less effective than muscarine.

Muscarine not only increases the incorporation of
$^{32}P_i$ into PI, but also increases incorporation of

TABLE I

Effects of Muscarine on Phosphoinositide Metabolism

Incubation Conditions	$^{32}P_i$ Incorporation	[^{3}H]Inositol Incorporation dpm/ganglion	[^{3}H]Inositol-1-P Formation
Control	510	9,517	377
Muscarine (100 μM)	4,230	21,232	9,430

Ganglia were incubated for 10 min at 37°C in medium containing $^{32}P_i$ (50 μCi/ml), or for 60 min at 37°C in medium containing [^{3}H]inositol (12.5 μCi/ml), in the presence or absence of muscarine. Phospholipids were extracted and separated by thin-layer chromatography; the radioactivity in the phosphatidylinositol spot was determined. Each point represents the mean of two determinations. In other experiments, ganglia were incubated for 120 min at 37°C in medium containing [^{3}H]inositol (12.5 μCi/ml) and were then incubated for an additional 5 min in medium containing 10 mM Li$^+$, in the presence or absence of muscarine. Inositol-1-phosphate was isolated by ion-exchange chromatography, and its radioactivity was determined. Each point represents the mean of three determinations.

[^{3}H]inositol into PI (Table I). Thus, muscarinic stimulation, like electrical stimulation (Burt and Larrabee, 1976), produces a greater increase in the incorporation of $^{32}P_i$ into PI than in the incorporation of [^{3}H]inositol. There are several possible reasons why muscarine has a smaller effect on [^{3}H]inositol incorporation than on $^{32}P_i$ incorporation. The incorporation of $^{32}P_i$ into PI reflects the net conversion of diacylglycerol to PI, by way of phosphatidic acid and CDP-diacylglycerol. While some [^{3}H]inositol is incorporated into PI during the synthesis of PI from diacylglycerol, this precursor can also be incorporated into PI by reversal of the CDP-diacylglycerol:inositol transferase reaction, or, possibly, by a distinct PI:inositol exchange enzyme. If the basal rate of incorporation of [^{3}H]inositol into PI is greater than the rate of net PI synthesis, then the stimulation of PI synthesis will cause a smaller percentage increase in [^{3}H]inositol incorporation than in $^{32}P_i$ incorporation. If this is true, then the incorporation of $^{32}P_i$ into PI is a better measure of PI synthesis than is the incorporation of [^{3}H]inositol.

The initial event in agonist-stimulated phospholipid turnover is thought to be the hydrolysis of phosphoinositides by an agonist-activated phospholipase C (Michell, 1975). According to this idea, the incorporation of radioactive precursors into inositol-containing phospholipids merely reflects the compensatory resynthesis of these lipids from diacylglycerol. Therefore, we studied the effect of muscarine on the phosphoinositide-specific phospholipase C activity in the SCG. In these experiments, we pre-labeled the inositol containing phospholipids by incubation of the ganglia with [^{3}H]inositol and then measured the formation of [^{3}H]inositol-1-phosphate ([^{3}H]IP) in medium containing Li$^+$ to inhibit the enzyme inositol-1-phosphatase (Hallcher and Sherman, 1980); [^{3}H]IP was isolated by ion-exchange chromatography, according to the method of Berridge et al. (1982). Ganglia incubated in the absence of muscarine form negligible amounts of [^{3}H]IP; the labeled phosphoinositides in the ganglion are apparently hydrolyzed at a very slow rate (probably no more than a few percent per hour). Muscarine causes an increase in the formation of [^{3}H]IP (Table I). This effect of muscarine occurs without a detectable lag and is linear for at least 5 min at 37°C. In 5 min, muscarine typically causes the accumulation of an amount of [^{3}H]IP equivalent to approximately 15 percent of the labeled phosphoinositides in the ganglion. This effect, like the stimulation of $^{32}P_i$ incorporation into PI, is essentially independent of extracellular Ca^{2+}. Agents that would be expected to raise the concentration of intracellular Ca^{2+} in the ganglion, such as high K$^+$, and the Ca^{2+} ionophore, ionomycin, cause only a small increase in the accumulation of [^{3}H]IP. Although muscarine causes an accumulation of [^{3}H]IP, it does not cause a decrease in the radioactivity of any of the phosphoinositides. The ganglia have a pool of [^{3}H]-inositol and evidently can maintain their content of labeled phosphoinositides when these substances are being degraded. Therefore, we cannot tell from these experiments which class of phosphoinositides is being hydrolyzed in muscarine-stimulated ganglia.

We have also assayed phospholipase C activity in homogenates of the SCG. In these experiments, we pre-labeled the phospholipids by incubation with [^{3}H]-inositol, homogenized the ganglia, and then measured the breakdown of phosphoinositides and the accumulation of [^{3}H]IP in these homogenates. When these homogenates are incubated at 30°C, the phosphatidyl[^{3}H]inositol phosphates are rapidly degraded and [^{3}H]IP accumulates. The phosphatidyl[^{3}H]inositol in these homogenates is stable. Thus, the ganglia appear to contain a phospholipase C activity that degrades phosphatidylinositol phosphates. This phospholipase C activity is Ca^{2+}-dependent and is stimulated by deoxycholate. This could be the enzyme whose activity is increased by muscarine. We do not yet understand the mechanism by which occupation of the muscarinic receptor leads to the activation of phospholipase C. Although the phospholipase C activity we have detected in ganglionic homogenates is Ca^{2+}-dependent, muscarine appears to activate this enzyme by some mechanism other than by increasing the influx of Ca^{2+} into the ganglion. There have not been any studies on the effects of muscarine on Ca^{2+} fluxes in the SCG. However, Brown et al. (1981) have reported that muscarine does not significantly affect the Ca^{2+}-activated K^+ current they have described in the ganglion. Thus, it is unlikely that muscarine raises the intracellular Ca^{2+} concentration in this tissue.

The biological significance of the muscarine-induced activation of phospholipase C in the SCG remains unclear. As mentioned above, activation of muscarinic receptors has many effects on the ganglion. Prominent among these are inhibition of the M current and activation of tyrosine hydroxylase. The breakdown of inositol-containing phospholipids in response to muscarine may mediate one of these other actions of muscarine. Alternatively, the activation of phospholipase C may be part of the dynamics of the muscarinic receptor and may not be related to the specific biological activities of muscarine. Experiments are currently in progress in our laboratory to investigate the relationship between muscarine-induced phospholipid turnover and the other biological actions of muscarine in the SCG.

Horwitz et al.

ACKNOWLEDGEMENTS

This research was supported in part by research grant HL29025 from National Institutes of Health. J.H. is the recipient of NRSA HL06701 from the National Institutes of Health.

REFERENCES

Berridge, M.J., Downes, C.P. and Hanley, M.R. (1982) *Biochem. J. 206*, 587-595.

Brown, D.A., Constanti, A. and Adams, P.R. (1981) *Fed. Proc. 40*, 2625-2630.

Burt, D.R. and Larrabee, M.G. (1976) *J. Neurochem. 27*, 753-763.

Hallcher, L.M. and Sherman, W.R. (1980) *J. Biol. Chem. 255*, 10896-10901.

Hokin, M.R., Hokin, L.E. and Shelp, W.D. (1960) *J. Gen. Physiol. 44*, 217-226.

Ip, N.Y., Perlman, R.L. and Zigmond, R.E. (1982) *J. Pharmacol. Exp. Ther. 223*, 280-283.

Jolles, J., Zwiers, H., Debber, A., Wirtz, K.W.A. and Gispen, W.W. (1981) *Biochem. J. 194*, 283-291.

Lapetina, E.G., Brown, W.E. and Michell, R.H. (1976) *J. Neurochem. 26*, 649-651.

Michell, R.H. (1975) *Biochim. Biophys. Acta 415*, 81-147.

DISCUSSION

Summarized by George Hauser

Ralph Lowell Labs, McLean Hospital
115 Mill St., Belmont, MA   02178

The three papers on changes in experimental diabetes elicited discussion relative to the decreased activity of Na/K ATPase. Dr. Gould (Inst. Basic Res., Staten Island) wondered whether other ATPases might be similarly affected and what effect stimulation might have on a nerve that contained no Na/K ATPase. Dr. Hawthorne had used a microsomal fraction rather than a whole nerve homogenate, which gave inconsistent differences, and therefore little non-Na-dependent ATPase was present. It was unlikely that Na/K ATPase was really zero, so that some response to stimulation was still possible. He also agreed with Dr. Perlman (Univ. Illinois, Chicago) that the decrease in inositol transport which required Na^+ might be secondary to the fall in Na/K ATPase activity and the consequent decrease in Na^+ gradient. However, the aldose reductase inhibitor sorbinil had not been added to see whether it might have an effect on inositol loss as Dr. Beyer (Pfizer Central Res.) suggested. The increase in Na^+ concentration within the diabetic nerve might be 30-40% estimated Dr. Greene replying to a question by Dr. Berridge, who wondered whether this would be the major cause of the defect in conduction velocity. The lower Na^+ equilibrium potential is, in fact, reflected in a decrease in the height of the action potential.

Inositol transport occurs in the Schwann cell whereas ATPase is in the axon and is probably related to the electrophysiological changes in diabetes. This distribution, Dr. Greene pointed out in response to a question by Dr. Gould regarding the locus of changes in diabetes, makes it likely that both structures are involved, especially since in myelinated fibers Na/K ATPase seems to exist in the node of Ranvier, both in the axon and the terminal vesicles of myelin. The node may be the functional unit which ought to be examined. Since Dr. McAfee (Beckman Res. Inst.) had wondered whether the deficit in the enzyme also occurred in non-neural tissues, Dr. Greene indicated that it had been at normal levels, for example, in kidney fractions two and four weeks after streptozotocin administration. Because of the complexity of the system, he had not performed experiments in media with low K^+ concentration which Dr. McAfee suggested would effectively block the Na/K pump.

The other related area which elicited questions and comments dealt with the changes in phospholipid levels and metabolism in diabetic rat nerves. In answer to a query by Dr. Gould, Dr. Hawthorne stated that all of the phospholipids examined other than phosphoinositides were at the same level in control and diabetic nerves although purified myelin had not been examined. Along similar lines, Dr. McAfee inquired whether sciatic nerve was special and whether the lipid findings could be gener-alized, occurring also in other nerves as well as in non-neural tissues. Dr. Eichberg felt that there was the suggestion that the deficit was axolemmal or at least non-myelin in location and that, therefore, unmyelinated nerves might show a greater effect. However, a smaller although significant increment in PIP_2 labeling was observed in vagus nerve as compared with sciatic nerve. This was consonant with the findings of Drs. McAfee, Perlman and Horwitz that no change in the production of inositol phosphates was seen in electrically-stimulated vagus nerve. Dr. Eichberg had not yet assayed for a possible loss in activity of PIP_2 phosphodiesterase which, Dr. Abdel-Latif suggested, could give rise to enhanced ^{32}P labeling. He did, however, speculate that the extent of the phospholipid changes might be related to the severity of the experimental disease. Animals appeared much sicker after intravenous than after intraperitoneal injections of streptozotocin and exhibited more pronounced changes.

Dr. Michell wondered whether constant specific radioactivity of ATP had been reached in the experiments described by Dr. Eichberg. Although Dr. Eichberg had not made this measurement, the fact that the effect was greater on PIP_2 than on PIP made it unlikely in his view that the explanation for the elevated incorporation lay in changes in the ATP pool. In endoneurial preparations such as those Dr. Hawthorne was using, equilibrium had undoubtedly been reached. He elaborate on the decrease in PC labeling which was apparent only when expressed as a percentage but not in terms of absolute incorporation level. The possibility of a feedback by choline influencing phosphoinositide metabolism which Dr. Greene had raised was an intriguing thought, but probably not supported by the evidence.

Dr. Gould commented that he had seen only extremely small effects in sciatic nerve, stimulated 24 h after ^{32}P or inositol was injected into the nerve bundle. There were slight increases in PIP but not PIP_2 labeling with either precursor. In the squid giant axon where label could be more readily introduced, the changes with stimulation were quite variable and very much smaller than in other systems.

Dr. Hawthorne remarked that in endoneurial preparations there seemed to be less polyphosphoinositides which he interpreted as the probable loss of a less well-labeled pool. Dr. Eichberg found it intriguing that substantial levels were still present under conditions favoring hydrolysis. In his own experiments very little polyphosphoinositides remained after incubation.

Dr. Sherman inquired whether the finding reported from Dr. Winegrad's laboratory in Philadelphia that a portion of the respiratory competence of the preparations was dependent upon arachidonate had been reported. Dr. Greene replied that in a single experiment in which an endoneurial preparation from diabetic animals was incubated with albumin-bound arachidonate no increase in resting O_2 consumption was seen. Possible relationship to a specific phosphoinositide pool was uncertain. Current investigations in Dr. Greene's laboratory center around the purification of a membrane fraction that manifests the defect. Efforts to then affect this defect by additives are planned.

Dr. Williamson introduced the idea that insulin might influence the phosphorylation state of regulatory proteins involved in the interconversion of phosphoinositides. Dr. Eichberg reacted favorably and recalled Dr. Gispen's model which points toward that possibility.

Dr. Howerton (Univ. Kansas) asked why there was a large difference in conduction velocity between the control groups for the two- and ten-week diabetic nerves. Dr. Eichberg explained that the experiments had been carried out at different temperatures which could account for this discrepancy. However, they would not explain the similarity of the values in the diabetic nerves at the two time points. The lack of a nerve conduction defect at two weeks could depend both on the conditions used and the particular nerve fibers measured, Dr. Greene remarked. There can be two classes of nerve fibers in the same fascicle, responding differently to diabetes, so that the placement of the recording electrodes would be critical. In some nerve fibers, changes can be seen within two weeks after administration of the drug.

In summary, the questions elicited further information on the nature of the changes seen in experimental diabetes. However, the precise relationship among altered inositol levels, transport and metabolism, Na/K ATPase and nerve conduction velocity as well as the exact locus of the deficit remain unexplained and in need of further experimental investigation.

The second part of this session dealt with the stimulation by muscarinic cholinergic agents of phosphoinositide metabolism and with cation effects in normal tissue from the CNS. Dr. Yandrasitz's presentation of synergistic effects of elevated K^+ concentration and ACh prompted the comment from Dr. Hauser that he had observed the same phenomenon in rat cerebral cortex mince accompanied by drastic reduction in PC labeling with ^{32}P. In Dr. Yandrasitz's synaptosomal preparation PC was virtually unlabeled, hence such a decrease would not have been noticeable. This caused Dr. Gould to remark that the enzymes required for PC biosynthesis had to be transported from the cell body to the nerve ending, that the movement, especially of cholinephosphotransferase, was very slow and that this might account for the poor labeling of PC in synaptosomes. Dr. Horwitz commented that in superior cervical ganglia prelabeled with [^{3}H]inositol high K$^+$ stimulates a Ca-independent release of inositol into the medium and that this could account for lower labeling of PI with [^{3}H]inositol as precursor as observed in Dr. Yandrasitz's system under depolarizing conditions.

Dr. Fisher reminded the audience that on the basis of his investigations with Dr. Agranoff in which neurotoxin and lesioning were employed, the effect in synaptosomes appeared to be postsynaptic and was probably located in dendritic fragments. Regarding the Ca sensitivity of the PI response he felt that Ca entry was not involved and extremely low concentrations, in the nM range, were required, so that EGTA pulling Ca out of the synaptosome would block the effect. Dr. Yandrasitz agreed that Ca entry might be a secondary phenomenon and emphasized that he had used conditions which would buffer Ca to about 10^{-6} M. The extent of the PI decrease concerned Dr. Fain who wondered whether it was as great as the 22% reported by Warfield and Segal. The reply was that because of the differences in incubation conditions the decrease in the present experiments was only half as great. Dr. Michell cautioned that labeling studies give only limited information and that chase experiments after labeling synaptosomes to equilibrium were required for the identification of the underlying events. This was indeed desirable, Dr. Yandrasitz agreed, but was difficult to do on synaptosomes which have no inositol transport and are unstable with time. He hoped to be able to do this type of experiment and examine polyphosphoinositides in particular.

Dr. Agranoff's paper also elicited comments on the Ca sensitivity of the polyphosphoinositide phosphodiesterase. Dr. Downes pointed out that inositol phosphates could be liberated at micromolar Ca as long as no Mg was present and the ionic strength was kept low. However, at physiological values for these two parameters Ca in excess of 0.1 mM was required. This moved Dr. Fain to speculate that this enzyme and Ca might regulate the breakdown of polyphosphoinositides but not of PI, especially in view of the ability of the Ca ionophore to mimic the effects. Although he did not try to bring in

Ca as a mediator, Dr. Agranoff felt that it might in reality be much more important than under the artifactual experimental conditions. He also stressed that the relative accessibility of PIP and PIP_2 to the enzyme might be a critical factor in their hydrolysis.

In conclusion, the nature of the initial reaction following muscarinic cholinergic receptor activation in the CNS remains uncertain as does the role of Ca and the phosphoinositide phosphodiesterases. Similarly, questions persist regarding the influence of K^+ depolarization on the ACh-induced phosphoinositide effect.

CONCLUDING REMARKS - A BRIGHT FUTURE

George Hauser

Ralph Lowell Laboratories, McLean Hospital,
Harvard Medical School, Belmont, MA 02178

It is difficult to make concluding comments at the end of such an excellent and exciting meeting. I think we are all mentally somewhat fatigued from the intensive activity that we have engaged in during the last three days, and I shall not tax you with a lot of additional scientific information. Obviously, I cannot thank the organizers, being one of them myself, but I do want to make special mention of the tremendous effort which John Bleasdale has had to exert owing to his special position as "Johnny on the spot" saddled with all the details of the local arrangements. We owe him and his wife a particular vote of thanks. It is, furthermore, a pleasure to acknowledge the invaluable support of the Chilton Foundation, represented by Mr. Andy Bell and Mr. Sam Winstead, which together with several pharmaceutical companies has provided the indispensable and generous wherewithal for this meeting.

The hospitality of the University of Texas Health Science Center at Dallas, Dr. Charles Sprague, President, was gracious, the energetic and steadfast support of Dr. Jack Johnston, Chairman of the Department of Biochemistry, vital, and both are greatly appreciated. Many thanks are also due everyone else who provided the necessary assistance to assure the smooth organization and running of the conference. If the interval between conferences on this topic continues to decrease at the same rate, the two earlier ones having been held in September 1968 and June 1977, we have less than 5 years until the next one in October-November 1988.

Much remains to be done. In this context it was especially gratifying to see so many younger investigators here. This assures the continuity of the field which has now entered a phase where a distinct payoff in terms of physiological function is in sight and is likely to be realized in the forseeable future. To speculate on the possible implications of knowing exactly how the binding of an agonist to a receptor is translated into a variety of intracellular events -- involving changes in phospho-inositide levels and metabolism, calcium mobilization, protein phosphorylation, cyclic GMP formation, and eicosanoid production -- all active areas with great poten-tial for significant advances -- would cause many of us to miss our flights home. There are clearly many possibil-ities for abnormalities and malfunction, a lot of room for genetically determined disease and hence much oppor-tunity for manipulation and modification of the systems to bring about changes necessary for correcting aberrations.

The possibility of exploring these fascinating aspects and the promise of a bright future for our field, together with that for all of mankind, could be obliterated if the trend towards nuclear armaments and their eventual use is not reversed. Ours is a scientific conference and political themes are seldom allowed to intrude at such meetings. Yet the threat of a nuclear holocaust is so ominous that an appeal for action seems justified at every opportunity. We must urge all governments with what-ever persuasive powers we can muster to stop and reverse the continued buildup of nuclear weapons which rob the world of social and other beneficial programs including medical research and endanger human existence itself. Continuous intensive lobbying against the insane policies of adding to the world's overkill capacity is needed whether inspired by the broadest humanitarian concerns or the narrowest self-interest. I hope that as caring individuals and informed scientists we shall make our voices heard.

To return to the specific topics of our discussion, for those unfamiliar with the earlier history of inositol as a constituent of phospholipids, a very brief review may be pertinent, covering a period of 50 years or so. It was 1930 when Anderson detected inositol in the phos-phatides of tubercle baccili (Anderson, 1930), but

Jordi Folch-Pi
Discoverer of inositol and inositol bisphosphate as constituents of mammalian phospholipids.
(Photo by G. Hauser at Dr. Folchi-Pi's retirement, October, 1977).

not very much fundamental advance occurred during the next 10 years. In 1939, at the beginning of World War II, Klenk described inositol in the phospholipids of soybeans (Klenk and Sakai, 1939) but it was not until 1942 that Jordi Folch, working with Woolley, identified inositol as a constituent of mammalian phospholipids, those in the cephalin fraction of beef brain (Folch and Woolley, 1942). He also discovered the polyphosphoinositides, reporting the isolation of inositol metaP$_2$ from a fraction of cephalin which was hydrolyzed with acid (Folch, 1946, 1949). The bisphosphate was identified as being the meta form by periodate oxidation -- a somewhat strange finding because this compound, rather than being the expected para-1,4-bisphosphate, would have had to be the 1,5-bisphosphate which is unlikely to be the one formed during hydrolysis or purification of either PIP or PIP$_2$. In any event, it was a number of years before the structures of all three phosphoinositides were elucidated (Brockerhoff and Ballou, 1961).

Seminal metabolic work on P turnover in phospholipids began in the 1950's by investigators still active in the field who identified phosphoinositides as having unique characteristics (Dawson, 1954; Hokin and Hokin, 1955) and the biosynthetic pathways for PI were clarified a few years later by Agranoff's group (Agranoff *et al.*,

651

1958) and by Paulus and Kennedy (1958). Soon afterwards in our laboratory we began to look at the incorporation of precursors (LeBaron et al., 1960, 1962) and at inositol biosynthesis (Hauser, 1963; Hauser and Finelli, 1963). Since then metabolic work has gone on at a steady pace, rapidly accelerating in the last 10 years to the level of activity shown by the interest and participation in this meeting, especially in the area of stimulus-induced changes in phosphoinositide turnover. Much of this was inspired by Michell's heuristic review (Michell, 1975) and has taken place in England.

In his classic book on inositol, Posternak wrote "(la) rôle biologique des phospho-inosites est encore très mal compris" (Posternak, 1962). The concluding speakers at the two previous conferences commented: "We do not yet understand all the details of the participation of phosphoinositides in physiological processes" (Hoffmann-Ostenhof, 1968) and emphasized that "There are still fundamental differences about the nature of the breakdown reaction" (Hawthorne, 1977). Hawthorne also paraphrased the title of an article by Agranoff and Bleasdale (1977) by saying "We are still not sure what it is trying to tell us".

And so we come to 1984 and these statements still apply in large measure, although some of the uncertainties have been resolved. Yet many questions remain and the lively interchanges at this meeting are testimony to the differences in opinions and interpretations of data. This is a healthy state of affairs. There is still much to be learned about the subjects we have been discussing -- I shall not review even the highlights of the meeting but some of the issues in need of resolution are worth listing:

What is the role of the high levels of free inositol, especially in the nervous system?
Are there pools of inositol? How are they regulated?
What are the details of the control of inositol uptake and phosphoinositide synthesis and degradation? Are there feedback loops?
Where does PI cleavage fit into the schemes of receptor-mediated phosphoinositide hydrolysis?
What is the relationship between receptor-mediated phosphoinositide turnover and the rapid disappearance of polyphosphoinositides after death?

How is receptor activation coupled to PIP_2 breakdown (what's in the black box)?
Is there a phospholipase D involved in a major way?
Is C kinase activity the main universal end result and what are its physiological substrates?
What is the relationship between calcium influx and calcium mobilization?
How and from what binding sites does IP_3 release Ca^{2+}?
What are the specific consequences of elevating the intracellular Ca^{2+} concentration by this mechanism?
To what extent is PI the source of released arachidonate?
Where does cyclic GMP fit into the scheme of things?

Even at our current state of knowledge, two additional general quotations from earlier reviews are pertinent and still valid: "Some of today's 'facts' may well be disproved by later work. By the same token, many observations that appear today no more than learned guesses may well be the opening leads to fruitful lines of work" (Folch and LeBaron, 1956) and "Il est certain que les efforts des biochimistes qui s'intéressent à l'inositol seront consacrés essentiellement à l'avenir à la solution des problèmes posés par le rôle biologique des phospho-inositides" (Posternak, 1962).

A myriad of specifics of this biological role await further clarification, and I predict that in the next few years substantial progress will be made in that direction leading to major advances in our understanding of the functioning and significance of a signal transmission system in widespread use by mammalian cells.

It is very exciting to be a part of this rapidly progressing field which promises to provide physiologically meaningful explanations for the intriguing observations on phosphoinositide metabolism that have haunted all of us for such a long time. Even so, it will be some time before, like Robin Irvine's dog, none of us will need to ponder the mysteries of inositol and phosphoinositide biochemistry because "'e knows it all". At any rate, this is no longer the esoteric subject that we almost had to apologize for, but one that has truly moved to one of the frontiers of biological science. It is pleasant to note that viewpoints which earlier seemed in frequent conflict are now reconciled, so that we can part as one happy family. Undoubtedly other stimulating controversies will arise.

On behalf of the organizers I want to thank every-one -- and the speakers in particular -- for participating in this scientifically and socially successful conference and for contributing ideas and inspiration for our future research. As we return to our laboratories and embark on the next round of experiments in preparation for another performance, I hope that we shall be filled with the imagination which, in Einstein's words, is more impor-tant than knowledge but which is bound to lead inex-orably to the knowledge necessary for the solution of the challenging but elusive problems still facing us.

P.S. In the months since the conclusion of the Chilton Conference, a very significant new connection with another area of biochemical research has emerged. Not only may protein kinase C be related to cell prolif-eration as Professor Nishizuka has suggested, but there is now evidence for the involvement of oncogene products in PI metabolism. Purified Rous sarcoma virus trans-forming gene product, pp 60^{v-src}, can phosphorylate diacylglycerol, PI and PIP (Sugimoto et $al.,$ 1984) and avian sarcoma virus UR2 gene product p68^{v-ros} can phos-phorylate PI (Macara et $al.,$ 1984), both leading to enhanced phosphoinositide turnover in rapidly prolifer-ating cells. This oncogene connection may have far-reaching implications for the study and control of malignancies and emphasizes the essential role phospho-inositides may play in physiological processes.

REFERENCES

Agranoff, B.W. and Bleasdale, J.E. (1977) in *"Cyclitols and Phosphoinositides"* (Wells, W.W. and Eisenberg, F., Jr., eds.), pp. 105-120, Academic Press, New York.
Agranoff, B.W., Bradley, R.M. and Brady, R.O. (1958) *J. Biol. Chem. 233*, 1077-1083.
Anderson, R.J. (1930) *J. Am. Chem. Soc. 52*, 1607-1608.
Brockerhoff, H. and Ballou, C.E. (1961) *J. Biol. Chem. 236*, 1907-1911.

Dawson, R.M.C. (1954) *Biochem. J. 57*, 237-245.
Folch, J. (1946) *Federation Proc. 5*, 134; and (1949) *J. Biol. Chem. 177*, 497-504, 505-519.
Folch, J. and LeBaron, F.N. (1956) *Canad. J. Biochem. Physiol. 34*, 305-318.
Folch, J. and Woolley, D.W. (1942) *J. Biol. Chem. 142*, 963-964.
Hauser, G. (1963) *Biochim. Biophys. Acta 70*, 278-289.
Hauser, G. and Finelli, V.N. (1963) *J. Biol. Chem. 238*, 3224-3228.
Hawthorne, J.N. (1977) in *"Cyclitols and Phosphoinositides"* (Wells, W.W. and Eisenberg, F., Jr., eds.), pp. 599-602, Academic Press, New York
Hoffmann-Ostenhof, O. (1968) *Ann. N.Y. Acad. Sci. 165*, 815-819.
Hokin, L.E. and Hokin, M.R. (1955) *Biochim. Biophys. Acta 18*, 102-110.
Klenk, E. and Sakai, R. (1939) *Hoppe-Seylers Z. Physiol. Chem. 258*, 33-38.
LeBaron, F.N., Hauser, G. and Ruiz, E.E. (1962) *Biochim. Biophys. Acta 60*, 338-349.
LeBaron, F.N., Kistler, J.P. and Hauser, G. (1960) *Biochim. Biophys. Acta 44*, 170-172.
Macara, I.G., Marinetti, G.V. and Balduzzi, P.C. (1984) *Proc. Natl. Acad. Sci. USA 81*, 2728-2732.
Michell, R.H. (1975) *Biochim. Biophys. Acta 415*, 81-147.
Paulus, H. and Kennedy, E.P. (1958) *J. Am. Chem. Soc. 80*, 6689-6690.
Posternak, T. (1962) *"Les Cyclitols - chimie, biochimie, biologie"*, Hermann, Paris.
Sugimoto, Y., Whitman, M., Cantley, L.C. and Erikson, R.L. (1984) *Proc. Natl. Acad. Sci. USA 81*, 2117-2121.

ABSTRACTS
OF
POSTER PRESENTATIONS

(arranged alphabetically by first author)

IMPLICATIONS OF THE METABOLISM OF INOSITOL PHOSPHATES IN PARO-
TID CELLS TO THE PATHWAYS OF THE PHOSPHOINOSITIDE EFFECT.

Debra L. Aub and James W. Putney, Jr. Med. Col. of Va./VCU,
Richmond, VA 23298.

Rat parotid acinar cells were used to investigate the time
course of formation and breakdown of inositol phosphates in response to receptor-active agents. In cells preincubated with
[^{3}H]inositol and in the presence of 10 mM $LiCl_2$ (which blocks
hydrolysis of inositol-1-phosphate (IP)), methacholine (10^{-4}M)
caused a substantial increase in cellular content of [^{3}H]IP,
[^{3}H]inositol-1,4-bisphosphate (IP_2) and [^{3}H]inositol-1,4,5-
trisphosphate (IP_3). Subsequent addition of atropine (10^{-4}M)
caused breakdown of [^{3}H]IP_3 and [^{3}H]IP_2 and little change in
accumulated [^{3}H]IP. The data could be fit to a model whereby
IP_3 and IP_2 are formed from phosphodiesteratic breakdown of
phosphatidylinositol-4,5-bisphosphate and phosphatidylinositol-
4-phosphate respectively, and IP is formed from hydrolysis of
IP_2 rather than from phosphatidylinositol. Consistent with
this model was the finding that [^{3}H]IP_3 and [^{3}H]IP_2 levels were
substantially increased in 5 sec while an increase in [^{3}H]IP
was barely detectable at 60 sec. These results indicate that
in the parotid gland the phosphoinositide cycle is activated
primarily by phosphodiesteratic breakdown of the polyphosphoinositides rather than phosphatidylinositol.

The possible involvement of phosphoinositide breakdown in the
control of pituitary hormone secretion

J.G. Baird, R.J.H. Wojcikiewicz, P.R.M. Dobson and B.L. Brown,
Department of Human Metabolism & Clinical Biochemistry,
Sheffield University Medical School, Sheffield, S10 2RX, UK.

Considerable evidence exists implicating calcium as a
primary mediator of the action of some of the hypothalamic
hormones on the anterior pituitary. However, information on
the role of phosphoinositide hydrolysis in prolactin, and
growth hormone secretion from normal, as distinct from tumour,
pituitary cells is sparse. We have analysed the effects of a
spectrum of modulators of secretion on the accumulation of
inositol phosphates in the presence of Li^+ in normal rat
anterior pituitary cells following preincubation with [^{3}H]-
inositol. Thyrotropin-releasing hormone (TRH) caused a rapid,
dose-dependent stimulation of inositol phosphate accumulation
with a similar dose response profile to that of prolactin
secretion. This effect of TRH was maintained in the absence of
added calcium and was only partially inhibited by the organic
calcium antagonists, methoxyverapamil and flunarizine. A23187
and increased concentrations of K^+ in the medium had little
effect on inositol phosphate production. Vasoactive intestinal
peptide had no effect and growth hormone releasing hormone only
stimulated breakdown at doses above 10 nM, which is the
maximally effective dose for GH secretion and cyclic AMP
production. Despite the evidence that dopamine causes a
decrease in ^{32}P-labelling of phosphatidylinositol, we have
not observed any effect of dopamine on inositol phosphate
accumulation either in the presence or absence of TRH. These
data indicate that TRH stimulated phosphoinositide breakdown
probably precedes calcium mobilisation. The action of dopamine
remains enigmatic.

659

Phosphatidyl Inositol Metabolism in Marine Salt
Secreting Epithelia

M.V. Bell, R.J. Henderson, C.M.F. Simpson and J.R.
Sargent

(Natural Environment Research Council, Institute of
Marine Biochemistry, Aberdeen, AB1 3RA, U.K.)

Phosphatidyl inositol (PI) in the osmoregulating,
salt secreting epithelia of marine teleosts and
elasmobranchs is rich in arachidonic acid, $20:4(n-6)$,
in contrast to the other phosphoglycerides that are
rich in the characteristically marine polyunsaturates,
$20:5(n-3)$ and $22:6(n-3)$. Agents that elevate intracellular levels of cAMP increase the activity of the
sodium ion pump in these epithelia and depress the
turnover of the phosphate and inositol groups of PI.
Increased intracellular levels of Ca^{2+} increase PI
turnover. The $20:4(n-6)$ but not the $18:0$ on PI also
turnovers rapidly and prostglandin formation from
$20:4(n-6)$ occurs. This formation is competitively
inhibited by $20:5(n-3)$. Mechanisms controlling the
sodium ion pump in these tissues are under
investigation.

Vasopressin, angiotensin II and α_1-adrenergic mediated inhibition of hepatic plasma membrane Ca^{2+} transport and its
relationship to phosphoinositide metabolism. P.F. Blackmore,
V. Prpic and J.H. Exton, Howard Hughes Med. Inst. and Dept.
Physiol., Vanderbilt Univ. Sch. Med., Nashville, TN
   There is much evidence which shows that vasopressin, angiotensin II and α_1-adrenergic agonists initially mobilize intracellular pools of Ca^{2+}, thereby raising cytosolic free
Ca^{2+} ($[Ca^{2+}]_i$). In order to maintain the elevated
$[Ca^{2+}]_i$, it has been postulated that Ca^{2+} influx from
the extracellular medium is stimulated and/or the plasma membrane Ca^{2+}-Mg^{2+} ATPase is inhibited. Rat liver plasma
membrane vesicles were isolated using isotonic medium and Percoll self-forming gradient centrifugation. The K_m for
Ca^{2+} transport into these inverted vesicles was 3 nM. The
Ca^{2+} transport was inhibited in a dose- and time-dependent
manner when livers were perfused with vasopressin. A maximum
50%/o inhibition was obtained within 3 min with 1 nM vasopressin. Maximally effective doses of angiotensin II and epinephrine inhibited Ca^{2+} transport by 21 and 26%/o respectively, while glucagon had no effect. A close correlation
between the inhibition of Ca^{2+} transport activity and the
stimulation of phosphoinositide hydrolysis induced by vasopressin, angiotensin II and epinephrine was observed, suggesting that the two processes are related. It is concluded that
inhibition of plasma membrane Ca^{2+} transport is involved in
the maintenance of elevated $[Ca^{2+}]_i$ by Ca^{2+}-dependent
hormones.

ACCUMULATION OF INOSITOL PHOSPHATES IN RAT SUPERIOR CERVICAL
SYMPATHETIC GANGLIA STIMULATED BY BETHANECHOL, VASOPRESSIN AND
HIGH K+-DEPOLARISATION

E.A. BONE, P. FRETTEN, S. PALMER, C.J. KIRK and R.H. MICHELL
DEPARTMENT OF BIOCHEMISTRY, UNIVERSITY OF BIRMINGHAM,
P.O. BOX 363, BIRMINGHAM B15 2TT, U.K.

Bethanechol, a muscarinic cholinergic agonist, induced an
accumulation of [3H]-labelled inositol phosphates in isolated
rat superior cervical ganglia prelabelled with [3H]-inositol.

[8-arginine]-vasopressin (AVP) induced a rapid, and much
larger, dose-dependent accumulation of inositol phosphates. The
first labelled products generated were inositol trisphosphate and
inositol bisphosphate, suggesting that a phosphodiesterase-
catalysed hydrolysis of phosphatidylinositol 4,5-bisphosphate
(and possibly of phosphatidylinositol 4-phosphate) initiates the
stimulated inositol lipid metabolism. The response to AVP was
inhibited by a V_1-vasopressin antagonist. Oxytocin was a much
less potent agonist than AVP.

The response to AVP was maintained in ganglia incubated in
media with reduced Ca^{2+} concentration and in the presence of
nifedipine, an antagonist of potential-sensitive Ca^{2+} channels,
suggesting that inositol phosphate production is a direct effect
of AVP. Incubation of ganglia with a raised extracellular K^+
concentration also caused a stimulation of inositol phosphate
accumulation, possibly due to release of intrinsic VP.

Phosphatidylinositol 4,5-bisphosphate hydrolysis may,
therefore, be important in vasopressinergic communication between
neurones in the rat superior cervical ganglion.

1321N1 ASTROCYTOMA CELLS: A MODEL SYSTEM FOR INVESTIGATING THE
ROLE OF PHOSPHOINOSITIDE HYDROLYSIS IN CALCIUM MOBILIZATION AND
DESENSITIZATION. Susan L. Brown and Joan Heller Brown, Univer-
sity of California, San Diego, La Jolla, Calif. 92093

Exposure of 1321N1 cells to carbamylcholine (CARB) causes
rapid (<30 sec), atropine-sensitive phosphoinositide (PhI) hydrolysis as measured by ^{3}H-inositol 1-phosphate (^{3}H-Ins 1P) formation. Acetylcholine also increases ^{3}H-Ins 1P formation, but
the muscarinic agonists oxotremorine (OXO) and pilocarpine have
little effect. The PhI response does not appear to desensitize
since the rate of ^{3}H-Ins 1P formation is constant for at least
90 min after treatment with CARB, and pretreatment (75 min)
with 100 μM CARB does not alter the PhI response. CARB also
stimulates $^{45}Ca^{2+}$ efflux and causes Ca^{2+}-dependent activation
of a cyclic AMP phosphodiesterase (PDE) in these cells; prolonged exposure to CARB leads to desensitization of PDE activation. If the PhI response is a mediator of these other molecular events, then OXO should be ineffective relative to CARB at
eliciting them. OXO and CARB cause equivalent Ca^{2+}-dependent
PDE activation and we are presently measuring Ca^{2+} mobilization
directly to determine of OXO and CARB are also equivalent in
this regard. OXO is poor relative to CARB at causing desensitization. This and our observation that the PhI response does not
desensitize suggest that the PhI response is related to the
phenomenon of desensitization. Further studies of the relationship between agonist effects on PhI hydrolysis, Ca^{2+} mobilization and desensitization in this homogenous cell line should
help define the physiological role of the PhI response.

INOSITOL(1,4,5)TRISPHOSPHATE EVOKES INTRACELLULAR Ca RELEASE IN LIVER CELLS. G.M. Burgess & J.W. Putney, Jr., Med. Col. of Va/VCU. Richmond, VA 23298. The increase in $[Ca^{++}]_i$ induced by Ca-mobilizing hormones in liver is due in part to release of Ca from intracellular stores. It is thought that a 2nd messenger is formed at the plasma membrane (P.M.) which diffuses to the intracellular organelles and releases Ca. One of the 1st effects of these hormones in liver is formation ($\leqslant$30s) of inositol(1,4,5)-trisphosphate [Ins(1,4,5)P$_3$], and as this is formed at the P.M. its ability to mediate intracellular Ca-release was tested.

Hepatocytes whose P.M.s. have been made permeable, when suspended in a cytosolic-type medium with $[Ca^{++}]$ buffered to 180nM (equivalent to resting $[Ca^{++}]_i$) take up ^{45}Ca in an ATP-dependent, DNP and oligomycin-insensitive manner, into a pool thought to be endoplasmic reticulum (ER). Addition of 5μM Ins(1,4,5)P$_3$ to these permeable cells with ^{45}Ca in apparent steady state caused a rapid release of 0.5nmol/mg protein cell associated ^{45}Ca. This approximates the amount of Ca released from intact cells by α-adrenergic stimuli. The concentration-dependence of the effect was tested, and the EC$_{50}$ value was ca. 0.25μM and was maximal at 5μM. In similar experiments, Ins(1,4)P$_2$, Ins1P, Ins cyclic P and inositol were without effect. If $[Ca^{++}]$ was set at 3.3μM to enable the mitochondria of the permeable cells to take up ^{45}Ca, Ins(1,4,5)P$_3$ was unable to cause ^{45}Ca release from this pool. This data suggests that Ins(1,4,5)P$_3$ may be a 2nd messenger linking receptor-activation to internal Ca release (from ER) in the liver.

FSH-MEDIATED ARACHIDONATE TURNOVER IN RAT SERTOLI CELLS. D.R. Cooper, B.F. Dickens and M.P. Carpenter, Okla. Med. Res. Fdn. and Dept. of Biochem., OUHSC, Okla. City, OK 73104

Sertoli cells (SC), somatic cells of the seminiferous tubules, are target cells for FSH action in the male. Germ cell development is mediated by FSH-SC interactions. The potential role of FSH on membrane lipid turnover and acylation patterns has not been previously investigated. Studies in this laboratory have shown that FSH stimulates prostanoid synthesis by the SC and suggest a potential role of FSH on membrane lipid turnover. FSH effects on SC lipid turnover were analyzed using SC prelabeled with [^{3}H] arachidonate (20:4) in primary culture from 21 day-old rats. There was a rapid uptake of [^{3}H] 20:4 (24% in 5 min); 89% of the incorporated [^{3}H] was found in the phosphatides. When FSH was added there was a decrease in label in phosphatidyl-choline (PC) and phosphatidylinositol (PI) after 1 min followed by increases in PI and phosphatidylethanolamine after 5 min. Concurrently, 25% of the [^{3}H] in neutral lipids appeared in the 1,2 diglyceride fraction. The FSH-mediated turnover of PI was confirmed using [^{32}P]-labeled cells. Within 2 min after FSH exposure, 50% less label was detected in PI than in lysoPI than in the control. These very early effects of FSH on SC implicate receptor-mediated turnover of 20:4 in PI and PC and suggest that phospholipase C and A$_2$ activation is FSH-dependent. Turnover of SC 20:4 correlates with stimulation of prostanoid synthesis by FSH.

662

DISSOCIATION BETWEEN THE STIMULATION OF PI LABELING BY VASOPRESSIN AND ANGIOTENSIN II AND THEIR METABOLIC EFFECTS IN RAT HEPATOCYTES. S. Corvera and J. A. García-Sáinz. Universidad Nacional Autónoma de México.

It is generally accepted that Vasopressin, Angiotensin II and alpha$_1$-adrenergic amines stimulate glycogenolysis and ureogenesis through a mechanism involving Ca^{++} ions. Phosphatidylinositol turnover seems to play a role in this mechanism.

In hepatocytes from hypothyroid rats, the stimulation of glycogenolysis and ureogenesis by vasopressin and angiotensin II is markedly diminished whereas the stimulation by alpha$_1$-adrenergic agonists is unaltered. However, PI turnover is stimulated by the three agents in hepatocytes from both control and hypothyroid rats. Thus in hepatocytes from hypothyroid rats there is a dissociation between PI turnover and metabolic effects: further insight into this problem is provided by the fact that the glycogenolytic effect of Ca^{++} ionophore A-23187 is also decreased by hypothyroidism. It appears then that insensitivity to vasopressin and angiotensin II in hepatocytes from hypothyroid rats is due to a decreased responsiveness to calcium signalling. The data further suggest that two mechanisms are involved in alpha$_1$-adrenergic actions: one which is Ca^{++} dependent and related to PI turnover, and another which predominates in hypothyroidism and is not shared by the vasopressor peptides.

REGULATION OF PHOSPHATIDYLINOSITOL METABOLISM BY LH AND GnRH. J.S. Davis. Dept. of Medicine, Univ. of South Florida College of Medicine; J.A. Haley Vet. Hosp., Tampa, FL 33612.

The response of rat granulosa cells to both luteinizing hormone (LH) and gonadotropin-releasing hormone (GnRH) involves alterations in phosphatidylinositol (PI) metabolism. Experiments were conducted to determine whether the effects of LH and GnRH occur by similar mechanisms. Granulosa cells were incubated for 1 h in medium containing ^{32}P in the absence (CTL) or presence of LH (1 µg/ml) or GnRH (100 ng/ml) and chased for 1 h in medium without ^{32}P or hormones. Aliquots (1.5×10^6 cells) from each group were further incubated (0-30) min in ^{32}P-free medium in the absence or presence of LH and GnRH. Zero time ^{32}P-PA levels were similar in all groups. Zero time ^{32}P-PI levels were increased in LH- and GnRH-treated groups (34% and 259%, respectively). In all groups, PI and PA labeling were not altered during subsequent incubations with LH. GnRH increased PA labeling within 2.5 min in all groups. In the CTL group, GnRH increased PI labeling (13%) at 30 min. In the LH-pretreated group, GnRH did not alter PI labeling. However, in the GnRH-pretreated group, GnRH-stimulated significant reductions in ^{32}P-PI levels at 10 min (16%) and 30 min (29%). The results indicate that LH and GnRH have both quantitative and qualitative differences in their actions on PI metabolism; 1) The effect of GnRH is many-fold greater than LH; 2) GnRH, but not LH, stimulates PI depletion; 3) GnRH-induced PI depletion involves a GnRH-sensitive pool of PI. Supported by NIH:HD-17939 and the Veterans Administration.

663

FREE FATTY ACIDS AND DIGLYCERIDES IN CONTROL, LAMINECTOMIZED, AND TRAUMATIZED CAT SPINAL CORD - Paul Demediuk, Douglas K. Anderson, and Lloyd A. Horrocks, Dept. of Physiological Chemistry, The Ohio State University, Columbus, OH 43210, and VA Med. Ctr., Cincinnati, OH 45220

Since lipolysis of phosphoinositides may play an important role in central nervous system trauma, the effect of surgical exposure of the spinal column (controls), laminectomy, and compression trauma on free fatty acid and diglyceride levels in the spinal cord of mongrel cats has been assessed. All samples were frozen in situ with liquid N_2 prior to removal. Control samples frozen with vertebrae intact and removed immediately following surgical exposure, demonstrated unexpectedly high levels of free fatty acids and diglycerides. Laminectomy further exacerbated this condition. Allowing laminectomized, anesthetized animals to equilibrate for time periods up to 120 min. resulted in large decreases in free fatty acids and diglycerides. Therefore all trauma samples have been taken from animals which have equilibrated for 90 min following laminectomy. Preliminary data indicated that 5 min. of compression trauma results in significant increase in free fatty acids and diglycerides. Arachidonate levels increased approximately 6-fold in these samples.

Supported by NIH Grant NS-10165

RECIPROCAL STIMULATION OF PHOSPHATIDYLINOSITOL-4-PHOSPHATE AND MYELIN BASIC PROTEIN PHOSPHORYLATION BY THE KINASES SOLUBILIZED FROM MYELIN. D.S. Deshmukh, S. Kuizon and H. Brockerhoff. Inst. for Basic Res., 1050 Forest Hill Rd., Staten Island, NY 10314.
We have shown that there is rapid metabolism of phosphatidylinositol-4-phosphate (PhIP) and phosphatidylinositol-4,5-bisphosphate ($PhIP_2$) in purified myelin, which contains the kinases and phosphohydrolases required for the turnover of the phosphomonoester groups of these compounds. In view of the abundance of polyphosphoinositides and myelin basic protein (MBP) and the presence of kinases for both substrates in myelin, the speculation of a connection between the phosphorylation of these compounds arises. We have, therefore, examined the mutual stimulation by MBP and PhIP of their phosphorylation by detergent solubilized myelin kinases. Pure myelin was isolated from young adult rats. The kinases were solubilized by treatment of myelin with Tris-Cl buffer (50 mM, pH 7.5) containing 10 mM non-denaturing Zwitterionic detergent 3-[-C3-cholamidopropyl)-dimethylammonio]-1-propanesulfonate (CHAPS). The specific activities of PhIP and MBP kinases of CHAP-solubilized myelin were increased by about 5-fold over whole myelin. When MBP and PhIP were present together, the rate of phosphorylation of each was increased about 6-8 fold. Such reciprocal stimulation was not observed by combination of other proteins or lipids. It appears that the phosphate turnover of MBP and of PhIP may be coupled in vivo (supported by grants NS14480 and NS14073 from NINCDS).

664

PHOSPHATIDYLINOSITOL CONTENT OF PLASMA MEMBRANES FROM PORCINE CORPORA LUTEA IS INCREASED BY LUTEINIZING HORMONE BUT NOT CYCLIC AMP. Michael J. Dimino, Raymond B. Allen, Jr., and Michael A. Neymark. Department of Biochemistry, Eastern Virginia Medical School, Norfolk, VA 23501

Luteinizing hormone (LH) treatment of bovine luteal cells increases the incorporation of ^{32}P into phosphatidylinositol (PI) (Davis et al., Endocrinology 109:469, 1981). To study the mechanism for these changes in more detail, we used an enriched plasma membrane preparation (as indicated by a 5-fold increase in 5'-nucleotidase specific activity) from porcine luteal tissue. The content of various phospholipids in these plasma membranes was determined by phosphorus analysis of lipids separated by two-dimensional thin layer chromatography of a chloroform/methanol extract. The concentrations of phospholipids (μg phosphorus/mg protein) obtained for unincubated, untreated plasma membranes were: PI 0.19, phosphatidylcholine 2.52, phosphatidylethanolamine 1.51, phosphatidylserine 0.25, phosphatidic acid 0.01. Treatment of plasma membranes with LH (0.8 μg/ml) for 5 min increased their PI content approximately twofold. No phospholipid changes were observed in plasma membranes incubated without hormone. In contrast to LH treatment, incubation of plasma membranes with 2 mM cyclic AMP for 5 min did not increase their PI content. These results suggest that the LH-dependent increase in the PI content of luteal plasma membranes is not mediated by cyclic AMP. (Supported by NIH Grant # HD 17589.)

MUSCARINIC RECEPTOR-LIGAND INTERACTIONS IN BRAIN AND THE ENHANCEMENT OF INOSITOL PHOSPHOLIPID TURNOVER. S.K.Fisher, J.C. Figueiredo and R.T.Bartus. Dept.CNS Research, Med.Res.Div. of American Cyanamid Co., Lederle Labs., Pearl River, NY 10965

Structural analogs of oxotremorine have been employed to examine the relationship between the binding of muscarinic agonists to muscarinic receptors in guinea-pig cerebral cortex and the enhancement of inositol phospholipid (IPL) turnover. Large differences were observed in the ability of the analogs to enhance IPL turnover, as measured by either the increase in labeling of phosphatidate (PtdA) and phosphatidylinositol from ^{32}Pi in a nerve-ending fraction, or by the stimulation of release of inositol phosphates (IP) from slices of cerebral cortex prelabeled with [3H]inositol. The quaternary N+ analogs, oxotremorine-M and its N-methyl acetamide derivative were the most efficacious analogs tested, being five to ten times as effective as oxotremorine. Receptor occupancy data obtained from the displacement of [3H]quinuclidinyl benzilate indicated that the more efficacious analogs interacted with both high and low affinity forms of the muscarinic receptor, whereas the less efficacious agonists bound to a single affinity form of the receptor. Dose-response curves for stimulated PtdA formation and IP release obtained in the presence of oxotremorine-M were predominantly correlated with occupancy of the low affinity form of the muscarinic receptor. These and previous results (J.Biol.Chem. 258, 7358) suggest that the differential effects of muscarinic agonists on IPL turnover in brain are directly related to the extent of agonist-induced conformational changes in the muscarinic receptor.

665

PHOSPHOINOSITIDE METABOLISM IN THE SQUID GIANT AXON. Robert M. Gould, Martha Jackson and Ichii Tasaki, Marine Biology Laboratory, Woods Hole, MA and Institute for Basic Research in Developmental Disabilities, 1050 Forest Hill Rd., Staten Island, NY 10314.

The axoplasm of the squid giant axon is a unique preparation for studying inositol lipid metabolism. Extruded axoplasm contains a select subset of neuronal proteins and organelles channeled to the axonal transport system. Among these proteins are enzymes of phosphoinositide metabolism (Gould, et al., J. Neurochem. 40 (1983) 1293-1299, 1300-1306).

By injecting lipid precursors into intact axons, it is possible to study the effect of electrical stimulation on axonal lipid metabolism. High frequency stimulation (50-100 Hz) for short (30 sec-5 min) or long intervals does not grossly alter the overall incorporation of ^{32}P (from orthophosphate or γ-labeled ATP) into phospholipid. Phosphatidic acid and phosphatidylinositol 4-phosphate are two of the dominent lipids formed by the axons. There is some incorporation of label in phosphatidylinositol 4,5-bisphosphate (PIP$_2$). Interestingly, when the glial cells are separated from the axon by microdissection, a large proportion of the PIP$_2$ is retained in these cells. Injected myo-inositol and acetate are selectively incorporated into axoplasmic phosphatidylinositol. With acetate labeling, the sheath contains little label in phosphatidylinositol and prominent label in phosphatidylcholine.
(Supported by grant NS-13980 from NIH.)

POSSIBLE REGULATION OF PHOSPHOLIPASE C ACTIVITY IN HUMAN PLATELETS BY TPI. G. Graff*, N. Nahas*, M. Nikolopoulou*, V. Natarajan+ and H. H. O. Schmid+. *Dept. of Biol. Chem., U. of Ill., Chicago, IL and +The Hormel Institute, U of Minn., Austin, MN

Cytosolic phospholipase C from human platelets was found to catalyze the Ca^{2+}-dependent degradation of phosphatidylinositol (PI), phosphatidylinositol-4'-phosphate (DPI), and phosphatidylinositol-4',5'-bisphosphate (TPI) at 150 μM to 5 mM Ca^{2+}. Both DPI and TPI inhibited the hydrolysis of PI (250 μM) in a concentration-dependent manner and, when incubated in mixture with PI, were competitive substrates for PI hydrolysis. Increasing the DPI/PI ratio to 0.3 resulted in preferential degradation of DPI over PI. TPI alone or in mixture with PI was a poor substrate. Increasing the TPI/PI ratio to 0.21 inhibited both PI degradation (≥95%) and the overall formation of 1,2-diacylglycerol (≥82%). TPI acted as a mixed type inhibitor with a K_i of about 10 μM. The K_a for Ca^{2+} in PI hydrolysis was increased from 5 to 180 μM when TPI (36 μM) was included with PI (250 μM) and optimum PI degradation occurred only at 4 mM Ca^{2+}. Since unstimulated human platelets had DPI/PI and TPI/PI ratios of 0.42 and 0.46, their phospholipase C activity may be completely inhibited. Changes in ^{33}P prelabeled phospholipids of intact platelets upon stimulation with thrombin indicated a transient decline in ^{33}P label in both TPI and DPI (15 sec) followed by an increase in [^{33}P]phosphatidic acid but no change in [^{33}P]PI. Our data suggest that DPI may be more important than PI in the formation of 1,2-diacylglycerol as potential precursor of arachidonic acid and that TPI may serve as modulator of phospholipase C activity. (PHS grants GM 28894, HL 24312, HL 08214 and NS 14304).

TIGHT ASSOCIATION BETWEEN RAT LIVER MICROSOMAL EPOXIDE HYDROLASE
AND PHOSPHATIDYLINOSITOLS. Martin J. Griffin and Ram B. Pala-
kodety. OK Med. Res. Fndn., Okla. City, OK.

Two forms of epoxide hydrolase (EH-A, pI 5.3 and EH-B, pI 6.8)
were resolved by chromatofocusing or by ion exchange chromatography. In vitro (γ-^{32}P) ATP treatment of microsomes and isolation of the enzyme forms showed the labeling of only EH-A.
Radioactivity from the enzyme could only be extracted with acidified chloroform/methanol (2:1) (C/M) and was found to comigrate
with authentic TPI. We found 3 ± 0.5 nmoles inositol per mol
of EH-A or B when the C/M extracted, purified forms were hydrolyzed, derivitized and analyzed by GLC. EH-A and B, labeled in
vivo with ^{32}P$_i$, were purified and first extracted with C/M until
all the radioactivity associated with loosely bound PLS was
removed. Acidified C/M was used to release the tightly bound
PI and these were analyzed by TLC. EH-B had 2 mol of PI and 1
mol of DPI per mol of enzyme and EH-A had 2 mol of PI and 1 mol
of TPI. The P$_i$ contents of EH-A and EH-B was about 5 for A and
4 for B. These two forms had similar amino acid composition, CN-
Br cleaved peptide patterns and were immunologically identical.
EH-A was found to be 12-fold less reactive towards styrene oxide
hydration than EH-B. The immunological reactivity of EH-A
towards anti-EH-B was about 5 times less as determined by ELISA.
The associated phosphatidylinositols may have a role in the modulation of enzymatic and immunological properties of epoxide hydrolase.

DEXAMETHASONE-INDUCED STIMULATION OF PI TURNOVER IN CUL-
TURED HUMAN EMBRYONIC FIBROBLASTS. R. I. Grove and
R. M. Pratt, Experimental Teratogenesis Section, LRDT,
NIEHS, NIH, Research Triangle Park, NC 27709.

The breakdown and resynthesis of phosphatidylinositol (PI) is thought to play a key role in the mechanism
of growth stimulation induced by certain growth factors
and other mitogenic agents. We have previously reported
that dexamethasone (Dex), a cleft palate-inducing glucocorticoid that inhibits cell proliferation in an established fibroblastic cell line derived from a human embryonic palate, stimulates the degradation and resynthesis of PI. We now report that the Dex-induced stimulation of PI synthesis appears to be independent of calcium
present in the medium and also occurs in the presence of
increased cAMP levels (8 bromo cAMP and theophylline).
Further analysis revealed that Dex caused a 40% decrease
in incorporation of labeled arachidonic acid into PI,
while stimulating incorporation into phosphatidylethanolamine and neutral lipids. The glucocorticoid antagonist
cortexalone inhibited the Dex effects on PI turnover and
cell proliferation. In addition, epidermal growth factor
(EGF), which prevents the Dex-induced inhibition of cell
proliferation, also prevents the Dex-stimulation in PI
turnover. These results suggest that Dex stimulates PI
turnover by a receptor mediated, extracellular calcium
independent mechanism and that the newly synthesized PI
has a decreased arachidonic acid content. In addition,
the cortexalone and EGF data are consistent with the
hypothesis that Dex inhibits human embryonic palate fi-
broblast proliferation by altering normal PI metabolism.

PHOSPHATIDATE LABELED IN MUSCARINICALLY-STIMULATED NERVE END-INGS IS PRIMARILY THE TETRAENE SPECIES. A.K. Hajra, L.A.A. Van Rooijen, E.B. Seguin and B.W. Agranoff, Mental Health Res. Inst. and Dept. of Biol. Chem., Neuroscience Laboratory Bldg., University of Michigan, 1103 E. Huron, Ann Arbor, MI 48109.

Labeled phosphatidate (PhA) formed following cell surface receptor activation, and associated with increased phosphatidyl-inositol (PhI) labeling, is believed to be produced by phos-phorylation of diacylglycerol (DAG) released in the phospholi-pase C-catalyzed hydrolysis of polyphosphoinositides. It is therefore expected that the fatty acid composition of resynthe-sized PhA should resemble that of the phosphoinositides, known to be primarily the 1-stearoyl,2-arachidonoyl species, although conflicting reports exist on this point. Nerve ending prepara-tions from guinea pig brains were incubated with carbachol in the presence of $^{32}P_i$. The PhA band obtained following TLC was extracted and methylated by diazomethane. Authentic 1-stearoyl, 2-[^{14}C]arachidonoyl glycerol-3-phosphate was used as a stan-dard. The non-polar dimethyl PhA species were separated accord-ing to their double bond content by argentation TLC, were lo-cated autoradiographically, and quantified by counting. Radio-activity comigrating with the standard tetraene species account-ed for 75% of the total muscarinically-enhanced $^{32}P_i$ content. This result is consistent with observed polyenoate distribution of [^{32}P]PhI formed in the same system (J. Biol. Chem. 249:1551). These observations indicate that the DAG remains intact in a closed cycle in which inositides are broken down and resynthe-sized. (Supported by NS 15413 and NS 08841.)

METABOLISM OF MOLECULAR SPECIES CONTAINING ARACHIDONATE, 20:4(N-6), or ADRENATE, 22:4(N-6), IN DIACYL-GPI OF MOUSE BRAIN. Hubert W. Harder, Yasuhito Nakagawa, and Lloyd A. Horrocks. Dept. Physiological Chemistry, The Ohio State University, Columbus, Ohio 43210.

Rigorous analysis of glycerophospholipid metabolism requires that lipids be separated into molecular species. We have used HPLC methods recently developed in our lab to study the metabolism of molecular species of diacyl-GPI that contain 20:4(n-6) or 22:4(n-6), and we have compared the results to those for diacyl-GPC and diacyl-GPE. Total IGP, EGP, and CGP fractions were obtained from mouse brain after simultaneous intracerebroventricular injection of [3H]20:4 and [14C]22:4; these fractions were reacted with phospholipase C and acetylated with acetic anhydride. The acetylated diradyl-glycerols were separated by normal phase HPLC into diacyl, alkylacyl, and alkenylacyl fractions; reverse phase HPLC was then used to separate the diacyl fractions into 15-20 peaks. Molecular species containing 20:4 or 22:4 were collected, quantified by GLC, and assayed for radioactivity, giving specific activity time curves for 3H for species containing 20:4 or 22:4 and for 14C for species containing 22:4. Molecular species containing 18:0, which constitute the largest fraction of 20:4 and 22:4 containing molecular species in brain in diacyl-GPI, -GPC, and -GPE, have a significantly slower metabolism than molecular species containing 16:0 or 18:1. The metabolism of 22:4 is significantly more active in diacyl-GPI than in diacyl-GPC or -GPE. Supported by NIH Grant NS-08291.

670

STIMULATION OF PHOSPHOLIPID (PL) PHOSPHORYLATION AND TURNOVER
BY PARATHYROID HORMONE (PTH) IN BASOLATERAL MEMBRANES (BLMs)
OF RENAL TUBULAR CELLS IN VITRO. Keith Hruska and Pedro
Esbrit. Jewish Hospital, St. Louis, MO. 63110

Synthesis and turnover of PLs in the phosphatidylinositol
(PhI) cycle are stimulated by PTH in isolated renal tubular
segments. These PLs stimulate calcium binding and translocation in renal tubular brush border cell membranes. When BLMs,
prepared by centrifugation of renal cortical homogenates in
Percoll, were incubated with 1 mM [γ32Pi] ATP, 1 μM Ca^{2+},
1 μg/ml calmodulin, and +/- b-PTH 1-84 (10^{-6}M), ^{32}Pi incorporation was stimulated into phosphatidic acid, PhI-P and PhI-P$_2$.
The effect of PTH was detectable by 15 sec and maximal between
1-5 min. At 5 min, PTH stimulated Pi incorporation was 5.7
(PA), 61.4 (PhI-P) and 11.4 (PhI-P$_2$) pmol/mg protein compared
to control of 3.8, 44.3, and 8.6 (p<.02). After 5 min,
dephosphorylation was predominant indicating turnover of the
phosphorylated PLs. PL phosphorylation or turnover was not
stimulated by incubation with 3'5' c-AMP (10^{-6} or 10^{-3}M).
In the presence of PTH, PL phosphorylation was not altered by
the absence of Ca^{2+} and calmodulin, but without PTH, PL
turnover was impaired by a zero calcium incubation. These
results were confirmed by release of ^{3}H-inositol phosphates
in tissues preincubated with ^{3}H-inositol. These data demonstrate stimulation of PL phosphorylation by PTH independent of
c-AMP and Ca^{2+}. These effects of PTH may be a mechanism for
modulation of Ca^{2+} binding and translocation in the BLM
directly associated with hormonal binding to the receptor.

THE PI-PHOSPHODIESTERASE IN BRAIN CYTOSOL AND SUBCELLULAR
MEMBRANES. Judith Kelleher and Grace Y. Sun. Sinclair Compara-
tive Medicine Research Farm and Biochemistry Department,
University of Missouri, Columbia, MO 65201.

A plasma membrane fraction enriched in [^{14}C]-arachidonoyl-
PI was used to assay the PI-phosphodiesterase (PI-PDE) activity in rat brain subcellular fractions. After incubation of
the membranes with [^{14}C]-arachidonate, ATP, Mg, CoA and a
small amount of lysophosphatidylinositol, the membranes were
washed twice with BSA and then heat-treated (80°C for 5 min)
to inactivate the enzymes. PI-PDE activity was assayed by
following the disappearance of labeled PI and the appearance
of labeled diacylglycerol (DG) after incubation of substrate
with brain subcellular fractions or cytosol. Appreciable PI-
PDE activity was observed in both membrane and cytosolic
fractions when incubation was carried out in the presence of
deoxycholate (1 mg/mg protein) + Ca^{++} (2 mM) or taurocholate
(4 mg/mg protein) + Ca^{++} (3.5 mM) + EDTA (1 mM), but not in
the presence of Ca^{++} alone. PI-PDE activity was highest in
the cytosolic fraction, but a substantial amount of activity
was also present in synaptosomes and the synaptic and non-
synaptic plasma membranes. Although activity of the enzyme in
the membrane fractions remained unchanged after lysing or
washing, treatment of the membranes with deoxycholate resulted
in solubilization of 50% of the enzyme. In the cytosol, the
DG formed is subjected to further hydrolysis by the DG lipase
to release the free fatty acids, but this type of conversion
is apparently not present in the membrane fractions.

RECEPTOR-MEDIATED MOBILIZATION OF CELL ASSOCIATED Ca^{2+} STORES IN HEPATOCYTES

C J Kirk and S B Shears, Department of Biochemistry, University of Birmingham, P O Box 363, Birmingham B15 2TT.

Activation of at least four separate hepatic receptors (α_1-adrenergic, V$_1$-vasopressin, angiotensin and ATP) is capable of increasing the intracellular concentration of Ca^{2+} via a mechanism thought to involve phosphatidylinositol 4,5-bisphosphate breakdown. Work from several laboratories, using Ca^{2+}-depleted hepatocytes and/or mitochondria isolated by differential centrifugation, has suggested that activation of these receptors may provoke the mobilization of Ca^{2+} from a mitochondrial pool.

We have incubated hepatocytes at physiological $[Ca^{2+}]$ and isolated mitochondria, within seconds of hormonal stimulation, by a technique that prevents re-distribution of Ca^{2+} across the mitochondrial membrane. Our results indicate that although α-adrenergic agonists and vasopressin mobilize Ca^{2+} from a cell-associated pool during the early stages of hormonal stimulation, mitochondria are NOT the source of this Ca^{2+}. In fact, vasopressin caused an accumulation of Ca^{2+} in the mitochondria of stimulated cells within 30 secs of receptor activation.

The possible role of phosphatidylinositol 4,5-bisphosphate breakdown in these events will be discussed.

CA^{2+} IONOPHORES AFFECT PHOSPHOINOSITIDE METABOLISM DIFFERENTLY THAN THYROTROPIN-RELEASING HORMONE. R.N. Kolesnick and M.C. Gershengorn, Dept. of Med., Cornell Univ. Med. Coll. NY, NY.

We have shown (J. Biol Chem 258: 227, 1983 and Biochem J, in press) that TRH decreases PtdIns-4,5P$_2$, PtdIns-4P and PtdIns, increases inositol-1,4,5P$_3$, inositol-14P$_2$, inositol-1P and diacylglycerol (DG), and mobilizes Ca^{2+} in GH$_3$ pituitary cells. To explore whether the effects of TRH may be mediated by Ca^{2+}, we studied the affects of A23187 and ionomycin. In cells pre-labeled with [3H]inositol, A23187 caused a decrease in [3H]-PtdIns-4,5P$_2$ to 74±4 % of control, [3H]PtdIns-4P to 62±13 %, and [3H]PtdIns to 88±1 %. A23187 increased [3H]inositol-1,4P$_2$ to 230±30 % and [3H]inositol-1,4,5P$_3$ was 124±25 %. By contrast, TRH increased [3H]inositol-1,4P$_2$ to 233±21 % and [3H]inositol-1,4,5P$_3$ to 251±23 %. When cells were exposed to TRH and A23187 simultaneously, the effect of TRH to increase [3H]inositol-1,4,5P$_3$ was not altered; hence, the inability to detect an increase in inositol-1,4,5P$_3$ after A23187 was not due to rapid conversion to inositol-1,4P$_2$. In cells prelabeled with [3H]arachidonic acid, A23187 caused an elevation in unesterified [3H]arachidonic acid to 150±10 %. By contrast, TRH did not increase unesterified [3H]arachidonic acid. Ionomycin caused changes similar to A23187 but of equal or greater magnitude. Hence, Ca^{2+} ionophores primarily stimulate the activities of a phospholipase C that acts on PtdIns-4P and of a phospholipase A$_2$ (and/or DG lipase), whereas TRH stimulates a phospholipase C that acts on PtdIns-4,5P$_2$ and does not stimulate release of arachidonic acid.

672

OXYTOCIN STIMULATES ARACHIDONIC ACID TURNOVER IN INOSITOL
TRIPHOSPHATE IN FETAL MEMBRANES AT HUMAN PARTURITION.

H.A. Leaver & A. Peatty, Department of Biological Sciences,
Napier College, Colinton Road, Edinburgh EH10 5DT, U.K., &
Eastern General Hospital, Edinburgh EH6 7LN.

The uptake and release of ^{14}C arachidonic acid into the
inositol triphosphate of fetal membranes and uterine decidua
were studied by incubating tissue obtained at Elective
Caesarian Section with ^{14}C arachidonic acid and ATP (7.5μM)
in the presence or absence of oxytocin (1 x 10^{-5}u/ml).
Inositol triphosphate was separated from all other
phospholipids on borate-treated high performance thin layer
chromatography plates. Within 5 min incubation, inositol
triphosphate incorporated 1-3.5% of tissue-bound ^{14}C
arachidonic acid. Oxytocin significantly (p<0.05) stimulated
both uptake and release of ^{14}C arachidonate from amnion,
chorion and decidua after 5, 10 and 30 min incubation. Our
results suggest that inositol triphosphate in the fetal
membranes represents a dynamic pool of arachidonic acid
which is sensitive to oxytocin mobilisation at parturition.
(Supported by the Scottish Home & Health Department).

Evidence for Phospholipase C-type Breakdown of Phosphatidylinositol and Subsequent de novo Synthesis in the Rat Parotid.

M. Lupu and Y. Oron, Tel Aviv University, ISRAEL

The incorporation of ^{32}Pi into phosphatidylinositol reflects de
novo synthesis , while the incorporation of ^{3}H-inositol reflects
both de novo synthesis and the exchange reaction. The ratio of
the incorporation of these two isotopes (tritium/phosphorus) into phosphatidylinositol should, therefore, reflect the relative
contribution of these two processes, provided the specific radio
activities of the precursors remain constant. Both alpha-ad-
renergic and cholinergic stimulation of rat parotid slices resulted in a dramatic drop in this ratio. This effect could be
either blocked, or reversed by appropriate blockers. Control
experiments rule out a change in the specific radioactivity of
the precursors. We conclude that an exchange reaction is active
in the intact tissue (hence the possibility of change in the incorporation ratio). The reversal of the biosynthetic pathway
following stimulation by neurotransmitters is not a viable hypothesis to explain the disappearance of phosphatidylinositol, but
rather a complete, phospholipase c-type of breakdown, followed
by de novo synthesis.

MYO-INOSITOL AND GLUCOSE IN RETINAL LAYERS OF NORMAL AND DIA-
BETIC RABBITS. Leslie C. MacGregor, Lauren R. Rosecan, and
Franz M. Matschinsky. Univ. of Pa., Philadelphia, Pa. 19104

Alterations in myo-inositol (MI) metabolism have been implicated in the pathogenesis of diabetic complications in neural
tissues. We have enzymatically measured the concentrations of
glucose and MI in freeze-dried samples of histologically defined
layers of retina from normal and from alloxan diabetic rabbits.
The absence of sucrose from retinal samples after i.v. sucrose
infusion demonstrated that the retinal permeability barriers are
functional in diabetics. In samples of both normal and diabetic
retinas, choroidal and pigmented epithelial layer glucose approximated serum glucose, with levels in other layers about half
that of serum. Diabetes did not affect serum MI levels. MI
values ($\bar{x}$ ± S.E.M.) in ocular samples of normal (N=5) rabbits
were (mmol/kg dry wt): choroidal vessel walls: 18 ± 1.4; pigmented epithelium: 12.8 ± 0.4; outer and inner segments: 8.7 ±
0.5 and 8.8 ± 0.8; outer nuclear and synaptic layers: 15.1 ± 0.3
and 21.9 ± 1.8; inner nuclear and synaptic layers: 18.3 ± 1.0
and 22.1 ± 0.8; ganglion/nerve fiber layer: 19.1 ± 0.4. MI was
significantly lower (p<0.02) in each layer of the diabetic
rabbit retinas (N=6), averaging 27% lower than in normals. In
particular, MI did not peak in the synaptic layers of the diabetic rabbit retina, as observed in normal retina. The specialized layers of the rabbit retina constitute a useful model in
which to study MI function in neural systems and deranged MI
metabolism caused by hyperglycemia.

STIMULATION OF PHOSPHATIDYLINOSITOL (PhI) AND PHOSPHATIDIC ACID
(PhA) METABOLISM IN RAT BRAIN CORTEX BY CARBACHOL (CCh), NORE-
PINEPHRINE (NE) AND K⁺. M. Dorota Majewska and George Hauser,
McLean Hospital (Harvard Medical School), Belmont, MA 02178.

In order to characterize the cholinergic and adrenergic
receptors involved in the stimulation of brain phospholipid
metabolism and to examine possible interactions with depolarization by K⁺, rat cerebral cortex mince was incubated with $^{32}P_i$
and suitable additions. Carbachol-enhanced incorporation into
PhI > PhA was blocked by atropine but not by tubocurarine,
implicating muscarinic receptors. Stimulation by NE of PhA >
PhI labeling was inhibited by prazosine and WB 4101, but not by
an α_2-blocker, yohimbine, or a β-blocker, sotatol, indicating
specific involvement of α_1-adrenergic receptors. NE and CCh
effects were additive. 50 mM K⁺ caused some elevation of $^{32}P_i$
incorporation into PhI and PhA, but drastic reductions in labeling of neutral phospholipids. These effects were insensitive to
either atropine or prazosine. CCh and K⁺ caused synergistic
incorporation increases in PhA, whereas stimulation of PhI
labeling was reduced. NE and K⁺ effects on PhA were essentially
additive. CCh partly reversed the negative K⁺ effect on neutral
phospholipid labeling. Thus, the receptor classes responsible
for enhanced PhA and PhI turnover are the same in cerebral
cortex as in other tissues. The findings also point to close
and complex interrelationships among agents inducing changes in
phospholipid metabolism in cerebral cortex as part of their
mechanism of action. (Supported by grants NS 06399, NS 19047
and RR 05484 from NIH and PCM 7824387 from NSF.)

EFFECTS OF THE VASOPRESSIN ANTAGONIST ON VASOPRESSIN-INDUCED PHOSPHOLIPID CHANGES AND ACTIVATION OF GLYCOGEN PHOSPHORYLASE

James S. Marks, Andrew P. Thomas, Rebecca Williams, Janette Alexander and John R. Williamson. Sandoz, Inc., East Hanover, New Jersey 07936 and University of Pennsylvania, Philadelphia, Pa. 19104

Recently, vasopressin-stimulated breakdown of hepatic polyphosphoinositides has been shown to be a Ca^{2+}-independent process which is sufficiently rapid compared to phosphorylase activation to enable it to play a role in the sequence of events leading to the mobilization of intracellular calcium. As a further test of the relationship between vasopressin-induced phospholipid changes and the activation of glycogen phosphorylase, in isolated hepatocytes, experiments were conducted using the vasopressin antagonist ([1-(B-Mercapto-B,B-cyclomethyleneproprionic acid),β-arginine]vasopressin). The addition of 1×10^{-7} M vasopressin to hepatocytes isolated from fed rats resulted in arapid activation of glycogen phosphory-lase and an equally rapid degradation of phosphatidyl-inositol 4,5-bisphosphate (TPI). In addition, the intracellular concentrations of inositol triphosphate (IP_3), inositoldiphosphate (IP_2) and phosphatidic acid (PA) increased. Additon of the vasopressin antagonist 30 seconds prior to vasopressin blocked the activation of phosphorylase,the degradation of TPI and the increase in PA, IP_3 and IP_2. Addition of the vaso-pressin antagonist 4 minutes after the addition of vasopressin caused phosphorylase to return to basal activity within 6-8 minutes as opposed to 35 minutes in the absence of the antagonist. In addition, the rates of TPI resynthesis and PA, IP_3 and IP_2 disap-pearance were all accelerated. These results suggest that the glycogenolytic effects attributed to vaso-pressin involve changes in the metabolism of the inositol phospholipids. Further experiments are required to confirm this relationship.

MUTUAL EFFECTS BETWEEN CDP-DG:MYOINOSITOL PHOSPHATIDYLTRANS-FERASE AND ATP-DEPENDENT, CALCIUM ION UPTAKE ACTIVITIES IN RAT LIVER MICROSOMES. S. McCune, M. Vogt, F. Procaccino and F. Slaby. George Wash. Univ. Med. Ctr., Wash., DC 20037

Both of the above activities can be expressed by microsomes incubated in medium containing equimolar concentrations of myoinositol (mI) and CMP (10-500 µM), 100 mM KCl, 5 mM NaN_3, 1 mM $MgCl_2$, 1 mM $MnCl_2$ and 12 mM histidine-HCl, pH 6.9. The separate addition of 500 µM $CaCl_2$ or 5 mM ATP has either no effect or a stimulatory effect on CDP-DG-dependent, [^{3}H]mI incorporation into PI; the combined effect, however, is an 80% inhibitory effect. The combined effect of calcium and ATP also inhibits the rate of the reverse reaction of the transferase by 50%, as measured by [^{14}C]CMP incorporation into CDP-DG in mI-free reaction mixtures. The rates of mI and CMP incorpo-ration neither affect nor are functions of the rate of ATP-dependent, calcium ion uptake. The magnitude of the combined inhibitory effect of calcium and ATP on [^{3}H]mI incorporation is a function of the calcium ion concentration, requiring a minimum concentration of 8-10 µM. Exogenous (100 µM) CDP-DG, however, inhibits active calcium ion uptake by 50%. The findings suggest that CDP-DG may inhibit active calcium ion uptake by the endoplasmic reticulum of hepatocytes during stimulated PI turnover, and that this is a mechanism by which the reactions regulating PI turnover may be associated with those processes regulating cytosolic calcium ion levels.

CHARACTERIZATION OF THE HORMONE-SENSITIVE PHOSPHATIDYL-INOSITOL POOL IN WRK-1 RAT MAMMARY TUMOR CELLS. <u>Marie E. Monaco</u>, Dept. of Physiol. & Biophys., New York Univ. Med. Ctr., and The VA Hospital, New York, N.Y. 10010.

We have recently demonstrated the existence of a discrete pool of hormone-sensitive phosphatidylinositol (HS-PI) within WRK-1 cells. Treatment of these cells with vasopressin (VP) results in a rapid increase in both the rate of synthesis and the rate of turnover of PI in this pool. Equilibrium labeling experiments indicate that this pool accounts for approximately 17% of the total cellular PI. We now report a further characterization of the HS-PI: 1. HS-PI is continuously synthesized and turned over even in the absence of VP, although at a much slower rate than when VP is present. Eight hours are required for an equilibrium situation to be achieved; that is, for 17% of the ^{32}P-PI to become hormone-sensitive. In contrast, if cells are incubated in the presence of VP for less than two hours, 60-90% of ^{32}P-PI is hormone-sensitive. With respect to turnover, prelabeled HS-PI slowly disappears even if VP has been removed. However, hormone-insensitive PI (HI-PI) is stable. 2. Incorporation of ^{32}P$_i$ into HS-PI in the absence of VP does not involve transfer of HI-PI to the sensitive pool. 3. Calcium ionophores can stimulate incorporation of ^{32}P$_i$ into PI; however, this PI is hormone-insensitive. In view of current data from other laboratories, we would like to suggest that the HS-PI is the precursor pool for synthesis of DPI and TPI.

CORRELATION BETWEEN THE DOSE-RELATED EFFECTS OF GENERAL ANES-THETICS ON PHOSPHOLIPID SURFACE TENSION AND BRAIN OXYGEN CON-SUMPTION.
E.M. Nemoto, Ph.D., M. Uram, M.D. and P.M. Winter, M.D. Dept. of Anes. and CCM, University of Pittsburgh, Pittsburgh, PA 15261

General anesthetics differ in their dose-related effects on cerebral metabolic rate for oxygen (CMRO2). Thiopental initially sharply reduces CMRO2 with a near maximal decrease at surgical anesthetic levels while halothane decreases CMRO2 linearly up to doses 10 times higher than required for anesthesia. We studied the dose-related effects of various inhalation anesthetics and thiopental on bovine brain phospho-tidylinositol monolayer surface tension as measured by a Cahn 2000 Electrobalance and dynamic surface tension unit with a Wilhemy plate. At concentrations up to 60 ug/cc, thiopental caused a rapid and linear increase in phosphotidylinositol surface pressure. However, the effect plateaued at about 3 dynes/cm at concentrations higher than 150 ug/cc. On the other hand, all inhalation anesthetics produced a linear increase in surface pressure with no sign of a plateau effect even at concentrations four times higher than that required for surgical anesthesia. There is an apparent correlation between the effects of general anesthetics on phospholipid surface pressure and CMRO2 at least for thiopental and halothane. The interaction of phospholipid monolayers (and presmably membrane) with thiopental differs from their interaction with inhalation anesthetics.

676

INDEPENDENCE OF CHOLECYSTOKININ-INDUCED PHOSPHA-TIDYLINOSITOL (PI) BREAKDOWN FROM EXTRACELLULAR CALCIUM AND MOBILIZATION OF INTRACELLULAR CALCIUM IN DISPERSED ACINI FROM GUINEA PIG PANCREAS.

S. Pandol, M. Thomas and M. Schoeffield, Dept. of Medicine, UCSD and VAMC, San Diego, CA 92161.

To determine the relationships between extracellular Ca, mobilization of intracellular Ca, PI metabolism and enzyme secretion in dispersed acini from guinea pig pancreas, we measured the ability of cholecystokinin-octapeptide (CCK-OP) and A23187 to cause amylase release, ^{45}Ca outflux, changes in cellular PI and phosphatidic acid (PA) and breakdown of ^{3}H-PI either with Ca (0.5 mM) in the incubation solution or with no Ca plus EGTA (0.2 mM). Both CCK-OP and A23187 increased amylase release and the increases were independent of media Ca for 5 minutes of incubation and dependent on Ca after 5 minutes. Both agents caused ^{45}Ca outflux (a measure of cellular Ca mobilization) occurring both in the presence and absence of media Ca. CCK-OP stimulated a 15-25% decrease in PI and a 2-fold increase in PA measured from 2.5 to 45 minutes of incubation that were independent of media Ca. In contrast, A23187 caused only a transient and Ca-dependent decrease in PI. A23187 stimulated a small Ca-independent increase in PA. CCK-OP induced a Ca-independent breakdown of ^{3}H-PI in acini prelabelled with ^{3}H-inositol while A23187 stimulated breakdown of ^{3}H-PI only in the presence of media Ca. A Ca-independent breakdown of ^{3}H-PI also occurred with acini incubated with both CCK-OP and A23187.

These results indicate that PI breakdown caused by CCK-OP is independent of extracellular Ca and not a result of its ability to mobilize intracellular Ca.

INSULIN BLOCKS ALPHA$_1$ADRENERGIC METABOLIC ACTIONS BUT NOT PI LABELING IN HEPATOCYTES FROM HYPOTHYROID RATS. C.K. Pushpendran, S. Corvera and J.A. García-Sáinz. Universidad Nacional Autónoma de México.

The stimulations of glycogenolysis and ureogenesis by vasopressin and angiotensin II were markedly diminished in hepatocytes from hypothyroid rats whereas the stimulation by epinephrine (through alpha$_1$-adrenoceptors) was unaltered. This suggested that alpha$_1$-adrenergic amines could act through two mechanisms; one shared with vasopressin and angiotensin II, and an alternative Ca^{++}-independent mechanism, which predominates during hypothyroidism and is not shared by vasopressin and angiotensin II. Previous reports have shown that in cells from euthyroid rats insulin partially inhibits alpha$_1$-adrenergic actions, whereas those of vasopressin and angiotensin II are unaffected. A possible explanation for these data was that insulin was interfering with the Ca^{++}-independent part of the alpha$_1$-adrenergic mechanism. As predicted, in hepatocytes from hypothyroid rats insulin completely abolished the effects of epinephrine on glycogenolysis and urea synthesis but it did not modify alpha$_1$-adrenergic mediated PI labeling.

677

5-HYDROXYTRYPTAMINE-STIMULATED PHOSPHATIDYL-
INOSITOL-4,5BISPHOSPHATE SYNTHESIS AND Ca^{2+}
GATING IN BLOWFLY SALIVARY GLANDS.
Karen Sadler, Irene Litosch and John N. Fain
Section of Biochemistry, Brown University,
Providence, Rhode Island 02912.

Blowfly salivary glands, previously exposed
to 10 μM 5-hydroxytryptamine incorporated 70% of
the total [^{3}H]phosphoinositide label into phospha-
tidylinositol-4,5bisphosphate (PtdIns-4,5P$_2$) and
14% into phosphatidylinositol-4phosphate (PtdIns-
4P) when allowed to recover in medium containing
3-5 μM [^{3}H]inositol. Subsequent addition of 5-
hydroxytryptamine produced an equivalent breakdown of the newly synthesized phosphoinositides
but no detectable $^{45}Ca^{2+}$ entry. Increasing the
medium inositol concentration to 300 μM produced a
14-fold stimulation of phosphatidylinositol (PI)
synthesis but only a 5-fold increase in PtdIns-
4,5P$_2$ synthesis. Full recovery of Ca^{2+} gating was
observed in glands incubated with 300 μM inositol
in which labeled PI constituted the major phosphoinositide. These results indicate that conversion
of PI to PtdIns-4,5P$_2$ occurs in blowfly salivary
glands and is secondary to an initial breakdown of
the phosphoinositides. Recovery of Ca^{2+} gating is
dependent on the restoration of both PI and
PtdIns-4,5P$_2$ to appropriate levels.

NOREPINEPHRINE STIMULATION OF PHOSPHATIDYLINOSITOL METABOLISM
IS ASSOCIATED WITH THE ALPHA$_1$-RECEPTOR SUBTYPE IN RAT BRAIN.
Darryle D. Schoepp, Sheila M. Knepper and Charles O. Rutledge,
Dept. of Pharmacology and Toxicology, School of Pharmacy, Univ.
of Kansas, Lawrence, KS 66045.

Norepinephrine (NE) has been previously shown to stimulate
phosphatidylinositol (PI) turnover by interacting with alpha
(α)-adrenergic receptors in brain. The present study was performed to determine the subtype of α-receptor which is associated with NE-stimulated PI turnover in rat brain. PI hydrolysis
was determined by incubating slices of cerebral cortex with ^{3}H-
myo-inositol in the presence of lithium ion (10 mM) and measuring the accumulation of ^{3}H-inositol phosphates (Berridge et al.,
Biochem. J. 206:587-595, 1982). NE and the selective α$_1$-agonist
phenylephrine (PE) both markedly stimulate the formation of ^{3}H-
inositol phosphates in a concentration-dependent manner. Maximal effect was observed at 10^{-4}M of either agonist (NE 626% of
control; PE 365% of control). The selective α$_2$-agonist, clonidine, did not significantly alter ^{3}H-inositol phosphate formation even at concentrations as high as 10^{-3}M. The β-agonist (-)
isoproterenol moderately increased ^{3}H-inositol phosphate formation to 140% of control at 10^{-4}M and 214% of control at 10^{-3}M.
The α$_1$-antagonist prazosin (IC$_{50}$ 0.036 μM) was 300 times more
potent than the α$_2$ antagonist yohimbine (IC$_{50}$ 10.7 μM) as an
inhibitor of NE (10^{-4}M) stimulated PI hydrolysis. These results
indicate that the α$_1$ but not α$_2$ adrenergic receptor subtype in
rat brain is coupled to PI hydrolysis. (Supported by American
Heart Assoc. Fellowship KSA-82-02, USPHS Grant NS 16364 and
General Research Support Grant 5606.)

678

PHOSPHOINOSITIDES AND ACETYLGLYCERYLETHER PHOSPHORYLCHOLINE
Shivendra D. Shukla and Donald J. Hanahan, Department of
Biochemistry, The University of Texas Health Science Center
at San Antonio, Texas, 78284.
Acetylglycerylether phosphorylcholine (AGEPC) or
platelet activating factor is a potent hypotensive agent
and its role has been implicated in inflammatory and
allergy reactions. Within 5 seconds of the mixing of [3H]
inositol labeled rabbit platelets with AGEPC (5 x 10^{-10}M)
a 25% decrease in the [3H]triphosphoinositide (TPI) was
evident. Treatment of these platelets for 5 seconds with
various concentrations of AGEPC exhibited a dose-related
decrease in [3H]TPI. The decrease in [3H]TPI was transient
since after 5 seconds the radioactivity in this phospholipid increased. Radioactivity in diphosphoinositide (DPI)
also increased. A 15-20% decrease in [3H]phosphatidyl-
inositol (PI) within 15 seconds and an increase in [3H]
lyso-PI also was observed when rabbit platelets were
exposed to AGEPC. These observations lead to the conclusion that phosphoinositides are closely associated with
the early processes of the interaction of AGEPC with the
rabbit platelets. Relevance of these findings in relation
to the cellular mode of action of AGEPC will also be
discussed. Supported by a grant from the American Heart
Association, Texas Affiliate.

RECEPTOR-MEDIATED CYCLIC GMP FORMATION IN CULTURED NERVE CELLS
INVOLVES ARACHIDONIC ACID RELEASE. R. M. Snider, M. McKinney,
C. Forray and E. Richelson. Mayo Clinic, Rochester, MN 55905.
Muscarinic or histamine H_1 receptors on neuroblastoma
(N1E-115) cells mediate cGMP formation. The hypothetical mechanism for this response is calcium channel activation, possibly
resulting from PI turnover. However, neuroblastoma cells
loaded with the Ca^{++} indicator, aequorin, did not increase
their light emission after challenge with carbachol or histamine but did in response to Ca^{++} ionophores or melittin.
Moreover, the cGMP time-course was much more rapid for receptor
agonist than for Ca^{++} ionophores, thus dissociating cGMP
stimulation from increased $[Ca^{++}]i$.
Thrombin stimulates cGMP in nerve cells in a similar manner as receptor agonists (Snider & Richelson, Science 221:566,
1983). The response of thrombin, carbachol and histamine were
all equally antagonized by quinacrine, ETYA and NDGA, suggesting that cGMP formation results from the release of arachidonic acid (AA) and its metabolism by the lipoxygenase enzymatic pathway. Guanylate cyclase is known to be markedly stimulated by oxidation products of lipids. Indomethacin had no
effect or slightly potentiated cGMP levels. Finally, we found
that receptor activation stimulates the release of [3H]AA from
prelabeled cells and this effect was blocked by quinacrine or
receptor antagonists. We hypothesize that receptor activation
leads to an increase in PLA_2 activity, release of AA which is
metabolized via lipoxygenase, and stimulates guanylate cyclase.
(Support: Mayo Fdn.; Grants MH27692, MH08823 and HL12186)

MODULATION OF PHOSPHOINOSITIDE LEVELS AND THE PATHWAYS OF
STIMULUS-INDUCED DEGRADATION IN MACROPHAGES

Sundler, R., and Emilsson, A. Department of Physiological
Chemistry, P.O.B. 750, University of Lund, S-220 07 Lund,
Sweden.

In mouse peritoneal macrophages 8.0 ± 0.3 and 5.7 ± 0.3 percent of
the inositol phospholipids consist of diphospho-(DPI) and tri-
phosphoinositide (TPI), respectively, as determined from their
radioactivity after 24 h labeling with [³H]inositol. Exposure
of the cells to Con A or Latex beads (both avidly endocytozed)
induced a shift from TPI towards DPI, with little change in
phosphatidylinositol (PI), while platelet activating factor
(which enhanced cell spreading) induced a shift from PI
towards DPI and TPI. A selective increase in DPI occured upon
exposure of cells to monensin. The latter would be consistent
with the known effect of monensin to specifically block membrane transport from the Golgi, since phosphorylation of PI is
very active in this organelle (Jergil and Sundler, J. Biol.
Chem. (1983) 258, 7960-7973.

A net degradation of inositol lipids was induced by zymosan
particles and by the ionophore A_{23187} in the presence of Ca^{2+}.
The former agent primarily induced deacylation of PI, while the
latter primarily enhanced the phosphorylation of PI to DPI and
a phosphodiesterase-catalyzed degradation with liberation of
inositol-diphosphate.

A COMPARISON OF THE EFFECTS OF CEREBRAL PROTECTIVE
DRUGS AND NONPROTECTIVE DRUGS ON MONOLAYERS OF PHOS-
PHOLIPIDS

M. Uram, M.D., E.M. Nemoto, Ph.D., P.M. Winter, M.D.
Department of Anesthesiology and Critical Care Med.
University of Pittsburgh, Pittsburgh, PA 15261

Various drugs including barbiturates such as
thiopental and the anticonvulsant, dilantin, attenuate
ischemic brain damage, whereas other anesthetics such
as ketamine and halothane do not. We evaluated the
effects of these drugs on phosphotidylinositol monolayer surface pressure over bicarbonate buffer using
the Cahn 2000 Electrobalance and Dynamic Surface Tension unit and a Wilhemy Plate. Dilantin, thiopental
and droperidol produced a progressive change in surface pressure but showed a definite plateau effect.
On the other hand, the ineffective anesthetics such
as ketamine and halothane produced a linear increase
in surface pressure. This difference between the
cerebral protective and nonprotective drugs on phospholipid monolayer surface pressure was not one of
magnitude but rather, the pattern of their effects.
These differential effects of the drugs on phospho-
tidylinositol surface pressure may account for the
differences in their cerebral protective effects.

680

AMINOGLYCOSIDE INHIBITION OF PHOSPHODIESTERATIC CLEAVAGE OF
POLYPHOSPHOINOSITIDES IN NERVE ENDING MEMBRANES. <u>Lucio A.A. Van
Rooijen and Bernard W. Agranoff</u>, Mental Health Res. Inst. and
Dept. of Biol. Chem., Neuroscience Laboratory Bldg., University
of Michigan, 1103 E. Huron, Ann Arbor, MI 48109.

We have recently described studies on a membrane-bound
Ca^{2+}-dependent phosphodiesterase in nerve ending preparations
which cleaves endogenous phosphatidylinositol phosphate (PhIP)
and phosphatidylinositol bisphosphate ($PhIP_2$) with the concomitant production of the water-soluble products, inositol bisphosphate (IP_2) and inositol trisphosphate (IP_3), respectively
(BBRC 112:919-926, 1983). We report here that the antibiotic
aminoglycosides neomycin and gentamycin, and to a much lesser
extent streptomycin, block degradation of both PhIP and $PhIP_2$.
The neomycin inhibition could be overcome by increasing the
concentration of Ca^{2+}. Under no conditions, however, could
the cleavage rates of PhIP and $PhIP_2$ be dissociated. The
competition between Ca^{2+} and neomycin may be of interest
in regard to the known neurotoxicity of the antibiotic. We
also find that nucleoside triphosphates specifically inhibit
the phosphatase activity that degrades IP_3 to IP_2. Since
there is indirect evidence that the inositol moiety is con-
served for reuse in the inositide cycle, and IP_3 has been
implicated as a possible intracellular messenger, this block
may have regulatory significance in stimulus-secretion mechanisms. (Supported by NS 15413.)

PHOSPHATIDYLINOSITOL METABOLISM IN RAT LIVER
PLASMA MEMBRANES - EFFECTS OF HORMONES AND
CATIONS.
Wallace, M.A., Poggioli, J., Giraud, F.,
Claret, M., Brown J. and Fain, J.N.
Section of Biochemistry, Brown University,
Providence, RI 02912; Depts. of Physiol.
Comparée et Physiol. de la Nutrition,
Université Paris XI, Orsay, France.

Phosphoinositide degradation occurs rapidly
when hormones that promote Ca^{2+} mobilization interact with their receptors. Subsequently, phosphatidylinositol (PI) synthesis is stimulated. The
enzymes responsible for PI synthesis are sensitive
to Ca^{2+}, and in this regard we have reassessed
their sub-cellular distribution. When rat liver
plasma membranes were incubated in media containing
5 mM Mg^{2+} and 10 nM free Ca^{2+}, addition of norepinephrine promoted a specific loss of PI in 60
sec. The effect was blocked by phentolamine.
Removing Mg^{2+} or reducing free Ca^{2+} to below 2 nM
also promoted PI loss. Phosphoinositides may be
degraded and their re-synthesis stimulated by loss
of Ca^{2+} from the membranes or phosphoinositide degradation may be part of the signal leading to Ca^{2+}
mobilization. It is also possible that hormone
stimulated Ca^{2+} flux and phosphoinositide degradation are simultaneous but independent events.

681

THE SECOND MESSENGER TYPE REGULATOR cAMP ANTAGONIST
IS A PROSTAGLANDYL-INOSITOL CYCLIC-PHOSPHATE (A regulatory function for inositol as molecular constituent of cAMP antagonist).

H.K. Wasner; Diabetes-Forschungsinstitut, Auf'm Hen
nekamp 65, 4000 Düsseldorf 1, FRG.

Parallel to the receptor mediated alteration of the
phosphatidylinositol metabolism, cAMP antagonist is
isolated from hepatocytes a few minutes after alpha
adrenergic- or insulin stimulation· (1981 FEBS lett.
133,260 - 264). Its regulatory properties are inhibition of cAMP-dependent protein kinases and activa
tion of phosphoprotein phosphatase (1975 FEBS lett.
57,60 - 63). By dephosphorylation/phosphorylation
reactions the cAMP antagonist activates the mitochondrial pyruvate dehydrogenase, inhibits the phos
phorylase, adenylate cyclases and ATPases.
The structural components of the cAMP antagonist are
prostaglandin E, inositol and phosphate. These have
been identified by degradation of the cAMP antagonist and mass spectrometric identification of the
fragments. Furthermore radioactive labelled components are incorporated into the cAMP antagonist.
From chemical derivatisation experiments a structure
for the cAMP antagonist can be inferred where the
phosphate is linked to an inositol and this inosi-
tol-phosphate to the prostaglandin.

SUPPRESSION OF PHAGOCYTIC FUNCTION AND PHOSPHOLIPID METABOLISM
IN MACROPHAGES BY PHOSPHATIDYLINOSITOL LIPOSOMES. Nabila M.
Wassef, Frits Roerdink, Earl C. Richardson and Carl R. Alving.
Department of Membrane Biochemistry, Walter Reed Army
Institute of Research, Washington, D.C. 20307.

Liposomes were opsonized by specific antibodies to a
liposomal antigen (galactosyl ceramide) plus complement. The
antibodies used (IgM) did not have an opsonizing effect alone,
however, in the presence of complement uptake and ingestion of
liposomes by mouse resident peritoneal macrophages was
enhanced up to 10 fold. Increased phagocytosis of complement-
opsonized vesicles was accompanied by increased phosphatidylinositol (PI) turnover in murine macrophages. However, when
PI was also present as one of the lipids in the opsonized
liposomes it caused reduced phagocytosis and reduced stimulation of endogenous PI turnover. These suppressive effects did
not occur with liposomes containing PI phosphate. When a
monoclonal IgM anti-PI phosphate antibody, that bound to liposomal PI phosphate, was substituted for anti-galactosyl ceramide antibodies for activating complement in the opsonizing
process, enhanced phagocytosis occurred normally, but
increased cellular PI turnover did not occur. Although PI
phosphate cannot replace PI for suppressing PI turnover, it
can be induced to be suppressive after binding to specific
anti-PI phosphate antibody. We conclude that ingestion of
complement-opsonized liposomes by macrophages, and complement-
induced turnover of cellular PI are separate but related
phenomena that can be independently modulated by the polar
group of liposomal PI.

682

INDEX OF ABSTRACT AUTHORS

INDEX

694